Ranjith Rajamanickam, Amit Sharma, Dhivya Ranjith, J. Paulo Davim (Eds.)

Machine Learning for Medical Applications

Advanced Mechanical Engineering

Edited by
J. Paulo Davim

Volume 14

Machine Learning for Medical Applications

Volume I: Computational Drug Discovery, Bioimaging, Smart Biomaterials

Edited by
Ranjith Rajamanickam, Amit Sharma, Dhivya Ranjith
and J. Paulo Davim

DE GRUYTER

Editors
Dr. Ranjith Rajamanickam
Rademics Research Institute
Coimbatore 641035
Tamil Nadu
India
ranjith.mecs@gmail.com

Prof. Dhivya Ranjith
Rademics Research Institute
Coimbatore 641035
Tamil Nadu
India
rdhivya1521@gmail.com

Dr. Amit Sharma
Professor
School of Computer Application
Lovely Professional University
Phagwara
Punjab
India
profamitsharma@gmail.com

Prof. Dr. J. Paulo Davim
Department of Mechanical Engineering
University of Aveiro
Campus Santiago
3810-193 Aveiro
Portugal
pdavim@ua.pt

ISBN 978-3-11-150318-9
e-ISBN (PDF) 978-3-11-150320-2
e-ISBN (EPUB) 978-3-11-150374-5
ISSN 2367-3796

Library of Congress Control Number: 2025936061

Bibliographic information published by the Deutsche Nationalbibliothek
The Deutsche Nationalbibliothek lists this publication in the Deutsche Nationalbibliografie; detailed bibliographic data are available on the internet at http://dnb.dnb.de.

Cover image: Sky_blue/iStock/thinkstock
Typesetting: Integra Software Services Pvt. Ltd.
Printing and binding: CPI books GmbH, Leck

www.degruyterbrill.com
Questions about General Product Safety Regulation:
productsafety@degruyterbrill.com

Contents

List of contributors

M. Parthiban
Computer Science and Engineering
SASI Institute of Technology and Engineering
Tadepalligudem 53410
West Godavari
Andra Pradesh
India

P. Krishnamoorthy
Computer Science and Engineering
SASI Institute of Technology and Engineering
Tadepalligudem 53410
West Godavari
Andra Pradesh
India

D. Amaravathi
Computer Science and Engineering
SASI Institute of Technology and Engineering
Tadepalligudem 53410
West Godavari
Andra Pradesh
India

K. Suresh
Computer Science and Engineering
SASI Institute of Technology and Engineering
Tadepalligudem 53410
West Godavari
Andra Pradesh
India

S. Karthikeyan
Computer Science and Engineering
SASI Institute of Technology and Engineering
Tadepalligudem 53410
West Godavari
Andra Pradesh
India

P. Prasant
Associate Professor
AIPH University
Bhubaneswar, Odisha
India
registrar@aiph.ac.in

Kiran Siripuri
Department of Cybersecurity and IoT
Malla Reddy University
Maisammaguda
Dulapally
Hyderabad
Telangana 500100
India

S.B. Mohan
Department Of Electronics And Communication Engineering
SA Engineering College
Chennai
Tamil Nadu
India

Venkatesan Hariram
PG and PhD Supervisor
Department of Practice of Medicine
Vinayaka Mission's Homoeopathic
Medical College and Hospital (A Constituent College of Vinayaka Mission's Research Foundation)
Deemed to be University
Salem 636308
Tamil Nadu
India

A.S. Vigneshwar
Ramco Institute of Technology
Rajapalayam
Sivakasi
Tamil Nadu
India

T. Sripriya
Department of ECE
Jeppiaar Institute of Technology
Chennai
Tamil Nadu
India

https://doi.org/10.1515/9783111503202-203

Praveen Talari
Department of CSE
Vignana Bharathi Institute of Technology
Hyderabad
Telangana
India

Amirthalakahmi T.M.
Department of ECE
SRM Institute of Science and Technology
Ramapuram
Chennai
Tamil Nadu
India

Senthilkumar B.
Department of Electronics and Communication Engineering
VSB Engineering College
Karur 639111
Tamil Nadu
India

T.M. Thiyagu
Department of Computer Science and Engineering
Vel Tech Rangarajan Dr. Sagunthala R&D Institute of Science and Technology
Avadi 600062
Chennai
Tamil Nadu
India

A. Aalan Babu
Department of Computer Science and Engineering, School of Computing
Vel Tech Rangarajan Dr. Sagunthala R&D Institute of Science and Technology
Avadi 600062
Chennai
Tamil Nadu
India

R. Selvakumar
Department of AIML
Saveetha Engineering College
Poonammallee High Road
Thandalam
Sriperumbudur
Tamil Nadu
India

D. Shobana
Department of Mechatronics
Rajalakshmi Engineering College
Chennai
Tamil Nadu
India

S. Kavitha
Department of ECE
Nandha Engineering College
Erode 638052
Tamil Nadu
India

P.R. Therasa
Department of CSE
R.M.K. Engineering College
RSM Nagar
Kavaraipettai 601206
Thiruvallur District
India

B. Yamini
Department of Networking and Communications
School of Computing
College of Engineering and Technology
SRM Institute of Science and Technology
SRM Nagar
Kattankulathur 603203
Chengalpattu District
Tamil Nadu
India

R.M. Dilip Charaan
Department of Computer Science and Engineering
Vel Tech Rangarajan Dr. Sagunthala R&D Institute of Science and Technology
Chennai 600062
Tamil Nadu
India

Janani G.
Department of Information Technology
RMD Engineering College
Chennai
Tamil Nadu
India

S. Sree Dharinya
School of Computer Science Engineering and Information Systems
VIT University
Vellore
Tamil Nadu
India

G. Parthasarathy
School of Computing and Information Technology
REVA University
Bangalore
Karnataka
India

P. Nagarajan
Department of Biotechnology
Vel Tech Rangarajan Dr. Sagunthala R&D Institute of Science and Technology
400 Feet Outer Ring Road
Avadi 600062
Chennai
Tamil Nadu
India

D. David Neels Ponkumar
Department of Electronics and Communication Engineering
Vel Tech Rangarajan Dr. Sagunthala R&D Institute of Science and Technology
400 Feet Outer Ring Road
Avadi 600062
Chennai
Tamil Nadu
India

A. Muniyappan
Centre for Computational Modeling
Chennai Institute of Technology
Chennai 600 069
Tamil Nadu
India

D.M. Kalai selvi
Department of CSE
RMD Engineering College
Chennai
Tamil Nadu
India

S. Ramkumar
Department of Electronics and communication engineering
Sri Eshwar College of Engineering
Coimbatore
Tamil Nadu
India

K. Suresh Kumar
Department of MBA
Panimalar Engineering College
Nasarathpettai
Varadharajapuram
Poonamallee, Chennai
Tamil Nadu
India

Ezhil E. Nithila
Department of ECE
Vel Tech Rangarajan Dr. Sagunthala R&D Institute of Science and Technology
Avadi
Tamil Nadu
India

Amit Sharma
School of Computer Application
Lovely Professional University
Phagwara
Punjab
India

G.L. Karthik
School of Biosciences Engineering and Technology
VIT Bhopal University
Kotrikalan Village
Ashta Tehsil 466114
Sehore District
Madhya Pradesh
India

Rakhi Kamra
Department of Electrical and Electronics Engineering
Maharaja Surajmal Institute of Technology
Janakpuri
New Delhi 110058
India

S. Praveena
Electronic and Communication Engineering
Mahatma Gandhi Institute of Technology (was a technological institution)
Gandipet
Hyderabad
Telangana
India

Spoorthi P. Shetty
Department of MCA
NMAAM Institute of technology
Karkala Taluk
Nitte 574110
Udupi District
Karnataka
India

N. Pragadish
Department of Mechanical Engineering
Rajalakshmi Institute of Technology
Kuthambakkam 600 124
Chennai
India

Ashish Verma
Department of Physics
School of Mathematical and Physical Sciences
Dr. Harisngh Gour Vishwavidyalay
Sagar University
Madhya Pradesh
India

K.N.V. Satyanarayana
ECE Department
SRKR Engineering College
China Amiram Bhimavaram
Andhra Pradesh
India

Vaishnavi V.
Faculty of Management
SRM Institute of Science and Technology
Kattankulathur 603203
Chengalpattu Dt.
Tamil Nadu
India

Meena G.
Faculty of Management
SRM Institute of Science and Technology
Kattankulathur 603203
Chengalpattu Dt.
Tamil Nadu
India

Santhanalakshmi
Faculty of Management
SRM Institute of Science and Technology
Kattankulathur 603203
Chengalpattu Dt.
Tamil Nadu
India

Anitha R.
Department of Electronics and Communication Engineering
Government Engineering College Palakkad
Sreekrishnapuram 678633
Kerala
India

K. Sivakumar
Department of Mathematics
Simats School of Engineering
Saveetha University
Thandalam 602105
Chennai
Tamil Nadu
India

J. Dafni Rose
Department of CSE
St. Joseph's Institute of Technology
Semmancheri, Chennai
Tamil Nadu
India

Pradeepkumar G.
KPR Institute of Engineering and Technology
Coimbatore
Tamil Nadu
India

T. Shanmugaraja
KPR Institute of Engineering and Technology
Coimbatore
Tamil Nadu
India

P. Janardhan Saikumar
Department of ECE
Audisankara College of Engineering and Technology
Gudur
Andhra Pradesh
India

M. Mahalakshmi
Department of Networking and Communications
SRM Institute of Science and Technology
Kattankulathur
Chennai
Tamil Nadu
India

S. Sujatha
Department of EEE
Sri Sairam College of Engineering
Bengaluru
Karnataka
India

S. Banumathi
Department of Computer Science and Engineering
Vel Tech Rangarajan Dr. Sagunthala R&D Institute of Science and Technology
Avadi, Chennai
Tamil Nadu
India

Purushothaman V.
Department of Electronics and Communication Engineering
Chennai Institute of Technology
Kundrathur, Chennai
Tamil Nadu
India

G. Jenifa
Department of Computer Science and Engineering (Artificial Intelligence)
Madanapalle Institute of Technology and Science
Madanapalle
Andhra Pradesh
India

R. Meenal
Department of Electrical and Electronics Engineering
VSB Engineering College
Karur
Tamil Nadu
India

A. Anandh
Department of CSE
Kamaraj College of Engineering and Technology
Virudhunagar
Tamil Nadu
India

R. Jennie Bharathi
Department of ECE
Saveetha Engineering College
Thandalam, Chennai
Tamil Nadu
India

M. Parthiban, P. Krishnamoorthy, D. Amaravathi, K. Suresh,
S. Karthikeyan

Blockchain technology to secure medical data sharing in machine learning applications ensure privacy and integrity

Abstract: The idea of blockchain provides unprecedented answers for brand new challenges with EHR structures with great disruption in the healthcare field. In this chapter an overview of the effects and possibilities of the uncontrolled growth of the blockchain technology in the acquisition, preservation, and dissemination of data in healthcare is described. After much research on the blockchain-based EHR systems, a number of interconnected applications that comprise data sharing, security, and patient-oriented EHRs has been developed. It was to mention that in the case of data sharing, and especially concerning the topic of privacy and security and where there is no ability to change data, stakeholders of healthcare facilities find blockchain an extremely effective tool for data sharing. Consensus process, smart contract, and encryption thus improve the security measures even further while giving the patients the most absolute control over their health information. It shifted to actual application scenarios and asked what blockchain means for blockchained healthcare systems and matter interaction and brought together numerous parties. The fluctuations witnessed within the landscape reveal what the future of blockchain adoption in the healthcare sector can entail in the right direction, how far the healthcare industry has come in blockchain, and how far it has to go in developing more practical and coordinated plans and visions to understand where blockchain can actually revolutionize the flow of the health initiative, favorable outcomes, and substantial end-user value proposition.

Keywords: blockchain, EHR, healthcare, data integrity, security, interoperability

M. Parthiban, Computer Science and Engineering, SASI Institute of Technology and Engineering, Tadepalligudem 534101, West Godavari, Andhra Pradesh, India, e-mail: parthimitit@gmail.com
P. Krishnamoorthy, Computer Science and Engineering, SASI Institute of Technology and Engineering, Tadepalligudem 534101, West Godavari, Andhra Pradesh, India, e-mail: krishnancse0206@gmail.com
D. Amaravathi, Computer Science and Engineering, SASI Institute of Technology and Engineering, Tadepalligudem 534101, West Godavari, Andhra Pradesh, India, e-mail: amaraduraipandian0@gmail.com
K. Suresh, Computer Science and Engineering, SASI Institute of Technology and Engineering, Tadepalligudem 534101, West Godavari, Andhra Pradesh, India, e-mail: koyyesuresh@yahoo.com
S. Karthikeyan, Computer Science and Engineering, SASI Institute of Technology and Engineering, Tadepalligudem 534101, West Godavari, Andhra Pradesh, India, e-mail: mr.kaarthik@gmail.com

https://doi.org/10.1515/9783111503202-001

1 Introduction

It is important and essential in health systems to be able to share medical data to encourage stakeholders' collaboration, to build quests, as well as to enhance patients' outcomes [1, 2]. Their acquisition is greatly complicated by many factors that cause challenges to their smooth flow [3]. The traditional ways of passing information based on the realm of hierarchical relational data patterns that in some cases are centered on consolidating data repositories for the distributed systems, face these challenges mostly from the aspect of modality of data integration [4]. Such barriers affect coordination and integration, which are requisite for effective caring and sharing of information in caregiving practices [5]. Besides these, growing concerns for data accuracy and privacy show lack of better methods that can protect personal medical information [6]. Because of the daily new cases surrounding hackers and unauthorized access to medical data that are reported, there is need to find new innovative solutions pertaining to the privacy-preserving health information exchange [7].

Privacy and data integrity are the two cornerstone features in health information that present standards governing the privacy of patient information [8]. Privacy allows the individual the discretion of who has access to his/her information on health; it also aims at preventing unauthorized disclosure of information [9]. Ensuring that patient's identity cannot be easily identified is important before people can be comfortable with the idea of placing their "faith" in a doctor-patient relationship, as well as in the healthcare field where private and personal details of an individual vary from one person to the other. Privacy should be well-protected more so since infringement of this right has extreme repercussions that go to the extent of causing devastating loss of reputation in addition to the legal consequences a person is bound to face as a result [10].

In similar respect, so as to guarantee that medical records are coherent, accurate, and reliable in the duration of one's lifetime, integrity of data has to be upheld [11]. Information that is inaccurate usually leads to wrong diagnosis or wrong treatment subjecting the patient to high risk [12]. Managers use more data in their firm's healthcare decisions, and consequently, data must be accurate and reliable for clinical decision purposes [13]. Interference with the data in any way negates the basis of evidence-based medicines and undermines the oft-proclaimed health system integrity [14].

Because such targets are closely related to healthcare organizations, it is essential to set adequate procedures regarding data accuracy and privacy in those enterprises [15]. This suggests UAV control through employing highly secure measures that reduce chances of data leakages and prohibition of other unfavorable associated access tools including, audit trails, access to limits, and encryption [16]. With respect to data security and integrity, blockchain technologies provide potential directions for creating a decentralized, non-tampering blockchain to handle efficient transactional records and data assurance [17].

Being a revolutionary innovation, blockchain technology is rapidly drawing attention for changing various fields including healthcare [18]. The foundational technology of Bitcoin blockchain developed into an immutable and shared database of records, which makes exchange safe and efficient [19]. This technology has developed in handling core domains of data operations, which enable its application in the healthcare sector [20]. Blockchain is used as an auditable and secure record of transactions and can provide for exclusive handling and sharing of protected health information [21]. It is capable of increasing data credibility and enhancing the protection of data information and also relieves the cumbersome work in healthcare network [22]. Blockchain technology is shown to be a contender that enhances some systems and opens up new avenues for imagination as healthcare firms struggled with challenges of dealing with huge amounts of patient data where privacy constraints were a problem [23].

This section reviews the precise requirements of multitudinous individuals such as the hospital administrators and technology vendors [24]. Currently there are discussions about the best data governance and the necessary integration needed to apply blockchain in the healthcare sector [25]. To this end and to develop everlasting approaches to the innovation of the application of blockchain solutions in healthcare, different experts with adequate training in healthcare management, information technology, law, and policy should come together. The implementation of blockchain in healthcare delivery and patient satisfaction requires expert knowledge.

2 EHR management systems

2.1 Fundamentals of blockchain-based EHR systems

Blockchain technology gained much attention in many sectors when it was first invented as a base of cryptocurrencies. Privacy and security issues in healthcare can be solved using blockchain technology. In simpler terms, blockchain is described as a decentralized and distributed register of safe and fully transparent transactions in a network with participants and protection against attacks on records. In block systems the users interact in a direct manner; every participant holds a copy of the ledger and data was safe with agreement protocols. One major disadvantage of having centralized databases is that unauthorized access to databases is possible based on different circumstances.

Figure 1 highlights the identified components of blockchain-based EHR systems that depict the relations between a patient and care providers as well as distinct technological ensembles such as EHR apps. There is potential for blockchain to transform some of the most important spheres of the healthcare industry including those of managing trials and overseeing the supply chains. In this discussion, the attributes of

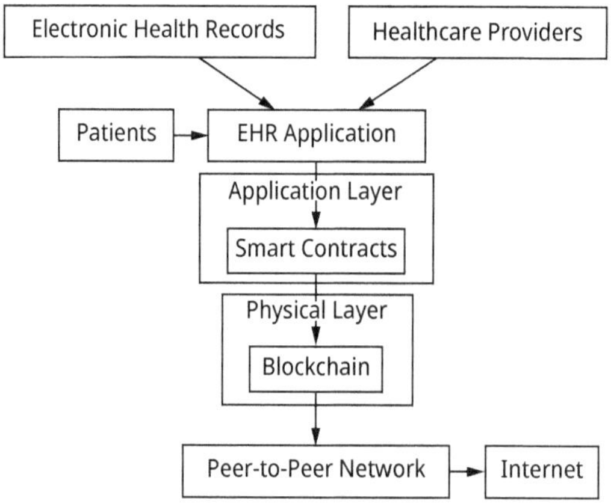

Figure 1: Fundamentals of blockchain-based EHR systems.

blockchain have been broken down to include immutability, and transparency; the implementation of every one of them in healthcare organizations improved the security of data, besides increasing the healthcare workflow and the patients' success rate. Due to its feature of smooth data exchange between stakeholders in healthcare, blockchain supports teamwork in care provision and also improves patient satisfaction.

2.2 EHR

The EHR format is an electronic record that consists of patient's full history, details of testing, and treatment plans. In EHRs, different characteristics become clear; EHRs provide safety in storage and sharing of information over the dangers that come with paper records that get lost and create confusion to all entire healthcare teams.

The EHR system outlined in Figure 2 organizes and links several items such as appointments and test results but also allows the healthcare professional and the individual to view and modify the record. At its core, it is an EHR system that stores patient health information in a structured way, which is backed by digital holdings for the users with the permission to access and modify them according to need. EHR systems often feature functionalities such as:

Patient demographics: The basic information that can identify a patient includes the patient's name, age, gender, phone number, and insurance.

Medical history: All past illnesses, surgeries, vaccines, medical and family history and any other record of the person.

Interaction between patients and healthcare staff is documented at the clinic as clinical notes, which are short descriptions of the interactions. The symptoms as well as an analysis of diagnoses and treatments have to be incorporated into the records.

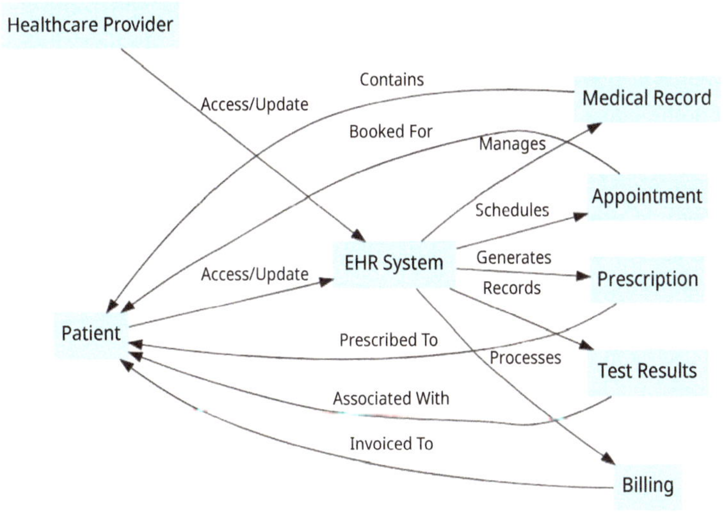

Figure 2: Basics of EHR.

Lists of combinations to be avoided, usage instructions, and dosing regimes are components of medication administration.

Results of laboratory and diagnostic tests: List of diagnostic investigations including imaging and other investigations updated by the treating doctors.

Appointment scheduling: Services that include the ability to schedule appointments with patient's visits, monitoring patient follow-ups, and notifying patients of scheduled visits.

Interoperability: Toward compiling and transmitting data between multiple cogent healthcare entities, unbroken dialogue and collaboration frameworks and methods continue to be established by multiple care providers and organizations.

EHR systems have several advantages including variations in patient data accessibility and processing time that have a positive impact on the clinical decisions and patient safety. About research and quality, EHRs provide potential definitive sources of data to drive constructive interventions in the field of health.

2.3 Overview of traditional EHR management systems

EHR systems have been essential for organizing flow and storing of the healthcare information. These frameworks are often centralized databases within a healthcare facility with specific servers that are outsourced. Traditional EHRs are those that existed prior to the time when digital means are standard for storing healthcare information that doctors can update using technology.

Traditional EHR systems can be defined by emphasizing on the unification of comprehensive patient's health information from various sources including clinical encounters and laboratory tests. It means that the entire medical data of the patient are accumulated in one place; they make care effortless and assist healthcare professionals in selecting superior therapies.

Data entry and management technology brings to the table of healthcare providers specific tools in documenting the visit of the patient and reviewing past records of a patient's health as well as in ordering certain tests and prescriptions. To increase the efficiency of the documentation and to ensure the required uniformity in guaranty records these systems can have basic interfaces and can be tailored according to different medical professional positions.

Figure 3 illustrates traditional Electronic Health Record (EHR) management systems, demonstrating how a patient's health information flows from data input and storage to output processes and security controls utilized by both patients and healthcare providers. These conventional systems offer functionalities such as access monitoring, security management, and data protection to prevent unauthorized access, legal violations (such as those under HIPAA), and to ensure the confidentiality of patient information.

While traditional EHR systems are still widely adopted and considered effective, they exhibit several limitations. One significant challenge is interoperability—the difficulty in sharing and integrating data across different Health Information Systems (HIS). This limitation often leads to information silos, restricting the seamless exchange of patient records and impeding coordinated care.

Interoperability issues: Interoperability issues in EHR systems are as such understated, which leads to the isolation of patient information and limits their integration with other HIS.

Data silos: Before the implementation of Share Everywhere EHR systems, patient health information was typically stored within the boundaries of individual healthcare organizations, often in proprietary platforms. This limited the ability to aggregate and share patient data across different organizations.

Centralization and single point of failure: This was so when EHR databases were centralized and it opened the doors for data leak and congestion of the service in a way that affects patient's information.

Vendor lock-in: As proprietary EHR systems are already implemented in healthcare entities, the latter have few choices at their disposal since they cannot tailor systems.

The traditional EHR systems play a great role in enhancing digital transformation of healthcare data together with increased support for clinical documentation. Some limitations limit this system's ability particularly in the areas of data communication and security. Blockchain technology has introduced a new approach to solving these concerns and enhancing safety and confidentiality of EHR systems in innovative ways.

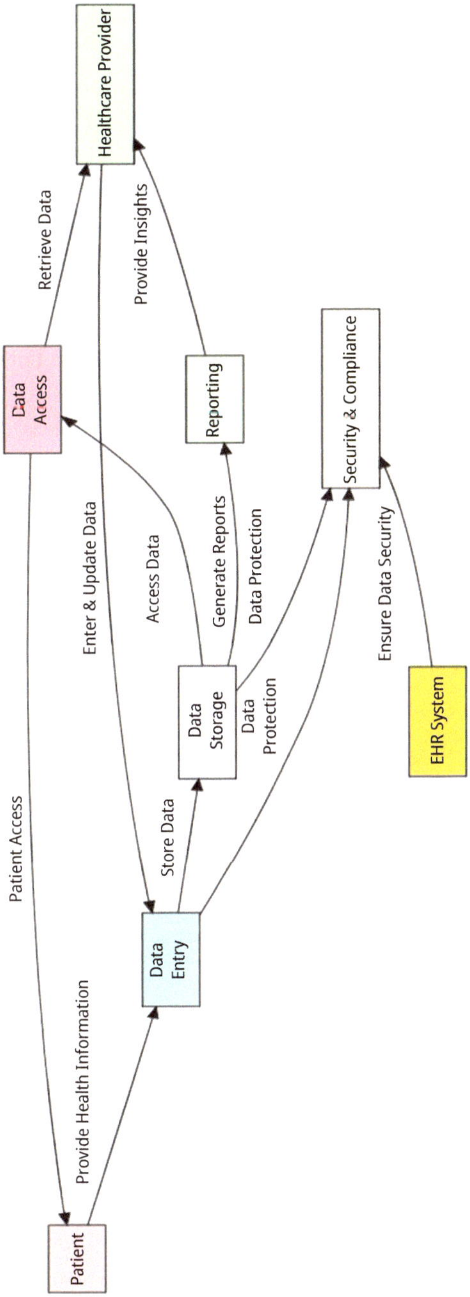

Figure 3: Overview of traditional EHR management systems.

2.4 Evolution of blockchain-based EHR solutions

A decentralized safe replacement to standard EHR systems came with the rise of blockchain EHRs. Because sharing of healthcare data required more secure and semantic communication, the idea of using blockchain technology came up in the 2010s.

Blockchain technology in its early stages of evolution mainly provided a means for supporting "transaction ledgers" and for securely creating and confirming digital currency such as bitcoins. Others began to realize how blockchain can preserve and handle special health information in addition to monetary applications.

As per many specialists, at an early stage of enhancement of blockchain-based EHR systems, all the emphasis was given to demonstrating projects to explore how useful blockchain was in healthcare. Although these initial attempts suffered from concerns of scalability and compliance, they helped showcase basic use cases of blockchain technology in terms of data security and the ability to share data.

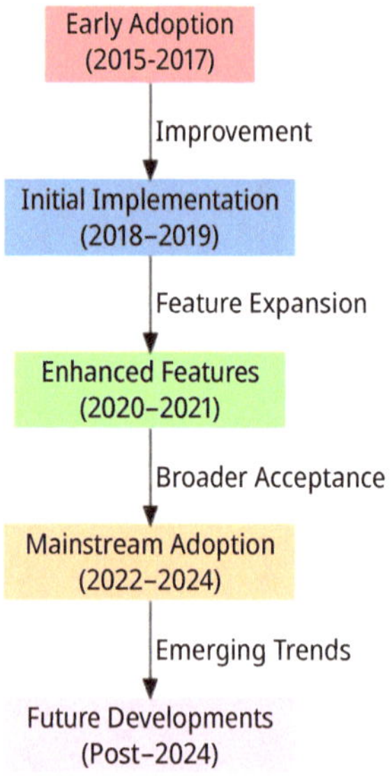

Figure 4: Evolution of blockchain-based EHR Solutions.

Solutions that integrated blockchain technology across a healthcare system came to the fore as the technology advanced and gained adoption. These solutions brought new technologies and functionalities in order to satisfy the demands of patients and healthcare organizations. All of these emerging trends address the issues of conventional EHR systems and open new possibilities for healthcare data handling using the advantages of the main characteristics of the blockchain: decentralization and cryptographic security.

2.5 Key milestones in the evolution of blockchain-based EHR solutions

Development of healthcare-specific blockchain platforms: Starts to develop blockchain solutions with an attempt to target the field of healthcare with different characteristics such as consensus algorithms and conditional contracts.

Integration with existing healthcare infrastructure: Blockchain-based EHR solutions are designed to interface with existing healthcare information systems to enhance interoperability, allowing diverse systems to exchange and share data while preserving privacy and ensuring data integrity (see Figure 5).

Standardization and regulatory compliance: To meet legal requirements related to regulations such as the GDPR and HIPAA, standardized guidelines have been developed for blockchain-enabled electronic health record systems.

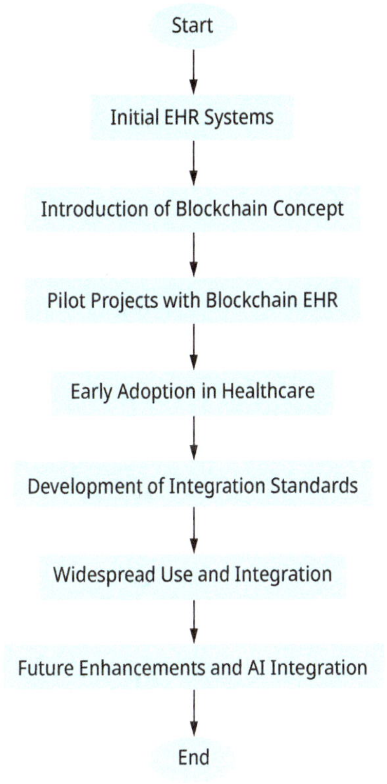

Figure 5: Key milestones in the evolution of blockchain-based EHR.

2.6 Key components and architecture of blockchain-based EHR systems

The characteristics of blockchain systems and the materials required for their implementation also guarantee the secure and compatible management of patient health information. An illustration of the components and design of the health record systems based on the blockchain paradigm is presented in Figure 6 while arguing about the patient's privacy and protection of their data. Figure 6 shows the basic building blocks and architecture of blockchain-based EHR systems, as well as an explanation of how blockchain was incorporated in order to make EHRs more secure and sharable for patients and clinicians.

Distributed ledger: Clinical data of the patients is safely stored in the blocks, which constitute a major component of the blockchain EHRS. The ledger provides data security in that it presents a clear and time-tagged view of transactions. EHR's state change happens with each transaction that creates a blockchain that links to the previous block cryptographically.

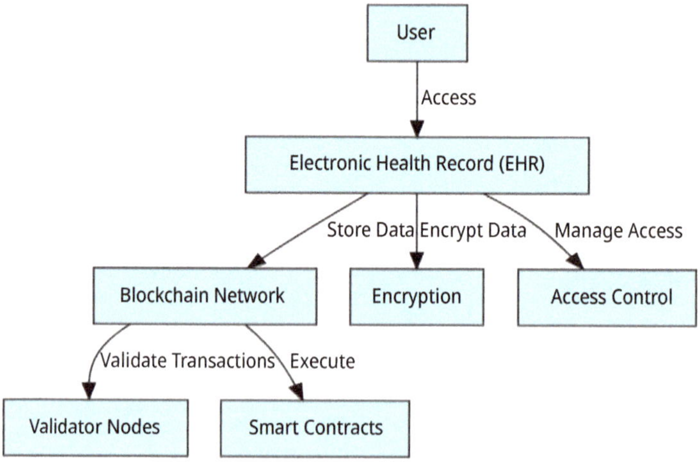

Figure 6: Key components and architecture of blockchain-based EHR systems.

Smart contracts: Smart contracts refer to terms that when fulfilled back agreements, perform actions without any supervision, and are written into the blockchain. It revealed that smart contracts control data access rights as well as other aspects of EHR systems based on the blockchain. Smart contracts mention who has the access to certain clinical information for purposes of data privacy and security.

Consensus mechanism: Protocols with the designation consensus mechanisms maintain consensus in blockchain systems and make users aware of the transaction legitimacy. BFT and PoS and PoW are known as consensus mechanisms. To ensure all the participants of the network are aware of the legal tender status of EHR transactions and security of the ledger in blockchain systems, suitable consensus methods are used.

Cryptographic hashing: A data block in the blockchain process in which we get a unique hash by applying cryptographic hashing techniques. These identities firmly link block elements safely and act as electronic labels of contents of the data. The members can confirm the trustworthiness of the blockchain and identify illegal modifications to the data using hashed values.

Decentralized network: Every node within a network established for the processing of blockchain records holds an updated version of the ledger. From the decentralized confirmation of this system, there are lesser chances of giving a point to be targeted when one weak point has been eliminated. In peer-to-peer networks, protocols have nodes that help them to link with each other and ensure data replication and synchronization within a network.

User Interface and Applications: Healthcare consumers connect to EHR systems on blockchain networks through user interfaces and applications tailored to the stakeholders of a particular healthcare provider. These interfaces allow users to navigate EHRs, review medical documents, manage access permissions, and conduct transactions. In addition to integrating with traditional healthcare institutions, these systems offer capabilities via web interfaces and mobile tools, enabling broader engagement and interaction.

3 Security and privacy in blockchain-based EHR systems

3.1 Importance of security and privacy in healthcare data

Sensitivity, reliability of patient information, security, and privacy turn into critical factors when managing health information. In the healthcare data management system, the statement, "secure and private" is an important aspect to enhance patient trust while adhering to government requirements and protecting patient data from exploitative use.

The two practices to prevent cyber threats and unwanted access into healthcare data are defense measures. Patient information such as personal identification details and personalized/routine treatment plans are stored by the healthcare firms attract high monetary value by the cyber criminals. The consequences of violation of healthcare data are severe: financial loss, damage to people's image, threats of identity theft, and possible harm to life of patients. To avoid situations where patient identity is disclosed proper measures such as encryption and firewalls must be working as required.

From Figure 7 below, it emerges that emphasis on security and privacy is critical to healthcare data. Collection, storage, and transmission of data in this sub-sector deserve the same level of protection as the information from unauthorized access. The ordinary citizens and consumers expect healthcare organizations and providers to protect their data correctly and honorably, so that they alone can decide who gains access to their personal health information. To ensure that patients continue to have confidence in them and work beneficially with doctors it is important to protect patients' information because it is legally and ethically right. The general public rarely presents themselves to seek medical assistance from doctors after a breach on their privacy has happened.

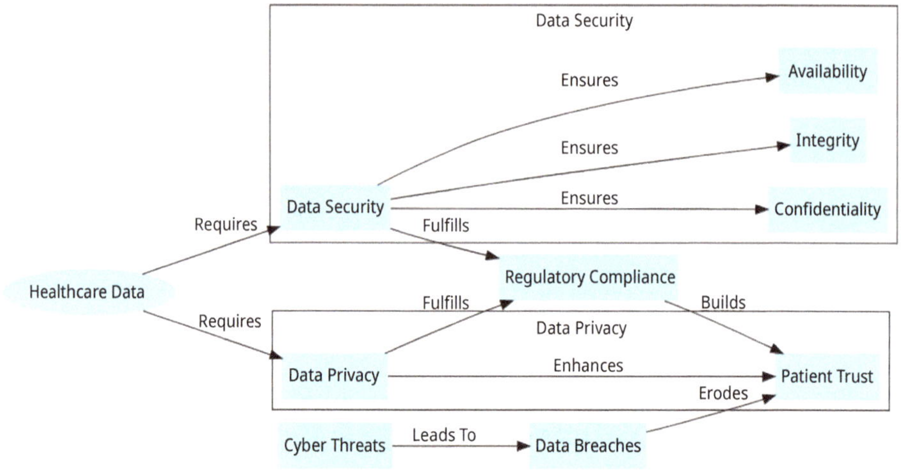

Figure 7: Importance of security and privacy in healthcare data.

The privacy of healthcare data is important in terms of general data protection regulations, for example, GDPR and health information privacy and accountability act (HIPAA). These norms define specific expectations on how patient medical data safety and privacy can be maintained along with activities as simple as data anonymization and as severe as breach notification. Should these regulations be flouted, severe consequences could develop with a legal aspect as well as the deterioration of the image of the healthcare providers.

In any healthcare data management to protect the privacy of the patients, in addition to the legal requirement, security and privacy are key to protect the patient details. The healthcare organizations have to implement both security and privacy measures. Including the security and privacy of data in the strategic perspectives of these organizations' lowers the levels of risks associated with data leakage and enables a high level of ethical and lawful practices in healthcare settings.

3.2 Security challenges in traditional EHR systems

To address requirements including the GDPR and HIPAA, of safeguarding healthcare-related information is required. Essential measures required for preserving patient health information are covered by regulations and include data encryption and limitation of access to these data. Failure to meet these legal obligations can attract heavy penalties, legal problems, and loss of reputation to the healthcare organizations.

In healthcare data management, to protect patient's identity and build and sustain patient confidence, and to meet regulatory requirements, both security and privacy are crucial. If the patient data were to be adequately protected, healthcare entities need to enforce security and privacy mechanisms. By ensuring protection of privacy during data handling processes, firms can reduce the threat of data violations concerning the patients' confidentiality.

Centralization: The conventional EHR systems rely on centralized databases resident on separate servers of the third-party vendors of the implementing healthcare facilities. Since most of the patient's information was stored in one point where it was easy to share and retrieve the information, the system was extremely prone to hacking, unlawful access, and data corruption. Many patients' electronic health records are at risk; just one successful attack on a central EHR system can enable fraudulent activity, identity thefts, and other vices.

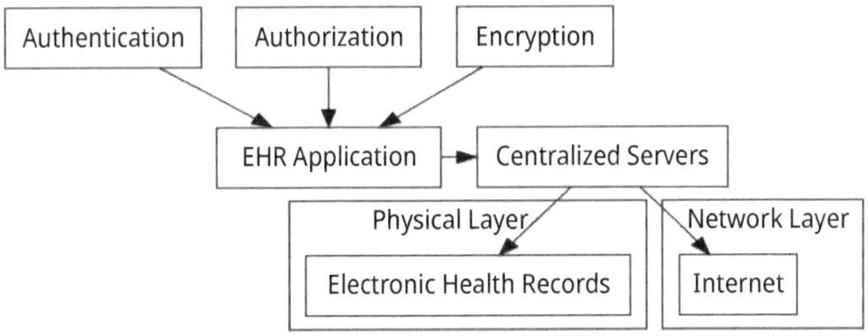

Figure 8: Security challenges in traditional EHR systems.

Insider threats: This threat arises from the fact that when workers, vendors, and contractors have access to personal patient information, they are a risk to healthcare businesses. There are inherent dangers in the different types of insider threats, both accidental and of the palindromic kind, in which the internal attacker aims to simply steal and corrupt patient data for personal reasons. Standard EHR structures are often deficient in strong auditing, and inside threats are often hard to detect and prevent.

Inadequate authentication and access controls: Lack of strong techniques in authentication and poor controls to access in the conventional EHR systems are SIAs. Some of the vulnerabilities include: password-based authentication was vulnerable to brute force attack, phishing, and password guessing and this would compromise user accounts and make it easy for unauthorized person to get access to the patient's record. Lack of proper access control measures exposed patient records to information invasions, modifications, and destruction by end-users who are not privileged to do so.

Vulnerabilities in legacy systems: These prove that most of the conventional EHR systems in use are based on older technologies with inherent software defects, and hackers have laid their eyes on them.

These vices were exploited by cybercriminals in order to compromise data and infuse damaging malignant software, by gaining unlawful access into EHR systems and confidential patient information. Correcting and upgrading the stand-in systems was complex, besides which they increased the security hazards.

Absence of data encryption: Some of the encryption methods supporting the protection of patient information in transit and at the time of storage are missing in the preexisting EHR systems. In the context of databases over the networks, the data remains unencrypted and is easily vulnerable to getting intercepted, methodologies allowed unauthorized access and access by the adversaries. In brief, without encryption, the privacy and confidentiality of the patient information are compromised since medical records are venerable to being lost, modified, and disclosed.

Numerous security flaws in traditional EHR systems jeopardize patient data's availability, confidentiality, and integrity. Healthcare companies must lay down strong security measures to address these issues, including encryption, access limits, auditing procedures, and recurring security assessments. Switching to blockchain-based, decentralized EHR systems presents viable ways to reduce security threats and improve data security, privacy, and integrity in the administration of medical records.

3.3 Enhancement of blockchain security in EHR management

Blockchain technology comprehensively eliminates the major shortcomings of typical EHR systems through several ways that enhanced security in EHR management. The adoption of blockchain technology eliminates the central administrator in managing EHRs, cryptographic methods, and consensus procedures enhance security. These benefits include:

Decentralization: A copy of the distributed ledger containing EHRRs is maintained on the file by each node in the chain of decentralized nodes of a blockchain. As a result of this decentralization there is a single point of failure, thereby reducing the risks of unauthorized access to EHR data. In blockchain, all nodes in the network contain a copy of the ledger and therefore even if the ledger of a particular node is compromised the blockchain remains sanctimonious.

In blockchain, there is a record of transactions that cannot be altered in any form, and for each block of data an albatross was formed with the previous block of data. Once a transaction was written to the blockchain, it cannot be altered without the cooperation of the majority of users in the network it serves. EHR data is guaranteed to be entirely traceable as well as unaltered due to its inherent feature of being unalterable by other parties.

Cryptographic security: To protect EHR data during its lifetime possession, blockchain adopts features of cryptology such as encryption, digital signatures, and hashing. Since every transaction performed on the blockchain is hashed, a unique ID can be obtained from it, which links it to the its previous block. It should be noted that only the specified users can send transactions to a blockchain using the digital signatures that check the validity of the transactions. Secure EHR data are used by applying encryptions to the data that are stored in the blockchain to increase security and privacy.

Consensus processes: It is on the basis of consensus processes that the users of a network can be effectively convinced to accept the legitimacy of transactions on blockchain. Security prevents immoral individuals from making any alterations to the chain by making sure that every node in the distributed system reaches a consensus of the status of the chain in question. Consensus procedures enhance the data stored in the blockchain EHR by strengthening participants' trust as mentioned below.

Access control and permission: include EHR data access control and permission rules that were developed using smart contracts which are self-executing contracts implemented on blockchain. To ensure that only the right people for the specific patient interact with information, smart contracts include rules and permission settings for EHR record use or modification. Smart contracts decrease the probability of unauthorized access leakage in blockchain-based EHR systems because of a robust access control system involved.

EHRs benefits from several features and approaches offered by the blockchain such as decentralization, immutable ledger, cryptographic security, consensus mechanism, and access control in enhancing the security of managing EHRs. Blockchain-based EHR solutions are secure due to the features that support patient confidentiality, integrity, and privacy, while providing a clear and secure way to store, maintain, and exchange patient health information.

3.4 Privacy-preserving techniques in blockchain-based EHRs

This aspect is crucial because the EHR systems developed based on blockchain technologies require explicit consideration of privacy preservation due to the fact that the data stored in EHRs is sensitive to unauthorized access and publication. There are several methods and approaches about how Blockchain in support of privacy to enhance the administration of EHRs for patients guarantee their privacy while allowing the parties to exchange data safely and cooperate. Among the most important privacy-preserving strategies used in blockchain-based EHRs are:

Data encryption: To avoid unauthorized access and breach of the patient's information, the information in EHRs on the blockchain is encrypted. By encrypting important patient identification and medical data such as diagnoses, treatments, EHR systems that are based on blockchain assure patient privacy. This disallows unauthorized users to access the granted data especially since they are encrypted. Access to the NMDB is tightly controlled such that only personnel with the appropriate credentials can access the data; the data is encrypted such that keys to such access are required to decipher the data in question.

Data minimization: To reduce the exposure of data to unauthorized parties, data minimization best practices limit the quantity of PII kept on the blockchain. Minimized to truly needed metadata hashes of the data, EHRs should remain off the blockchain. This method keeps data on the blockchain while preserving data privacy and it also allows far less exposure of patient-specific information as a result.

Zero-knowledge proofs (ZKPs): ZKPs are cryptographic methods that let the prover demonstrate to the verifier that a statement is true without disclosing any further information. ZKPs can be used to confirm the legitimacy of transactions of EHR data in the context of blockchain-based EHRs without revealing the underlying patient data. This makes it possible for numerous parties to validate and share data on the blockchain in a safe and private manner.

Differential privacy is a strategy for strengthening privacy that tries to safeguard individual privacy by adding noise to aggregated data, all the while enabling useful analysis of the data. Differential privacy can be used in blockchain-based electronic health record systems to aggregate and anonymize patient data prior to its storage on the blockchain. This method helps prevent patient reidentification while maintaining the quality and usefulness of the data for analysis and study by hiding individual contributions to the dataset.

Enhancing privacy: Smart contracts used on blockchain technology can impose access limitations and privacy-maintaining guidelines for electronic health record information. Smart contracts that improve privacy, for instance, include guidelines for managing consent, exchanging data, and anonymizing health records. These contracts ensure that patient privacy wishes are honored and that data is exchanged safely and openly among authorized parties by automatically enforcing privacy regulations stored in the blockchain.

3.5 Compliance with healthcare regulations (e.g., GDPR, HIPAA)

One strategy of creating and implementing blockchain-based EHR systems with patient data confidentiality and privacy guarantees is by the legislation of healthcare. The two popular legal structures for managing privacy of health information are called the General Data Protection Regulation (GDPR) that operates in the European Union and the Health Insurance Portability and Accountability Act (HIPAA) that exists in the United States of America. Subsequently, patient's identity and the information provided have to be protected and for organizations involved in the provision of healthcare and technology firms, these rules mean that they need to have appropriate security and privacy measures in place.

GDPR: The General Data Protection Regulation is a comprehensive data protection law that establishes strict standards for the collection and processing of personal data belonging to individuals within the European Union (EU) and the European Economic Area (EEA). Healthcare organizations and technology vendors are obligated under GDPR to ensure the lawful, fair, and transparent processing of Protected Health Information (PHI) and other EHR data. This includes obtaining explicit patient consent before the collection and use of personal data, ensuring that consent is specific and informed, implementing appropriate organizational and technical safeguards to prevent unauthorized access, and notifying relevant authorities of any data breach without undue delay and no later than 72 hours after becoming aware of it. Additionally, healthcare entities must ensure compliance with GDPR provisions regarding individuals' rights, such as the right to access, rectify, or erase their personal data.

HIPAA: HIPAA is a federal statute in the USA that deals with how healthcare providers, health plans, and other entities that are bound by it should handle PHI. Thus, for ensuring confidentiality, integrity, and availability of PHI, including EHR data, HIPAA lays down stringent security and privacy measures. To curtail the access, use of PHI by parties other than the individual, the covered organizations and the business associates need to develop administrative, physical and technical safeguards. These covered procedures include risk assessments, audit trails, encryption, and access restrictions. HIPAA mandates that covered organizations get patient consent before using PHI for reasons other than treatment, payment, or operations. Patients must also be able to view their own health information by requesting it.

Healthcare organizations and IT companies developing and implementing blockchain-based EHR systems must adhere to GDPR, HIPAA, and other pertinent healthcare legislation. Serious consequences, such as fines, legal ramifications, and reputational harm arise from breaking these restrictions. As a result, businesses need to make sure that their blockchain-based EHR systems follow the guidelines and specifications specified in these laws, giving patients' health information security, privacy, and rights top priority. Healthcare organizations foster trust and confidence among patients, regulators, and other stakeholders about the integrity and confidentiality of their EHR systems by exhibiting compliance with GDPR, HIPAA, and other legal frameworks.

4 Interoperability and data sharing in blockchain-based EHR systems

4.1 Interoperability challenges in healthcare data exchange

Interoperability is often considered to be the necessary precondition for efficient healthcare data sharing and cooperation and is defined as the ability of different information systems and devices to exchange and utilize data seamlessly. The achievement of interoperability on healthcare data exchange brings a lot of challenges that hinder the easy sharing and incorporation of EHR across different systems and institutions.

Lack of standardization: Lack of standard formats, protocols, and vocabulary usage for describing and exchanging EHR data is one of the biggest issues with healthcare data interchange. To extract, transfer, and analyze data, healthcare industry often uses proprietary coding and formats, which result in conversion issues. Failure to standardize results in formation of data silos, clutters data duplicates, and reduces effective delivery of healthcare services by compromising the integration of EHR systems.

Fragmented EHR systems: Conventional healthcare organization adopted multiple EHR systems provided by various vendor solutions based on the organization's requirements. Currently, there are huge differences in EHR systems regarding their data models, data interfaces, and offered functionalities, and thus data exchange between different systems is a challenging task. Lack of integration of EHR systems also means that the data gathered in the systems remain in some sense isolated from information in other systems.

Data governance and consent management: HL7 standards in care consent management give protocols and policies on the use and sharing of patient's health information regarding healthcare data exchange. For example, the development of uniform data governance and consent initiatives across a number of disparate healthcare institutions is difficult because of the different regulations, privacy protection laws, and institutional guidelines in various settings. The current situation of data management and consent in different systems and patients' identity hinders integration and raises questions concerning data privacy and patients' privacy rights.

Technical integration complexity: Since EHR systems need to interoperate with other HIS like LIS, imaging systems, and electronic prescribing systems, we are faced with some technical issues of data formats, protocol, and interfaces. Multiple and distinct architectures and the use of legacy systems, proprietary APIs, and others, also contribute to the technical integration problem, raising development time and costs, at the same time. Sustaining compatibility as systems grow and develop, with the ever-transforming technological environment raises a number of cognate issues to healthcare organizations.

Semantic interoperability: Semantic interoperability that allows data to be shared as well as understood by receiving IT systems and applications is crucial in guaranteeing the meaningful and consistent sharing of EHRs. To get semantic interoperability there is need to use common coding schema, vocabularies, ontologies, and data type to encode clinical terms and their association. Semantic incompatibilities of terminologies, coding discrepancies, and semantic vagueness inhibit semantic integration, subsequently causing data variation, mistakes, and misunderstanding.

The main hurdles to achieving efficient interoperability in exchange of data in the sphere of healthcare include lack of unified data models, absence of unity in the EHR systems being used, and data management problems that give rise to integration problems and semantic misinterpretations. These challenges call for increased working together by members in the healthcare system to establish and implement interoperability roadmaps, data management policies, technical integration solutions, and semantic convergence. Those interoperability challenges make healthcare organizations ensure that data exchange is done effectively, care coordination is enhanced, and patients' outcomes are improved in blockchain-based EHR systems.

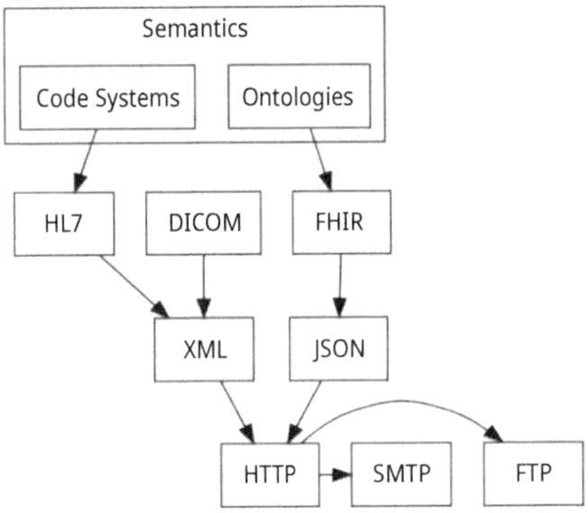

Figure 9: Interoperability challenges in healthcare data exchange.

Figure 9 illustrates the intricacies of interoperability challenges in healthcare data exchange. It delineates key components essential for achieving seamless data exchange: Standards like HL7, FHIR, and DICOM serve as foundational frameworks, dictating data representation. Data formats, namely XML and JSON, determine the structural organization of healthcare data, influencing its exchange. Semantics, represented by code systems and ontologies, play a pivotal role in ensuring accurate interpretation across systems. Meanwhile, transport mechanisms like HTTP, FTP, and SMTP facilitate the movement of healthcare data between systems. This comprehensive visualization highlighted the multifaceted nature of interoperability hurdles, encompassing standardization, data format, semantics, and transport, imperative for effective healthcare data exchange.

4.2 Role of blockchain in enabling interoperability

Blockchain technology has potential to help with healthcare data exchange interoperability issues and make data sharing across different EHR systems easier. Blockchain networks' distributed and decentralized architecture, along with its cryptographic security characteristics, provide a number of benefits for improving interoperability in the transmission of medical data:

Decentralized data management: Distributed ledgers holding patient health data are maintained in duplicate by every member of decentralized networks of nodes, which are the foundation of blockchain-based EHR systems. Because of its decentralized design, peer-to-peer communication and cooperation between healthcare professionals, patients, and other stakeholders was made possible with the requirement for a central authority to permit data transmission. Blockchain lessens dependency on centralized repositories and eases interoperability issues brought on by incompatible EHR systems by decentralizing data management.

Consensus methods: In order to reach a consensus among network users on the legitimacy of transactions, blockchain depends on consensus methods like PoW. Ensuring that every node in the network has a shared knowledge of the blockchain's current state, consensus methods help to build confidence and agreement on the accuracy of EHR data. Blockchain improves interoperability by encouraging confidence between many parties and enabling safe data sharing by offering a shared and unchangeable record of transactions.

Smart contracts built into blockchain hold the potential for enabling self-execution of exchange and interaction protocols of data contained in multiple EHRs. Specifically, smart contracts conceptualize policies governing the access, modification, and sharing of EHR information, while guaranteeing compliance with interoperability standards and data management policies in various systems. Smart contracts become effective means for increasing the speed and integrity of data exchange between participants in healthcare systems through the automation of exchange processes and compliance with standards for interoperability.

Data standardization and semantic interoperability: Blockchain-based EHR systems can take advantage of standard structure and exchange formats, and standard terminology and knowledge representation systems for semantically seamless use and interchangeability of EHR data. The use of common data understandings and common languages enables EHR systems built on the application of blockchain technology to effectively share and exchange these patient details without a high probability of data incompatibilities, errors, and misunderstandings. Interoperability is further promoted in this case by standardization combined with blockchain technology, both on the technical and the semantic levels of sharing health information.

Patient-controlled data sharing: EHR systems built on blockchain technology provide patients the ability to have control over their personal health data and grant access to just certain physicians approved organizations. For privacy, security, and consent management, patients are allowed to provide access to specific EHR data to specific parties by signing them digitally with digital signatures and cryptographically with keys. Blockchain allows for fine-grained management of read/write access to specific data fields and places data ownership back in the hands of the patient, which emphasizes both patient sovereignty and patient centricity while simultaneously promoting high levels of system integration.

Blockchain has a central function in unlocking the interoperability feature of health information exchange by integrating decentralization, consensus, smart contract, and information convergence with patient-directed data accessibility. As a result, with the common key issues of interoperability addressed and the use of blockchain technology in facilitating a decentralized, secure, and transparent data exchange within blockchain-based EHR systems, it is about time that the healthcare system within this connected environment delivers value and improves outcomes for patients.

4.3 Standards and protocols for interoperable EHR systems

Interoperability within EHR systems entails compliance with standard data and structural interoperability, communication, as well as the relationship between EHR systems. A number of standards and protocols have been evolved, which offer standards and guidelines for representing, exchanging, and integrating EHR systems. Some of the key standards and protocols for interoperable EHR systems include:

Specifications that relate to the exchange, combination, sharing and access to electronic health information are referred to as HL7 (Health Level Seven) standards. IT infrastructure of healthcare operates on HL7 standards, which offer data formats, communication formats and interaction standards for clinical care, public health, and administrative healthcare transactions, among others. HL7 standards are applicable to the management of EHR information as well as facilitate integration of information from one HIT system with another. Some of these standards are HL7 Version 2 (V2) and HL7 Version 3 (V3).

FHIR which stands for Fast Healthcare Interoperability Resources is a relatively modern interoperability standard unveiled by HL7 that employs Web-based technologies to allow timely and seamless exchange of health information. Examples of these technologies include RESTful APIs (Application Programming Interfaces). EHR continuation is the defined structure of the data, namely patient identification data, observations derived from the patient, prescribed medications, etc., based on the FHIR resources using a concept of the module and extension. FHIR's basic and fine-grain features make it ideal for interoperability of EHR systems to enable developers to construct light weight as well as interoperability applications that are capable of acquiring EHR data.

DICOM (Digital Imaging and Communications in Medicine): DICOM is a standard for the communications and storage of medical images and related information throughout the healthcare systems. DICOM encompasses a full solution for protocols, data formatting, and methods of communication for integrated sharing, storage, and retrieval of images in various imaging specialties and healthcare organizations/centers. DICOM facilitates exchange of medical imaging data in a way that allows them to interoperate with EHRS; it guarantees availability of diagnostic images in the overall ambience of the healthcare systems.

IHE (Integrating the Healthcare Enterprise) profiles: IHE is a strategy for working with "clusters" of higher-level standards as well as defining guidelines and interoperability profiles for implementation. This defines various HL7 and DICOM use cases and how they have been implemented and exist to support various interoperability information exchange use cases such as lab result exchange, transferring medical images, and querying patient demographic. Through specification of connectivity profiles and supporting templates IHE ensures effective data exchange and integration of EHRs.

SMART on FHIR: SMART (Substitutable Medical Applications, Reusable Technologies) on FHIR was an open framework for developing cross-platform healthcare applications operating on the FHIR framework. SMART on FHIR includes specific principles and prescriptions for designing and building loosely coupled components that integrate with EHRs through FHIR interfaces. SMART on FHIR delivers through web-based technologies and applying open standards, empowers developers to build applications that are integrated with EHR systems to advance solutions for health information exchange and interoperability.

4.4 Blockchain-based solutions for seamless data sharing

Blockchain can provide new approaches to enable EHR data sharing, solve problems of interoperability, and promote more effective exchange of both electronic and personal health information. Some solutions utilizing blockchain technology have been proposed to improve data exchange between different EHR systems; these solutions all take advantage of unique characteristics of the blockchain platform to foster interoperability and cooperation. As shown in Figure 10, blockchain-based solutions for data sharing by integrating blockchain technology into the solution environment is illustrated for better understanding of the associated benefits such as security and efficiency of data sharing between involved parties in the healthcare system. Some of the key blockchain-based solutions for seamless data sharing in EHR systems include:

Decentralized data exchange platforms: The opposite of centralized middlemen, the decentralized data exchange based on the blockchain helps patients, healthcare organizations, and other parties to exchange and get required EHR data securely. These systems allow for dissemination of data among systems with privacy, integrity, and consent by incorporating blockchain's dispersal of data, the cryptographic principles, and smart contract. Decentralized data exchange platforms work on the principles of multiple data systems and do not require intermediaries and a central hub to share data and allow EHR systems and healthcare facilities to exchange information.

Federated learning and data sharing networks: On the basis of blockchain technology, federated learning and data sharing networks enable cooperative studying and sharing of EHR data together with preserving data confidentiality and privacy. In federated learning localization, several decentralized nodes have a part of EHR data, and collaboratively train a machine learning model. When data is aggregated and analyzed, cryptographic approaches like differential privacy and homomorphic encryption safeguard patient privacy. Blockchain guarantees the integrity and traceability of model modifications and data contributions. Federated learning networks facilitate interoperability and knowledge exchange between academics and healthcare organizations by providing safe and private data sharing.

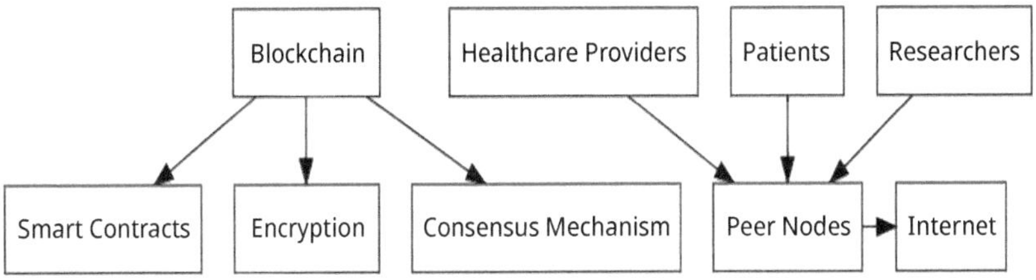

Figure 10: Blockchain-based solutions for seamless data sharing.

Patient-controlled data sharing platforms: These allow the ownership and control of personal health data and allow health and research recipients access to them only when appropriate. These blockchain-based systems allow clients to grant different parties access to specific EHR information while ensuring data privacy and control of access by use of digital signature and keys. Systems that patients control also means that patient data can be easily provided and accessed across multiple EHR systems and across the various settings where patients seek care.

Interoperable health information exchanges (HIEs): Interconnected HIEs leveraging information technology consult blockchain to enable a secure and standardized exchange of electronic data across greater healthcare systems. Smart contracts enable efficient handling of permission-related processes and contracts between several parties; blockchain ensures the availability and credibility of EHR data transferred through exchange implementation. Blockchain-based HIEs enable the sharing of information across care givers and patients, co-ordination of care between different providers and systems, as well as cooperation between patients through an interoperable sharing platform.

Standards-based blockchain networks: Such networks leverage standard and recognized norms and procedures for conglomerate connectivity, for instance, HL7 FHIR, for the conversions and the reconciliations of EHR information among different blockchain networks and healthcare systems. Standardized blockchain networks ensure compatibility, interconnectivity, and interchangeability of data across the healthcare systems since they conform to set data format, messaging, and integration standards embraced by the various networks that make up a standards-based blockchain. This promotes interoperability and data sharing across varied EHR systems and stakeholders.

4.5 Case studies of successful interoperable EHR implementations

Examples of the effective use of EHR systems connected through an interoperable network with the help of blockchain technology are described, and such cases indicate promising areas of the application of blockchain in the healthcare domain. These case descriptions demonstrate that conceptualization and strategy development are critical to creating viable and successful models of interoperability and data sharing among many stakeholders in healthcare systems. Some notable examples of successful interoperable EHR implementations include the following:

Medicalchain happens to be an application that is run through the blockchain platform and enables patients to store their electronic health records securely and share securely with other parties like doctors and researchers. Through using of the blockchain technology, Medicalchain guarantees data security and nondisclosure of patient's health records while creating an auditable trail. The EHR data was electronically shared by permissioned contracts, which include provision of consent for the authorized healthcare providers to access the data. Medicalchain's interoperable EHR solution resolves health stakeholders working in unison by ensuring proper communication and thus patient care is enhanced.

Coral health: Coral Health was an EHR platform designed on blockchain characterized by seamless integration of records from one ecosystem to another. Coral Health offers a patient health record repository through blockchain, making it easier for patients, payers, and providers to share EHR data. Patient identity and data security are also maintained during the exchange of data through cryptographic techniques employed on the platform. Coral Health EHR Solution was an integrated EHR system where the sharing of data, open integration, and team approach enhance quality and operational performance of care.

Nebula Genomics: Nebula Genomics is the company that is built on blockchain technology to use it for genomic data sharing and analysis. Nebula Genomics operates on the blockchain, which is used to establish a safe and trustworthy market place for genomics data where people can sell their genomics data to researchers, pharmaceutical companies, and other healthcare providers. Blockchain maintains confidentiality and data authenticity of individuals' genomes, while the smart contracts bring efficiency in decision-making for sharing of data and reward for contributors. With Nebula Genomics' interoperable platform, all individuals, researchers, and organizations involved in genomics and personalized medicine can work together to unearth more discoveries and develop new related technologies that push the boundaries of healthcare.

5 Smart contracts for access control and governance

5.1 Introduction to smart contracts in blockchain technology

Smart contracts contain the terms of the contract straight up in the code. This shall mean that in the implementation of the contractual provisions, enforcement shall be done via a blockchain network without the intermediaries. An important part of the blockchain infrastructure is smart contract capable of performing agreements, business logic, and transaction programmatically and in a decentralized manner.

In its simplest form, smart contracts are self-executing programs that reside on blockchain, which execute automatically once predetermined conditions are met. These conditions are often called triggers and are coded into a smart contract and occur on their own when activated by outside events. Although smart contracts are initiated by the user, the set actions are executed without the user's direct interjection and with no human interference.

One feature of smart contracts is its potential to facilitate customizable and credible cross-party transactions. Smart contract results of those executions are transparent and can be audited as well since they occur on a blockchain, which is inherently open and immutable. Therefore, today there is no need to have middlemen and other reliable parties for supervision and verification of the transactions, which reduces the costs, time, and risks associated with the traditional methods of contract execution.

The use of smart contracts can automate a large spectrum of agreements and transactions ranging from banking, real estate, supply chain, as well as healthcare. Since blockchain is an emerging technology to support health information sharing, smart contracts play a decisive role in the consent process, governance, and access rights in blockchain-based EHR systems. Smart contracts ensure that only authorized parties have access to patient health information and that data sharing happens in accordance with set rules and permissions as also describe data sharing agreements, automate consent management procedures, and enforce access control measures.

5.2 Application of smart contracts in healthcare

Smart contracts are a true revolution in the progressively developing field of healthcare owing to the numerous opportunities created for the efficient management of data, its protection, and access. Smart contracts that leverage on blockchain technology safely are transparent and automatically execute protocols, agreements, and transactions. This ultimately fiddles the effectiveness, responsibility, and reliability of the healthcare structures. Among the most important uses of smart contracts in the medical field are:

Smart contracts are used to handle patient permission for the sharing and use of EHR data, as well as to enforce access control measures. Healthcare businesses guarantee that sensitive patient information is only accessed by authorized persons by encoding consent preferences and access rights into smart contracts. Smart Contracts enhance security, privacy, and data protection laws because they automate consent management procedures and provide patients with centralized authority over who receives their health information and when.

Using an example represented in Figure 11, one can see how smart contracts work in building a healthcare system and eradicating the need in many intermediaries that concentrate on such procedures as patient consent, billing, information exchange.

Claims Processing and Insurance Verification: There is potential to effectively automate processes such as claims and insurance verifications in the field of healthcare thanks to smart contracts that can facilitate insurance claims decision making and settlement. Through the use of smart contracts, the insurer and healthcare providers are thus provided with the means of automating the process of checking the validity, admissibility, and number of claims to be paid against the entire policies, coverage, and reimbursement rules are issued. Smart contracts eliminate some of the challenges of using traditional claims processing approaches by enabling clear and accurate transaction tracking.

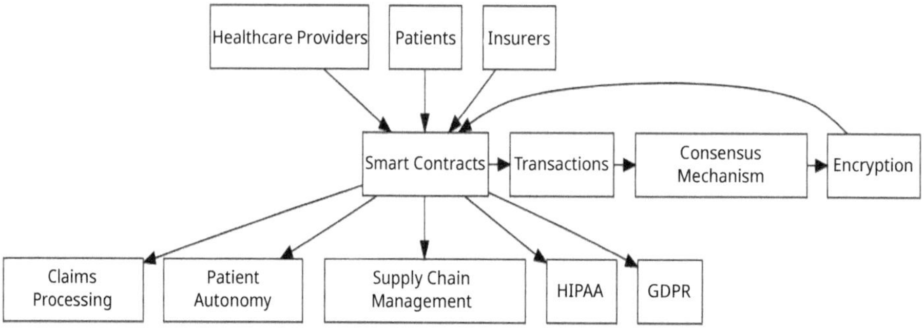

Figure 11: Application of smart contracts in healthcare.

Supply chain management and drug traceability: Smart contracts can offer a positive impact by helping track the origin and ensuring the genuineness and quality of pharmaceutical products right from the forecasted supply chain network. Smart contracts also offer immutable record-keeping of the entire supply chain process, used in manufacturing, various shipments, and dispensing through contract documentation on a blockchain network. Smart contracts check the compliance of the product and its authenticity and regulatory compliance, hence reducing the risk of fake drugs, drugs diversion, and fraudulent practices.

Clinical trials and research collaboration: Smart contracts as apply for clinical trials and interdisciplinary research can automate the contract negotiations and data share and be rewarded between researchers & participants and sponsors and researchers. This paper identifies six potential benefits of smart contracts in clinical trials: reformulating trial protocols, consent documents, and data sharing agreements into smart contract formats empowers researchers to simplify trial subject enrollment, consent process, and data gathering. Smart contracts enable research task tracing and documentation thereby providing data, compliance, replications, and audit trails.

Healthcare payments and revenue cycle management: Smart contracts have the potential for enhancing HC payment and revenue cycle by eliminating paper-based processing, self-executing billing, invoicing, as well as payment reconciliation procedures between the providers, payers, and patients. We posit that smart contracts in healthcare can reduce the time spent on payment transactions and improve accuracy since payment terms, fee schedules, and reimbursement rules submit themselves to logical and mathematical analysis. Smart contracts result in clear accountable recording of payment transactions to ensure that they reflect actual transaction and improve the overall trust for the payment transactions within the health smart environment.

5.3 Role-based access control (RBAC) in EHR management

Another access control technique employed in utilizing access to delicate healthcare data is the widely used RBAC controlling access to EHRs based on roles of people in an organization. RBAC prevents unauthorized personnel from accessing certain data and, at the same time, provides complicated information to employees, since access rights and privileges are delegated in relation to users' jobs. RBAC is one of the main foundations in healthcare data management and is critical for ensuring data availability, confidentiality, and data integrity of EHR data.

In RBAC systems, access control is typically defined based on three main components:

Roles: Organizational job descriptions and positions are reflected in the roles represented in the study. Examples of roles in healthcare organizations can include physicians, nurses, and even administrators and laboratory technicians. Each role is linked with some of privileges or rights that describe the functions, which the user of the particular role can make within the EHR system.

Permissions: Permissions define the functions that a user is allowed to perform within the EHR system. The activities it encompasses are: patient records such as view, add, modify; certain modules and reports and analysis. Roles are permissioned by the role requirements and the concept of least privilege, which grants a user the minimum level of privileges on the network that is needed for them to do their work.

Access control lists (ACLs): These data stores are role permission tables that enumerate the right of users in the EHR system. Users define who can perform and which operations can be updated, allowing creation of highly detailed access control rules consistent with the organizational and/or regulatory standards.

RBAC offers several benefits for EHR management, including:

Simplified administration: In contrast with the traditional attribute-based access control (ABAC), which focuses on specific persons, RBAC means that rights and privileges are assigned to the roles and thus can be adapted rather easily when giving out access control rules. Thus, there is a relative lack of paperwork and reporting, and granting or revoking the access rights of the employees who switch their jobs in the company become less problematic.

Enhanced security: By limiting the number of people with certain predefined roles and responsibilities from accessing vital healthcare information it enhances security by implementing RBAC. Insider threat and unauthorized access functions are also controlled in RBAC that guarantees patient privacy and confidentiality and least privilege concept works in this model.

5.4 Implementing fine-grained access control with smart contracts

Fine-grained access control allows for exact control over who access what data actions inside the system. It is the granular regulation of resource access based on certain qualities. To ensure patient confidentiality and privacy and that only authorized persons access sensitive healthcare data, fine-grained access control must be implemented in the context of EHR management. Because they are decentralized and programmable, smart contracts provide an effective way to establish granular access control in EHR systems.

By utilizing smart contracts, access control restrictions are enforced according to a multitude of criteria, including user roles, organizational hierarchies, patient preferences, and contextual elements. Healthcare companies automate the implementation of fine-grained access control policies, guaranteeing that only authorized individuals have access to particular EHR data under predetermined conditions, by encoding access control rules into smart contracts and putting them on a blockchain network.

Using ACLs control matrices (ACMs) to specify the permissions and access privileges of certain users inside the EHR system is one method for integrating fine-grained access control with smart contracts. These ACLs can be updated dynamically by smart contracts on the blockchain in response to modifications in user roles or permissions. The accompanying smart contract checks the user's identification and credentials against the access control list when the user seeks access to a certain resource or a specific activity inside the EHR system. It then grants access in accordance with preset criteria.

Another way of achieving fine-grained access control with smart contract is to apply ABAC. In ABAC models access control decisions are made based on assignments of attributes, conditions related to the user, resource, other contextual factors, and its environment. Smart contract can assess these attributes into the contract, with flexibility to change the permission profile depending on the situation. For example, a smart contract should only allow a specific user to gain access to a specified patient's file if that user has the right credentials to access patient and client information if he/she is given the authority by the patient and only if the access is compliant with the requisite regulatory laws.

By implementing fine-grained access control with smart contracts, healthcare organizations achieve several benefits, including:

Transparency and auditability: Smart contracts ensure the access control decisions and transactions are always consigned and can help organizations monitor the user's access to EHR data in real time. Recording of access control events on the blockchain ledger strengthens the efficiency of smart contracts since it improves the levels of both accountability and trust in records related to access management.

5.5 Governance models for blockchain-based EHR systems

Management structures and structures of decision-making in block chain EHR systems are used in governance models to provide guidance on the manner in which the operation, maintenance and development of such a system is to be dealt with. Good corporate governance helps to maintain accuracy; has justice; takes into account the goals and the position of organizations, laws, and shareholders. Different governance models are implemented for blockchain-based EHR systems and these models are distinct from each other and have certain strengths and weaknesses. Some common governance models include:

Consortium governance: Consortium governance model involves two healthcare organizations, institutes in managing the operations of a consortium blockchain network for EHRs. In most cases of consortium blockchain, there is formation of the consortium itself as the governing body that oversees the management, operation, and decision-making of the blockchain network. There are mechanisms through which consortium members are involved in the governance process including by voting and consensus and other decision-making processes. There is integration of the different parties through use of the consortium governance models, which enables the structures in a blockchain to be collective, interoperable, and jointly managed.

Community governance: Community governance models select the participation of a larger and organized community of stakeholders in the governance of kinds of blockchain-based EHR systems, including the healthcare providers, patients, developer, and researcher. Governance communities include the open source, contributors for the development, the decentralized autonomous organizations (DAOs), and community steering that directs the design, development, and the steering of the blockchain network that permits stakeholders to participate in. Community governance frameworks involve openness and the development of effective involvement of participants in the decision-making process about the management model of the EHR system.

Federated governance: Consortium and community types of governance are distinguished as well but with the focus on the idea, which lets several independent entities retain control over their nodes within the overall federated blockchain network. Federated governance allows organizations to remain independent and sovereign over their data and action while engaging in cooperation activities at the level of federation. Overlapping mechanisms include the creation of the governance structures, policies, procedures, and interfaces for management, communication, and consensus among the federated members.

Hybrid governance: There is a blend of consortium/community/federated/hierarchical governance models where different governance models are combined in line with the strengths required to suit the needs of blockchain-incorporated EHR systems – introducing hybrid governance models that provide for structuring the required tiers of governance at the consortium, community, organizational, levels of the blockchain network. This is because hybrid systems offer flexibility, customization, and opportunities for organizations of every structure and in different jurisdictions to adapt and meet the needs of all stakeholders and legislation.

6 Data integrity and auditability in blockchain-based EHR systems

6.1 Importance of data integrity in healthcare data management

In fact, the consistency and dependability of the patient information as it cycles through its life is the reason for data integrity, which is an important function in healthcare data management. At this juncture, it is equally important for the quality, reliability, and accessibility of healthcare data in view of EHR systems from which patient care, clinical decision-making, and compliance can be derived.

Patient safety: A number of functions of the healthcare setting can be closely related to the dependency on data integrity to ensure patient safety and standard of care. Healthcare staffs diagnose medical conditions, advise on appropriate treatment, and arrive at evidence-based clinical judgments based on accurate and reliable healthcare information. On this note, pulling off wrong prescriptions, untoward incidents, and poor patient care outcomes ensue from a lack of information of the patient. It can also be seen how healthcare companies can eliminate patient risks and enhance the level of quality of care offered, by maintaining data integrity.

Clinical decision-making: EHRs are used in healthcare settings to analyze patient data for clinical decision, patient care and management, and evaluating treatment regimes. Data quality contributes to the completeness, consistency, and currency of EHR data to support clinicians in making decisions based on patient information anytime and at any point of need. Better quality and consistency of the data supplied to these decision-making processes enhance the practice's effectiveness and accuracy, enhancing patient outcome, reducing heath care medical errors, and promoting the delivery of care.

Continuity of care: Continuity of care requires data from previous encounters, regardless of the provider, therefore data integrity was critical. People visit different doctors and specialists, and all those doctors take part in shaping the patient's treatment plan. Data consistency helps guarantee the quality of information about patients, including their medical history, diagnoses, medications, and test results and their availability to other authorized caregivers in other care settings. Recording creates orderliness and reliability, and hence facilitates proper care given to the patients, cooperation between other shareholders in the healthcare sector, and communication.

Research and analytics: Rarely does one find such a high-level statement of the necessity for the integrity of heath care data for research, analytics, as well as population health management. EHR data has been used in the study of disease prevalence, for examining the outcomes of implemented treatment, as well as in development of strategies to enhance the quality and efficiency of the healthcare system. The accuracy, reliability and validity of data provide assurance for reliable research findings and analytics to support informed decision-making and to develop comprehensible healthcare policies.

6.2 Challenges in ensuring data integrity in traditional EHRs

Incorporated into conventional EHR systems, there are several obstacles that threaten data integrity and that can lead to doubts about the patients' records' trustworthiness, accuracy, and reliability. Some of the key challenges include:

Data centralization: To store and manage patient health records, traditional EHR systems frequently rely on centralized databases. Data corruption, loss caused by a single point of failure makes centralization a danger to data integrity. The key problems in using a central design are coordination, rerouting, and backup, which means that data could be compromised not only when the system was down, but also due to hardware failure.

Lack of transparency: Standard EHR systems are essentially opaque and do not readily provide audit trails of records where modifications have been made. Since there are no obvious audit trails it was very hard to determine the genuineness, origin, and sanctity of EHR data. Batching entry, feature variety, and various modes of documentation lead to lack of standardization, system error, and a lack of metadata about modifications, thus posing threats to the validity and reliability of data.

Limited interoperability: Problems of interfacing in traditional EHR systems restrict the communication and transfer of information concerning the patient between the healthcare organizations and systems. Most of the data is in incompatible formats and applications use proprietary standards and have data in different silos, which makes integration and correlation of data a nightmare. Inadequate data exchange specifications and interface connection limit interconnectivity of EHRs, thereby affecting the quality of information shared and coherence of patient care.

Data security risks: Legacy traditional EHR are vulnerable to many factors of the data insecurity such as hacking, exposure, and even cyberattacks. Lack of strong authentication, poor control of access, and limited encryption mean that EHR data can be compromised by being released or changed. Independent threats include other risk sources, which are internal adversaries, motivated actors, and personnel mistakes whereby insiders expose the confidentiality, integrity, and availability of patient data.

Limited data validation: Typical models of EHRs do not have sufficient features for data validation to maintain the correctness, comprehensiveness, and coherence of patients' records. Traditional methods of data entry, pre-implementation data validation rules, and an overall limited data validation scale increase the chances of data entry error, omission, and inconsistency. Lacking automated validation components in data quality checks, such problems in EHR data remain unseen, thus impairing the complexity, quality, and general soundness of the data.

6.3 Ensuring blockchain data integrity and immutability

One of the promising strategies to ensure data consistency and verified inalterability of entries in EHR systems is applying the blockchain mechanism. The methods listed below show how blockchain accomplishes these goals:

Decentralized and Distributed Ledger: Blockchain stores data across multiple nodes in a network, forming a decentralized and distributed ledger. This structure is achieved through a chain of blocks, where each block contains a cryptographic hash of the previous one, creating a tamper-evident link between blocks. This decentralized nature enhances data integrity and resistance to tampering, as no single entity has full control over the data a key advantage of blockchain technology.

Cryptographic hashing: Blockchain employs cryptographic hash algorithms to generate unique digital fingerprints, known as hashes, for each block of data. These hashes are determined based on the block's content, making it virtually impossible to alter the data without affecting the hash. If any modification occurs, the block's hash changes, alerting the network to potential tampering thereby maintaining trust and immutability across the system.

Consensus procedures: The additional transactions flowing into the blockchain, their sequence, and legitimacy are confirmed by consensus procedures, which form the basis of block chains. There is agreement mechanism such as proof of work (PoW), proof of stake (PoS), and practical byzantine fault tolerance (PBFT) that ensures that node members are in agreement regarding the current state of the blockchain. Interested parties cannot manipulate the information that was stored in the systems since the blockchain negates such actions by allowing only the nodes in the system to approve such changes.

Immutable data records: The beauty of the blockchain technology is that once data is entered on the blockchain, it cannot be changed. New blocks are added to the blockchain at the core, in a sequential chain and are nearly impossible to manipulate without consensus of the entire network to the historical records. The article explains that these characteristics guarantee confidentiality of data and the EHR network's reliability since any alteration of records is easily identified by other members of the network.

Transparent audit trails: Blockchain makes it possible to guard and tag all business transactions and even changes to the information in the EHR. It maintains a visible log of who accessed or modified the data provided by the timestamp and digital signature inherent in every block of a blockchain transaction. This allows stakeholders to verify the authenticity and origins of data, which enhance accountability, traceability, and credibility of the EHR data.

Data encryption and access controls: These EHR systems normally utilize encryption and access controls, which belong to reliable structures, to secure the patient's data. Database information was in encrypted form with cryptographic keys to help maintain the privacy of the data. Smart contracts' distributed ledger system-based access control limits the ability of users to access EHR data and thus increases security, data accuracy, and privacy.

6.4 Auditing and traceability features of blockchain-based EHR systems

High auditability and accountability afforded by blockchain-based EHR systems enhance responsibility, transparency, and credibility of patient health information management. Figure 12 shows how the use of blockchain in EHR systems provides guaranteed data integrity and transparency in records through a blockchain technology approach. The auditing and traceability capabilities of blockchain-based EHR systems are highlighted by the following features:

Clear and unchangeable audit trails: It is also important to know that thanks to blockchain, all transactions and change of data in EHRs are recorded, being immutable and easily traceable. Each and every bitcoin transaction was linked with every other transaction with some type of digital signature, and the blocks formed are immutably arranged chronologically.

Time-stamping: Blockchain attaches a date and time stamp for each transaction thus giving a chronological sequence to activities recorded throughout the blockchain. Various EHR actions performed on the EHR data must be accurately quantified and sequenced to assist with post-incident analysis and coverage for a range of issues including forensic, compliance, and reporting. Time-stamping helps healthcare organizations verify the origin and pedigree of information within EHR and monitor and audit all activities of interest, which could include, but is not limited to detecting an attempted unauthorized access of EHR data.

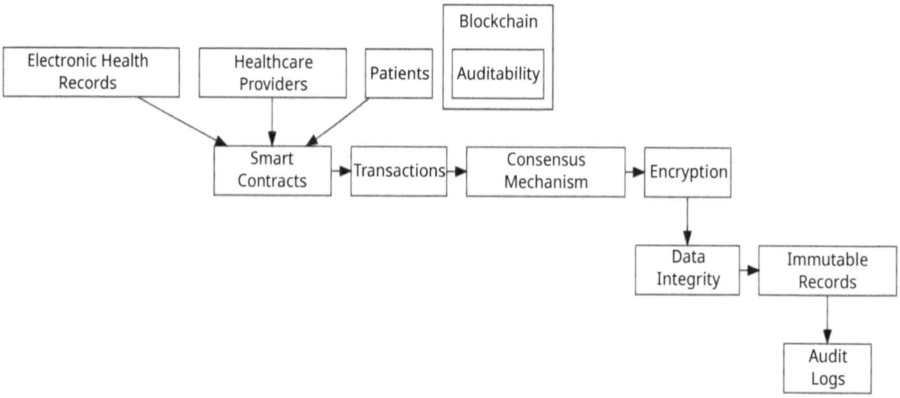

Figure 12: Data integrity and auditability in blockchain-based EHR systems.

Cryptographic signatures: Blockchain transactions are basically digital signatures that incorporate the identities of the involved parties and confirm the validities of the transactions. Cryptographic signatures help guarantee that anyone tampering with the transactions featured in the audit trail is going to be caught red-handed, which in turn makes the whole system trustworthy and reliable. This is further enhanced using the various steps in a block linked to the digital signatures of the participants, rendering the blockchain EHR systems adept at tracking actions of the doers involved, which discourages prevaricators.

Access control and permissioning: The method of implementing EHR using blockchain employs the use of access control, which involves assigning special assets to decision-making instances including smart contracts and permissioned blockchain to control data access of patient's health information. Privacy and security provisions are about the conditions and people who can read, write, update, EHR information. By enforcing access control measures based on preestablished guidelines and permissions encoded into the blockchain, smart contracts make sure that only those with the proper authorization access confidential patient data. Permissioned blockchain networks lower the possibility of unwanted access by limiting participation to recognized and reliable institutions.

Alerts and real-time monitoring: Features such as real-time alert and notification on activities such as malicious attacks and other anomalies detected in the blockchain network are provided by blockchain EHR systems. Monitoring tools and analysis algorithms residing in the automated system scan the blockchain data for gory abnormalities, variations from the norm. Real-time alerting informs administrators/specified user of any strange activities to provide timely action in relation to threats to data confidentiality and integrity.

Compliance audits and regulatory reporting: EHRs are alleviated from it through the use of blockchain EHR systems that offer expansive records of all the transactions and modifications made to the EHR information. In this aspect, the auditors check the records in the blockchain to assess the firm's compliance with regulations like the HIPAA and the GDPR as well as its internal policies and procedures. By using blockchain as audit trails, audits are made easier and cheaper, and show organization's stance on data and information authenticity, security, and compliance.

6.5 Real-world examples of data integrity assurance in blockchain EHRs

The effective application of blockchain-based EHR systems to guarantee data integrity and auditability in the healthcare industry is illustrated by a number of real-world examples:

Medicalchain: Medicalchain is a blockchain-based EHR system that aims to improve patients' care and patients' data protection and privacy through decentralized and patient-controlled records. By means of blockchain, the platform enables patients to store and transfer their records securely to insurers, researchers, and healthcare stakeholders. Through a technique known as encryption the records of patients' health are secured and every operation conducted on the medical chain is recorded permanently on the blockchain database, further enhancing its credibility. Users can easily improve privacy and confidentiality since make their own decisions regarding their data, sharing it with healthcare providers sparingly. In the Medicalchain, the EHR data is arranged and secured through the blockchain system, which provides clear tracks of audit that can make it respond to the responsibility and confidence of the patient interaction records.

Solve.Care: Solve.Care is the healthcare ecosystem deployed on blockchain to facilitate payments, administrative and care programs through apps and smart contracts. This way, the platform captures patient-related activities, treatment, and diagnostic procedures, as well as administrative transactions related to funding and reimbursement for EHR management. Take Care of Itself.Wallet application provides an interface to securely view and manage a patient's health records; providers can see the records and contribute toward the care plan in a "traceable" fashion. Solve. Care's blockchain solution for EHR makes sharing, reconciliation, and verification of data more reliable and effective, supporting patient-centered outcome delivery and healthcare administrative practice.

Hu-manity.co: Hu-manity.co was built using blockchain technology that allows customers to own, control, and sell their health data. It employs the concept of blockchain whereby records made for patients' consent and agreements made for sharing of data are sacrosanct and can be audited for accuracy and authentication of patients' data. #My31 app on Hu-manity.co plans on enabling users to reclaim their data control of health information and blockchain contracts ensure that third-party data users shall only access the data that users have granted permission for, through data usage compensation agreements on the online platform. Hu-monition's EHR solution is developed on blockchain technology and allows patients to own personal health data and be engaged in data-driven projects while data remains trustworthy and safe.

Coral Health: Coral Health is a healthcare platform that uses blockchain in a way that enables patients, researchers, and healthcare providers securely and accurately share patient data. The platform achieves protection of data and non-repudiation by applying blockchain technology to construct a secure, tamper-proof registry of patient's EHRS. By holding the rights to the records' privacy settings and access control, patients consolidate and distribute their data to various parties with Coral Health's EHR solution built on the blockchain. For this reason, Coral Health integrates blockchain to create transparency, security, and trusted data management in healthcare for collaboration in care delivery and research.

The practical use of blockchain-based EHRS for the purpose of guaranteeing data credibility and historical transportability in the sphere of healthcare is outlined. Thus, utilizing blockchain technology, these platforms offer their users clear, unalterable, and patient-centered solutions concerning the proper handling of EHR, facilitating the issues of interoperability, patient privacy, and trust.

7 Conclusion

This chapter examined how blockchain technology is transforming healthcare, especially in the field of electronic health records. After a thorough analysis of blockchain-based solutions, the revolutionary effects on data security, administration, and interoperability in the healthcare environment have been clarified. Healthcare stakeholders effectively manage the complex difficulties of data sharing while maintaining data integrity and protecting patient privacy by utilizing the decentralized and irreversible characteristics of blockchain technology. Consensus processes, smart contracts, and encryption strengthen security protocols while giving patients more control over their medical records. By dismantling organizational walls and enabling smooth data sharing across various healthcare organizations, blockchain promotes interoperability. This study demonstrated how blockchain technology has the enormous potential to spur innovation, better patient outcomes, and improve healthcare delivery. To fully realize the revolutionary potential of blockchain technology, joint efforts and strategic initiatives will be crucial as the healthcare industry continues to change.

References

[1] Saif, S., Biswas, S., & Chattopadhyay, S. (2020). Intelligent, secure big health data management using deep learning and blockchain technology: An overview. Deep Learning Techniques for Biomedical and Health Informatics, 187–209.

[2] Singh, S., Rathore, S., Alfarraj, O., Tolba, A., & Yoon, B. (2022). A framework for privacy-preservation of IoT healthcare data using federated learning and blockchain technology. Future Generation Computer Systems, 129, 380–388.

[3] Kumar, R., Kumar, P., Tripathi, R., Gupta, G. P., Islam, A. N., & Shorfuzzaman, M. (2022). Permissioned blockchain and deep learning for secure and efficient data sharing in industrial healthcare systems. IEEE Transactions on Industrial Informatics, 18(11), 8065–8073.

[4] Vatambeti, R., Krishna, E. P., Karthik, M. G., & Damera, V. K. (2024). Securing the medical data using enhanced privacy preserving based blockchain technology in Internet of Things. Cluster Computing, 27(2), 1625–1637.

[5] Le Nguyen, B., Lydia, E. L., Elhoseny, M., Pustokhina, I., Pustokhin, D. A., Selim, M. M., . . . & Shankar, K. (2020). Privacy preserving blockchain technique to achieve secure and reliable sharing of IoT data. Computers, Materials and Continua, 65(1), 87–107.

[6] Jin, H., Luo, Y., Li, P., & Mathew, J. (2019). A review of secure and privacy-preserving medical data sharing. IEEE Access, 7, 61656–61669.

[7] Zou, R., Lv, X., & Zhao, J. (2021). SPChain: Blockchain-based medical data sharing and privacy-preserving eHealth system. Information Processing and Management, 58(4), 102604.

[8] Alzubi, O. A., Alzubi, J. A., Shankar, K., & Gupta, D. (2021). Blockchain and artificial intelligence enabled privacy-preserving medical data transmission in Internet of Things. Transactions on Emerging Telecommunications Technologies, 32(12), e4360.

[9] Zhang, H., Li, G., Zhang, Y., Gai, K., & Qiu, M. (2021). Blockchain-based privacy-preserving medical data sharing scheme using federated learning. In Knowledge science, engineering and

management: 14th International Conference, KSEM 2021, Tokyo, Japan, August 14–16, 2021, Proceedings, Part III 14, (pp. 634–646). Springer International Publishing.
[10] Wang, M., Zhang, H., Wu, H., Li, G., & Gai, K. (2022, May). Blockchain-based secure medical data management and disease prediction. In Proceedings of the fourth ACM international symposium on blockchain and secure critical infrastructure, (pp.71–82).
[11] Khan, A. A., Laghari, A. A., Shafiq, M., Cheikhrouhou, O., Alhakami, W., Hamam, H., & Shaikh, Z. A. (2022). Healthcare ledger management: A blockchain and machine learning-enabled novel and secure architecture for medical industry. Human-centric Computing and Information, 12, 55.
[12] Wang, L., Meng, L., Liu, F., Shao, W., Fu, K., Xu, S., & Zhang, S. (2022). A user-centered medical data sharing scheme for privacy-preserving machine learning. Security and Communication Networks, 2022.
[13] Rahman, M. A., Hossain, M. S., Islam, M. S., Alrajeh, N. A., & Muhammad, G. (2020). Secure and provenance enhanced internet of health things framework: A blockchain managed federated learning approach. IEEE Access, 8, 205071–205087.
[14] Zheng, X., Mukkamala, R. R., Vatrapu, R., & Ordieres-Mere, J. (2018, September). Blockchain-based personal health data sharing system using cloud storage. In 2018 IEEE 20th international conference on e-health networking, applications and services (Healthcom), (pp. 1–6). IEEE.
[15] Zhang, D., Wang, S., Zhang, Y., Zhang, Q., & Zhang, Y. (2022). A secure and privacy-preserving medical data sharing via consortium blockchain. Security and Communication Networks, 2022.
[16] Rajput, D. S., Sharma, S., Tiwari, S. K., Upadhyay, A. K., & Mishra, A. (2020). Medical data security using blockchain and machine learning in cloud computing. In Mathematical modeling and soft computing in epidemiology, (pp. 347–374). CRC Press.
[17] Huang, H., Zhu, P., Xiao, F., Sun, X., & Huang, Q. (2020). A blockchain-based scheme for privacy-preserving and secure sharing of medical data. Computers and Security, 99, 102010.
[18] Ramani, R., Mary, A. R., Raja, S. E., & Shunmugam, D. A. (2024). Optimized data management and secured federated learning in the Internet of Medical Things (IoMT) with blockchain technology. Biomedical Signal Processing and Control, 93, 106213.
[19] Chen, X., Ji, J., Luo, C., Liao, W., & Li, P. (2018,December). When machine learning meets blockchain: A decentralized, privacy-preserving and secure design. In 2018 IEEE international conference on big data (big data), (pp. 1178–1187). IEEE.
[20] Neelakandan, S., Beulah, J. R., Prathiba, L., Murthy, G. L. N., Irudaya Raj, E. F., & Arulkumar, N. (2022). Blockchain with deep learning-enabled secure healthcare data transmission and diagnostic model. International Journal of Modeling, Simulation, and Scientific Computing, 13(04), 2241006.
[21] Chakraborty, S., Aich, S., & Kim, H. C. (2019, February). A secure healthcare system design framework using blockchain technology. In 2019 21st International Conference on Advanced Communication Technology (ICACT), (pp. 260–264). IEEE.
[22] Das, S., Karmakar, S., & Saha, H. N. (2022). Secure digital health data management in internet of things using blockchain and machine learning. In Digital health transformation with blockchain and artificial intelligence, (pp. 23–45). CRC Press.
[23] Liu, H., Crespo, R. G., & Martínez, O. S. (2020, July). Enhancing privacy and data security across healthcare applications using blockchain and distributed ledger concepts. In Healthcare, (Vol. 8, No. 3, pp. 243). MDPI.
[24] Tan, L., Yu, K., Shi, N., Yang, C., Wei, W., & Lu, H. (2021). Towards secure and privacy-preserving data sharing for COVID-19 medical records: A blockchain-empowered approach. IEEE Transactions on Network Science and Engineering, 9(1), 271–281.
[25] Salim, M. M., & Park, J. H. (2022). Federated learning-based secure electronic health record sharing scheme in medical informatics. IEEE Journal of Biomedical and Health Informatics, 27(2), 617–624.

P. Prasant, Kiran Siripuri*, S.B. Mohan, Venkatesan Hariram, A.S. Vigneshwar

AI-powered sensors and devices for sustained health tracking

Abstract: The steady progress made in artificial bio interfaces and sensing systems has strong impacts on the development of continuous health monitoring systems in biomedical sensor technology solutions. In this chapter, the authors focus on enhancing biomedical sensors and devices using AI and underlining the need to incorporate complex algorithms and real-time analysis of large datasets. Some of the focal areas described include the utilization of smart sensor nodes and machine learning for diagnostic prognosis, and the optimization of data credibility. As the chapter seeks to discuss the future of health monitoring technologies in coming up with an improved advanced remote health monitoring system, it discusses the challenges in sensor calibration, data processing, and real-time feedback. The future prospects and challenges of personalized medicine and preventative care resulting from the direct application of AI to patient care are also described in relation to the potential for dramatically changing patient management and treatment algorithms.

Keywords: AI-driven sensors, biomedical optimization, continuous health monitoring, machine learning, predictive diagnostics, real-time data analytics

1 Introduction

The integration of AI into biomedical sensor technology marks a transformative shift in the domain of health monitoring [1]. As healthcare systems evolve toward more personalized and preventive models, the demand for sophisticated, real-time monitoring solutions has intensified [2]. AI-driven biomedical sensors are at the forefront of this revolution, offering unprecedented capabilities for continuous health surveil-

***Corresponding author: Kiran Siripuri**, Department of Cybersecurity and IoT, Malla Reddy University, Hyderabad, Telangana, India, e-mail: siripurikiran@gmail.com

P. Prasant Associate Professor AIPH University Bhubaneswar, Odisha India, e-mail: registrar@aiph.ac.in

S.B. Mohan, SA Engineering College, Department of Electronics and Communication Engineering, Chennai, Tamil Nadu, India, e-mail: drsbmohan@gmail.com

Venkatesan Hariram, Department of Practice of Medicine, Vinayaka Mission's Homoeopathic Medical College and Hospital, A Constituent College of Vinayaka Mission's Research Foundation Deemed to be University, Salem 636308, Tamil Nadu, India, e-mail: venkathompath@gmail.com
https://orcid.org/0000-0003-2237-7071

A.S. Vigneshwar, Ramco Institute of Technology, Rajapalayam, Tamil Nadu, India, e-mail: viigneshs@ritrjpm.ac.in

https://doi.org/10.1515/9783111503202-002

lance [3]. These advanced systems leverage AI algorithms to analyze vast amounts of data with high precision, providing actionable insights that were previously unattainable with conventional methods [4]. The convergence of AI and biomedical sensors is reshaping how health conditions are monitored, diagnosed, and managed, thus driving significant advancements in patient care and clinical outcomes [5].

The development of intelligent sensor platforms is a pivotal aspect of this transformation [6]. Modern biomedical sensors equipped with AI can continuously track and assess a wide array of physiological parameters, such as heart rate, blood glucose levels, and respiratory patterns [7]. These sensors are designed to operate seamlessly in real-time, delivering immediate feedback that is crucial for managing chronic conditions and detecting anomalies early [8]. By incorporating advanced machine learning algorithms, these sensors can also adapt and improve their accuracy over time, learning from individual patient data to refine their predictive capabilities [9]. This adaptability enhances the reliability of the data collected, leading to more informed and timely medical interventions [10].

Machine learning plays a critical role in the optimization of AI-driven biomedical sensors [11]. Through the application of sophisticated algorithms, these systems can identify patterns and correlations within large datasets that may not be evident through traditional analysis methods [12]. Predictive diagnostics, enabled by these machine learning techniques, allow for early detection of potential health issues before they manifest into severe conditions [13]. This proactive approach not only improves patient outcomes but also reduces the overall burden on healthcare systems by preventing the escalation of chronic diseases [14]. The ability of AI to predict and preemptively address health concerns is a key advantage of modern biomedical sensors [15].

The deployment of AI-driven sensors is not without challenges [16]. Ensuring the accuracy and reliability of sensor data is paramount, particularly in critical health monitoring scenarios [17]. Calibration of sensors, data processing efficiency, and real-time feedback mechanisms must be meticulously designed and validated to maintain high standards of performance [18]. Addressing these challenges requires ongoing research and development to refine sensor technologies and algorithms, ensuring they meet the rigorous demands of clinical applications [19]. By overcoming these obstacles, AI-driven biomedical sensors can achieve their full potential in enhancing patient care and health monitoring [20].

The implications of AI-driven biomedical sensors extend beyond individual patient management to broader public health outcomes [21]. By facilitating more precise and personalized health monitoring, these technologies contribute to the advancement of preventive care strategies [22]. The insights gained from continuous health monitoring can inform public health initiatives and shape policies aimed at improving population health [23]. As AI continues to evolve, its integration into biomedical sensors likely leads to further innovations, ultimately revolutionizing how healthcare is delivered and experienced [24]. The future of health monitoring is poised for significant transformation, driven by the ongoing advancements in AI and sensor technology [25].

2 Overview of biomedical sensors and devices

2.1 Types of biomedical sensors and devices

Wearable sensors represent a transformative advancement in personal health monitoring, offering continuous, noninvasive measurements of various physiological parameters. These devices have become integral to both personal fitness and clinical applications, providing real-time data that can significantly enhance health management and preventive care.

There are two most common forms of wearable sensors, which include smartwatches and fitness trackers and basic traditional watches with additional sophisticated health sensors. These gadgets usually entail including heart rate sensors, step counter, sleep pattern and physical activity sensors. Most of the models use GPS for the tracking of outdoor physical activities and then offer further statistics such as caloric consumption and stress. Different types of biomedical sensors and devices that are wearable, implantable, and diagnostic are depicted in Figure 1, where the various sensors and devices are designed to measure a diverse physiological parameter. These devices collect data and using some algorithms, guide the user on their general fitness and health; they are helpful to both the regular fitness fanatic as well as the chronically ill.

Skin-mounted patches and adhesive sensors are aimed for skin on-surface wear where it monitors different health parameters continually. These sensors are mainly employed for observing cardiac activity and blood glucose levels of dehydration. The commonly used adhesive patches are thin and flexible to prevent weighing the patient down and also to allow the patient to go about his/her normal business without much disturbance. For example, ECG adhesive monitors the heart rhythms for weeks, while glucose adhesive monitors the blood sugar levels in an unobtrusive manner. It is also convenient that the data from these sensors can be delivered to mobile devices for real-time analysis and display of results.

Already, we have smart clothes that have sensors placed to the cloth so that people do not need extra devices to monitor their health. These garments incorporate a range of underlining sensors to monitor factors such as heart rate, respiratory rate, body temperature, and muscle activity, among others. Examples of smart clothing are sportswear that help improve performance and recovery from diseases, and other apparels that help monitor long-term diseases. Technical innovations in textiles also in the sphere of sensors and the shrinking of the devices let smart clothing become much more comfortable and versatile, so they are worn not only during sports activities, but can also be used in healthcare.

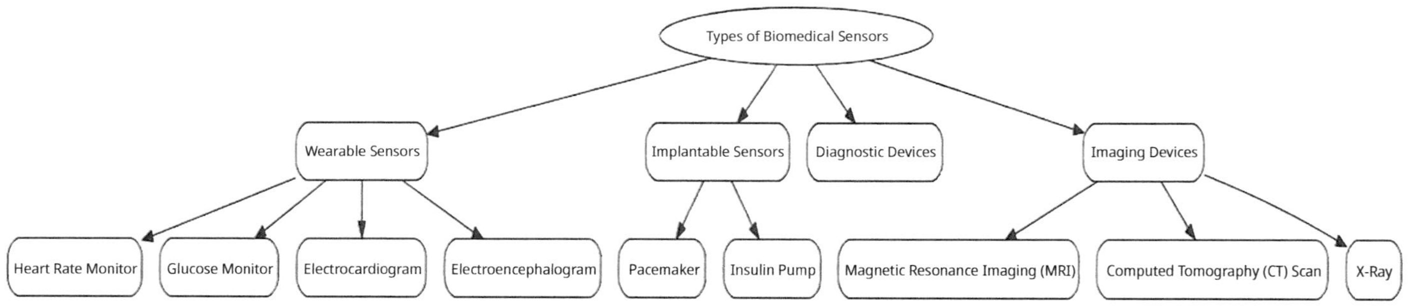

Figure 1: Types of biomedical sensors and devices.

Wearable ECG monitors are considered as unique gadgets primarily intended for constant registration of the activity of the heart. Wrist band, chest harness of other wearable forms, and wearable ECG monitors help to regulate heart problems as they involve monitoring arrhythmia, heart rate variability, and other aspects of cardiac structure and function. Consequently, the constant data gathered by those gadgets can be examined for flagging of possible cardiovascular disorders with a view to taking suitable remedies on time alongside handling heart-related complications.

CGM devices are designed to be worn by patient with diabetes for tracking blood glucose levels at a constant rate during everyday activities. These devices generally comprise a small sensor that is inserted under the skin, which checks the glucose level in the interstitial fluid. The data is then transmitted to a wearable receiver, thereby giving users an instant feedback on the glucose values and trends. The benefits of CGM are that it can improve the overall diabetic management since patients can adjust the doses of insulin more accurately, monitor patterns of glucose level, and use fewer fingerstick tests.

2.2 Implantable sensors

Implantable sensors are sophisticated biomedical devices intended to be implanted within the human body to deliver permanent constant surveillance and control of many physiological elements. The type of data that such sensors provide is highly accurate, stable over extended periods, and in real-time, which serves the patient with chronic diseases and the conditions that require comprehensive medical treatment.

Pacemakers and cardiac monitors are among the most widely used implantable sensors, designed to manage and monitor heart health. Pacemakers are small electronic devices implanted in the chest to regulate heart rate, especially in patients suffering from conditions like atrial fibrillation or heart block. Several implantable sensors, as illustrated in Figure 2, have been developed for continuous health monitoring and data recording.

These sensors can monitor the heart's electrical activity and continuously analyze heart rate and rhythm. For instance, modern pacemakers can provide detailed, real-time feedback on cardiac activity, enabling healthcare providers to adjust treatment plans based on accurate, up-to-date information. Additionally, portable yet invasive devices like implantable loop recorders offer long-term ECG monitoring and can effectively detect arrhythmia-related abnormalities over extended periods, which is highly beneficial for preventive care and diagnosis.

Insulin pumps and glucose sensors are important diabetes technology tools that allow for constant check and accurate regulation of blood sugar levels. Please note that insulin pumps work similar to an ICD but instead of being implanted inside the body, are worn like a beeper, and regulate insulin conveyance to the body through a slim pipe inserted just beneath the skin, and the insulin can be regulated according to

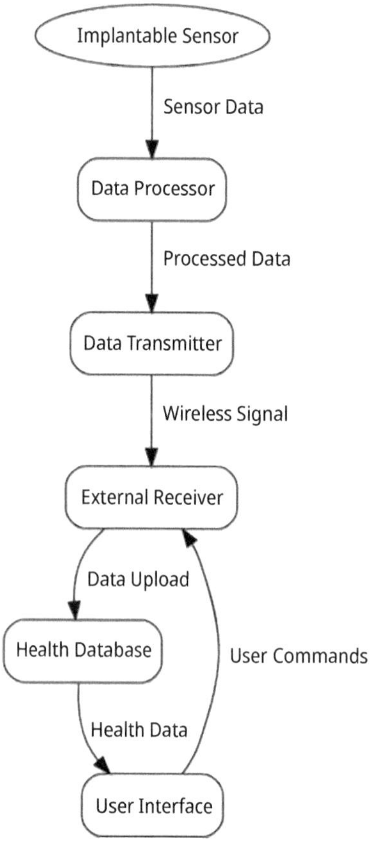

Figure 2: Implantable sensors.

the measurements of the blood glucose level. It is discreet and comprises of sensors placed subcutaneously, which monitor glucose levels of the interstitial fluid and display data to the insulin pump monitor. These devices can essentially be synced to deliver insulin automatically and provide better means to regulate glucose levels and lessen the fingerstick tests required, and with better glycemic results.

Intracranial pressure (ICP) probes are implantable devices for monitoring the pressure of the skull in patients with diseases of intracranial dynamics, including traumatic brain injury. These sensors are most commonly located in the brain system and their role is to continuously record the ICP. Given the valuable information from the ICP monitors, the treating physicians and surgeons can go ahead with intervention measures such as changing medication dosages that are intended to alter brain pressure in cases affecting the patient.

Cochlear implants are advanced technological tools developed to help people diagnosed with severe to profound sensorineural hearing loss. These implants have an external auditory processor and an internal receiver-stimulating electrode that is fixed in the cochlea of the patient. The external processor gathers sound source and converts these into electrical signals that can be received by the internal electrodes. These electrodes thus interfere with the auditory nerve so as enable the user to have an understanding of sound. Cochlear implants refer to a developmental assistive technology that enhances the process of hearing and communication for those with hearing loss.

Neurostimulators can be surgically inserted into the body to specifically adjust the activity of the neural tracts to serve a therapeutic role. Some of these devices are deep brain stimulators and spinal cord stimulators, which are employed to manage multiple neurology disorders. Pulse generators are placed directly in the brain through surgical procedures and used to provide electrical signal to control dysfunctional brain signals involved in ailments like Parkinson's disease and essential tremor. Spinal cord stimulators are surgical devices placed in the epidural space of spinal cord and deliver electrical impulses to the patient's body to control chronic pain by altering the pain's message en route to the brain. Both kinds of neurostimulators offer designed therapies that adapt, depending on the reaction of the individual.

2.3 Remote monitoring systems

Tele monitoring systems is one of the latest innovations in the field of medical technology since it allows the assessment of the conditions of the patient, together with the determination of what actions to take without having to physically check up on the patient. These systems use different types of sensors and communications technologies to gather and transmit health information to the healthcare givers to enhance effective and timely management of patients' health.

Telemonitoring devices are developed to be positioned at the patient's home environment to enable endless regular tracking of the patient's health status without constant referrals to the healthcare facilities. These include weight measurement through a digital scale, home blood pressure monitoring, and oximeter for blood oxygen levels. Such devices, when combined with mobile application, help patients to monitor their health over a period and to share the acquired information with physicians for remote evaluation and treatment.

Telemedicine is an innovative approach that allows the patient and/or healthcare provider to consult almost everything through the Internet. Some of these platforms comprise common amenities like video calls and messaging, and access to records of the health institution, among others. Using such technologies, patients can get medical consultation, examination, and even diagnosis without having to visit a healthcare facility. This makes it possible to enhance the use of telemedicine by connecting the telemedicine work with remote monitoring devices in order to give a view of the patient, in addition to enabling better virtual care.

Telemonitoring of blood pressure is an advanced tool used by patients to record and convey their blood pressure levels to healthcare givers without direct contact.

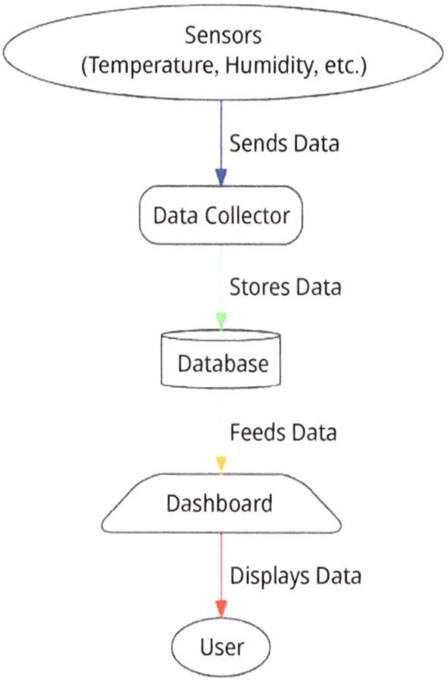

Figure 3: Remote monitoring systems.

These monitors include automatic cuff inflators and digital displays that measure and display systolic and diastolic blood pressure. They represent key components of remote monitoring systems, including sensor mechanisms, communication methods, and real-time data transmission and analysis for clinical oversight. Data is wirelessly transmitted using Bluetooth communication technologies, allowing healthcare providers to track blood pressure trends, adjust medications, and manage hypertension without requiring in-person visits. Such continuous monitoring enables early detection of potential health issues, leading to more effective and timely hypertension management.

Digital stethoscopes improve traditional auscultation by transforming sounds into signal data that can be documented and examined. These devices are sometimes incorporated with wireless systems in order to forward heart and lung sounds to other sites for analysis. Specifically, by using the technology of the digital stethoscopes, one can control a patient's cardiovascular and respiratory systems from remote, which means that doctors can make more precise diagnosis and apply necessary measures, when needed. The recorded sounds are listened to by other practitioners, which enhances the assessment process as well as the multidisciplinary management of the patient's healthcare.

Environmental/ambient sensors are responsible for measuring factors around a patient's environment that impact health, such as temperature, humidity, and quality of air. These sensors are always employed, together with other types of remote monitoring devices, in order to offer a broad perspective of the patient's home environment. For instance, sensors can be used to identify variations that exist within rooms that endanger the respiratory system of patriotic citizens. The information that can be obtained can serve as a basis for identifying areas that need change in the patient's environment and their general condition.

2.4 Diagnostic and imaging sensors

Sensors used in diagnostics and imaging are one of the precursors of modern medicine as they give the needed information on the internal functioning of the human body and its changes. The features of these sensors allow for the diagnosis of diseases, determination of the most appropriate strategies of treatment, and assessment of the outcome of intercessions. The use of imaging devices has improved overtime through technological improvement, making them precise and more versatile.

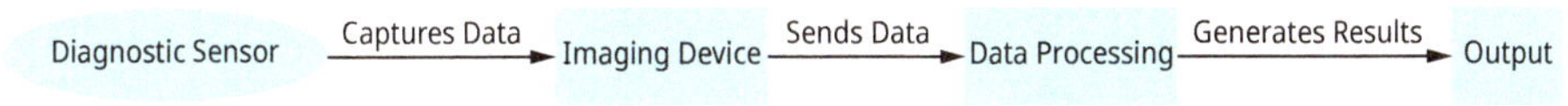

Figure 4: Diagnostic and imaging sensors.

MRI and CT are two modern radiation diagnosis facilities that help doctors to visualize deeply settled pathologies in detail. MRI involves the use of strong magnetic fields and radio frequency waves to produce a detailed picture of the soft tissue areas of the body, including the brain, muscles, and organs. Figure 4 shows some of the diagnostic and imaging sensors used in today's advanced ICT medical technologies for the purpose of improving accuracy and early detection of diseases as an underlying goal. Computed tomography (CT scanners) delivers cross-sectional images by applying X-ray and provide close views of bones, organs, and blood vessels. MRI and CT imaging give crucial information while diagnosing ailments that include tumors, fractures, internal bleeding syndromes, and neurological problems.

Ultrasound uses sound waves of high pitch to present the internal structures of the body in real-time. This noninvasive imaging modality is commonly used for studying soft tissue areas, including the abdomen, pelvis, and fetus during pregnancy. As for ultrasound imaging, it is appreciated owing to its noninvasiveness and mobility and the possibility to deliver real-time information. The technology consists of different forms, B-mode for structural evaluation and Doppler ultrasound for evaluating vessels' functionality. Ultrasonic imaging devices have improved over the years – the quality of images captured has improved and new features have been added to the devices to enable more specific uses such as echocardiography and ultrasound musculoskeletal system.

For determining which of the medical imaging techniques is most beneficial, PTA 1 Proposed Technology OCT, a noninvasive imaging tool based on light-wave tomography of biological tissues at resolutions up to 10 µm. OCT is used mostly in ophthalmology applications that help in the diagnosis of various diseases through imaging the retina and the anterior segment of the eye, including macular degeneration, glaucoma, and diabetic retinopathy. The technology has also been used in other medical specialties such as cardiology and dermatology because it allows for high-resolution imaging of tissue characteristics.

Endoscopic cameras are particularly developed for observing the cavities and organs of the human body using minimally invasive techniques. These cameras are placed on a flexible endoscope inserted into the body via a natural duct or surgical opening. In endoscopy, fiber optic cameras transmit pictures of the digestive tract, respiratory system, and other body cavities to monitors for further examination, diagnosis, and treatment of various diseases. Modern types of endoscopic equipment, including high-definition and narrow band-imaging, have improved the diagnostic and therapeutic capacities of these instruments.

Electromyography (EMG) sensors are devices that record the electrical activity of muscles and nerves, and so are useful to diagnose neuromuscular disorders. EMG sensors can be used to measure the strength of a muscle, identify abnormalities in electrical activity, and determine the condition of a nerve. Information acquired from EMG testing plays a role in identifying diseases, including muscular dystrophy, peripheral neuropathy, and carpal tunnel syndrome. Recent technological developments carried out on EMG include portability and wearable technology, where EMG has been advanced is and used to give a broad basis for the evaluation of muscles and nerves.

2.5 Therapeutic devices

Therapeutic devices are technologies used to diagnose and combat different ailments and illnesses. Such devices have significant value in various aspects of patient care since they are used in providing optimized treatment schedules, boosting individual patient recovery, and facilitating continuing healthcare in the community. Therapeutic devices have undergone tremendous developments over the last few decades and have increased the prospects of medical procedures.

Drug delivery systems are complex structures intended for the delivery of drugs to the patient's body and tissue in a calibrated and precisely measured manner. These systems include the infusion pump that gives a continuous delivery of such things as insulin and antibiotics agents. The other forms of advanced drug delivery involve biodegradable implants, where the material used releases the drug in a controlled manner at an agreed timetable without the need for further injections. Examples of systems include the microneedles and transdermal patches, making drug delivery less invasive, thereby improving patient satisfaction. These systems are designed to work as efficiently as possible in terms of the intended therapeutic effects without composing with its side effects, and give the best results possible for the kind of treatment needed by the patient.

There are physiotherapy and rehabilitation aids that are meant to supplement the physiotherapy process for patients with musculoskeletal condition that requires physiotherapy. These devices include the robotic rehabilitation systems aimed at assisting motor rehabilitation and repeating motion tasks. Other examples include exoskeleton wearable assists for locomotion and posture, and therapeutic ultrasound for tissue repair and pain relief. Through sensors incorporated in the rehabilitation equipment, patient's progress can be closely observed and the treatment adapted to the requirements of a particular client.

Obstructive sleep apnea is treated using Continuous Positive Airway Pressure (CPAP) machines as a method of controlling pauses in breathing during sleep. CPAP devices provide ongoing pressure of air through a facial mask in the nose and avoid the occurrence of apneas. Today's CPAP devices have features include the auto-titrating devices that adjust the pressure, depending on the feedback given at that time, and the capability of connecting the devices to other devices for monitoring and sharing of information. Different therapeutic devices intended for specific medical treatments and improving patient's conditions are presented in Figure 5. As we have highlighted, CPAP use enhances sleep quality, thereby minimizing the necessity of additional sleep and the disease related to it and cardiovascular disease.

Electrostimulation products are used to deliver electrical signals to challenged nerves for medical treatment. Such devices include TENS that is used in order to lessen pain by manipulation of nerve contractions, and NMES used in muscle remodeling and building up. Neuro stimulators include spinal cord stimulation (SCS) and deep brain stimulators (DBS), together with implants that are used as treatments that involve electro stimulation in specific areas of the nervous system, mostly for the management of chronic pain and certain neurological disorders. Employing different parameters of electrical stimulation results in the development of a specific therapy plan to meet needs of the patient, and improve the treatment outcome.

Medical man-made organs implants are sophisticated devices used for the treatment of ailing organs. Such devices are mechanical ventricular pumps (left ventricular assist device) that help patients who have severe heart failure, with mechanical support. Artificial kidneys (dialysis machines) replace the function of filtering blood in cases of renal failure and artificial lungs (extracorporeal membrane oxygenation systems) supply limited respiratory support to patients with critical respiratory illnesses. Artificial organs are in the process of providing hope and means between organ failure and transplantation as well as significantly enhancing the quality of life of those suffering from critical organ breakdown.

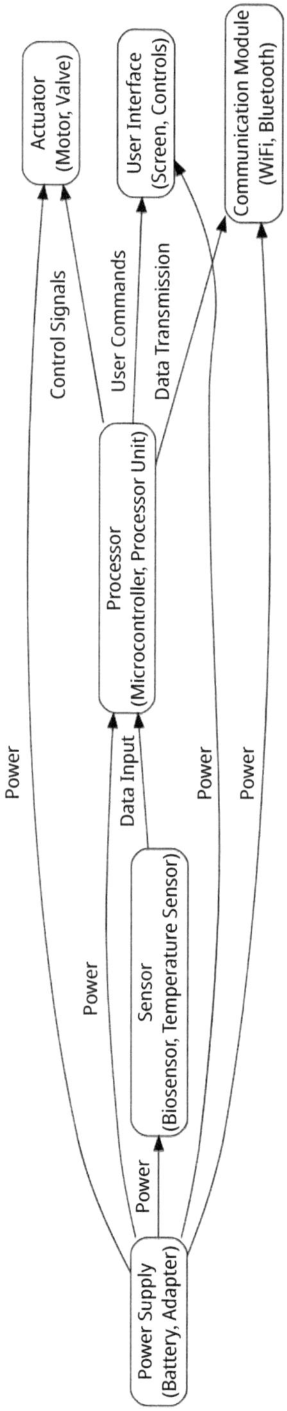

Figure 5: Therapeutic devices.

3 Functionality and applications

3.1 Vital signs monitoring

Supplementary to what has been discussed, the recording of the vital signs is a core part of the working process since it involves essential information of a patient's body functions. This practice makes it possible to spot trends that in turn point toward a medical condition that needs treatment, as well as management of chronic illnesses. Significant development exists in sensor technology, thereby strengthening the efficiency of the vital signs monitoring contributing to the positive outcomes of the patients as well as enhancement of the healthcare systems.

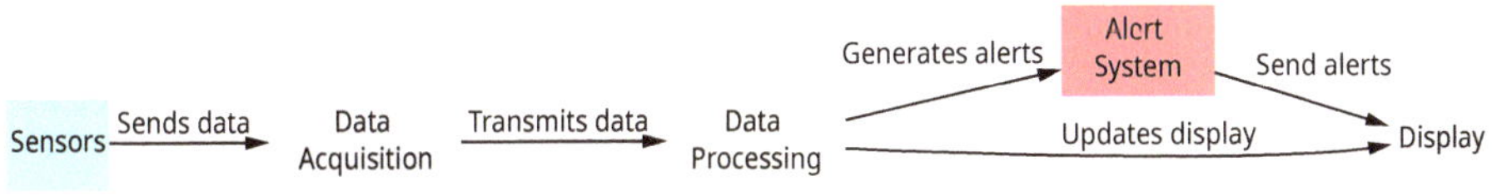

Figure 6: Vital signs monitoring.

Hear rate and ECG are two important parameters in diagnosing cardiovascular disorders. Pulse oximeters convey the number of pulse repetitions in one minute, giving details of the pattern and efficiency of the heart. Smartwatches and chest straps provide prolonged input and analysis of the wearer's heart rate. ECG, on the other hand, records electrical activity of the heart such as heart rate irregularities, lack of blood supply to the heart muscles, and other heart-related ailments. The complex ECG devices help in diagnosing and monitoring the various heart conditions as well as offer detailed evaluations, apart from maintaining the long-term records of the patient.

Tracking of the pressure within the blood vessels is relevant in evaluating the overall heart health and treatment of ailments like hypertension. Dialysis monitors track blood pressure automatically and give accurate systolic and diastolic values to enable frequent checks and effective treatment. That is why real-time patient monitoring with the digital interface of the patient's essential vital signs such as pulse, blood pressure, and SpO_2 are illustrated in Figure 6. There is a need to monitor SpO_2 using pulse oximeters as a means of assessing respiratory compromise and, specifically, hypoxemia. A pulse oximeter operates through the use of an infrared light, which is used to estimate the percent of oxygenated Hb and is very useful for monitoring respiratory diseases as well as oxygen therapy.

Pulse rate measurement is a significant means of assessing respiratory health as well as general health status. By observing the respiratory rate over time, the client is easily able to identify any abnormalities such as tachypnea, which is indicative of a number of underlying health problems. Smart clothing as well as remote monitoring systems, accompanied by respiratory sensors, are able to monitor an individual's breathing and give immediate feedback. Respiratory rate data is critical in chronic conditions like COPD, asthma, and sleep apnea illnesses.

Temperature measurement is one of the first features of the HM process to assess the ability of the body to regulate temperature and to detect fever. Current temperature sensors are incorporated into wearable, smartwatch bands and devices, and connected to remote controllers and thermometers. This type of a sensor provides continual temperature readings, which enable users to identify febrile diseases and all other diseases that relate to temperature. New technologies in temperature measurement also help combine the temperature readings with other parameters of a patient state, which is valuable.

Multiparameter monitoring systems utilize information from various invasive/ non-invasive vital signs sensors to give the overall picture of a patient. These systems integrate the readings of pulse, blood pressure, respiratory rate, and temperature into one platform for easy health evaluation and monitoring of trends. Multiple parameter devices are frequently employed in patient care and in telemonitoring applications for those who have a chronic illness. These systems improve the accuracy of the assessments of the health status further by integrating data from the different sensors.

3.2 Chronic disease management

COPD stands for chronic obstructive pulmonary disease, which is one of the chronic diseases that requires the active monitoring of a patient and in offering regular treatment. On analyzing the role of management of chronic diseases, the researcher noted that it impacts patient's QOL, risk of complications, and healthcare spend. Biomedical sensing and, particularly, the technological developments of the related portable devices have greatly improved the diagnosis, follow up, and management of chronic diseases with more precision, and in real-time.

Another area that has seen continuous monitoring technologies being important is in the management of diabetes. CGM systems provide constant monitoring and tracking of blood glucose levels and provide the patient's real-time information of their glucose levels. These systems involve a tiny strip of glucose sensor that is placed under the patient's skin to record the glucose level in the interstitial fluid; the information is relayed to an electronic monitoring mobile application The World Health Organization's comprehensive framework for chronic disease management, which involves; a) Patient data, b) Treatment plans, and c) intimately connected real-time monitoring to deliver the best results is presented in Figure 7. The combination of CGM with insulin pumps has allowed for closed-looped insulin delivery systems, which adjust the dose of the insulin intake, depending on the level of glucose, hence improving glycemic control and minimizing the use of fingerstick tests. In glucose control, these technologies are useful due to the adjustment that allows for timely appropriate changes and minimization of both hyperglycemia and hypoglycemia incidences.

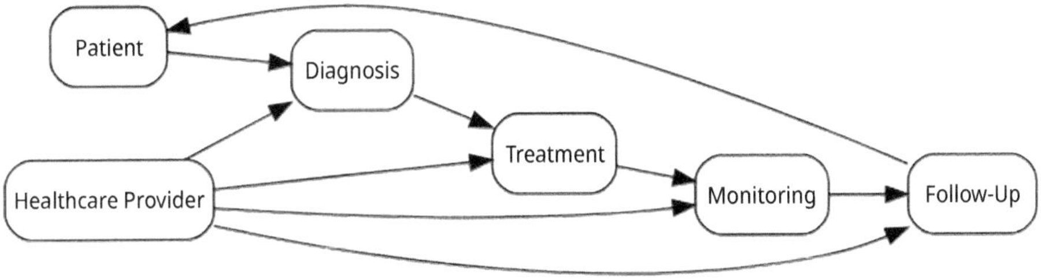

Figure 7: Chronic disease management.

Cardiac care management is of high importance to maintain control over health conditions like hypertension, heart failure, and arrhythmias. Wearable ECG and pocket blood pressure measurement equipment give continuous details on the performance of the heart and blood pressure to recognize early anomalies. Trans-telemer IT and implantable cardiac monitors are used to monitor abnormal rhythms and different cardiac problems, including arrhythmia, which need intervention. These devices help to control cardiovascular illnesses by offering necessary information needed in the alteration of medications and changes in habits to prevent worsening of situations and adverse effects on patients.

Neurological disorders like epilepsy, Parkinson, and multiple sclerosis are handled with improvements to the monitoring technologies. Smart clothes and integrated circuitry deal with motor coordination, seizures, and other signs occurring in neurological disorders. For instance, wrist-worn accelerometers can get data on tremors in Parkinson's disease patients, which is beneficial to update the therapy. In the same way, seizure detection gadgets can provide caregiving informants and other medical decision makers with an insight into the deterioration, resulting in the need for a change of medication dosage. These technologies improve the care of patients with neurological disorders by offering real-time data on the management of the diseases for better quality patient outcomes.

Respiratory healthcare management entails managing diseases that include asthma, COPD, and sleep apnea. Monitoring devices are portable spirometers and peak flow meters that enable patients to monitor their lung function and symptoms, simultaneously. In the case of sleep apnea, CPAP devices with tracking capabilities offer information of the apnea episodes and the effects of therapy. Further, wearable sensors are primarily designed to continuously assess respiratory rate and oxygen saturation levels to help treat, assess the efficiency of the interventions, and inform patients and therapists about emergent complications. These monitoring tools enhance the effective management of respiratory ailments through providing nurses with tools to intervene appropriately and fashion out individualized patient care procedures.

Serum creatinine and electrolyte levels are paramount in CKD patients with dialysis. Smart and bio-accepted devices like gems and chips can monitor some aspects such as blood pressure, fluid, and electrolyte ratios which, are significantly important to renal health. Dialysis machines that support intricate monitoring systems guarantee appropriate treatment by constantly analyzing and drawing parameters from real time. These technologies are useful in taking care of chronic kidney disorders by facilitating adjustments of the patients' treatment regimes, assessment of complications, and general patient care.

3.3 Preventive and predictive health

Preventive and predictive health categories of technology minimize the chances of somebody coming down with a particular health issue before the situation becomes hopeless. These technologies, which involve using imageries and data acquisition devices, sensors, data collection, and analytics, and backed up by AI computing power, provide solutions for effective health management, early disease identification, and proactive health management. The incorporation of these tools into healthcare can in fact be observed as having the ability to alter the course of ailments and diseases that would otherwise constantly plague humanity.

Diseases diagnosis has a massive impact on successful outcomes of a treatment and on the overall percentage of people's life, immediately after being affected by various health conditions. One of the main functions of the diagnostic sensors and imaging equipment is to detect diseases in their infancy. For instance, wearable electronic devices that record physical signs include heart rates blood pressure discrepancies that indicate additional problems. Likewise, screening devices that involve biomarkers and genetic information can always point out people with high risks of acquiring diseases like cancer and cardiovascular conditions. Early detection helps our bodies, lives, and overall health situation to improve through early intervention, and hence better prognosis.

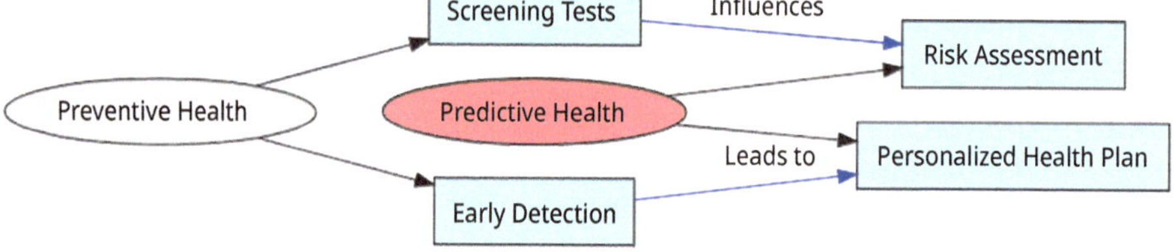

Figure 8: Preventive and predictive health.

The data collected from wearable sensors, EHR, and genomics give an understanding of how susceptible an individual is to develop particular diseases known as risk assessment and prediction models. Statistical analysis and prognosis tools are used to identify patterns and connection between data to give a customized risk profile. In Figure 8, there is a depiction of how preventive and predictive health strategies operate together, with the aim of improving the probabilities of early diagnosis- and patient-orientated intercession for superior patient outcomes. For example, algorithms can estimate one's probability of becoming a diabetic, depending on the height, exercise regimen, and genes. These models can help choose the methods for prevention, according to the risks, and can include everything from changing a patient's habits to prescribing certain medications.

Smart gear gadgets are those that are developed to track the number of steps taken, the amount of sleep obtained, and other factors that define health fitness. Information like how many steps a person has taken, his/her heart rate, and the quality of the sleep each day give a lot of information about a person's health. The information is frequently used to deduce specific recommendations for the subject regarding physical exercise, sleep quality, and overall health. The everyday application of wrist-worn fitness trackers encourages healthy behavior, help in the achievement of individual training objectives, and advance a chronic disease-prevention framework.

Environmental exposures, in this case, entails keeping track of things like quality of air, temperature, and the humidity level, which affect the health of an individual. CO sensors, SO_2 sensors, NO – NO_2 – NOx sensors, air quality monitors, AQI sensors, PM2.5 Matter 2.5-micron sensors, Ozone sensors, VOC sensors, and Radon sensors can determine the concentration inside environments and can track exposure risks in respiratory diseases and cardiovascular diseases. It is equally possible to determine the relationships between environmental conditions and the status of people's health, which allow individuals and healthcare professionals to avoid frequent exposure to various hostile environment conditions and diseases.

Comprehensive health and indicative data are generated from the data harvested from all health monitoring gadgets and probabilities' model. As analytic interfaces, personalized health platforms offer guidance based on specific measures of health, genetic data, and behavior. These include information on nutrition, physical activity, use of medications, and how to be more active. This offers the possibility of achieving tailored health intelligence to reduce health risks and improve preventive health planning, according to people's needs, choice, and susceptibility.

3.4 Data acquisition and analysis

Data collection and interpretation, therefore, form the cardinal steps in enhancing the use of biomedical sensors and devices. HISs are the processes whereby information regarding a patient's health is gathered, analyzed, and used for the purposes of diagnosis and treatment planning as well as further management of the person's health. Over the years, new technologies of data acquisition and analysis have boosted the potential and utility of biomedical devices.

The raw data collected through sensor networks is processed using various signal processing techniques to enhance its quality and usability. These methods include filtering, which helps eliminate noise and preserve signal integrity; amplification, which strengthens weak signals; and transformation techniques, which convert signals into more analyzable forms. One widely used method is Fourier analysis, which transforms signals from the time domain into the frequency domain to help isolate and remove unwanted noise. Additionally, feature extraction techniques are applied to identify key characteristics within the data.

In certain advanced applications, more sophisticated signal processing techniques are required to handle nonstationary signals and complex noise patterns. These scenarios often leverage computational intelligence methods, such as machine learning and adaptive filtering, to improve signal interpretation. Effective signal processing is essential for ensuring that data acquired from biomedical sensors is accurate, reliable, and suitable for clinical decision-making.

Figure 9: Data acquisition and analysis.

Data fusion and integration means the use of data collected from different sources in order to come up with a holistic view of a patient's status. This process involves the combining of data collected from numerous types of sensors, including wearable, implantable, and imaging. Figure 9 gives an overview of the data acquisition and analysis process and the links between raw data, data preprocessing, and analytical outcomes. Some measures like multi-sensor fusion and data fusion algorithm are applied in the integration of different nonhomogeneous data to enhance the accuracy and comprehensiveness of health data. Another area of data integration is where sensor data is combined with EHR to give a broad picture of the patient's condition and medical background. It improves diagnostic assessments because the presence of multiple data sources makes the assessment's conclusion more accurate and reliable.

Real-time data acquisition refers to the process of acquiring sensor data as it is collected, and processing it instantaneously. This capability is critical for the patient-facing application that needs real-time interaction and response, such as the CGM and cardiac arrhythmia application. Some details of the real-time data acquisition systems include the use of high-speed data acquisition hardware, fast communication methods of data transfer, and low latency data analysis in the acquisition of data. The given benefit of constant observation of physiological data highlights the fact that potential risks are identified right away and the course of treatment can be immediately changed for a patient's sake.

Historical data analysis and trend monitoring involve examining data collected at multiple time points to identify patterns, detect changes, and assess overall health status. Monitoring techniques such as time-series analysis, statistical modeling, and feature extraction are used to uncover trends in longitudinal data (i.e., repeated measurements over time). In addition to time-series methods, cross-sectional analysis can be employed to evaluate sequential data and identify both major and minor health anomalies. By analyzing historical data, healthcare professionals can gain insights into disease progression, treatment outcomes, and other important clinical indicators. This information enables healthcare managers and providers to improve patient care, make informed decisions, and proactively plan interventions by anticipating future health events.

Sensor calibration involves establishing sensor calibration coefficients by comparing it with a reference sensor, while validation aims at obtaining calibration equations that translate the sensor output to biomedical variables of interest. Calibration is the process of setting existing sensor parameters to the specific degree of existing standards. The other process is validation that refers to the process of checking the existing sensor on how it performs against predetermined benchmarks. The continuous high-quality performance of the system requires a calibration and validation process due to factors such as drift and environment changes, as well as inter-individual differences. Calibration, validation, and quality assurance guarantee the integrity and reliability of sensor information and in turn enhance the quality of diagnostic and therapeutic interventions.

3.5 User interaction and feedback

Involvement and feedback from the users of biomedical sensors and devices is very important in the design and functionality of biomedical sensors and devices. These aspects make the devices usable and are designed keeping in mind the needs of the patient as well as the healthcare provider for meaningful reports from the collected data. Consultation of the users increases the level of compliance to treatment programs and their direction, and consequently the overall health status of the users.

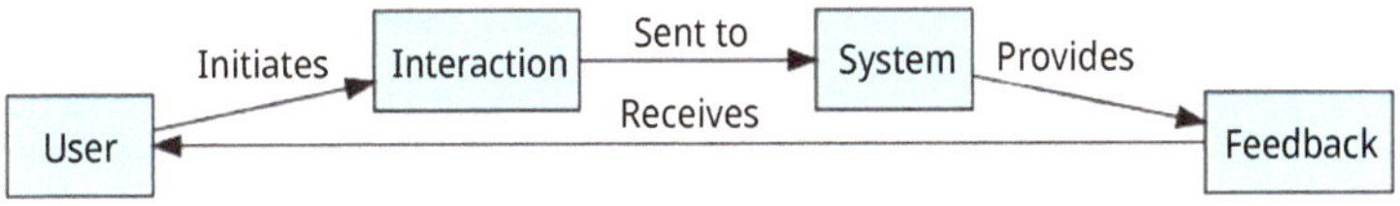

Figure 10: User interaction and feedback.

People interfaces and display systems are critical in determining the functionality of biomedical devices. Complex instrumentations of contemporary gadgets involve graphical touch screens, display interfaces, and voice acknowledgment systems for interaction. Open platforms should be easily navigable for similar data presentation in the form of graphs that use colored alerts. These technologies allow patients to easily interpret their health parameters, device settings, and other functionalities, thus making the experience using this device more interesting.

There are a few more important things apart from notifying users on the upcoming changes in their state of health, in the gadget wear. Biomedical devices include different types of alarms, including visual alarms such as flashing light, sound alarms such as alarms beeps, and vibration alerts, which help in intimating the users, among other things. The process of engagement and providing feedback of the users are shown in Figure 10, where the data flow demonstrates how the users' input is employed to modify the system and improve the user experience. It is also possible to set these alerts, depending on conditional preferences, in addition to the alerts being incorporated into mobile applications in order to facilitate remote notification. Alerts assist the users to respond appropriately to their health condition through the device, which enhances safety and treatment compliance.

Mobile and web applications improve client engagement as health information and performance of the device become readily available. These applications help the users to monitor their basic health indices; monitor daily, weekly and monthly trends; and get feedback, depending on their value. Some of the features include the ability to produce graphical representation of facts and figures as well as trends and real-time report that assist a user in the management of his/her status health. Applications help patients interact with their healthcare providers by sharing data and getting consultations when the patient is already managed and is still under care.

Connection with Personal Health Records (PHRs) represents a significant shift in how patients engage with their health and wellness. It enables the integration of health data collected from biomedical devices with an individual's electronic health history. This seamless coordination provides a comprehensive view of a patient's current condition and medical background, thereby supporting more informed decision-making by healthcare providers. Patients benefit from having all their health records consolidated in one accessible place, allowing them to easily review and update their information as needed. Furthermore, this integration promotes continuity of care by aggregating data from multiple healthcare providers, ensuring that all relevant information is available and accurately captured throughout clinical treatment.

It is important to understand user experience and usability design aspects to determine how burg0eoning biomedical devices address the needs for a range of user groups. Usability design deals with the physical, cognitive, as well as the affective mode of interaction with the device. Some of the factors are portability, size, and shape of device, and accommodated by the device's interface simplicity, practicality of handling of special needs, and user training. Managing feedback mechanisms such as surveys and usability testing fosters a view of getting to know the future problems and challenges. Addressing these concerns regarding devices' designs ensures that device developers develop wearable solutions that do not only work but also provide comfort to the patient, hence ensuring that the patient uses it consistently.

4 Technological advancements

4.1 Miniaturization and wearability

Size reduction and incorporation into clothing and accessories are two critical technical advancements in the biomedical sensor and device technology that have boosted their application and usability. These have resulted in new developments in smaller, more comfortable, and highly functional devices that can be worn and used for monitoring one's health all day and even at night without inferring with the patient's daily routine.

MEMS technology has played a critical role in miniature biomedical sensors. MEMS devices combine micromechanical and microelectronic systems into small chips to design miniature sensors with considerable sensitivity and selectivity. These systems are used in a number of areas such as pressure-sensitive transducers, accelerometers, and gyroscopes, which are particularly relevant for the monitoring of activity levels, heart rate, and motion. MEMS also has significant impact on miniaturization of sensor devices so that one can wear several of them on the body, expanding the intended usage of the wearable devices.

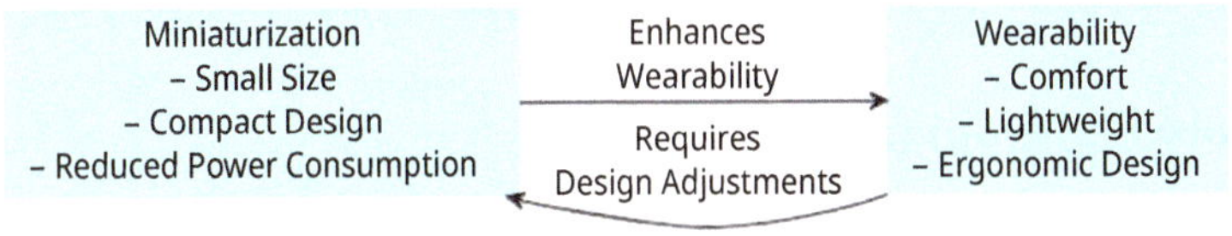

Figure 11: Miniaturization and wearability.

Flexibility and stretchability of electronic components offer a vital improvement in the wearable biomedical technologies. These technologies involve the integration of flexible materials and forms of designs that enable the working electronic components to adapt to the shape of the human body and any forms of movement without affecting its performance. Figure 11 reflects the future evolution of miniaturization and wearability, the progression of technology toward smaller and more comfortable to wear, and even integrated into clothing. Wearable health monitors developed from flexible electronics are smart patches and bands that comfortably fix to the human skin. Stretchable electronics, on the other hand, are used in devices that have to be functional when stretched in smart clothes. These impacts improve the wearability of wearable devices and increase their readiness for constant use.

Portability of physiological sensors has also been enhanced by miniaturization of biomedical sensors to produce implantable and ingestible devices for constant internal monitoring. For the monitoring of glucose level in diabetes patients of cardiac rhythm, implantable sensors as real-time monitors inside the human body are developed. Wearable, implantable sensors, typically encased in a biocompatible material, can be ingested to check gastrointestinal health and absorption of drugs. These devices take advantage of this concept in that they offer useful data on internal health conditions without any requirements for external fittings interferences.

Technological developments in power management have played an important role, especially in the miniaturization of biomedical devices. New solutions in power supply, micro batteries, energy harvesting technologies, and low-power electronics have provided ways to build devices to have longer power operation times and small sizes. Power supply methods like kinetic energy, thermal energy, photo voltaic, and radio frequency that are inherent in the human body can be used to supply power for wearable and implantable devices, thus easing the frequent battery replacement of the devices used in continuous monitoring systems.

The development in miniaturization and wearability has made drastic changes in terms of comfort for the users as well as general appearance. Contemporary wearables are portable, inconspicuous, and ornamental, and therefore acceptable to be worn at workplaces and other social settings. This involves the creation of slim and sectional form and the application of Asiatic materials that do not affect the skin, such as wrist bands that people wear as a fashion accessory, and eye ware and clothing that users have adopted; user comfort is maintained since health monitoring is done without causing any inconvenience.

4.2 Connectivity and communication

Communication, networking and connectivity play central roles in the development of biomedical sensors that facilitate flow of information, real-time coordinated operations, and in exchange of information between machines, humans, and physicians. The relationships and connections between nodes and subsystems within the system are depicted in Figure 12 Stylized flow diagram: Nodes are only accompanied by numeric references, and only the most significant connections between the nodes are shown. These technologies have made a broad range of enhancements to biomedical devices to enhance patient health management and results.

Accessibility and compatibility are interrelated to wireless networking technologies for biometrics sensors and devices, and a number of mobile applications. Some of the most familiar technologies are Bluetooth, wireless fidelity, and cellular technology, each with its characteristics suited to specific application. Bluetooth is commonly employed in transmitting data and short distance communication between the wearable devices and the smart phones. Wi-Fi gives larger coverage and higher data transfer rate, and is used in cases where devices need perpetual data streaming to home-based healthcare systems. Internet networks such as 4G and 5G support remote monitoring by enabling the devices to broadcast data to distant locations, hence enabling tracking of health information, regardless of the place.

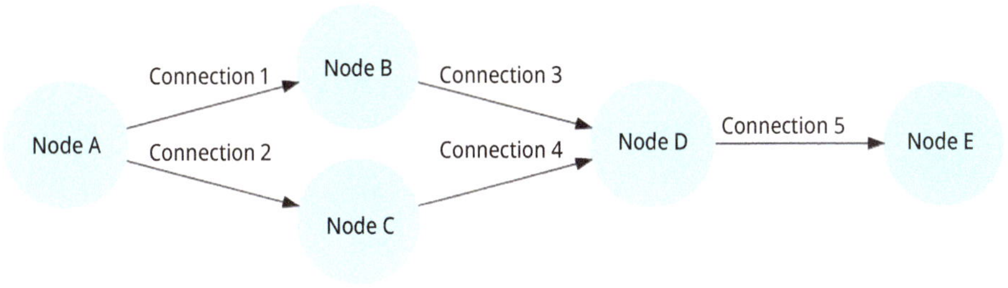

Figure 12: Connectivity and communication.

The adoption of the biomedical devices with the IoT structure helps expand their capability by linking the biomedical devices to the wider IoT framework. IoT objects have the ability to interface with other related healthcare objects, including EHRs, smart home, and telemedicine applications. Such integration enables the overall health monitoring, and information exchange to facilitate real-time update and efficient enhanced and coordinated treatment. IoT enhances sophisticated capabilities, including self-monitored notifications, statistical analysis, and possible health state information, due to the given levels of innovative interaction in healthcare.

There is a need to store and analyze large volumes of data derived from biomedical sensors and devices, and here cloud computing solutions come in handy. Electronic health records are most effectively stored and managed with cloud solutions that offer the possibility of safe data storage and data availability for use. Cloud platforms that support proper data analysis help in making valuable conclusions concerning tendencies and patterns, as well as techniques to apply in their organization. Such solutions include data visualization on aspects that include a patient's health status, and the capability of monitoring and predicting specific conditions with different levels of precision by employing machine learning and artificial intelligence in predicting different conditions.

Security and privacy of data emerge as chief concerns when the biomedical devices are connected and are in a state of communication. The protection of the privacy of health-related information requires the proper administration of its security measures such as coding, identifying, and transmission of data. Secure transmission and storage of information are offered by encryption, and authentication secures the user access to the devices. Due to the legal rules in the context of data protection acts – like HIPAA in the United States and in the majority of European Union countries – the privacy of patients has to be taken care as well as their confidence in the given institution. Security management measures prevent individuals, who are unauthorized, to access the information.

This paper aims to demonstrate that making the communication process between the biomedical devices and systems more integrated, calls for interoperability and standardization. The structure of communication and the use of format allow the products, being devices from different manufacturers, to cooperate with each other. There are currently HL7 and FHIR perspectives that include the necessary requirements and procedures for interoperability in healthcare systems. Standardization promotes ubiquitous device connectivity to wider health networks, as well as improves data interoperability, and boosts individual and team integrated care processes in different forms of health settings.

4.3 Materials and manufacturing

There are nearly universal synergies between the progress of new materials and manufacturing technologies with state-of-the-art biomedical sensor and devices developments. Therefore, the choice of the right materials and integration of innovative technology production processes are critical factors when defining the performances of devices, their robustness, and biocompatibility. The above discussed technologies enhance the development of improved effective biomedical gadgets.

The use of biomaterials is considered to be pivotal in biomedical devices in terms of their performance and sterility. These are specifically developed materials, including biocompatible polymers, metals, and ceramics for use with the intended application, with properties of strength, flexibility, and ability to integrate with the biological tissues. Some of the most used polymers are polyurethane and silicone in implants prosthetics because they are flexible and do not interfere with the fluids of the body. Based on the results obtained, a brief summary of materials and manufacturing processes used in the design of high-throughput screening for new medical materials is presented in Figure 13. Cohesive materials such as titanium and stainless steel are preferred for mechanical parts of implants for their strong and cleat antagonism to corrosion. Alumina and zirconia ceramics, due to their outstanding hardness and wear resistance, are applied in joint replacements and dental implants.

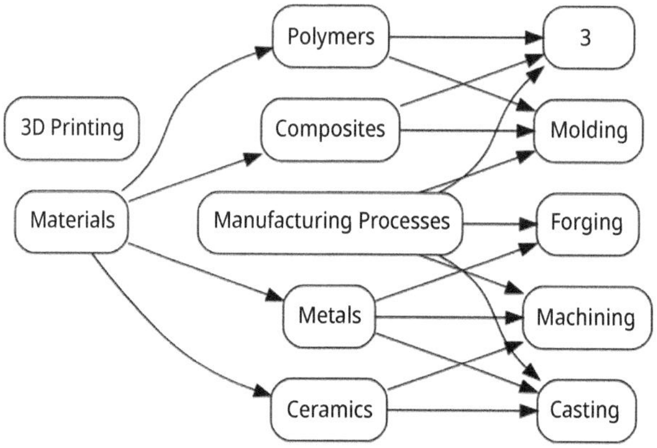

Figure 13: Materials and manufacturing.

Two important enabling technologies are miniaturization and microfabrication, which play an important role in the further development of biomedical sensors. MEMS and nanotechnology are in the forefront in these developments, thus making it possible to incorporate both the sensors and actuators within single highly effective devices. MEMS allow for manufacturing of miniaturized elements in the form of sensors and actuators with high sensitivity and accuracy, as used in self-monitoring and diagnosing devices such as blood glucose monitors and health monitoring wearable devices. Nanotechnology can help in the integration of nanoscale sensors as well as other devices that exhibit better performance features; higher sensitivity and selectivity being examples.

Among the advanced manufacturing technologies, the so-called additive manufacturing has recently gained a status of a revolutionary technological tool in biomedical devices fabrication. This technique makes it possible to build thin-walled structures with close contours and unimpeded through-holes with high accuracy, and detail additive technologies such as 3D printing are used to produce implants, prosthetics, and surgical instruments, which have the required dimensions identical to the individual patient's needs. Materials utilized in 3D printing for biomedical applications are biodegradable polymers, metals, and ceramics. The capacity to rapidly model and produce individualized gadgets promotes overall effectiveness in device development and the benefits experienced by the patient.

The advancement of stretchable and flexible electronics has led the way for the prolonged biomedical use of wearable and implantable applications. These technologies involve the use of new material, which is then processed in new methods to make devices that can bend, to fold around the body and also bend with the motions. Flexible electronics are generally based on materials such as organic semiconductors and conductive polymers because they are both flexible and offer conductivity. Examples of such metamaterials-based applications are wearable health monitoring systems, bendable sensors for physiological signal acquisition, and smart clothing with inbuilt electronics. The enhancement of flexible electronics increases comfort for the end-users and device performance in biomedical application.

Manufacturing quality and reliability are very critical in ensuring that biomedical devices meet certain safety and performance standards when used. Modern production operations minimize manufacturing variation by utilizing technology such as automated inspection system, statistical quality control, and failure analysis. This involves controlling how institutions and their accessories get contaminated in a very high standard, to minimize this aspect's effect on the devices. Compliance with the requirements, as prescribed by the FDA and ISO, is critical in achieving devises' reliability and guaranteeing that it is suitable for use in clinical practice.

4.4 Enhanced sensor capabilities

Several developments in the biomedical field have made the sensors more efficient in terms of functionality and utility. Advancements in physical sensors have led to better measurement, higher sensitivity, and greater versatility, advancing the general quality of health check and discovery.

Technological developments in the recent past have seen quite remarkable enhancements regarding sensitivity and precision of sensors. The current biosensors are also very sensitive to pick on any small changes in physiological indices like glucose level, pulse rate, and blood pressure, among others. High sensitivity of the method helps to reveal the initial signs of the disease due to new biomarkers as well as infinitesimal shifts in physiological processes, which were not observed before. For instance, the improved glucose sensors can determine the glucose concentration in interstitial fluid with relatively low calibration, necessary for diabetic treatment. This added precision offers greater capacity in diagnosis and aids in the management and planning on the treatments offered.

The integration of advanced sensors has been made possible by miniaturization of components, which has also had the effects of permitting the creation of less obtrusive and less uncomfortable devices. When the current high technologies are applied, the enhanced sensor functions, indicated in Figure 14, are obtained, so that the data accuracy and operating efficiency are improved. New developments in LIGA and nanotechnology have enabled the development of compact sensors to be incorporated into wearable and implantable system, without significant reduction in the sensing capabilities. Multiplexing, inclusion of several sensors within one hardware, has also grown increasingly popular, making it possible to monitor quite a few different physiological signs at the same time. For instance, fitness trackers have evolved to integrate both heart rate monitor, ECG, and blood oxygen sensors, and data from all these devices can be obtained from a single wearable device.

Different paradigms in the utilization of the outcomes of the sensor data interpretation and the role of the improved data processing and analytics have grown significantly in their importance. Real-time biomedical signals are monitored through sensors with the help of enhanced algorithms and trends in machine learning. These technologies allow for far reaching analysis of data, pattern recognition, and health trend prognosis. For instance, machine learning algorithms can filter ECG signals in order to diagnose atrial fibrillation and probable cardiovascular incidents for users and their caregivers. There is also improved data analytical processing, which offers real-time monitoring and feedback, while modifying treatments and management actions.

Facilitating the integration as well as the functionality of biomedical sensors has been made possible by enhanced connectivity and communication technologies. Many of today's sensors can integrate Bluetooth, Wi-Fi, and other communication means, which means that data can be transferred straight to a mobile device and cloud space health system. This integration allows operating on a single platform; the data is shared

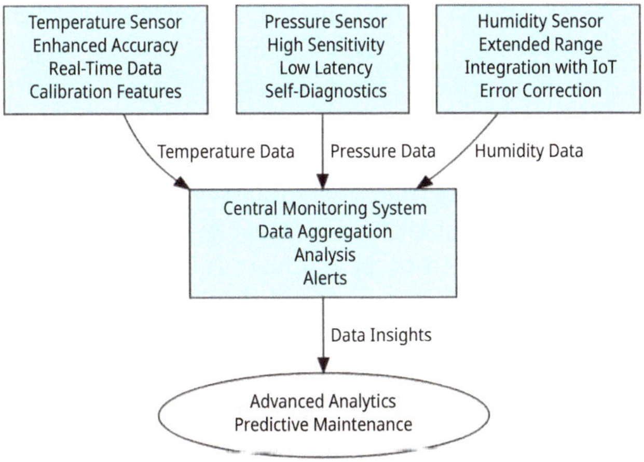

Figure 14: Enhanced sensor capabilities.

in real-time and enables remote monitoring and telehealth solutions. The enhanced communication technologies allow early interventions and grants authorized healthcare providers access to critical patient data, regardless of their physical location and status, in general, and upgrades the productivity of the care process.

Recent material engineering developments have made possible the detection of biosensors with enhanced biocompatibility, and increased their sturdiness. Emerging technologies include the use of flexible electronics, materials that can dissolve in the body after performing a certain function, and protective coatings that reduce sensor interaction with body tissues, instead giving them better interaction, while reducing inflammation. For example, stretchability allows the development of thin plane, form-fitting sensors for long-term skin-worn applications. Better biocompatibility also means that there are few issues with the use of such sensors in implantable devices, hence improving the lot of patients.

4.5 Regulatory and compliance advances

The authors propose that certification/review and compliance/follow-up are critical to the expansion and application of biomedical sensors and apparatus. These advancements guarantee that brand-new technologies are secure, biomedical, and up to par, all at the same time helping promote innovative growth and admissions into new markets. The specifics of specific instructions and trends in the sphere of regulatory and compliance are shown in Figure 15. Current change events and trends in the regulatory system and compliance techniques have helped to respond to the implementation of improved technologies, including AI and real-time data.

The development of biomedical sensors and devices and their regulation have undergone some reforms, which have deliberately sought to minimize the approval procedures. Expedited access programs have been launched by the U S Food and Drug Administration, known as Breakthrough Devices Program, and by the EUA European Medicines Agency, known as PRIME (Priority Medicines) scheme, to fast track the review and approval of innovative technologies. They exist to quicken the advancement and deployment of such devices that cry out for improvements in patient outcomes, and to portend a shortened timeframe in the delivery of promising innovations.

Because the number of biomedical devices with AI and cloud features is on the rise, the legal requirements for data protection are even higher. Adherence to rules like the HIPAA in America and the GDPR administration in Europe are compulsory to protect patient information. These regulations specific standards on how data is to be encrypted, how access to such information is to be controlled, and how user consent is to be obtained before their sensitive health information is created and used. They have allowed the patients to trust their doctors as well as fulfill legal requirements of certain standards.

Introducing artificial intelligence and machine learning technologies into biomedical devices has called for standard measures to regulate them. Agencies are making efforts to define new guidelines that can cover difficulties involved in the application of AI-driven devices, including algorithm explanation, verification, and testing, for their performance. For instance, the FDA has come up with guidance in the assessment of algorithm-based SaMD, where the evaluation criteria include accuracy, additional sturdiness, and functionality, regarding consistent enhancement. This guide is created to provide manufacturers with practical procedures that are important in developing smart technologies that are safe to use and are capable of continued high performance.

Innovations in the current regulations have led to a focus on post-market evaluation and compiling of actual life data in relation to the performance and safety of biomedical devices after they have been released for public use. Modern regulatory agencies are demanding more and more post-market data for evaluating overall device performance and for searching out possible problems and outcomes. This approach of using real-world evidence is done through registries, user feedback, and continuous post-marketing studies. Follow-up evaluation contributes to the review of particular devices in their context of use and overall assessments of their safety and effectiveness, and identifies improvements to the devices, post market surveillance studies, and regulations.

The process of global synchronization of regulatory requirements is progressing to enable the global deployment of biomedical sensors and devices. The IMDRF and the ISO are in the process of harmonizing their standards in different areas to enhance the approval procedures within the various regions. It is also important to note that harmonized standards decrease regulatory restraints, and also facilitate the understanding of compliance requirements for manufacturers, allowing for the export of technologies on an international level, with ease. These initiatives are designed to

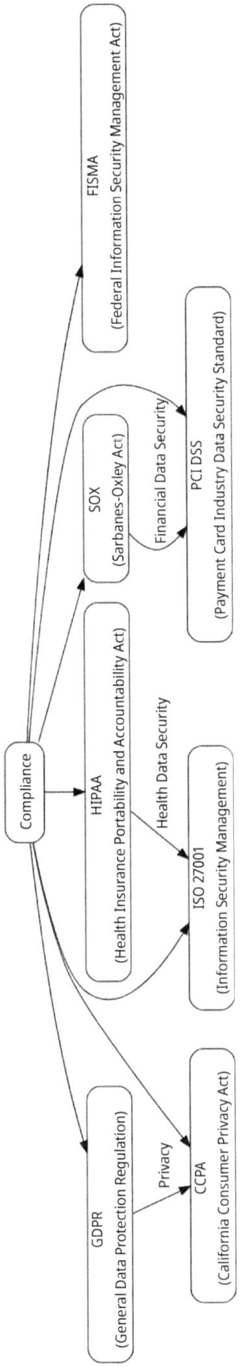

Figure 15: Regulatory and compliance advances.

establish a far less variable environment for regulating biomedical devices, as well as facilitate international cooperation and promote the global distribution of superior biomedical devices.

5 Challenges and limitations

5.1 Accuracy and reliability

Precision is the core characteristic that defines efficiency and clinical applicability of BM sensors and gadgets. When using these devices, it is very important that they would give accurate measurements and consistently provide correct results to allow correct diagnosis, therapy, and patient protection. Some of the factors due to which one can't get highly accurate and reliable results are the factors that can affect the success of such technologies and their acceptance in clinical practice.

Variability in sensor performance remains one of the main issues affecting accuracy and reliability of measurements. Biomedical sensors are prone to many enhancements such as environmental influence, user influence, and damage to sensors. For instance, changes in temperature with humidity and electromagnetic interfere affect the performance of the sensors, leading to deviation from true values. Monitoring the sensors' performance at different conditions is necessary through meticulous periodic checks, profound control procedures, as well as keenly constructed design, to minimize interferences from external environment conditions on the sensors.

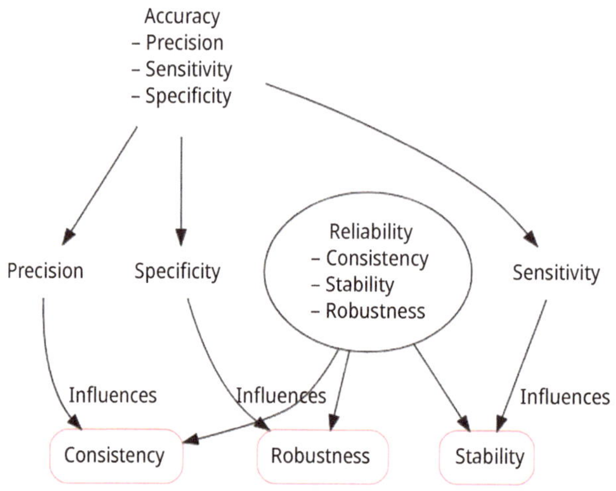

Figure 16: Accuracy and reliability.

Calibration and validation are fundamental techniques for enhancing the credibility of biomedical sensors. While calibration means the modification of its output to optimized known values, validation focuses on the implant's performance, compared to set standards. Low calibration and low validation can cause wrong measurements, which interfere with the clinical decisions made. Routine calibration and elaborate validation procedures are crucial to check the device's consistency throughout its usage and across a number of applications. And, it still remains a major issue to create simple and effective calibration processes as well as validation techniques for various sensor types.

When it comes to biomedical devices powered by AI, the information analysis capability is critically reliant on the algorithms. In machine learning and artificial intelligence, the different algorithms have to be trained with big data sets, where the sample represents the population properly. Formation of improper algorithms due to the training information is another problem; it leads to poor efficiency and reliability of the device. In Figure 16, the A&U of the proposed method shows that the method is accurate and reliable and that it provides the exact and consistent result. This must be followed by subsequent checks, diagnostics, and optimization of the algorithm's performance in case of shortcomings, when improving the accuracy of the analysis and interpretation of data. Perceived algorithm opacity mainly affect trust with regards to diagnostics and therapeutic devices that are based on AI capabilities.

This paper also highlights that the actual way in which the user engages himself / herself with the devices, and can greatly affect the precision and reliability of the biomedical sensors by following the recommended procedures of the equipment. This normally arises out of improper handling of the devices and usage of the equipment, which results in erroneous measurements that are dysfunctional of the equipment used. HRM control risks relate to the possibility of improper use of the technological devices and charging of wrong credits; the use of correct procedures in the operation of the devices needs to be understood and practiced by the users. Also, making some input/output devices friendly to users and giving clear instructions to use such devices reduces the rate of user's effects to degrade the performance of the devices.

Stability and product endurance are essential for biomedical devices; these are significant for these products that are designed for long-term use. As time progresses, the performance of sensors is affected by either mechanical wear or deterioration of the material. These changes can cause the device to become less accurate and less reliable, thereby depriving it of improvement and making it gradually worse each day. Sustaining component standards and employing sound engineering solutions to cater to the actual use span of a device is another criticality in maintaining its reliability during its functional life.

5.2 Data privacy and security

Security and privacy of personal information is a major issue in the use of biomedical sensors and devices, especially since much of the information is in digital format. Most importantly, the management recognizes privacy of health information as key to patients' confidence, meeting various legislation standards and guarding patients' data against cyber vulnerabilities.

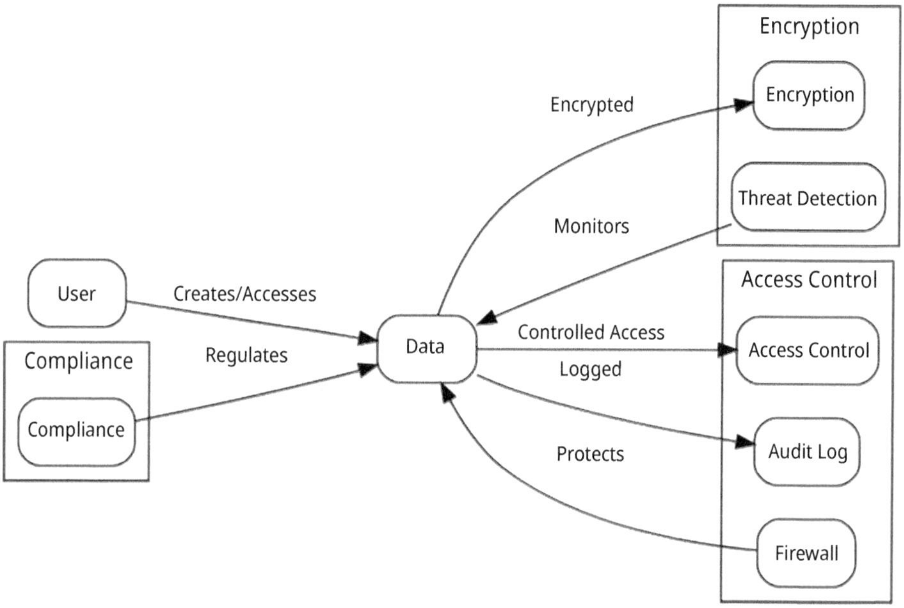

Figure 17: Data privacy and security.

Many biomedical sensors and devices are communication-enabled and send data over the wireless network. This transmission poses potential risks and could lead to data interception as well as unauthorized access. In Figure 17, we present aspects related to data privacy and security that involve the main factors and approaches to security of data. Cryptographic solutions such as AES are used to protect the data in transit to various destinations. Encryption makes sure that in the limited time, the data is encoded and it cannot be intercepted by interested parties. It can also provide a secure connection for the transferred data, as well as an ability to update encryption methods in response to the appearance of threats and risks.

The issue of storing and managing health data is an American concern of privacy and security. Biomedical devices and the associated systems have to guarantee that the data is stored and processed adequately, being compliant with privacy requirements. Organizations in various jurisdictions, including the United States through

HIPAA, and the European Union via GDPR, require data protection standards on security measures like access control and anonymization of data, among others, standards that have enforced the use of secure storage solutions. It can be mentioned that a proper access solution involves the use of multifactor authentication mechanisms and role-based access, which prevent exposure to unauthorized access to data.

Adherence to the principles of data privacy act help in defending information pertaining to health. Most countries have health data protection standards that bar any data use for purposes not mentioned in the consent form; the US has the HIPAA act and Europe, the European General Data Protection Regulation. Implementation measures are the use of privacy policies, periodic assessments, and staff sensitization on measures of protecting and securing data. Businesses also need to draw guidelines for handling data breaches, which consist of notice and response steps in the case of a data privacy violation.

The general aspects of biomedical sensors and devices require that the integrity and accuracy of biomedical data are preserved. Data integrity compromises stem from problems such as those related to software and data tampering. In order to obtain valid and accurate health data, high levels of validation and verification as well as system audits have to be incorporated. Further on, the use of mechanisms that record data through reverse logistics and track changes offer an audit trail that can be used when investigating and resolving data integrity issues.

A lot of biomedical sensors and devices place high dependence on third-party storage processing services. Third-party risks are considered as the risks involved with the evaluation of security and compliance of other affiliated partners. Business entities should legally bind the parties involved in handling of data and frequently conduct checks on the third-party service providers. It reduces the risks associated with data sharing and outsourcing because every third-party service provider has an obligation to observe common standard and follow regulatory standards.

5.3 User acceptance and compliance

This points out that in the biomedical sphere, user acceptance and compliance are key success conditions in the successful use of sensors and devices. Maintaining huge numbers of user engagements and their strict following has always been a desirable goal due to technological development. The nature and perception of these barriers can further impact the effectiveness of these devices in health-related results, as understanding the above aspects is critical to determine the strategies toward enhancing user compliance.

The escalating rate of integrating technologies requires users' education and training as it is the basis for high acceptability and usage compliance of biomedical devices. For the device to be used, users are to give consistent inputs. To get the corresponding appropriate outputs, there is the necessity to learn about the purpose, the functionality, and the operation of the device. Since the device is complex and new, there needs to be further training through instructional manuals, online lessons, and demonstration so that the device is incorporated into the user's daily activities. Real-time support and constant resources also ensure that any challenges encountered by the users are dealt with, while at the same time, reminding them of correct and appropriate device utilization.

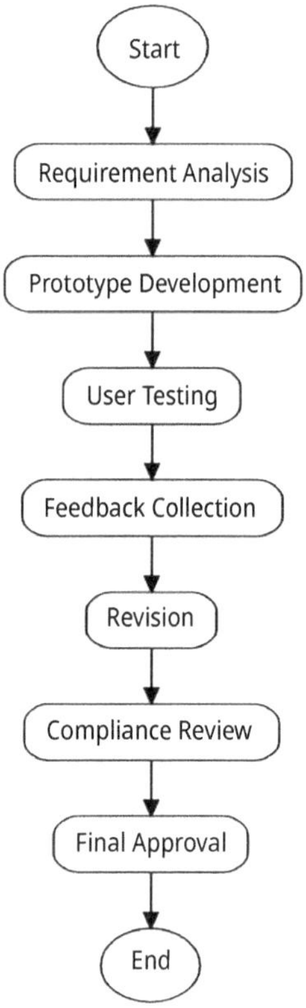

Figure 18: Data privacy and security.

The comprehensible and colorful design of biomedical devices considerably influences the users' acceptance. Products therefore need to be integrated with interfaces that consumers can easily manipulate. Controls should be easily understandable and the products should be designed to support different user capabilities. People do not use complicated devices, as required, because their is not convenient. The measures for the protection and safeguard of the data are depicted by the Figure 18, which shows the various levels and techniques that are used in order to safeguard the unauthorized access and breaches of sensitive data. The first advantage refers to the ability to identify problems within the design process, considering the needs of the users, who also submit complaints about the product. Products that are functionally adaptable, ergonomically appealing, and visually promising would likely be accepted and utilized repeatedly.

Issues relating to privacy and security pose great influence to the acceptability of biomedical devices, especially those that entail information through mobile technologies. It is also important that users feel that their data is handled and protected from unwanted access hacks. Accurate data management procedures, strong encryption procedures, and a transparent privacy parasol are prerequisites to garner the trust of the users. Adherence to legal requirements like GDPR and HIPAA is also useful to discharge responsibilities regarding protection of users' privacy to boost the confidence of the device users.

But in this case, the extent and the cost of the biomedical devices can affect the modem acceptance and compliance of such gadgets. The high costs consequently affects the usage of devices within specific demographics, and as a result, obstructs its usage across the general public. Other costs, such as those that are recurrent by way of maintenance and cost of consumables, are also important factors affecting usage by users. Some of the challenges include cost of accessing devices, insurance, and/or subsidies for a device, which can be scaled to meet the user's need at a cheaper price.

Biomedical devices cannot be utilized properly if they cannot be easily fit into the lifestyles of the users. Acceptance rates are lower for devices that demand radical alteration of existing activities of the users. This integration can be improved by standardizing devices according to the lifestyles of the users as well as by providing clear and optimizable feedback. Simple functions like auto report generation, low interference from users, and adaptability with other technologies like smart phones help integrate easily and ensure compliance.

5.4 Cost and accessibility

Pricing and ease of access are the main barriers toward popular adoption and deployment of state-of-art biomedical sensors and devices. These affect the deployment of technology across the regions and the population in general and hence the overall efficiency of health services.

These critical biomedical sensors and devices are developed and produced through heavy capital investment. Technical promotion costs associated with the creation, development, and testing of new technologies are relatively high. Figure 19 gives an indication of the cost associated with the accessibility and reach of resources. Expenses such as the cost of manufacturing, quality assurance, as well as several regulatory compliances add to the cost of these devices. All these factors can increase the cost for both end consumers and healthcare organizations, thus barring those with narrow means from accessing the devices.

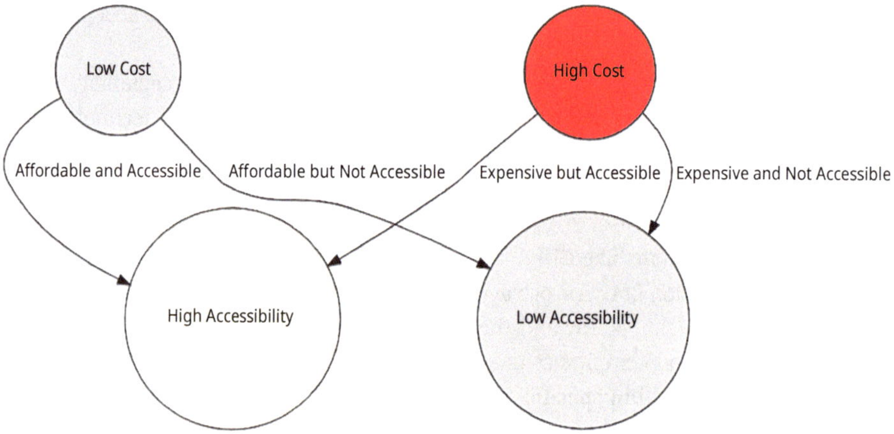

Figure 19: Cost and accessibility.

Provision of insurance cover and reimbursement policies are key determinants of the extent to which biomedical devices are available to the public. Some of the popular and the most recent sensors and devices, which are still regarded as hi-tech, are not included in insurance. Another factor is reimbursement rate; this is a factor that distinguishes one region provider from another and which impacts the cost of these technologies to the patients negatively. Restriction of insurance frequently leads to the situation when the patient has to pay for the services personally; this, in turn, plays a role in discouraging the patient from using advanced devices and, therefore, not reap the related advantages.

Extensive variations in the geographical distribution of healthcare facilities influence the availability of biomedical sensors and devices. While in some areas, factors such as institutional, human, and technical resources remain a significant constraint in the coverage and implementation of advancement technologies, especially to populations in developing nations. A small percentage of individuals in those regions do not get the same benefit of advanced diagnostic and therapeutic technology due to the lack of access to current sophisticated devices. These geographic disparities mean that specific efforts must be made to begin building better healthcare systems and to extend the technological options available.

In the deployment of advanced biomedical sensors and devices, healthcare givers benefit from competency in their profession. Due to the developmental nature of these devices, together with the appropriate interpretation and handling of data, clinicians must continue learning and receive backup. Education and information services are critical to making sure that practitioners can utilize the implemented technologies successfully. In return, due to the costs and difficulties involved in arriving at a level of complete training and support, these devices are not yet popular.

Economic constraints also influence specific patient's access to the new biomedical sensors and devices. Patients with few resources would be hard put to afford such gadgets and the cost of sustaining them. The use of devices during treatment means an additional cost, which is financed by the patient himself, as also the expense related to the care of devices, apart from the cost of repair, which is also additionally paid by the patient himself. Economic barriers can be addressed through case approaches between healthcare providers, policy makers, and device manufactures to ensure that the latest technologies are provided to all stakeholders, including the needy and the indigent.

5.5 Technical and operational issues

It can be illustrated that technology and operation factors are the major concerns in biomedical smart sensors and related devices manufacturing. Solving these issues act as a precondition for achieving satisfactory levels of device stability, efficiency, and consumer enjoyment. There are a number of technical and operation challenges within these technologies affecting the application and uptake of these solutions.

Accuracy and reliability of the sensors that are used in developing biomedical devices are always questionable. Biosensors should be accurate in order to detect physiologic parameters such as temperature, heart rate, and blood glucose. The variations of the sensors can be caused by aspects such as environmental, users, and degradation of the devices. These problems call for frequent calibration and maintenance for the system to offer a standard performance. Future developments include improved method of calibration and the invention of sensors with self-calibration in a bid to minimize the number of manual adjustments.

The data fusion derived from multiple sensors and devices present tremendous technical difficulties. Mobile biomedical devices, in particular, require integration with EHR systems as well as other healthcare applications and technologies. The interoperability of the systems is a significant challenge to ensure that the whole picture of patient health condition is represented. It is important to establish unified formats and communication protocols in order to make integrations and consequently data from various sources can be combined and analyzed. Solving the issues of interaction involves the action of manufacturers of devices, application developers, and healthcare institutions to set universal norms.

Battery management is always a major issue in the use of wearable and implantable biomedical systems because of their need for long-time use. Short battery life affects the device's functionality and, often, user adherence, especially for devices that need to be used for extended periods continuously. Battery technology, electronic parts efficiency, and efficient power use are central to prolong battery longevity and lessen the need to charge batteries. This has prompted researchers to look for other solutions of power supply, including using power from body movements, to improve device sustainability.

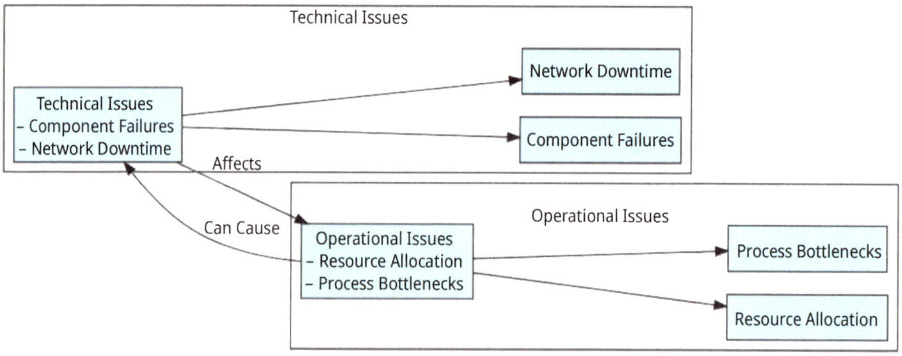

Figure 20: Technical and operational Issues.

Another main operational concern of biomedical devices is the need to safeguard the security and privacy of health data because some of the devices collect and relay such data wirelessly, while others relay them in protected formats. Having measures in the form of excellent encryption methods, proper means of transmitting information, and safe controls are essential in protecting patient's information from erosion. Regulation like HIPAA and GDPR have to be followed strictly so as to ensure that data privacy and security are adhered to. In Figure 20, system performance and operation are tabulated technically and tactfully, where improvements and optimizations are required. It thus becomes important that initiatives to improve and adopt measures toward the protection of users' data should continue toward avoiding raising further concerns.

User training as well as user compliance are very important in ensuring that biomedical devices run as expected. Users have to be trained well in its use, interpretation of data, as well as handling of alerts actions. Lack of training means that people do not use the gear in the right manner, making their resources less effective and posing health dangers. Using friendly interfaces, incorporating full instructional documentation, and offering supporting services enhance the specifics of using the devices as well as in making necessary adjustments to their algorithms. Further, recognizing user worries and adapting the device for the users' feedback could improve usability and maintain device usage.

6 Future directions

6.1 Emerging technologies

Advanced technologies that are on the rise are a clear determinant of biomedical sensors and devices that are on display because they act as the development engine for biomedical health monitoring systems. Developments in this area hold potential to improve the accuracy, speed, and coverage of diagnosis methods and treatment approaches in medicine. Of the many subfields of engineering, several of them are expected to bring a more profound influence in the area of biomedical devices with the advancement of technology.

With the help of AI and ML, biomedical sensors and devices have become more versatile and efficient to develop. The use of these technologies allows handling a massive amount of health data to identify patterns and potentially unusual occurrences, otherwise not visible to the naked eye. AI algorithms can help in effective execution of predictive analysis, customization of treatments, and improving device functionality, based on the results obtained from the real-world data, making progressive changes. Emerging technology in the AI diagnostic tools, including image recognition and natural language processing, are predicted to enhance the technologies embedded in devices and assist in clinical decision-making.

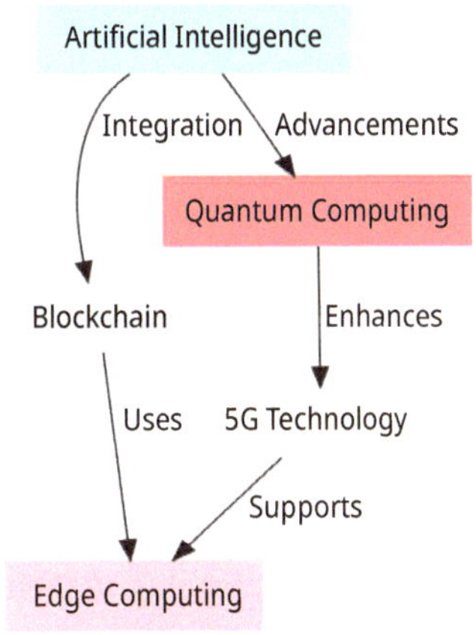

Figure 21: Emerging technologies.

Nanotechnology is preparing a whole field of nano sensors that are expected to be very sensitive and specific in detecting biochemical alteration on the molecular level. These can be used as biosensors, for signs of early disease, metrics of biomarkers, and as drug delivery systems. Development of nanoparticles and nanowires offers the opportunity to develop highly sensitive and selective sensors suitable for integration into wearable devices. The capability of doing real-time molecular diagnoses can be useful in enhancing and optimizing human health.

IoMT is a nascent technology, which bridges medical devices and systems through a common network, the internet, to share real-time data and perform remote monitoring. IoMT relates to a broader class of associated technology such as body wearables, remote monitoring equipment, and connected medical tools. Appendix 4 outlines the interconnection as well as the role played by emergent trends within several industries and their depiction of the future, as shown in Figure 21. The integration of IoMT allows devices to talk to each other, the patient, and healthcare givers, enhancing care co-ordination and customization for patients. Advanced solutions for IoMT technology seek to streamline device communication, data exchange and integration and, patient monitoring.

Wearable biosensors have now become important in life-extending technologies, which facilitate tracking of health conditions. Such devices, which can be attached on the body of the individual or into fabrics, are used for monitoring of various physiological activities of the human body, including heart rate, blood glucose level, and respiratory movement. New developments in fabrics, composites, and coating, have improved wearability and bio sensitivity of wearable biosensors. Such devices afford the means by which individuals can always monitor their health status and take necessary action, based on the data acquired.

There is an escalating growth in remote therapeutics and telemedicine due to emerging and existing technologies in communications and virtual care. Thus, telemedicine stations facilitate consultations and diagnostics, as well as in defining a treatment strategy; therapeutic tele-devices involve the subject in a direct, individual interaction. Advancements in Virtual Reality (VR) and Augmented Reality (AR) technologies are improving accessibility of and support for remote therapeutic interventions. The use of telemedicine, in conjunction with diagnostic and therapeutic equipment, is to introduce new solutions that are important for the development of medicine, increase patients' attendance, and contribute to better management of chronic diseases.

6.2 Personalized and precision health

Precision health and person-centered care are new approaches that integrate superior technologies to deliver unique medical solutions for patients. More importantly, it focuses on reaching out to individuals with tailored health solutions, according to their genetic makeup, environmental influences, and personal lifestyle, hence providing better solutions to clients. Biomedical sensors and devices remain a critical enabler of personalized and precision health since they offer highly specific and individualized health information.

Molecular diagnostics and genomic analysis are rapidly advancing fields in healthcare, enabling the development of personalized treatment strategies tailored to individual patients. Thanks to progress in human genome research, scientists have identified numerous genetic markers associated with various diseases, including cancer, cardiovascular conditions, and rare disorders. Artificial Intelligence (AI) can leverage biomedical sensors that connect an individual's genomic information with physiological and biomarker data. This integration allows for personalized and proactive healthcare management plans based on a patient's genetic makeup and current health status, supporting more precise and effective interventions through advanced health informatics.

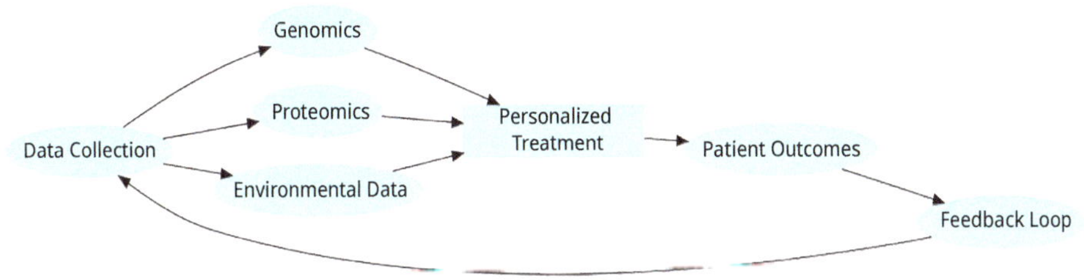

Figure 22: Personalized and precision health.

One strength of precision health pertains to the sensitivity to patient characteristics that the administration of symptom-relieving measures allowed. Popular AI-enabled technologies and interfaces can process data from various sources, including genetic makeup, medical history, and current health status, and suggest the right course of treatment. In particular, Figure 22 reveals how strategies of personalized and precision health movements have embraced the way of medical treatments therapies for patients, and improved the overall result. For example, smart pills for drug delivery enable modifications in dosages and the timing of administered drugs based on high-frequency feedback from wearable devices that measure the amount of drug in circulation and the body's physiological signs. Likewise, owing to the information documented from motion sensors and activity trackers, rehabilitation programs that would be individualized can be developed, based on the functional limitations of an individual, and the recovery process can be enhanced.

Integrated with biomedical sensors and AI algorithms, predictive analytics provides an essential function for precision health: the assessment of risk and disease progression. By comparing data with the current model of a human body and closely related biomarkers parameters' increase/decrease rates, one can predict individuals with high probability of developing certain health conditions. This allows for timely interventions that include screening and changing of habits to reduce disease risks. Proper risk assessment utilizing Big data analytics allows for optimized distribution of resources as well as increasing the efficacy of the general utilization and prevention steps and therapies.

Monitoring of real-time health assists in expanding the management of health conditions in a dynamic way. Smart sensors and gadgets can constantly gather data on patient's symptoms, mobility, and environment, which help to monitor the state of a person's health accurately. This capability allows for the logistical updates to a patient's treatment and dietary plan according to current status of health. For instance, systems using CGMS linked with AI can control the insulin pump for setting rates, in accordance with continually acquired glucose levels, and wearable technologies can give feedback on the movement and activity, and recommend changes to enhance fitness and health.

The person-centered environment and lifestyle data improve the applicability and effectiveness of implemented health interventions. AI sensors are able to track different aspects in the environment that affect an individual's health, including the kind of air, temperature, and humidity. If this data is combined with lifestyle information like diet, exercise, and stress levels, it gives a better picture of the determinants of health. Proband's specific advice and approaches can be derived from this extensive information, which has a beneficial effect on chronic illnesses and general health quality.

6.3 Interdisciplinary research and collaboration

Appropriate interdisciplinary research is fundamental to the enhancement of AI-driven biomedical sensors and fundamental device technology. These technologies are relatively complex and their problems need more than one discipline set of knowledge and skills to solve them. Such synergistic approaches of researchers, engineers, clinicians, and data scientists are paramount to advancing and integrating such technologies for ambulatory tracking.

The freedom of engineering and biomedical sciences appears to be the critical for progress of biomedical sensors and devices. It is the engineer's knowledge in the conception and engineering of subtle sensors, materials, and systems, and the knowledge of biomedical scientists in physiological aspects and medicine. Interdisciplinary collaborative research in this field aims at developing sensitive, accurate, and biocompatible sensors. Technologies like wearables and implantable involve both engineers and biomedical scientists in their development and therefore enhancing the development of high-quality health monitoring technology.

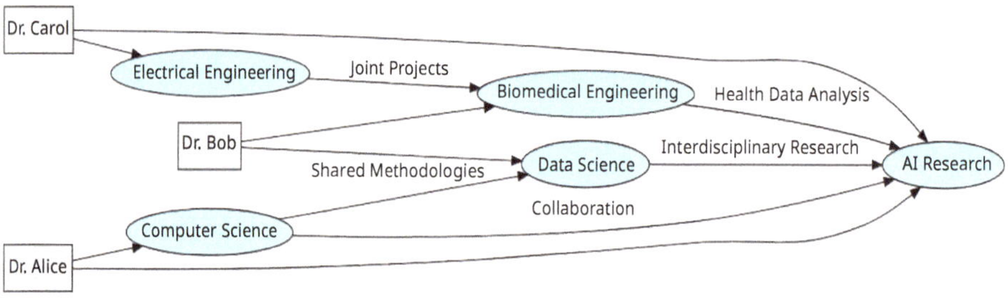

Figure 23: Interdisciplinary research and collaboration.

Applying of AI and machine learning in biomedical devices has an enormous dependence on the partnership of data scientists and clinicians. Data scientists devise ways of analyzing health data, including set algorithms and models, while clinicians offer understanding of patient situations. As indicated in Figure 23, an interconnection of interdisciplinary research and collaboration leads to a future advancement of innovative solutions with application in various disciplines. Collaborative work in this area is needed to develop intelligent systems to forecast health events, develop individualized treatment, and offer better diagnostics. Integration makes sure that to the AI models, the real data is fed back into the models and the models are tweaked to meet working requirements that are possible in real settings of healthcare.

In the case of biomedical technology, encouraging cooperation between the industry and academic institutions remains crucial to improve technology research. Industry partners offer real-life exposure, tools, and market perspective; academic institutions offer up-to-date research, ideas, and expertise. Common work includes cooperative research measures, technological exchange, and construction of preliminary versions and preliminary studies. These linkages help to make ready research results for commercial application and guarantee that innovative technologies are well developed before they are released into the market.

Regulatory compliance of biomedical sensors and device design require collaboration between the regulatory bodies and policy makers. Working with the regulatory agencies is useful in deciphering the complicated legal structure and bringing forward new technologies that are safe for use. Members in policy-making have responsibilities for drawing up regulations that concern the evolution and implementation of biomedical devices. Through collaboration, funding, and better policy-making, regulators, researchers, and other stakeholders in the medical field can develop, evaluate, and approve new medical technologies while addressing emerging issues contributing to the safe and ethical certification of such technologies.

Integrated curricula and interdisciplinary training are vital to create an interprofessional climate and prepare competent personnel for the job that involves several disciplines. Engineering sciences, data sciences, and healthcare put in information literacy-based educational projects in universities help the next generation biomedical technology researchers and professionals. This makes it possible for other stakeholders to develop a better understanding of the different aspects of device development and also to foster a better team. Consequent effort to promote cross-disciplinary training is beneficial in fostering more changes and in responding to the emerging needs in the area of healthcare.

6.4 Global health and accessibility

Global healthcare needs and biomed devices availability are two more important domains for the further evolution of biomedical sensors and instruments. This is required because making available the advanced health technologies for the various populations and different parts of the world is critical in turning around health timelines and closing gaps in health services. Based on these considerations, it is possible to state the following steps that should be taken for the future developments of this field: technology should become more accessible and its cost should be lowered; it is also important to provide equal access to technology to people of all countries.

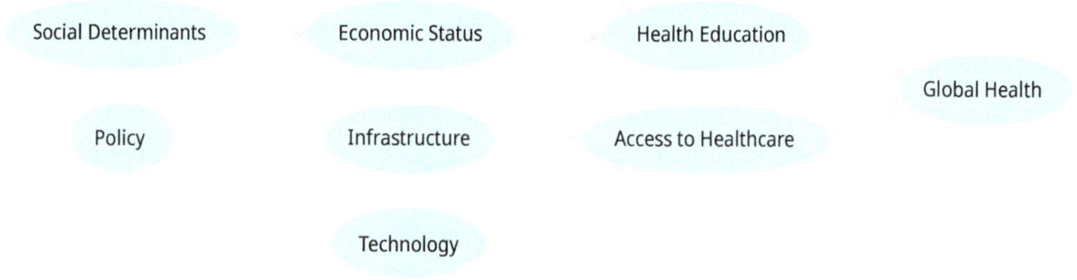

Figure 24: Global health and accessibility.

Closing the digital divide is therefore a core critical success factor to achieving the goals of global health. Thus, many regions, especially LMICs, are unable to access more sophisticated biomedical technologies because of lack of amenities such as roads, very low internet access, and lack of skilled personnel. Figure 24 further elaborates the world's health and access of people, distinguishing the demand for universal healthcare and access to products in different parts of the world. Developing such devices involves understanding the real-world constraints of the environments in which the device is to be deployed, and then developing low-cost, rugged, and easy-to-use devices that effective even in such environments. Likewise, attempts at enhancing the digital environment, such as the extension of the internet connection or proper use of information and communications technology, are critical in making certain that socially excluded populace is not left behind in profiting from the extensions of modern-day technologies in the delivery of healthcare.

Cost effectiveness is another strategic factor that would enable making biomedical sensors and devices available to the larger society. Many healthcare technologies, due to their high costs, are only limited to the cash-rich health sectors and regions. To this, new strategies such as flexibility in device design, standardization and integration, and outsourcing with local manufactures can be employed to make the phones more affordable. Solutions that can be easily extended for the given purposes are significant as well in order to achieve the implementation of technologies for different contexts of the healthcare system.

Special consideration is made for the desire to implement inclusive design to allow biomedical devices to be utilized by as many people as possible from diverse backgrounds with disabilities. This encompasses the design of devices that can be operated by individuals of low health literacy, low physical mobility, and low technological competence. Usability includes additional components, which include interfaces that can be personalized, and the support of multiple languages, as also devices that are variable and interactive, among other factors. In this way, accessibility problems can be revealed during the design phase through interaction with patients and healthcare providers as users and participants.

This paper shows that collaboration and partnerships are critical and are at the core of progress toward global health and accessibility. Working with IOs, governments, INGOs, and CBOs help to create and deliver biomedical technologies, where there is inadequate access to advanced technologies. Partnership areas include research partnerships; technology development and dissemination; and strengthening of capacity to address and respond to health needs at the local level. Such partnerships can assist in negation of barriers to access and enable augmentation of the benefit that is occasioned by technology on special populations across the world.

Ethical and regulatory are were significant, where these principles help to achieve compliance with the international regulatory frameworks to implement biomedical devices within different countries, culture, and values systems. Due to this, appropriate ethical standards must be developed for new emerging technologies to guarantee that their use is proper and fair. There is a need to overcome the differences in the required regulatory frameworks to achieve the needed safety and efficacy of devices as well as their compliance with the peculiarities of the local healthcare systems. With these considerations, developers and regulators of these biomedical technologies can shape public conduct toward their precise, reasonable, and fair usage.

6.5 Regulatory and ethical considerations

As the technology in biomedical sensors and devices gets more advanced, more effort should be applied in the regulation and addressing ethical concerns in relation to its use. It therefore becomes clear that future work in this field shall have to contend with these regulations and has to answer the following ethical challenges, which includes data management, patients' self-determination, and influence of advanced technologies within the health sector.

The exponential growth and development of new biomedical technology, such as the use of AI and ML, requires new legal frameworks to counter new emerging issues. New guidelines must be defined to incorporate the AI-related characteristics, including algorithm disclosure, learning capability, and performance improvement capability of the devices. The future rules probably revolve around setting rules for formal AI verification to prove the work successfully and take the right and proper steps with all the different segments of population. It is the collaboration between legal implementing bodies, and industry and academic publishers, as well as researchers to come up with robust responsive and adaptive regulations and guidelines that can capture technological changes while at the same time protecting patients privacy and ensuring disease efficacy.

Since there is a growing number of biomedical devices that are connected and produce/detect data, protection of this information and resources is crucial. Since devices capture and transmit health information from individuals, quality measures to prevent intrusions and security breaches are necessary. The future considerations include fortification of encryption measures, improvement in data anonymization procedures, and compliance with legal frameworks, including GDPR and HIPAA. The creation of rules for the effective exchange of data and creating an interoperability environment must be crucial for restoring public confidence in patients and protecting their data.

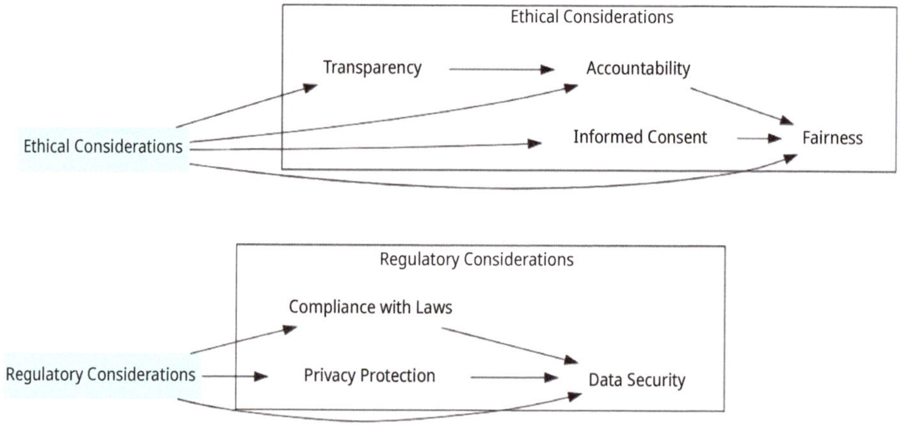

Figure 25: Regulatory and ethical considerations.

Based on the analysis of the relationship between AI and machine learning in biomedical devices, several issues are identified as follows: Some of the anticipated concerns include algorithmic bias in the diagnosis and recommendation of the intervention process and the capacity of the public to understand AI-based decisions. Figure 25 summarizes the regulatory & ethical considerations essential to decide the application of high throughput screening technologies for new-age medical materials along with indicating high level of safety and ethical norms that are to be followed. It would be crucial that future work aims to increase understanding of how to prevent AI bias, guarantee fairness of AI models, and keep AI decision-making transparent. Standard protocols need to be set out on how such AI-driven devices can be developed, deployed, and used to create a balance to ensure patient safety and fairness in the practice of the technology.

The tension between social entrepreneur and patient self-advocacy is an important foreground when considering new forms of biomedical technology. With the devices being capable of monitoring and analyzing personal health data, it is important to be sure that the patient is fully informed and capable of making an informed decision about their engagement in the monitoring programs/data exchange. Issues for future research should include definition of proper consent procedures, patient education with regard to ways that the data can be used for their benefit, and making sure that patients have control over their health information. It remains equally important to respect patient's self-determination while pursuing the use of such technologies in order to further remaining ethical.

The development of biomedical sensors and devices should also consider problems that refer to the availability of advanced gadgets to all the population groups. When it comes to contemporary further development in medical technologies, there are crucial differences that only aggravate existing healthcare discriminations. Further work should be oriented toward finding how, what, and what not make these technologies effective in operating in areas with scarcity of such essentials. This comprises the concept of costs, innovation of cheap technologies, and formulation of regulations promoting reasonable inclusive healthcare technology.

7 Conclusion

The incorporation of AI in continuous health monitoring through biomedical sensors represents a major conceptual shift in the future of healthcare delivery. This chapter has provided insight into how AI-powered sensors are transforming the healthcare industry by improving patient monitoring and management. Although these technologies are still emerging, they have the potential to revolutionize health monitoring by offering enhanced accuracy, timely updates, and personalized information about patients' health conditions.

The transformation to intelligent sensor platforms is driven by advances in technology that leverage sophisticated algorithms to continuously monitor numerous physiological parameters with high precision. Continuous, real-time feedback from AI-enabled sensors enables healthcare providers to make faster, more accurate decisions an essential factor in effectively treating chronic diseases and acute conditions. This capability is fueling the shift from volume-based healthcare models to value-based care, where prevention, rather than treatment, becomes the central focus of healthcare systems.

Machine learning techniques are essential, where the best of the functions of these sensors are enhanced. Machine learning in both analyzing large data sets and recognizing details pinpoints to more accurate diagnostic predictions as well as individualized healthy living approaches. This innovation enhances the scalable effectiveness of the unit care of individuals and impacts the general healthcare systems. What makes such algorithms used widely is that they can be modified continually to enhance their sensor precision as well as forecasting efficiency, making the overall technology ever-evolving to accommodate new medical advancements.

A number of issues still remain in the use of AI-driven biomedical sensors, as mentioned in the above discussion, Data verification, calibration problems, and continuous real-time processing are three major elements requiring further research and technological advancements. To address these issues, it is critical to unlock the full benefits of AI sensors and achieve high levels of system dependability in a clinical environment. More commitment to technological development and extensive procedures of testing should be central to these concerns and the progress of the field.

As the future unfolds, there exist ample possibilities in having biomedical sensors to transform the aspect of personalized health and human care. As technologies that support more personalized, predictive, and preventive medicine, these devices are well positioned to revolutionize how health status is detected and addressed around the world. That is why it appears that these advancements in AI as well as the development of sensors have further possibilities of development, creating an even more effective health monitoring. Alvino et al. earlier pointed out that as the field continues to advance, it is expected that biomedical sensors driven by artificial intelligence play a more central role in defining the future of healthcare, thus enhancing the quality of the systems and consequently the health of patients internationally.

References

[1] Junaid, S. B., Imam, A. A., Abdulkarim, M., Surakat, Y. A., Balogun, A. O., Kumar, G., . . . & Hashim, A. S. (2022). Recent advances in artificial intelligence and wearable sensors in healthcare delivery. Applied Sciences, 12(20), 10271.

[2] Ahmadi, A. (2024). Digital health transformation: Leveraging AI for monitoring and disease management. International Journal of BioLife Sciences (IJBLS), 3(1), 10–24.

[3] Qi, J., Yang, P., Min, G., Amft, O., Dong, F., & Xu, L. (2017). Advanced internet of things for personalised healthcare systems: A survey. Pervasive and Mobile Computing, 41, 132–149.

[4] Diaz-Flores, E., Meyer, T., & Giorkallos, A. (2022). Evolution of artificial intelligence-powered technologies in biomedical research and healthcare. Smart Biolabs of the Future, 23–60.

[5] Khang, A. (Ed.). (2024). AI-driven innovations in digital healthcare: Emerging trends, challenges, and applications: emerging trends, challenges, and applications.

[6] Kasowaki, L., & Deniz, B. (2024). Revolutionizing big data: AI algorithms and predictive analytics (No. 11685). EasyChair.

[7] Kasowaki, L., & Ozan, A. (2024). Artificial intelligence in big data: Transforming insights through advanced algorithms (No. 11684). EasyChair.

[8] Zaid, A., & Milson, S. (2024). Precision analytics: AI-driven solutions for big data challenges (No. 11895). EasyChair.

[9] Zaki, H., Shayea, I., Saoud, B., Azmi, M. H., El-saleh, A. A., Saad, S. A., & Alankhi, M. (2024). Advancing patient outcomes: the intersection of biomedical engineering, medical devices, and AI. EasyChair Preprints, 12174.

[10] Vermesan, O., & Friess, P. (Eds.). (2013). Internet of things: Converging technologies for smart environments and integrated ecosystems. River publishers.

[11] Albahri, O. S., Albahri, A. S., Mohammed, K. I., Zaidan, A. A., Zaidan, B. B., Hashim, M., & Salman, O. H. (2018). Systematic review of real-time remote health monitoring system in triage and priority-based sensor technology: Taxonomy, open challenges, motivation and recommendations. Journal of Medical Systems, 42, 1–27.

[12] Appelboom, G., Camacho, E., Abraham, M. E., Bruce, S. S., Dumont, E. L., Zacharia, B. E., . . . & Connolly, E. S. (2014). Smart wearable body sensors for patient self-assessment and monitoring. Archives of Public Health, 72, 1–9.

[13] Ahmed, Z., Mohamed, K., Zeeshan, S., & Dong, X. (2020). Artificial intelligence with multi-functional machine learning platform development for better healthcare and precision medicine. Database, 2020, baaa010.

[14] Boyce, M. B., Browne, J. P., & Greenhalgh, J. (2014). The experiences of professionals with using information from patient-reported outcome measures to improve the quality of healthcare: A systematic review of qualitative research. BMJ Quality & Safety, 23(6), 508–518.

[15] Solovchuk, D. R. (2024). Advances in AI-assisted biochip technology for biomedicine. Biomedicine & Pharmacotherapy, 177, 116997.

[16] Kriegel, H. P., Kröger, P., & Zimek, A. (2009). Clustering high-dimensional data: A survey on subspace clustering, pattern-based clustering, and correlation clustering. Acm Transactions on Knowledge Discovery from Data (Tkdd), 3(1), 1–58.

[17] Rasool, S., Husnain, A., Saeed, A., Gill, A. Y., & Hussain, H. K. (2023). Harnessing predictive power: Exploring the crucial role of machine learning in early disease detection. JURIHUM: Jurnal Inovasi Dan Humaniora, 1(2), 302–315.

[18] Clarke, J. L., Bourn, S., Skoufalos, A., Beck, E. H., & Castillo, D. J. (2017). An innovative approach to health care delivery for patients with chronic conditions. Population Health Management, 20(1), 23–30.

[19] Dogheim, G. M., & Hussain, A. (2023). Patient care through AI-driven remote monitoring: Analyzing the role of predictive models and intelligent alerts in preventive medicine. Journal of Contemporary Healthcare Analytics, 7(1), 94–110.

[20] Sahu, M., Gupta, R., Ambasta, R. K., & Kumar, P. (2022). Artificial intelligence and machine learning in precision medicine: A paradigm shift in big data analysis. Progress in Molecular Biology and Translational Science, 190(1), 57–100.

[21] Yao, Y., Han, T., Yu, J., & Xie, M. (2024). Uncertainty-aware deep learning for reliable health monitoring in safety-critical energy systems. Energy, 291, 130419.

[22] Darwish, A., & Hassanien, A. E. (2011). Wearable and implantable wireless sensor network solutions for healthcare monitoring. Sensors, 11(6), 5561–5595.

[23] Giordano, M. R., Malings, C., Pandis, S. N., Presto, A. A., McNeill, V. F., Westervelt, D. M., . . . & Subramanian, R. (2021). From low-cost sensors to high-quality data: A summary of challenges and best practices for effectively calibrating low-cost particulate matter mass sensors. Journal of Aerosol Science, 158, 105833.

[24] Cox, S. M., Lane, A., & Volchenboum, S. L. (2018). Use of wearable, mobile, and sensor technology in cancer clinical trials. JCO Clinical Cancer Informatics, 2, 1–11.

[25] Wang, C., He, T., Zhou, H., Zhang, Z., & Lee, C. (2023). Artificial intelligence enhanced sensors-enabling technologies to next-generation healthcare and biomedical platform. Bioelectronic Medicine, 9(1), 17.

T. Sripriya, Praveen Talari, Amirthalakahmi T.M., Senthilkumar B., T.M. Thiyagu

Development of AI-driven biomedical sensors and devices optimization for continuous health monitoring

Abstract: Due to the rapid evolution of artificial intelligence (AI), current biomedical sensor technology has experienced the evolution of exciting concepts and continuous progress in nonstop health monitoring. The data given in this chapter provide an effective focus on AI driven biological sensors, their growth, incorporation, and impact on society in the modern-day healthcare industry. The AI technologies, including machine learning and edge computing are described as the enablers of sensor performance for real-time data processing and individualized healthcare applications. Finally, the principles of AI-integrated sensors along with examples and case studies from clinical practice and conducted research are presented in the chapter and numerous examples of successful AI applications in chronic diseases, diagnostics accuracy, and remote patient control are provided. It also explores the major issues, such as data privacy issues, integration of different systems, and issues of compliance. Students can derive specific knowledge for discovering the potential of AI in the improvement of biomedical sensors and explore the future developments in the biomedical field from this chapter.

Keywords: artificial intelligence, biomedical sensors, continuous health monitoring, machine learning, edge computing, remote patient monitoring

T. Sripriya, Department of ECE, Jeppiaar Institute of Technology, Chennai, Tamil Nadu, India, e-mail: sripriya.t@jeppiaarinstitute.org

Praveen Talari, Department of CSE, Vignana Bharathi Institute of Technology, Hyderabad, Telangana, India, e-mail: praveent738@gmail.com

Amirthalakahmi T.M., Department of ECE, SRM Institute of Science and Technology, Ramapuram, Chennai 89, Tamil Nadu, India, e-mail: amirthatm@gmail.com

Senthilkumar B., Department of Electronics and Communication Engineering, VSB Engineering College, Karur 639111, Tamil Nadu, India, e-mail: professor.senthilkumar@gmail.com

T.M. Thiyagu, Department of Computer Science and Engineering, Vel Tech Rangarajan Dr. Sagunthala R&D Institute of Science and Technology, Avadi 600062, Chennai, Tamil Nadu, India, e-mail: t.m.thiyagu@gmail.com

https://doi.org/10.1515/9783111503202-003

1 Introduction

Biomedical sensors are a rapidly growing subfield of the broader sensor technology domain, with advances in material science, including miniaturized and networked devices within the last few decades [1]. The first biomedical sensors that were developed were large, non-intelligent, and applicable only to narrow functions such as glucose detection and estimation of heart rate [2, 3]. Most of these sensors needed frequent calibration and maintenance; deploying these sensors into everyday healthcare practices was thus difficult [4]. Due to innovations in modern technology, there are now better and less-intrusive sensors [5, 6]. It has become possible to have "chronic tracking" of many physiological parameters in real time due to incorporation of current sensors into wearables and implants [7, 8]. Such a development advances the applicability of using health data and enhances the reliability and consistency of health data for constant health check-ups, which have now become beneficial and significant in the advanced areas of present-day healthcare services [9].

As for biomedical sensors, their application of AI has greatly enhanced both their abilities and impacts on health monitoring [10, 11]. In the context of Big Data and their processing and analysis, several AI technologies are more relevant, including machine learning, and deep learning [12]. Employing these advanced techniques, AI-based sensors can determine features and interconnections in the collected information that escape the traditional approaches [13]. This capability enables enhanced and timely assessment of a patient's health status and subsequently the possibility to diagnose previously unidentified problems as well as shape individualized care plans [14]. It is also able to enhance the operation of sensors since they learn from patient needs and thus enhance the efficiency of health monitoring devices [15]. Consequently, artificial intelligence drives substantial progress in the development of biomedical sensors from enhancing the human body monitor for discrete, occasional checks to a method of constant health monitoring [16, 17].

Although great advancement has been achieved with biomedical sensors powered by the AI technology, there are still some challenges that need to be addressed fully to enhance the capabilities and application of biomedical sensors [18]. One of the main challenges is to guarantee the correctness and reliability of measurements provided by the sensors in a number of diverse real-life scenarios [19]. Pertinent factors like sensor calibration, the level of noise within the signal, and issues with interference can have a direct bearing on the quality of data collected and this explains why further research toward better sensor designs as well as noise reduction methods [20, 21] is needed. The practical issues arising from a multiple sensor system, which is combined into a health monitoring system, are technical challenges [22]. Three key challenges had to be addressed in a Tetra system and the development of formats for data fusion and system interfaces and interfaces represented a major contribution in solving them [23]. Many patients were concerned whether they could share their personal health data as regulating privacy and security of these sensitive data was a significant challenge towards compliance of legal requirements [24].

Based on case and clinical studies, practical application and outcomes of the biomedical sensors that have been advanced by artificial intelligence are brought to light. For instance, with a gadget like the Dexcom G6, a CGM that can go for about 10 days without needing a new sensor, patients' glycemic control has been enhanced as well as the frequent incidences of hypoglycemia are detected. Non-wearable AI-driven ECG monitoring has shown enormous potential in identifying early-stage atrial fibrillation hence saving lives [25]. Telemonitoring systems like the Biofourmis system are effective in managing chronic diseases through early identification of the risks of worsening of a condition and/or frequent hospitalizations. Such cases present AI-hinged sensors as enablers that contribute to improving diseases control, diagnostics, and even telemedicine. The experiences from these case studies serve to improve the growth of biomedical sensor technologies as a continuous process.

In the future, the authors predict that biomedical sensors operated by AI are going to develop and progress in the field even further. It is suggested that the means of enhancing the sensors together with the advancement of healthcare can be brought to the next level with new technologies like edge computing AI algorithms. Edge computing reduces response time while accelerating response time by processing data on the sensors. Nowadays, emerging fields of AI as Explainable AI and Federated learning open new possibilities for succeeding model reveal. The important elements are the integration of these technologies with biomedical sensors, which form the next generation of health monitoring systems that are more accurate, personal, and efficient. Biomedical sensors that would be fueled by AI are going to be increasingly important in deciding affairs of healthcare as long as concepts continue to be expanded in terms of research and development.

2 Overview of biomedical sensors

2.1 Definition and purpose of biomedical sensors

The name "biomedical sensors" is used in the detection and measurement of several varieties of physiological variables in the body. These sensors convert biological signals to electrical signals in order that applications in medicine can process, analyze, and interpret the data. Figure 1 presents the definition and use of biomedical sensors for measuring physiological parameters for diagnosing diseases and for individualized treatment. Biomedical sensors were initially designed to deliver accurate and real-time data with significant health parameters including heart rate and rhythm, blood glucose level, body temperature, blood pressure, and neural signal.

Biomedical sensors as data acquisition tools are insufficient; also, critical to enhancing patient care and thereby medical science. These sensors can promote early condition identification and early supervision of relevant health issues, all of which help to effectively control diseases. The ability to follow up continuously is especially helpful in various chronic disorders such as diabetes, hypertension, and arrhythmias where constant attention needs to be paid to prevent complications.

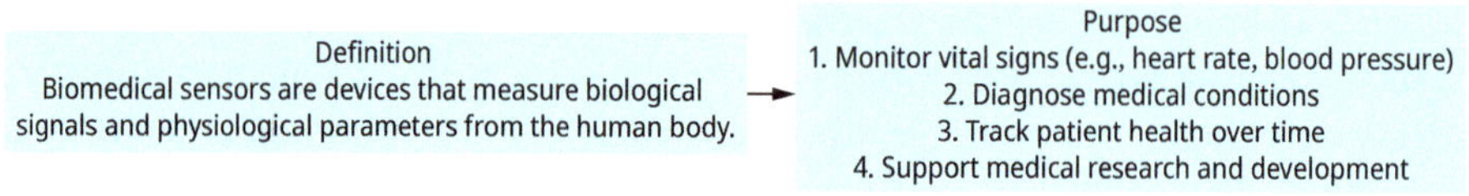

Figure 1: Definition and purpose of biomedical sensors.

Because biomedical sensors provide data that is utilized to customize treatment plans to each patient's needs, they help personalize healthcare. This individualized strategy improves patient outcomes and lowers healthcare costs by facilitating more effective and efficient healthcare delivery. Biomedical sensors offer useful data in research settings that can spur advances in medical science and the creation of novel medications and treatments.

2.2 Historical development and milestones

Biomedical sensors experience prominent evolution stages that have changed notions about healthcare and medical diagnostics. Biomedical sensors have their roots in the early part of the twentieth century when simple tools were used to assess fundamental physiological signs. Early sensors were large and required to be operated manually, which greatly discouraged their use and precision.

The main turning points for biomedical sensors can be traced back to the 1950s with the invention of the electrocardiogram (ECG) and the electroencephalogram (EEG). These devices marked a significant advancement in the ability to diagnose and monitor the electrical activity of the heart and brain, respectively. They highlighted the critical intersection of electronics and medicine, driving innovations in health monitoring technologies and diagnostic methods (see Figure 2).

Figure 2: Historical development and milestones.

A breakthrough in the 1970 s and 1980 s in the development of sensor technology includes the solid-state sensors as well as miniaturization process. An important way to understand EP analysis was through the presentation of its historical evolution and main stages; Figure 2 shows the timeline of these events. The increase in sophistication of devices enabling continuous monitoring of physiological parameters is attributed to the advancement within silicon-based sensor technology. During this period, the first wearable sensors were developed allowing patients to monitor their status outside the clinic.

The advancement of wireless technology in the 1990s along with integration of microelectronics to biomedical sensors was a major step forward. With the evolution of wireless communication, data could be collected remotely and transmitted in real time; this provided further utility in fields of biomedical monitoring both in clinics and at home. A technologically related invention that perhaps originated from this time period was the development of the glucose monitoring appliances used by diabetes patients to monitor and record their blood sugar levels.

As seen, biomedical sensors went through another revolution in the early 2000s through the integration of digital technology and the Internet of Things (IoT). The available and affordable smart phones and wearable technologies expanded the possibility of developing comfortable and effective health monitoring systems, which can monitor and analyze multiple physiological properties over time. Activity monitoring devices with sensors for heart rate, physical activity, and sleep improved due to these advancements and are thus used by consumers to monitor health parameters. Such devices include fitness trackers and smartwatches.

AI broke out into biomedical sensors over the most recent decade and has actualized the new age of health checking. AI and machine learning techniques have also been applied in analysis of data received from sensors making the results much more profound to develop models and healthcare intercessions. Present AI-based sensors have improved capabilities of real-time monitoring and data analysis, identification of anomalies during disease progression, and individual health management considered as a progression in biomedical sensor development.

In sum, the progress of biomedical sensors in the historical evolution can be described as consisting of constant advancement from the initial mechanical structure to the application of a modern AI system. All of these milestones have continued to build the competence of biomedical sensors by broadening their use and enhancing the effect of the technologies in the domain of health and treatment.

2.3 Types of biomedical sensors

Biomedical sensors are defined as multifunctional devices for tracking different organisms' functions since every type is effective in particular contexts. It is illustrated in Figure 3 that biomedical sensors are categorized by the nature of their working process as electrochemical sensors, optical sensors, and mechanical sensors, as well as their intended uses for diagnostics. It is necessary to identify the kinds of biomedical sensors in order to comprehend their functions in constant health surveillance and improvement based on IT systems and AI.

2.3.1 Electrochemical sensors

Electrochemical sensors are used to identify and quantify concentration of targeted chemical compounds using an electrochemical process. Such sensors are central in glucose monitoring for diabetic patients, which measures glucose concentrations in blood fluid. The electrical reactions that occur in the electrochemical cell are found to mimic the concentration of the targeted analyte, in this case, glucose, which is accurately measured. Such sensors are important in that they are extremely sensitive and selective and are often used in the monitoring of long-term illnesses such as diabetes.

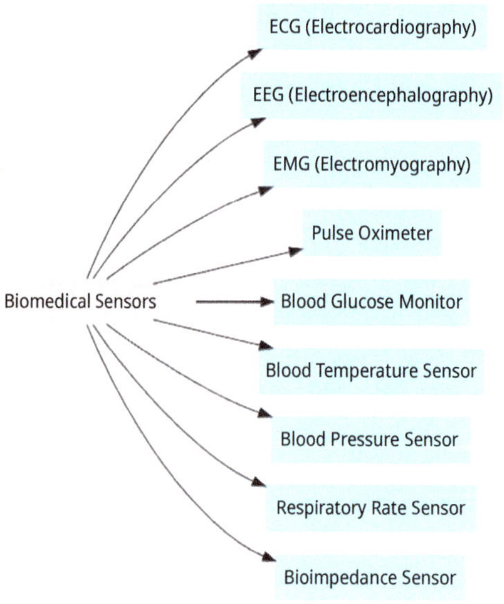

Figure 3: Types of biomedical sensors.

2.3.2 Optical sensors

Optical sensors use the properties of light together with special techniques to quantify related physiological variables. One such device is the pulse oximeter, which measures light absorption and reflection in order to estimate blood oxygen levels. With regard to the variable absorption of light by oxygenated and deoxygenated hemoglobin, these biosensors have enough features for the assessment of respiratory and cardiovascular conditions. Optical sensors are not intrusive in nature and can be incorporated into useful wearable gadgets used in tracking the rate of human breathing as well as the oxygen level in the blood.

2.3.3 Temperature sensors

Temperature sensors are meant to detect changes in physical temperature that are an early sign of many diseases. These sensors are associated with wearable gadgets such as smart and wearable devices to monitor changes in body temperature. For instance, in fever monitoring, the thermal sensors give real-time data to produce indicators of febrile conditions thus enabling early medical intervention. These sensors have to be accurate and respond fast to ensure proper health monitoring is achieved.

2.3.4 Pressure sensors

Physiological pressure measurement devices measure different forms of pressure including blood pressure. For estimating blood pressure, which as a variable fluctuates over time, pressure sensors are applied in devices used for continuous health monitoring, to track trends in cardiovascular health. These sensors operate on the change in pressure and turn that change into an electrical signal to compare pressure and look for signs of health problems.

2.3.5 Bioelectrical sensors

Bioelectrical sensors in turn record electrical signals produced by endogenous physical functions. ECG and EEG are prominent examples, which are used to monitor the activity of the heart and the head, respectively. ECG sensors are attached to the body in an attempt to pick up electrical product of the cardiac contractions and assists in the diagnosis of arrhythmias and other cardiac abnormalities. EEG involves acquiring of brain wave patterns by use of sensors to diagnose conditions such as epilepsy and sleep disorders. These sensors are basic in the determining and treating of various cardiac and neurological conditions.

As such, all kinds of biomedical sensors have been developed to address certain concerns using diverse technologies in order to yield relevant health information. These sensors become smarter when connected to AI technologies, meaning that the data received is real-time, precise, and broader health data.

2.4 Key components and technologies

Biomedical sensors are intelligent gadgets developed to quantify and screen features of health, and the results rely upon different crucial segments and essential advances. Figure 4 shows key components and technologies of high importance in improving electrical performance in sophisticated systems. Awareness of these elements is important for comprehending the utilitarian innovation in BM sensing.

2.4.1 Sensing elements

Sensing elements are the critical parts of the BCI system and are used to detect and quantify physiological signals. These elements depend on the type of the sensor and the parameter of interest to be sensed. Common sensing elements include:

Electrochemical cells: Applicable in sensors for tracking chemical compounds, for instance, glucose in the human body. These cells produce electrical signals when an interaction with the analyte takes place chemically.

Optical sensors: Use light to quantify things such as amount of oxygen in the blood. These sensors depend on principles such as absorbance and reflectance of light in the processing of information.

Thermocouples and thermistors: Used in temperature sensors for thermal measurements to detect fluctuation in body temperatures by converting heat into electricity.

Pressure transducers: Determine pressure, which is the force acting on any surface, for example, a liquid substance. In biomedical applications these are fundamental for tracking blood pressure and other pressure-related signals.

2.4.2 Signal transduction mechanisms

Detection of physiological signals translates into meaningful outputs that are made possible by signal transduction mechanisms. This process involves several devices:

Transducers: Transduce the physical changes assessed by the sensing elements into electrical signals. Such applications include piezoelectric sensors for pressure and capacitive sensors for displacement.

Amplifiers: Amplify the electrical signals generated by the transducers so as to obtain higher amplitudes of these signals for subsequent procedures.

Analog-to-digital converters (ADCs): Convert the boosted constant analog signals into a form suitable for the treatment in computational systems. ADCs are important in translating sensor data into information that can be incorporated into digital health monitoring systems.

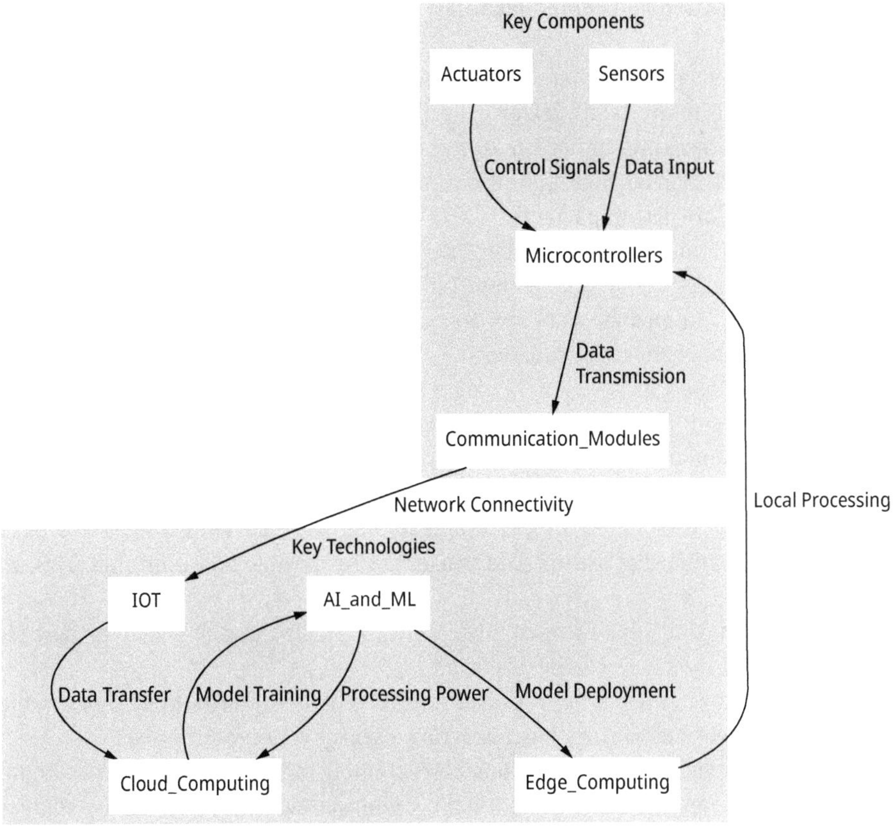

Figure 4: Key components and technologies.

2.4.3 Power supply and energy management

Management of power supply and energy is critical to biomedical sensors particularly in applications such as wearables and implantable devices. Key considerations include:

Battery technologies: Structural merits including compactness and durability are achieved by incorporating new forms of battery technologies including lithium-ion and are flexible.

Energy harvesting: Ambient power using such devices like thermoelectric generators for powering the sensors for longer periods, thus minimizing the requirement for battery replacement.

2.4.4 Data processing and communication

If we capture the data then there is a big challenge of how to process this data and how it would inform the health sector. This involves:

Embedded processing units: The processing components within the context that are responsible for signal conditioning and sensor data preliminary evaluation, such as filtration, etc., are also important.

Wireless communication modules: Facilitate the sharing of information in the form of sensor data through Bluetooth or Wi-Fi to other devices such as smartphones.

Data storage: Immediate data storage solutions including on-board memory for holding sensor data before transmission.

2.4.5 Interface and user interaction

The interface and user interaction components that were developed for the project warrant confirmation that sensor data was easily retrievable and utilizable. This includes:

Display units: Units that have an interface with people through which the current state of their health notices can be displayed.

User input mechanisms: A graphical interface made up of buttons that enable users to control various settings for calibrating a sensor collection routine.

Software applications: Applications on the mobile administrate and coordinate the visualization, analysis, and integration of information with others in the context of HM systems.

Such components and technologies give an insight as to how biomedical sensors work, as well as the key enabling technologies behind them. All these elements would be very central to the design, operation, and use of modern biomedical sensors in the enterprise of continuous monitoring of human health status.

2.5 Importance in modern healthcare

Biomedical sensors are an essential component of contemporary medicine due to the fact that they allow monitoring of the human physiological data in real time. These sensors are both the preventative and the diagnostic elements of healthcare and provide several fundamental benefits that dramatically affect healthcare delivery and outcomes.

First, biosensors in biomedical enable the diagnosis of diseases at an early stage via consistent tracking of health markers. It means that through this continuous monitoring system, the healthcare provider is able to notice any pathophysiologic changes in the patient and this is done before the clinical indicators begin to show up. Screening is highly important in the early identification and a lot could be done when diseases like cardiovascular diseases or diabetes is picked up in its early stage.

Real-time data given by biomedical sensors is contributing to the concept of personalized medicine. These sensors enable healthcare practitioners to adapt treatment schedules to the need of every patient because personal data from patients is constantly collected and assessed. Thus customization reduces the odds of episodes and stays in the hospital because it enhances the efficacy of treatment and brings better ways to handle chronic diseases.

That biomedical sensors enable telemedicine and remotely monitored care makes healthcare more effective. Due to the home monitoring the patient can get better and easier treatment instead of having to visit hospital frequently. This improves the availability of healthcare and lessens the pressure from general illnesses on the health facilities for consumers from other regions.

Also biomedical sensors are critical for enhancing safety outcomes for patients and serving better clinical goals. The sensors that measure the vital signs like blood oxygen levels and pulse rate for instance, give a signal to the healthcare providers, of the changes in the health state of the patient and trigger an appropriate reaction and action. When the patient is actively managed, adverse consequences are averted, and necessary treatment can always be provided on time.

Further, the use of biomedical sensors with EHR systems helps optimal health informatics. The continuous flow of data from the sensors to EHR systems enables health trends to be monitored, better diagnostics to be made, and care givers to effectively and efficiently coordinate their efforts.

Real-time, constant, and noninvasive biomedical sensors are vital in today's healthcare environment because of their potential to contribute to individualization and distance treatment, patient protection, and data control. All of these are integrative to enhance the approach towards healthcare improvement as well as the realization of improved health status.

3 Evolution and advancements in sensor technology

3.1 Early biomedical sensors: limitations and challenges

It ought to be appreciated that the evolution of biomedical sensors commenced with the rudimentary sensors, which had a number of numbing constraints and hitches that diminished their credibility and feasibility within the health institutions. Early sensors embraced in these applications were mainly devised for determining unique variables including pulse rate and glucose level, and were afflicted by several inherent problems.

The first observed drawback in the evolution of biomedical sensors was the size and intrusiveness of the sensors themselves. Some of these sensors necessitated coming into direct physical contact with the body in an invasive process that could be uncomfortable for patient. For instance, older models of glucose monitors required the use of a blood sample for evaluation; getting a blood sample was painful and inconvenient for diabetic patients who needed to take several tests every day.

The other major factor that emerged is the requirement for constant calibration and other maintenance activities. The use of the early sensors in healthcare needs was cumbersome as they had to be periodically readjusted for them to be accurate; this made them unsuitable for real-life use. This constant adjusting not only amplified the operational demand but also added some human errors and variability in the process of measurement.

The issues of accuracy and reliability of early biomedical sensors were also raised. Most of these sensors were vulnerable to noise by other influences such as variability in the environment. Unfortunately, this limited their chances of giving accurate and constant measures that help them influence effective diagnostics of many health issues.

Data transmission and integration were other issues noted by authors in the article. Sensors in the early stages of development were found to be lacking in sophisticated data transmission qualities and therefore the data was to be recorded and analyzed manually. This process was very much slow and bulky since it involved much manual work, which could have confounded the analysis on health changes and the subsequent response.

Another limitation was the small number of parameters that early sensors could measure. The initial sensor systems were designed to monitor only a limited set of physiological indicators, which restricted their ability to provide a comprehensive overview of a patient's condition. Consequently, this narrow focus required the use of multiple separate instruments to track different parameters, greatly complicating the screening process.

These limitations and challenges of early biomedical sensors highlighted the need for improvements in technology to enhance the effectiveness and interconnectivity of these devices. Many of these issues have since been addressed through advances in materials science, sensor miniaturization, and wireless communication technologies, which underpin the development of modern noninvasive healthcare sensors.

3.2 Advances in materials science

Nanotechnology and material engineering have played central roles in the progress of biomedical sensors by bringing qualitative changes in their functionality, performance capability, and versatility. Such advancements refer to the emergence of new materials, enhancement of new material processing methodologies, and advancement in material application in sensors.

Some of the major progress made include development of biocompatible media for implantable sensors that improve on safety and functionality of sensors. Biocompatibility means that a material does not provoke negative responses when it is in contact with living tissue. New technologies introduced are enhanced polymer, hydrogels, and bioactive coating that reduce inflammation and tissue rejection, thus enhancing the lifespan and efficiency of the implantable sensors.

In sensor material, nanotechnology has also played a crucial role in developing nanoscale sensors that have higher sensitivity along with specificities. The unique properties of nanomaterials include, for instance, high surface area-to-mass ratios, and equally remarkable electrical and thermal conductivity. Some examples of such materials are carbon nanotubes, graphene, and metal nanoparticles. Such characteristics facilitate the realization of very highly sensitive sensors that enhance the ability to detect even the slightest perturbation in the human physiological system.

Flexible and elastic sensors used in wearable applications that map the shape of the body and its movements are a development that can be attributed to the material innovations. Substrate materials including elastomers, conductive fabrics, as well as stretchable electronic inks can all be used to fabricate wearable sensors that are comfortable to wear and elastic enough to perform and survive mechanical stretching. They improve the practical utility and comfort for the wearer that is associated with wearable health monitoring devices.

Modern composite material together with microfabrication technologies has created compact sensors that are miniature in design but highly efficient. MEMS technology makes it possible to fabricate miniaturized sensors with all elements arranged internally, thus enabling nonintrusive and long-term monitoring of health.

Also, the progress made in the use of smart materials with tunable characteristics has formed a base for biomedical sensors. Such materials can change their characteristics in any environment, for instance, if the temperature of the solution is altered. All the above ways of adaptability help in putting the sensor in a position where it can deliver right and contextually correct health data.

3.3 Miniaturization and nanotechnology

Miniature techniques with nanotechnology have brought serious changes in the strategy of biomedical sensors and in their efficiency. Thus, these technological advancements have made pivotal roles to revolutionize the features of the sensors along with the drawbacks that were present earlier in the sensor capabilities, such as size, invasive nature, and sensitivity.

Miniaturization means reducing the size of biomedical sensors and yet being able to maintain the functionality of the sensors. By this improvement, portable and wearable sensors that fit in wearable/implantable systems have been designed and produced. The advantages of miniaturized sensors include: 1) the patients' comfort level is enhanced, 2) there is reduced intrusion, and 3) convenience has improved. For instance, telemetry glucose sensors can be incorporated in the small-size continuous glucose monitoring devices that can be worn on the body in order to monitor glucose level changes in near-real time and without much discomfort.

In biomedical sensors, nanotechnology refers to the incorporation of materials and structures that are at least one dimension less than one nanometer in order to enhance the performance of sensors. Taking advantage of the advantageous properties of nanoparticles that include increased surface area-to-volume ratio and higher reactivity, it becomes possible to develop very sensitive and selective sensors. Nanowire, nanotube, and nanoparticle have unique properties, which enhance sensitivity and specificity of the sensors to detect various physiological parameters and biomarkers.

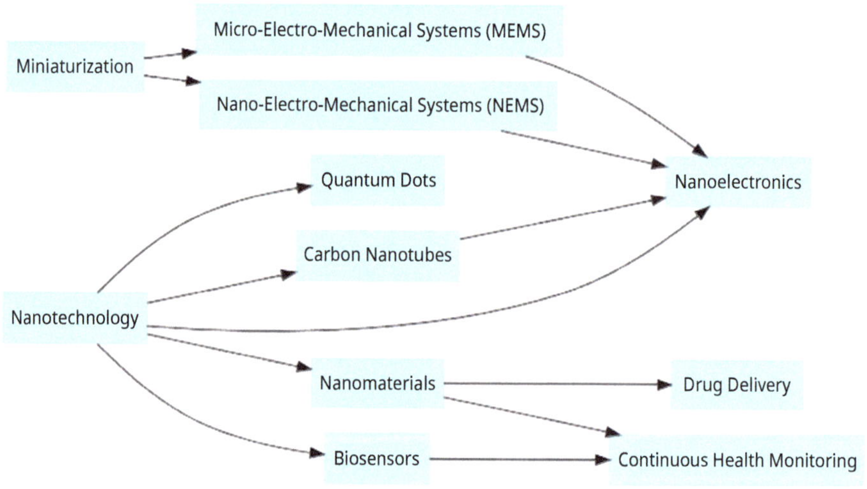

Figure 5: Miniaturization and nanotechnology.

Nanotechnology also enables formation of nanoscale sensors that can be used for small portable wearable gadgets. These nanoscale sensors have the potential of achieving much higher resolutions and therefore improve measurement of physiological parameters. For example, the use of nanoscale sensors would make it possible to measure the changes in biomarkers associated with diseases at the early stage.

Miniaturization and nanotechnology have led to availability of sophisticated sensing techniques including the nano sensors for molecular diagnostic and imaging. These innovations push the envelope to provide improved accuracy and individualized care by enabling examination of biological specimens and constant observation of cellular and molecular activity.

Miniaturization and nanotechnology innovations have boosted the functionality of biomedical sensors with increased performance, minimized invasive approaches, and highly sensitive monitoring systems. Figure 5 depicts the progress of miniaturization and nanotechnology and the benefit of their integration of detail of the devices on the highest level. These technologies remain as drivers of innovation in design of sensors in healthcare and if implemented could revolutionize the healthcare system.

3.4 Wireless and wearable sensor technologies

The advancements in the wireless and wearable sensors systems are a step up evolutionarily in the biomedical sensors and have revolutionized how health data is collected. Such capabilities have effectively solved many weaknesses associated with conventional wired and static sensors, expanding new prospects for constant and concealed health surveillance.

Wired sensors have posed insecurities as patients can only undergo collection of their health data if connected through some wires. The data collected by these sensors is wirelessly communicated to external instruments using radiofrequency, Bluetooth, Wi-Fi, and other wireless technology interfaces. It also improves the feasibility measures of collecting data, real-time observation, and prompt execution of health information. The use of wireless sensors is especially seen in cases where subject mobility was an issue – in this case, during physical activity they leave their beds better monitored than when stationary, therefore giving a better picture of the overall patient health.

Wearable sensors such as smartwatches and fitness trackers are some of the essential everyday technologies normally used to manage our health. These sensors are always evaluating a profile of numerous physiological parameters and personal adornments are expected to be worn, often in clothes. Examples of biological parameters include; body temperature, respiration rate, heart rate, and physical activity. Due to enhanced surface area-to-volume ratio and higher reactivity, nanoscale materials provide an opportunity to develop highly sensitive and selective sensors, including nanowires, nanotubes, and nanoparticles; it is believed that the sensor used in detection of different physiological indicators and biomarkers can benefit from these nanomaterials. What can be seen in Figure 6 is a range of wireless and wearable sensor technologies for real-time health status and data acquisition. Because wearables are nonintrusive, they can be worn for extended periods and data collected more often without invading the lives of the patients.

Miniaturization of sensors along with the incorporation of advanced functional materials has continued to improve the efficiency of wireless and wearable types of sensors. Sensors today have received upgrades in the form of battery life, durability,

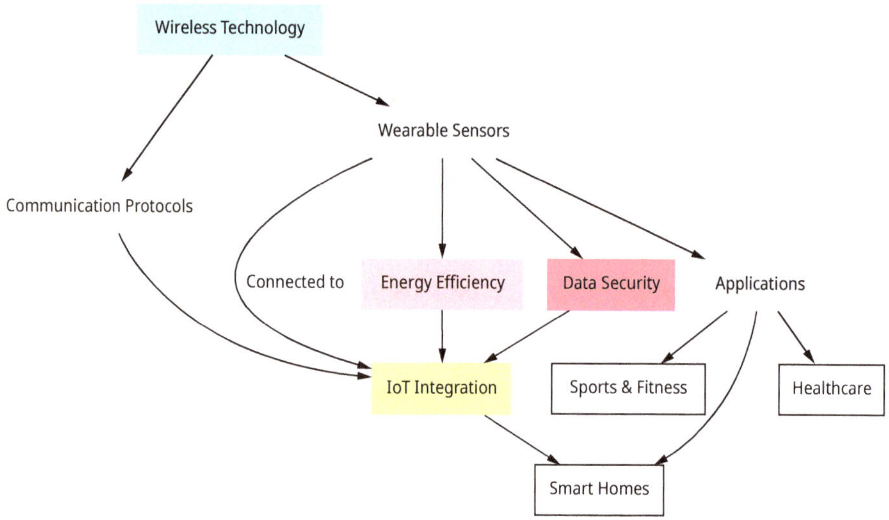

Figure 6: Wireless and wearable sensor technologies.

and the comfort in using them. Some of the factors that have led to development of wearable sensor devices include enhanced physical form of the sensors and availability of microfabrication technologies, which enable manufacturing of thin flexible sensors that conform to body contours.

Another improvement is the incorporation of wearable sensor and machine learning and artificial intelligence concepts. These are supplemented by the following technologies resulting in more advanced data analysis for input such as predictive analytics and real-time interpretation inputs. Smart wear with AI elements can detect potential diseases, note varying changes in physiological characteristics, and provide valuable information, which ensure better preventive healthcare and approach.

These and other wireless and wearable sensors also enable easy interfacing with mHealth applications and EHR systems. This connectivity also provides real-time monitoring of aspects of health, same data update in different applications and system, and organized care. Consumers can get the latest health information, engage in distant patient management, and make decisions based on detailed information.

It is noted that further development in wireless and wearable sensor technologies has enhanced the biomedical monitoring research field. These technologies allow flexibility and integration, which enhance the health processes management, patient-centered care, and better patient health status.

3.5 Integration with mobile health (mHealth) applications

The consolidation of biomedical sensors with mHealth applications is among the biggest developments in healthcare technology. The integration of these systems has changed the way that health data is gathered, evaluated, and used, promising new prospects to advance patient care and health.

Biomedical sensors when integrated with mHealth apps enable the recording of health information in real time and its immediate relay to mobile devices. Modern people with diabetes just scan a couple of buttons on their smartphone to monitor the changes of their heart rate, blood sugar, and physical activities. Their availability ensures that people take personal responsibility every time they need the health of a certain body organ to be enhanced by practicing new lifestyle changes in treatment regimens.

In addition to that, the mHealth applications also offer to boost the performance of biomedical sensors since they are able to carry out advanced analysis and visualization. By these applications, the users can get detailed analysis of the trends, graphs, and alerts based on the sensor readings. It helps the patients including the client and the healthcare providers to grasp complex data, hence comprehend better patterns that in appropriate decision-making when it comes to health.

The consolidation of biomedical sensors with mHealth applications is among the biggest developments in healthcare technology. The integration of these systems has changed the way that health data is gathered, evaluated, and used, promising new prospects to advance patient care and health.

4 Importance of AI in continuous health monitoring

4.1 Role of AI in data processing and analysis

Continuous health monitoring systems require AI in order to analyze and interpret data. The amount and quality of health data generated by biomedical sensors is improved through the management and analysis of Big health Data with the help of AI integration.

This suggests that there is need for algorithms to be used so as to handle the complexity and vast data generated by chronic health monitoring systems. These algorithms are especially important when they are based on machine learning and deep learning; Figure 7 demonstrates how preprocessed data using AI techniques such as machine learning algorithms helps in mining important data features from large and complex dataset. Such algorithms can handle big datasets and provide features and information that are not easily identifiable using more traditional analytical methods.

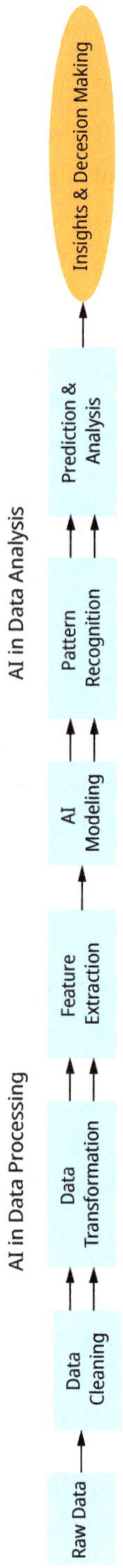

Figure 7: Role of AI in data processing and analysis.

Another important activity of applying AI in data processing is real-time analysis of the data collected by sensors. Currently, AI algorithms can analyze flowing data from multiple sensors to identify various issues, patterns, and deviations from typical healthy values. This real-time analysis is important in alerting the patients as well as healthcare givers thus maintaining close follow-up on any health complications likely to arise.

AI employs more advanced statistical and computational methods on data to increase the accuracy of the interpretation of the data. Stationary datasets, with many records are used to teach machine learning algorithms for pattern and correlation generation for diagnostic and prognostic purposes. Through such techniques, the models learn with time and are also adaptive to new information and this results in more improved health evaluations.

AI also facilitates ways to fuse and integrate data from multiple sensors. Information is obtained continuously from various sensors administrators in chronic health monitoring systems such as pulse rate, blood glucose level, and motion. All these multisensor data can be combined by appropriate AI algorithms and provide a full picture of a patient's health state. AI ensures that health information covers the entire picture as each source of information is considered.

There has been an acceleration of emerging data processing solution based on artificial intelligence for personal health management. Thanks to analyzing health information of an individual a patient, an AI solution is capable of indicating specific patterns and biomarkers unique to a particular patient. It makes appropriate health advice and treatment suggestions with solutions that can increase the impact of health interventions and advance the health of patients.

4.2 Enhancing accuracy and precision of sensors

It is worth emphasizing that AI has developed as an important supporting factor in improving the efficiency of biomedical sensors that are used in punctual health monitoring applications, more specifically, improving the level of accuracy that is provided by existing sensors. Through the combination of AI algorithms with sensors, some major issues concerning data credibility and measurement inaccuracy that are prevalent in the present system are solved, resulting in better heath evaluations.

AI polishes biomedical sensors through altering the manner in which data is processed to enhance the level of accuracy obtained. Among the machine learning strategies that are used to analyze the data of the sensors and identify those patterns, which are not apparent, support vector machines and neural networks are perhaps some of the most well-known. These algorithms can pick slightly different patterns owing to their training on huge databases; such differences suggest errors in sensors. Due to the real-time error detection and correction features, measurements are more accurate and more trustworthy.

AI adapts calibration of sensors by enhancing the accuracy of the sensing systems. Current and conventional approaches of calibration involve manual adjustments and calibration, which have to be done frequently and can be integrated with AI whereby the calibration is made automatic and efficient. The AI algorithms can dynamically remodel the parameters of the sensors depending on the real-time dataset received hence making alternative means of ensuring that these sensors preserve their precision over time and other conditions.

Another advantage of AI is elimination of noise as well as interferences, which affect the data that sensors are feeding. The noise that results from a sensor include unwanted objects/components such as noise, and in order to eliminate such an aspect, signal processing is used as filter it out. Due to the increase, signal-to-noise ratio is useful in the sense that AI increases the recognition of gathered data to be accurate to monitor health.

With the help of AI, it is possible to merge and combine data received from different sensors, which increases the accuracy of general measurements. AI can work through different types of sensors, optical, electrochemical, and bioelectric, therefore integrating such data allows much more accurate estimation of health parameters. This enables cross validation of measurements and reduces the possibility of such measurement having wrong reading at any given time.

Another area where sensor accuracy has also improved through predictive analytics powered by artificial intelligence is security. With the help of integration of history and real-time data, the AI identifies problems in the operation of the sensors. This provides a predictive capability that can be monitored and addressed making sure that the sensor does not have a false reading and ensuring that the calibration is constantly spot on.

4.3 Real-time health monitoring and alerts

The integration of AI into the systems of continuous health monitoring has provided increased reliability to the real-time analysis of health data and timely alert generation. The implementation of technology backed by artificial intelligence paves way to new horizons in automated physiological changes and yields numerous points of versatility to the patient as well as the healthcare technician.

Integrating AI technologies allows constant monitoring of collected health data from unique biomedical sensors and gives immediate and constant notification of important vital signs like heart rate, blood pressure, and sugar levels. Such systems simply analyze tremendous quantities of data at a rather fast rate and identify any variation from the standard metrics related to the patient's health status. In this context, Figure 8 shows the real-time health monitoring system, as well as the health alert system, showing how real-time data acquisition and health analysis lead to real-time health alerts for events related to health. This real-time capability requires timely

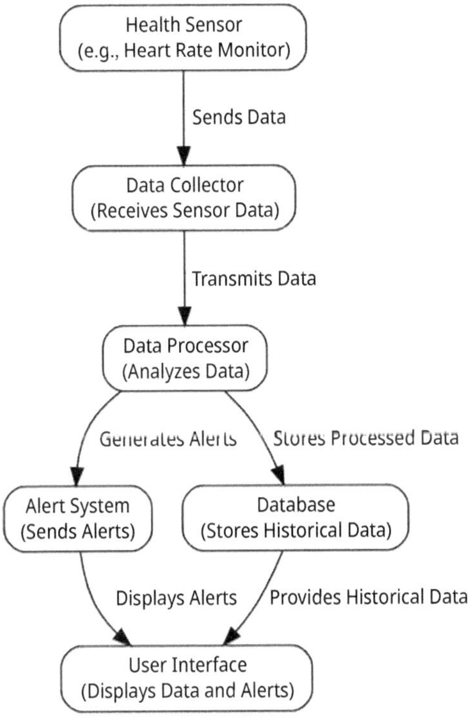

Figure 8: Real-time health monitoring and alerts.

monitoring of possible health complications so that appropriate medical treatment can be given without delay to avoid further complications.

Smart health monitoring systems also favor the use of AI algorithms in decision-making and elaborate alert systems. Potential models, which can be trained using multiple parameters, are cardiac arrhythmias, hypoglycemia, and hypertensive events. When such patterns are detected, the system sends the alarms to the practitioners. These alerts are also intended to be tangible and send clear information about the issue it has marked, including severity level. It means that no situation can catch one off guard so there are prompt measures that call for modification of drugs by consulting a doctor.

Other mechanisms, which emancipate the use of artificial intelligence, include real-time monitoring systems as a way of advancing the personalization of healthcare. AI works by taking the patient data of each person and processing it in a repetitive manner to find various patterns to each patient's overall condition and even risk factors for creating many tailored alerts. This level of personalization enhances alert usefulness, making them more probable to be effective and timely based on the recipient's health situation.

Modern health monitoring systems have been designed to employ artificial intelligence to enable consistent and efficient real-time interaction between patients and healthcare givers. Notification alerts from such systems can be embedded in health apps, hence enabling the patient to receive a notification on his/her smartphone device. This actual access to health information improves patient interaction and fosters a culture of choice for individuals at the required time.

The ability of AI in handling real-time data and performing analytics is also useful in lightening the burden of the healthcare providers. Notifications and data processing can be done independently from the monitoring human intervention, freeing them for the more important duties, as well as patient healthcare. It does this not only in the sense of increasing the speed at which healthcare is delivered, but also in raising the general quality of care.

4.4 Predictive analytics for early disease detection

In this contusive health monitoring, the major role of predictive analytics is to identify possible health issues before they turn into major medical diseases, and this is where AI comes in. The proactive approach that allows people to go to a doctor as soon as something is indicated in the data would not have been possible without sophisticated data analysis tools that translate big amounts of health data into insights.

Machine learning-based predictive analytics processes historical and current biometric data collected from biological sensors. These algorithms are quite capable of identifying correlations and patterns that suggest that illnesses began early. As an example, prognostic models can capture slight changes in vitals or movement, which precede the signs of chronic disorders such as diabetes or respiratory diseases.

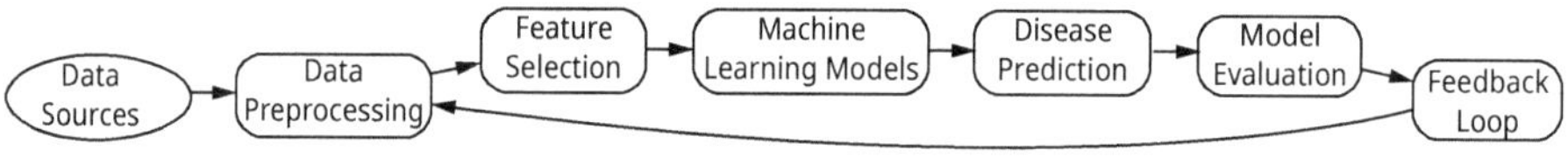

Figure 9: Predictive analytics for early disease detection.

One of the main benefits of such an approach is the fact that predictive analytics allows for early alerts about various threats to one's health. Some models, which can analyze data over time, provide reminder alerts when they conclude that the optimized health patterns are absent. In Figure 9, an early disease warning system shows how predictions offer advanced data models in creating awareness of possible health before reaching the critical phase. It enables early interferences to be made and prevents circumstances that worsen diseases and, therefore, patients' fates. For instance, analytics in healthcare can predict a cardiovascular event through monitoring fluctuations in heart rate variability and blood pressure and timely treatments can be carried out with AI in healthcare.

Apart from improving the care of individual patients, predictive analytics also improves the management of groups of people. A large volume of data yielded from different patients helps predictive models detect tendencies and possibilities that occur in large groups of people. Such information is helpful in formulation of public health strategies, outlining focused approaches to programmers and distribution of resources toward those ends. For instance, it is possible to use predictive analytics in regards to flu spread prognosis groups that need to be vaccinated.

Thus, the capability of employing predictive analytics for early disease detection has received a boost from integrated multiple data sources. By integrating the data gathered from sensors with EHRs, genomics, and behavioral parameters the overall evaluation of the health risks is more accurate. This approach enhances the general dependability estimates and also assists in the implementation of tailored healthcare measures.

The steady enhancement of the AI technologies also leads to the evolution of predictive modelling. With growing amounts of data and when enough data is gathered and processed, machine learning algorithms can be made more accurate with time. Thus, this iterative process keeps predictive analytics high on early disease diagnosis and makes it a cornerstone in patient health management.

Continuous health monitoring is the next big improvement in predictive analytics as it allows health professionals to anticipate health concerns by using predictive analytics. In this way, even by applying the predictive analytics tools within the frame of AI, early disease diagnosis, prevention, therapeutic monitoring, and results, as well as population, health gains are promoted.

4.5 Personalization of health monitoring and treatment

AI in continuous health monitoring has brought about personalized continuous health monitoring, enhancing monitoring and subsequently treatment strategies to the need of the individual patient. It is done by the use of AI algorithms where the information generated from different biomedical sensors can be used to better target solutions to manage health and illnesses.

It also improves the level of personalization through the use of AI to study large volumes of people's health data, which is obtained from sensors, among other methods. Other applications of machine learning are used in clinical diagnosis by picking out features about a person's health that are not noticeable from normal observations. Thus, through doing so, the monitoring parameters and even therapies can be matched to an individual's health conditions, life perceptions, and genetic propensities.

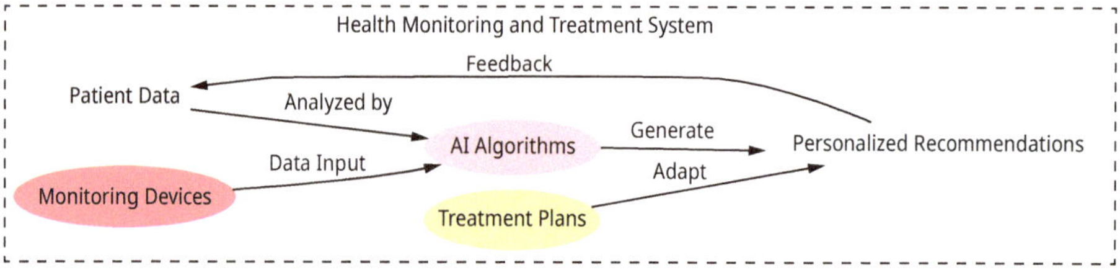

Figure 10: Personalization of health monitoring and treatment.

Among the major advantages of the AI application in the fields of PHM is the proposed model scenario's flexibility to address real-time variations in the patient's state. Figure 10 provides an overview about the possibility to individualize health monitoring and treatment according to the preferences of the patients. Such parameters are remotely perceived from wearable devices as well as other sensors, and AI algorithms constantly learn from the acquired data and refine the monitoring thresholds, and consequently the treatment plans. This real-time adaptability increases the efficiency of rendered treatment and thus draws higher results in patient rehabilitation.

AI makes health predictions that improve real-life management via tailored healthcare plans. AI is able to forecast possible diseases and estimates the probability of contracting certain ailments using past and current data. This predictive capacity, therefore, means that potential problems can be managed very well by taking preventive measures especially based on the risk factors that have been identified in the case of the individual. For example, patients with a genomic profile of susceptibilities to heart diseases receive suggestions on preferable changes in their living habits as well as prognosis of the possible effects of certain medications.

These personalization concerns themselves do not stop at truly customized application but apply to therapeutic suggestions as well. AI can automate processes of dosing and selecting the proper treatment for a patient because the AI system analyzes the data and finds out which medications and in what doses give the best result. It makes therapies more effective and side effects fewer, resulting in improved general health of the patient.

AI equally fosters patient-centered communication whereby through health applications the patients are able to receive individualized feedback and education. These applications provide specific health tips and message how the patient has to continue following the plan that has been set by identifying with the patient's information and goals. Such measures engage patients in an active way, promote better compliance with treatment regimens, and give them a sense of "usefulness" in managing their healthcare.

Mobile AI in chronic disease management, particularly in continuous health monitoring, delivers personalized monitoring, proactive anticipations of future health status, and tailored care packages. This approach enhance not only the impact of care interventions but also brings about an individual-oriented model of care that enhances health literacy among the patients.

5 Key AI technologies used in biomedical sensors

5.1 Machine learning algorithms

Machine learning algorithms form a core component of development and optimization of biomedical sensors for increasing the accuracy, effectiveness, and usability of the devices. These algorithms utilize techniques that enable the modeling of various health related data with a view to finding patterns as well as making informed forecasts, which facilitate correct health check and control.

The most employed supervised learning algorithms in biomedical sensors are used in classification and regression. They include algorithms that rely on labelled data to make sense of new sets of data and arrive at a certain decision. For instance, supervised learning models estimate blood glucose level from the past data collected using sensors to help diabetes patients. Supervised learning methods are support vector machines (SVM), decision trees, the ensemble methods of random forests, and gradient boosting machines.

Exploratory techniques are widely used to find and extract useful patterns within health data without any prior knowledge. *K*-means and hierarchical clustering together bring similar data points into tight knit groups and make the patterns in patient data easily identifiable. For instance, unsupervised learning can discover distinct clusters of patients with the same health needs and thus the clustering provides the necessary basis for customizing the approaches to patients.

Neural networks, which belong to the broader category of the machine learning, employ multiple layers of mathematics to map and interpret data issues. Computer vision is usually performed by special deep learning algorithms, like convolutional neural networks (CNN), which are best at analyzing image-based data like MRI scans. Figure 11 shows the differences and relationships between various machine learning categories and their potential fields of use, such as supervised learning and reinforcement learning. Recurrent neural networks (RNNs) and their special form, the long short-term memory (LSTM) is recommended for processing sequential data including data from wearable sensors such as time-series. Making use of deep learning models a number of features are capable of being automatically extracted from raw data, thereby improving the ability to detect even small information patterns as well as anomalies.

There is improved deployment of reinforcement learning algorithms that enhance operations in online real-time decision-making. These algorithms learn new skills from their environment through receiving rewards after interaction. It should

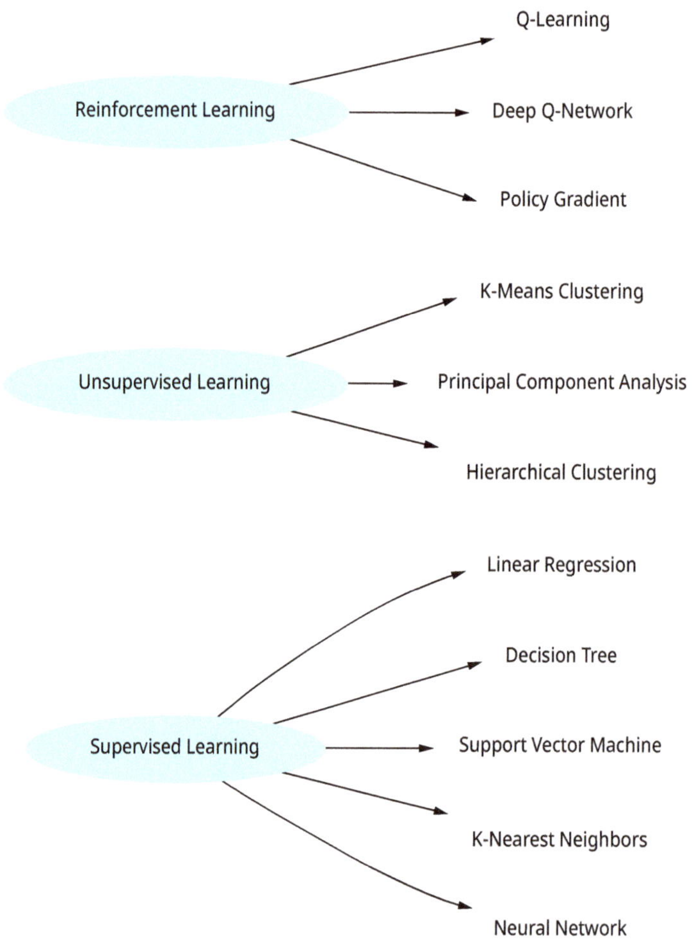

Figure 11: Machine learning algorithms.

be noted that reinforcement learning is applied to the context of biomedical sensors for the purpose of proposing modifications to a treatment approach based on patient responses in order to improve overlong health outcomes.

Abnormality detection algorithms are used for finding out that external conditions are shifting overall health away from normal. Anomalies are identified by using such methods as isolation forests, autoencoders, and statistical methods when it comes to sensor data. Some of the actionable and patient safety concerns include the ability to identify early signs of an abnormality, including irregularities in the heart rate.

Biomedical sensing, which involves the use of sensors in the diagnosis of diseases, is made possible by machine learning algorithms, which can perform complex computations, identify patterns, as well as make future predictions. With the help of these algorithms, biomedical sensors give better, reliable, and personalized health assessment and help promote better results of patients and the effectiveness of healthcare delivery.

5.2 Deep learning and neural networks

Neural networks and deep learning form core aspects of AI that have hugely improved the efficiency of biomedical sensors. These technologies play a role in analyzing stringent health data and finding patterns in sensor data to support continuous health monitoring research.

Deep learning is a category of machine learning that utilizes neural networks, which consists of many layers, therefore called "deep." As stated earlier, deep learning algorithms are well-suited for interpreting high dimensional signals from biomedical sensors, wearable bioelectrical sensors, and image sensors. These algorithms are particularly helpful in the large and complex datasets of health monitoring because they can learn characteristics from raw data and do not need feature engineering.

CNNs and RNNs in particular, have been found to yield the greatest results when it comes to analyzing biomedical sensor data. Figure 12 specifically presents deep learning and neural networks in terms of their hierarchical design and connection, which work on improving data analytics. CNNs excel at detecting spatial relationships in data and can used, for example, in image analysis of regions of the human body and spatially localized particularities in data originating from sensors. RNNs, including LSTM networks, perform well in handling sequence data, which is crucial in tracking time-series data with sensors such as electrocardiogram (ECG).

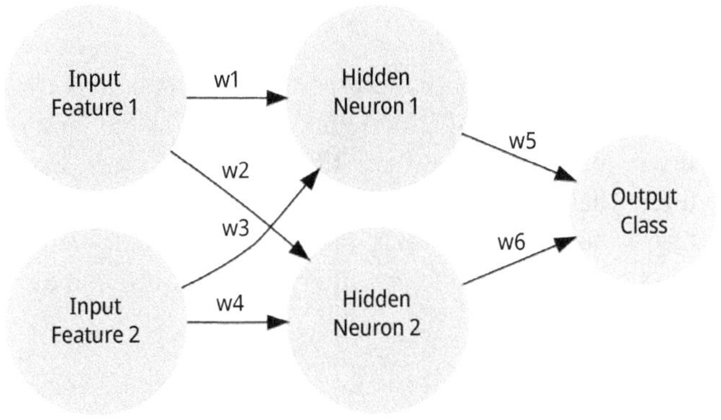

Figure 12: Deep learning and neural networks.

Deep learning improves the capabilities of biomedical sensors for the outcomes like real-time alarms, prognosis, and health alerts for an individual. For example, deep learning models are used to read nonstandard forms of vital sign pattern configurations, which indicate some health complaint. These models are designed to detect minor variations from normal patterns thereby alerting and ensuring timely intervention.

Advanced learning improves the health diagnosis procedures because it allows more accurate detection of the pattern and sorting. For instance, deep learning algorithms can identify tumors, distinction between benign and malignant tumors, and presurgical preparation using medical imaging. The fact that several of these algorithms enhance their performance in each iteration of learning and adaptation guarantees that diagnostic models are up-to-date and accurate.

Neural networks also play an important role in using the various sensor data to get the overall picture of the health state. As a consequence, deep learning models, which can take inputs from many different types of sensors at once, can help physicians see the overall health status of the patient and provide a context for the data from the wearable sensors and the environment, as well as the medical records.

5.3 Signal processing techniques

Signal processing is critical for the functionality and performance of AI-integrated biomedical sensors in their basic form. These techniques are indispensable to get useful information from raw data collected by the sensors as these are usually contaminated by noise, artifacts, and many other types of distortions. Improvement also occurs in accuracy in data due to efficient signal processing leading to increase in performance of AI and efficiency of health monitoring systems.

5.3.1 Noise reduction and filtering

Noise reduction is a significant component in signal processing that requires filtering out of interference and improving the quality of information produced by the sensors. Examples are digital filtering techniques like low-pass filter, high-pass filter, band-pass filter, and band-stop filter. They are used to reduce those frequencies that are related to noise while passing through frequencies of the signal of interest. Nonstationary noise is also addressed by means of adaptive filtering techniques that alter their parameters according to signal properties.

5.3.2 Signal denoising techniques

EMD and wavelet transform are two distinctive denoising techniques applied in signal processing used for identifying important signatures from noisy signals. From a signal, the wavelet transforms split up a signal spatially and temporally to different scales, thus enabling developers to remove noise from such a signal. Empirical mode decomposition, in a way, de-noises complex signals by unwrapping it to recover components dubbed intrinsic mode functions.

5.3.3 Feature extraction and selection

Feature extraction is the process of selection and filtering down of various characteristics, which defines the architecture of the signal relevant in health monitoring. Fourier and time-frequency analysis are employed to transform the signal to the frequency domain in which features like amplitude and frequency are measurable. Other techniques in machine learning also assist in feature selection because, through using certain algorithms, the importance of each feature to the results and, therefore, which ones are most appropriate to use in precision and classification are considered.

5.3.4 Data fusion and multisensory integration

Multifactor data fusion methods integrate information from various sensors in order to build a picture of a patient's condition. Techniques like Kalman filters and Bayesian fusion are used to amalgamate information acquired from different sources of data to get over the possible drawback of individual sensors and to tend the accuracy of health appraisal. One grants the anti-abnormal operations performances as high as the other by fusing information from multiple sensors for improving the reliability of the health monitoring systems.

5.3.5 Real-time signal processing

Only applications that require real-time response and input processing require real-time signal processing. Other techniques such as digital signal processing (DSP) algorithms and fast Fourier transform (FFT) enable fast analysis of the data gathered through the sensors pointing to real-time monitoring and alarm production. The application of real-time processing ensures the identification of changes in a patient's condition and timely intervention, thus improving the effectiveness of combating health tactics.

5.4 Data fusion and multisensor integration

Data fusion and multisensor integration are two effective innovations that have added more accuracy and inclusiveness to the concepts of biomedical sensors in monitoring health systems. Figure 13 shows the use of the data fusion techniques into the multisensor systems to improve the efficiency and effectiveness of information gathered in systems environments. Such technologies interconnect several sensors and produce a less arbitrary picture of a patient's condition, making the identification of health status and subsequent intervention more accurate.

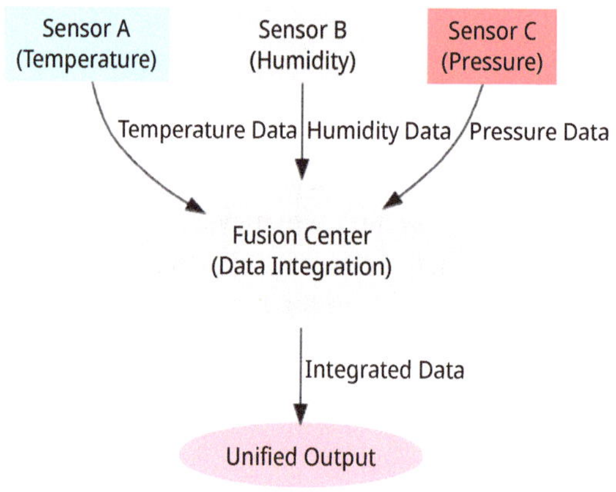

Figure 13: Data Fusion and Multisensor Integration.

Multisensor fusion refers to the ability to merge data acquired from different sensors into a single and coherent database. This process improves the quality and reliability of health information because different sensor modalities are utilized to their full potential in the system. For example, integrating data obtained from electrochemical sensors, which count glucose levels with optical sensors, which measure blood oxygen saturation offers a more holistic view of a patient's metabolism and respiration. It enables better diagnosis and treatment since the two features are integrated into one system.

Multisensor integration generalizes this idea as it compiles data from a wide range of different sensors, all of which measure different physiological values. It further allows for the detection of several condition markers, which makes for better evaluation of overall health status of the patient; for instance, a wearable device with temperature-sensing, acceleration, and heart rate to detect the thermos regulation mechanisms, physical activity, and cardiovascular state of the wearer simultaneously. This form of data collection is particularly useful where disease processes affect many physiological systems in serving the multifaceted health needs of clients.

It makes an important contribution to data fusion and multisensor integration by using intricate algorithms to process and analyze fused sensor data. Fusion algorithms and ensemble methods are used to amalgamate data generated from heterogeneous sources and problems associated with it including redundancy, noise, and inconsistency. Thus, these AI-techniques increase the reliability of the health information integrated with feature fields since they can detect possible errors in data input and as well as gaps where data is scarce.

It has been seen that using fused sensor data, AI algorithms can analyze patterns and correlation and offer a better estimation of an individual's health. For instance, analyzing data from a Fitbit and a smart glucometer makes it possible to develop insights concerning the correlation between physical activity and glucose, with suggestions on how to promote the best results. The capacity to identify these patterns improves decision-making in the use of Big Data and practical benefits for patients and care givers.

It also enhances real-time monitoring and the adaptive health management of data from several sensors. When equipped with AI, such systems can always process data from different sensors and detect abnormalities when they occur. Such real-time application is critical in monitoring fluctuating health states and in giving anticipatory care.

5.5 Edge computing and AI at the sensor level

The incorporation of edge computing with AI at the location of the sensors is a major step forward for biomedical sensors as it provides efficient analytical capacity and data evaluation right at the origin. This approach helps solve main issues of health monitoring that include latency in the data acquisition, low bandwidth, and security.

Edge computing relates to the accommodation of computational assets nearer to the point of data collection, for example, within the biomedical sensor edge device. Here, data processing occurs at the network edge eliminating dependency on physical servers and thus enhancing decision-making by reducing latency. This is especially important in monitoring health sessions, which require real-time response to such conditions so that intervention and diagnosis can be made on time.

When used at the sensor level, embedded AI algorithms allow certain analysis and decision-making to be performed directly on the sensor. Local real-time processing means that the analyzed data from sensors can be interpreted instantly, for example, for anomaly detection or pattern recognition. Figure 14 shows how edge computing and AI are deployed at the point where the sensors are located to allow true-time analysis and decision-making. For example, a wearable sensor comprising AI can indefinitely and autonomously measure physiological parameters, observe fluctuations, and give out alarms without needing other systems. This capability is appropriate for

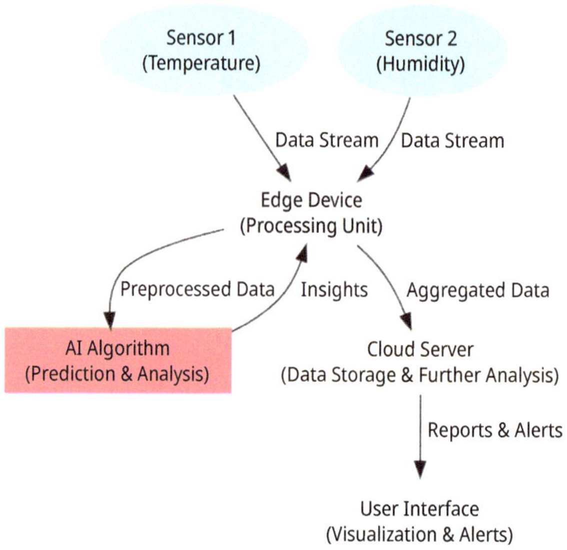

Figure 14: Edge computing and AI at the sensor level.

the applications where quick answers are required, e.g., in the case of abnormal heart rhythm, for preparation for the critical event.

Both edge computing and AI benefit from the integration in a way that improves data privacy and security. This way, the approach ensures that a large amount of health information is processed at a local level, thus significantly reducing the amount of information transmitted over the network – and vulnerability to unauthorized access. Furthermore, edge computing makes it possible to apply encryption and other security measures closer to the sensor level and this also provides extra protection to patient records.

Lee establishes that edge computing enables the deployment and flexibility of AI-based biomedical sensors. Integrated processing functionality of sensors enables them to manage big amounts of data and advanced calculations without referring to the central system due to the currently limited network capacity. Such scalability is important for sustaining higher demands of the persistent and multidimensional health check.

6 Case studies and applications

6.1 AI-enhanced wearable health monitors

Smart health watches with AI integration can be regarded as one of the most revolutionary developments in health control and improvement empowered by AI, which allows for constant, precise evaluation of multiple strategic physiological indices using AI and no external professionals. They provide solutions, which are much better than conventional health monitoring devices.

Wearable health monitoring with the help of AI uses different sensor technologies including accelerometers, gyroscopes, optical sensors, electrochemical sensors, and various others to capture health data. Due to the application of the AI algorithms to this data, these devices can harness this kind of functionality as well as pattern detection and analytics. For example, with the help of AI, wearable devices are capable of tracking essential signs like pulse frequency, blood pressure, as well as oxygen level and analyze the tendencies to define pathologic states health problems.

Another area that research has identified use of AI enhanced wearable monitors is in personal health management and specifically among those with chronic diseases like diabetes and cardiovascular diseases. For instance, glucose wearables have AI to evaluate blood glucose levels and continuously give feedback to these individuals. It can also forecast changes in blood glucose level depending on past records as well as lifestyle advice on appropriate meal plans and in dosage of medicines for certain behaviors. This is a proactive approach that is efficient in management of glucose levels and the prevention of complications.

Several wearable devices designed for cardiovascular disease monitoring can assess heart rate variability, ECG signals, and physical activity to help identify early signs of conditions such as atrial fibrillation. These devices are capable of analyzing long-term data and can alert users to abnormalities in heart rhythm, prompting further medical evaluation. This enables earlier diagnosis of cardiovascular diseases in outpatients and helps healthcare providers adjust treatment plans accordingly.

In the fitness and wellness niche, AI wearables provide users with customized fitness advice based on user activity history, exercise, sleep, and other aspects of their well-being. These devices then have the option of recommending particular workout schedules, assessing how the body is recovering, and advising on how best one can perform to get maximum results and minimize chances of developing an injury. The application of AI makes it easier to track the achievement of fitness goals and variability of results, thus improving the efficacy of trainer's personalized training plans.

Another use is in telemonitoring, where AI-integrated wearables allow patients with chronic diseases to be monitored constantly. These devices send and receive information regarding the patient's health to and from the healthcare givers in real time thus allowing doctors to monitor and treat their patients from the comfort of their homes. As AI algorithms can recognize early important changes in health status, they can produce alerts and ensure timely medical interventions with overall positive impact on the patients' well-being and the number of admissions to hospitals.

Technological advancements in wearable health technology enhanced by AI make it easier to monitor one's health consistently and in real-time, receive tailored feedback and get preventative advice. These devices could be considered as defining the modern healthcare systems, to the extent that disease control, enhanced fitness, and better, more iso-personalized care is underwritten by them.

6.2 Implantable sensors for chronic disease management

It should also be noted that implantable sensors have become one of the key technologies for the effective monitoring and control of chronic illnesses as they supply invaluable constant information to the healthcare system and the patient. These implantable sensors offer critical information on internal body circumstances and assist in prevention and control of chronic illnesses including diabetes, cardiovascular diseases, and neurodegenerative diseases.

A well-known use of implantable sensors is in healthcare as regards diabetic patients. Frese style CGM systems use implantable sensors in order to note periodic measurements of blood glucose concentration. These sensors wirelessly send glucose information to other devices and help patients track their glucose levels constantly and fine-tune their insulin treatment. It is useful in keeping blood glucose levels near the desired optimal values and significantly decrease chances of the development of hypoglycemia and so optimize diabetes care.

In the case of cardiovascular disease, implantable sensors have the most important task of tracking the heart rate and diagnosing arrhythmias. Some ICDs and pacemakers have sensors that record electrical activity in the heart and check for abnormality. These sensors can administer therapeutic shock rates independently, thus applying early reparative measures to ambiguous rhythms and minimizing sudden cardiac deaths. The complementation of AI with these sensors goes further by applying prediction on the existing problems and prevents them from escalating.

Parkinson's disease is one of the neurological disorders that is managed through implantable sensors for the tracking of motor signs and responsiveness to medication. Figure 15 shows different implantable sensors for chronic disease management. Real-time tracking can also be used and DBS systems contain implantable sensors that can monitor the severity of the symptoms and adjust the stimulation parameters for improving therapeutic results. Such an approach of treatment enhances the patients' symptoms control and enhances overall quality of life of patients with Parkinson's disease.

Similar implantable sensors are also considered for application in other chronic diseases like chronic pain and inflammatory diseases. Such sensors can be used for

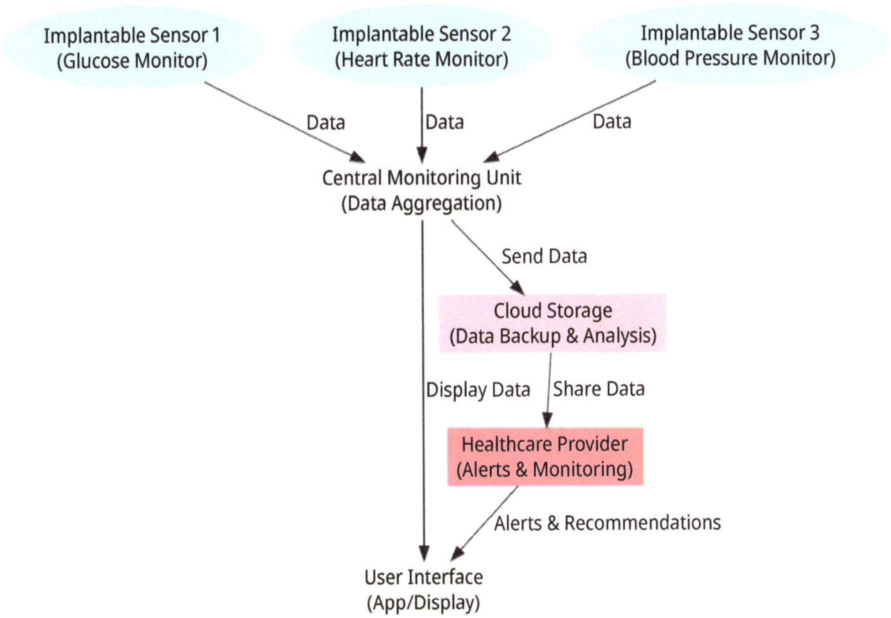

Figure 15: Implantable sensors for chronic disease management.

measuring biomarkers related to activity of the disease and are useful for treatment planning and evaluation of therapeutic outcomes.

Integrated sensors are also used to enhance the use of telemedicine and remote monitoring. These sensors can share information with the doctors, and patients' progress could be constantly monitored, and adjustments can be made to therapeutic regimens without repeated face-to-face consultations. This remote capacity ensures that the healthcare professionals are in a position to access new information to make better decisions besides enhancing patient convenience.

6.3 Remote patient monitoring systems

RPM devices involve integrated biomedical sensors and communication technology and are recognized as an indispensable aspect of modern medical practice to provide constant interphase vigilance without direct clinical control. Through these technologies, remote data collection, transmission, and analysis are accomplished, and it has some benefits in patients as well as healthcare workers.

RPM systems monitor numerous health parameters and physiological signs employing different biomedical sensors. These sensors are often integrated into wearable technology devices, for example, smart wristbands and watches, fitness trackers, and into home healthcare devices, for example, glucose and blood pressure meters. RPM systems are excellent in presenting an all round view of a patient's health status by periodically measuring factors such as pulse rate, blood pressure, blood sugar level, and oxygen level, among others.

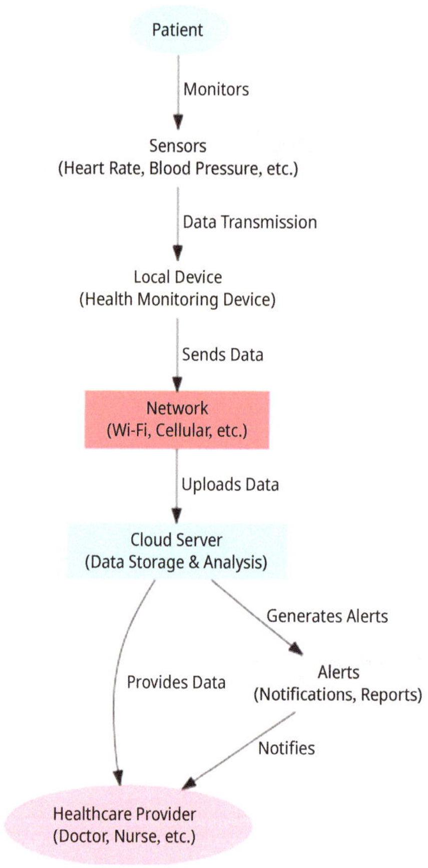

Figure 16: Remote patient monitoring systems.

One of the advantages of RPM systems is its ability to offer proactive management and early treatment for chronic illness. Diabetic patients can, for instance, use glucometric devices that allow medical practitioners to get values from patients undergoing treatment. This helps to be able to change the further program of work and make adequate decisions within a short time if there are signs of deterioration. In a similar way hypertension patients enjoy the convenience that comes with constant monitoring of blood pressure since it simplifies management of prescriptions and avoidance of complications among patients with cardiovascular diseases.

Also, RPM systems are required to enhance the patient's engagement and compliance with various treatment regimens. Such solutions help people to become more engaged in their healthcare process through using the own health data shared through mHealth applications. The remote patient monitoring system is described in section IV-A; a complete block diagram with wearable sensors, data transmission modules, and cloud-based analytical components are shown in Figure 16. In these applications, feedback involves patient-tailored advice, which makes patients change their habits and focus on following a treatment regime, adopting better lifestyles, and assuming more responsibilities for their treatment.

These systems have a lot of advantages: the number of visits of doctors and other caregivers during direct patient consultations is reduced and this can potentially decrease the load on medical centers and patients. This is from the view point of the healthcare professional. From the process of continuous monitoring, providers can therefore tweak care plans and come up with better judgments. This method leads to enhanced diagnostic and therapeutic processes as well as enhanced impact of care.

Several examples of RPM systems show that they are useful in many areas of medicine. For example, COPD patients managed through telemonitoring have had enhanced control of exacerbations and consequently enrollment reduction in hospitals. Likewise, RPM systems for individuals recovering from surgery have also indicated early signs of postsurgical complications and reduced rehabilitation duration.

Alongside the benefits, there are still challenges to RPM systems general adoption. Thus, implications of new technologies have to be identified and met, which would allow ensuring the efficient and moral use of technology; concerns such as data protection, safety, and integration into the existing framework of healthcare should be considered. In addition, ongoing investigations are necessary to enhance RPM systems' precision, reliability, and relevance.

6.4 AI in telehealth and telemedicine

Telehealth and telemedicine are defined as the use of modern telecommunication technologies that allow the healthcare stakeholders to attend to the needs of the patients, no matter where they are located as well as to provide medical professionals, physicians, patients, and other stakeholders with valuable information in a more efficient and effective manner – thanks to the integration of AI in telehealth and telemedicine. Telehealth platforms can harness these technologies to improve diagnostics and patient engagement and customize care for those who use telehealth services thereby revolutionizing the delivery of care.

Telemere and telehealth are enhanced by AI through the ability to enhance data analysis and decision support. To make better decisions, healthcare professionals rely on the results of the analysis of patient information, including EHRs, medical images, and PROMs that are carried out by ML algorithms and NLP. For instance, diagnostic instruments can "read" images of the patient's scans to look for signs of disease like cancers and then report back to radiologists who attest its accuracy of up to 95%. This capability helps maintain a fast and accurate diagnosis, which is of great help to clinicians in making the right decisions.

The other major area where AI has made a big impact in the area of telehealth is in the creation of virtual health assistants and chatbots. Such devices can interact with patients in order to obtain symptoms data, propose preliminary diagnosis, and reveal the actions to be taken. These virtual assistants filter patient complaints and provide the information on illnesses to avoid overload of healthcare officials. This is especially beneficial in the event that the user is in a remote area where access to medical practitioners can be very rare.

In telemedicine, AI is also used to identify patients' needs and provide appropriate follow-up service. Analytical models for predictive assessment work through patterns in records to identify potential sickness risks and to formulate individual treatment plans. Using wearable sensors, AI for instance, tracks patients with chronic diseases such as diabetes and alerts the patient in real time. Preventive measures in healthcare are promoted and the management of such diseases is enhanced by the targeted approach.

The impact of AI in telehealth is in data integration and data interoperability to improve the efficacy of the platform. AI systems can combine patient data not only from mHealth applications, wearable devices, and home monitoring applications, but also data from the patient's EHR. Such an approach enhances continuity of care since care providers are compelled to work as a team in delivering optimum care and the resultant care plans are based on a coordinated perception of the patient's ailment.

AI also has an impact on the technical side when applied to telehealth with regard to access and equity questions. Since teleconsultations offer the possibility to overcome language barriers, telehealth platforms can reach populations and communities that hardly have access to relevant healthcare services. This helps spread the knowledge of medical professionals and also helps achieve fairness with medical services.

6.5 Success stories and clinical trials

Several success stories in connection with AI integration with biomedical sensors exist, and multiple significant clinical trials have been made proving the effectiveness of such technologies.

A new success story is the creation of AI-based wearable technology for constant blood glucose monitoring. Abbott Diabetes Care and Dexcom are leading glucose sensors that utilize AI algorithms to give out glucose level readings in real time. Some of these devices are being tested with a level of efficacy that has reduced incidences of hypoglycemia and enhanced glycemic control in diabetics. For instance, the Dexcom G6 continuous glucose monitoring system has been tested scientifically and clinically, showing that it can measure glucose levels correctly and help patients with diabetes to manage it better.

Another is AI in wearable ECG monitor application in early identification of atrial fibrillation (AF). Smart watches of today like the Apple Watch Series 4 and Series 5 even have intelligent algorithms that can work in real time to analyze information from the device's ECG sensors that can detect signs of AF. Application of the devices in early identification of AF has been evidenced through clinical studies and enhanced patients' management. Several researches have presented that AI-based ECG monitoring can effectively identify AF with high sensitivity and specificity, which is helpful information to avoid stroke and other cardiovascular events.

The application of AI-coupled biosensors in the area of remote patient monitoring has borne fruit, especially for managing chronic illness. For example, the Biofourmis platform includes wearable sensors and AI that help control conditions in patients with heart failure. As with any HEP platform, the studies revealed that the platform's predictive analysis can predict cases of worsening and hospitalization effectively thereby eradicating overwhelming episodes of the disease. Frequently, the utilization of remote patient monitoring and the possibility to change the treatment plans based on the gathered information improves patient's quality of life and decreases the rate of ER visits.

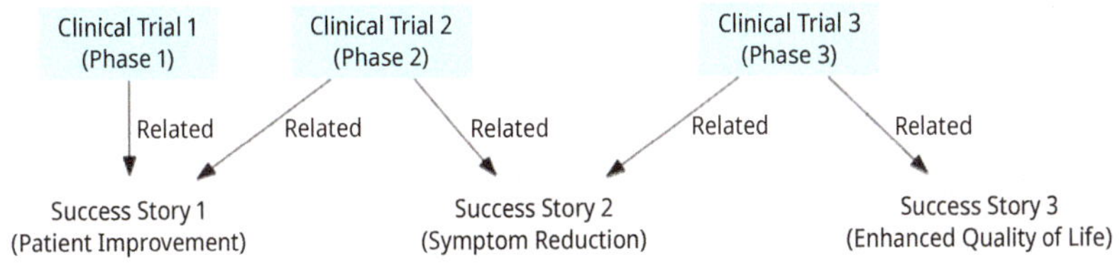

Figure 17: Success stories and clinical trials.

AI-driven sensors are equally effective in another field of health, mental health. Other endeavors like the Mindstrong Health platform ensure they use AI to gauge information from smartphones and wearables based on cognition and emotion. The proposed method has been applied in clinical practice and proved to provide accurate estimation of change in mental health conditions like depression and anxiety, along with identification of recommendations for therapeutically relevant further actions. Figure 17 captures a few facts and current clinical trials that show different kinds of updates and innovations in medical science and therapeutic approaches that have cropped up recently. AI applications in monitoring mental health can be useful in increasing the patient's compliance with the recommended therapy regimes.

These success stories coupled with clinical trials enrich the theme of the paper on the huge commendable progress achieved in the application of AI in biomedical sensors. They show how these technologies can be valuable tools in reaching said goal, by helping to increase the diagnostic precision of the clinician, and ultimately, improve overall patient treatment. New study in this area remains active and constructs upon these accomplishments, to create new innovative treatment and expand the utilization of these approaches in various healthcare-related fields.

7 Conclusion

The utilization of AI with biomedical sensors can be viewed as providing a creative and qualitative advancement in the field of health monitoring systems by providing higher levels of accuracy, effectiveness, and patient-centered approach in patient care. In this chapter, the complexity of the biomedical sensors that have been developed under the influence of AI has been discussed besides their historical development and the technology that has been applied in enhancing their operations. As described, advanced sensors and AI integration have transformed CHM into a system with superior possibilities of real-time analytics and personalized health self-maintenance.

It is now possible to collect and analyze large amounts of data with incredibly high accuracy using biomedical sensors due to the new advancements in AI technologies, including machine learning and edge computing. These advances facilitate early detection of health abnormities to allow efforts to be taken to start treating the afflictions as well as optimal treatment plans to be implemented. The possibility of establishing interaction between patients and AI algorithms based on the patient's need, as well as the ability of the algorithms to forecast heath issues, is a step up from simple monitoring and tracking. Consequently, AI-driven sensors offer the specificity and practical steps necessary to optimize disease control, minimize hospitalizations, and increase patient satisfaction.

Examples of AI applied in biomedical sensors and examples of case studies and clinical trials shown in this chapter make it clear that the presence of AI in biomedical sensors is quite helpful. Some successful deployments include using continuous glucose monitoring systems for the management of chronic diseases as well as AI incorporated into ECG monitors. These success stories illustrate how AI possesses the capability to fill healthcare gaps in delivery and also this paper has shown how AI can improve patient care through choice of care delivery models.

The combination of AI and biomedical sensors has faced some inevitable issues. Challenges are its proximity to data accuracy, privacy, and integration, which continue to constitute areas for further research and development. Solving these problems is critical to provide high-level sensors and integrate them into modern healthcare systems. The problems of ethical use of these technologies and their compliance with the legislation and regulations should be of great concern in protection of patients' information.

In the future, the application of AI biomedical sensors will see more innovations. Health monitoring systems can be taken to a new level by employing new technologies like advanced models of AI. This evolution has persisted and continued to produce higher efficiency, integration, and usability of the technologies that define the future of healthcare.

References

[1] Tovar-Lopez, F. J. (2023). Recent progress in micro-and nanotechnology-enabled sensors for biomedical and environmental challenges. Sensors, 23(12), 5406.

[2] Ates, H. C., Nguyen, P. Q., Gonzalez-Macia, L., Morales-Narváez, E., Güder, F., Collins, J. J., & Dincer, C. (2022). End-to-end design of wearable sensors. Nature Reviews Materials, 7(11), 887–907.

[3] Kazanskiy, N. L., Khonina, S. N., & Butt, M. A. (2024). A review on flexible wearables-Recent developments in non-invasive continuous health monitoring. Sensors and Actuators A: Physical, 114993.

[4] Antunes, R. S., Seewald, L. A., Rodrigues, V. F., Costa, C. A. D., Jr, G., L., Righi, R. R., . . . Campanatti, G. (2018). A survey of sensors in healthcare workflow monitoring. ACM Computing Surveys (CSUR), 51(2), 1–37.

[5] Villena Gonzales, W., Mobashsher, A. T., & Abbosh, A. (2019). The progress of glucose monitoring – A review of invasive to minimally and non-invasive techniques, devices and sensors. Sensors, 19(4), 800.

[6] Do Amaral, C. E. F., & Wolf, B. (2008). Current development in non-invasive glucose monitoring. Medical Engineering & Physics, 30(5), 541–549.

[7] Darwish, A., & Hassanien, A. E. (2011). Wearable and implantable wireless sensor network solutions for healthcare monitoring. Sensors, 11(6), 5561–5595.

[8] Mathew, A. A., Chandrasekhar, A., & Vivekanandan, S. (2021). A review on real-time implantable and wearable health monitoring sensors based on triboelectric nanogenerator approach. Nano Energy, 80, 105566.

[9] Din, I. U., Almogren, A., Guizani, M., & Zuair, M. (2019). A decade of Internet of Things: Analysis in the light of healthcare applications. Ieee Access, 7, 89967–89979.

[10] Manickam, P., Mariappan, S. A., Murugesan, S. M., Hansda, S., Kaushik, A., Shinde, R., & Thipperudraswamy, S. P. (2022). Artificial intelligence (AI) and internet of medical things (IoMT) assisted biomedical systems for intelligent healthcare. Biosensors, 12(8), 562.

[11] Jin, X., Liu, C., Xu, T., Su, L., & Zhang, X. (2020). Artificial intelligence biosensors: Challenges and prospects. Biosensors and Bioelectronics, 165(112412).

[12] Ngiam, K. Y., & Khor, W. (2019). Big data and machine learning algorithms for health-care delivery. The Lancet Oncology, 20(5), e262–e273.

[13] Perera, Y. S., Ratnaweera, D. A. A. C., Dasanayaka, C. H., & Abeykoon, C. (2023). The role of artificial intelligence-driven soft sensors in advanced sustainable process industries: A critical review. Engineering Applications of Artificial Intelligence, 121, 105988.

[14] Castaneda, C., Nalley, K., Mannion, C., Bhattacharyya, P., Blake, P., Pecora, A., . . . Suh, K. S. (2015). Clinical decision support systems for improving diagnostic accuracy and achieving precision medicine. Journal of Clinical Bioinformatics, 5, 1–16.

[15] Alshamrani, M. (2022). IoT and artificial intelligence implementations for remote healthcare monitoring systems: A survey. Journal of King Saud University-Computer and Information Sciences, 34(8), 4687–4701.

[16] Junaid, S. B., Imam, A. A., Abdulkarim, M., Surakat, Y. A., Balogun, A. O., Kumar, G., . . . Hashim, A. S. (2022). Recent advances in artificial intelligence and wearable sensors in healthcare delivery. Applied Sciences, 12(20), 10271.
[17] Shajari, S., Kuruvinashetti, K., Komeili, A., & Sundararaj, U. (2023). The emergence of AI-based wearable sensors for digital health technology: A review. Sensors, 23(23), 9498.
[18] Wang, C., He, T., Zhou, H., Zhang, Z., & Lee, C. (2023). Artificial intelligence enhanced sensors-enabling technologies to next-generation healthcare and biomedical platform. Bioelectronic Medicine, 9(1), 17.
[19] Gu, L., Jia, D., Vicaire, P., Yan, T., Luo, L., Tirumala, A., . . . Krogh, B. H. (2005, November). Lightweight detection and classification for wireless sensor networks in realistic environments. In Proceedings of the 3rd international conference on Embedded networked sensor systems (pp. 205–217).
[20] Picaut, J., Can, A., Fortin, N., Ardouin, J., & Lagrange, M. (2020). Low-cost sensors for urban noise monitoring networks A literature review. Sensors, 20(8), 2256.
[21] Jesus, G., Casimiro, A., & Oliveira, A. (2017). A survey on data quality for dependable monitoring in wireless sensor networks. Sensors, 17(9), 2010.
[22] Philip, N. Y., Rodrigues, J. J., Wang, H., Fong, S. J., & Chen, J. (2021). Internet of Things for in-home health monitoring systems: Current advances, challenges and future directions. IEEE Journal on Selected Areas in Communications, 39(2), 300–310.
[23] Bleiholder, J., & Naumann, F. (2009). Data fusion. ACM Computing Surveys (CSUR), 41(1), 1–41.
[24] Thapa, C., & Camtepe, S. (2021). Precision health data: Requirements, challenges and existing techniques for data security and privacy. Computers in Biology and Medicine, 129(104130).
[25] Fahimi, B., & Beikmohammadi, S. (2023). Role of artificial intelligence in atrial fibrillation management: A comprehensive review.

A. Aalan Babu, R. Selvakumar, D. Shobana, S. Kavitha, P.R. Therasa

Design and development of AI-driven biomedical sensors and devices and their optimization for continuous health monitoring

Abstract: Thanks to the developments in the biomedical sphere, sensor technologies has advanced rapidly, thus making the monitoring of continuous and real-time physiological parameters possible. This chapter reviews continuous health monitoring biomedical sensors created through the use of AI, the major advancements, and the trends. It goes further and outlines the key kinds of sensors such as the optical, chemical, and mechanical types, and their uses, all the while explaining the innovations bringing about change. The chapter also analyzes the main characteristics of hybrid as well as multimodal sensors that are composed of various sensing modes for the purposes of providing complete monitoring solutions. Particular attention is paid to the prospecting long-term monitoring solutions needed for geriatric care, as well as issues that are related to chronic illness management in elderly people. The discussion includes considerations of the applicability and affordability of sensor technologies, the two aspects that often define the universality and implementation of these technologies. By outlining these aspects in considerable detail, the chapter is intended to present useful information for the further development of biomedical sensor for enhancing health treatment results.

Keywords: biomedical sensors, AI-driven technology, optical sensors, hybrid systems, geriatric care, cost-effectiveness

A. Aalan Babu, Department of Computer Science and Engineering, School of Computing, Vel Tech Rangarajan Dr. Sagunthala R&D Institute of Science and Technology, Avadi, Chennai, Tamil Nadu, India, e-mail: aalanbabu@gmail.com
R. Selvakumar, Department of AIML, Saveetha Engineering College, Poonammallee High Road, Thandalam, Sriperumbudur, Tamil Nadu, India, e-mail: sachein.pretty@gmail.com
D. Shobana, Department of Mechatronics, Rajalakshmi Engineering College, Chennai, Tamil Nadu, India, e-mail: shobana.d@rajalakshmi.edu.in
S. Kavitha, Department of ECE, Nandha Engineering College, Erode 638052, Tamil Nadu, India, e-mail: gskkavitha@gmail.com
P.R. Therasa, Department of CSE, RMK Engineering College, RSM Nagar, Kavaraipettai, Thiruvallur District 601206, Tamil Nadu, India, e-mail: therasa.peter@gmail.com

https://doi.org/10.1515/9783111503202-004

1 Introduction

The environment in the healthcare sector is changing dramatically as a result of discoveries in biomedical sensor technologies [1]. These innovations are transforming continuous health monitoring by giving real-time, accurate information about the physiological signs [2–4]. Biomedical sensors are defined as a series of devices used to measure the human body's diverse parameters [5]. This chapter explores the options of designing and developing AI biomedical sensors for patient treatment improvement and as efficient instruments in health treatment [6, 7]. The emphasis is on the major developments of the project, the use of AI for optimizing the characteristics of the sensors, and their capabilities and the role of these technologies in long-term health monitoring [8].

Biomedical sensors fall under optical, chemical, and mechanical types, depending on the type of health parameter that is to be monitored [9]. For example, optical sensors use light to measure oxygen levels, glucose level, and concentrations of other biomarkers in the blood. Biochemical sensors quantitate specific chemicals present in a solution; for instance, glucose, using electrochemical reactions [10–12]. Mechanical sensors respond to the physical alterations in such values as strain; these can provide valuable information on joints' movements, gait, and muscles activity [13]. These sensor types are also described in detail in this chapter, together with an overview of the manner in which they operate, the fields that they are aptly suited for, and the modern technological improvements that have been made with for their precision and performance [14].

Hybrid/multimodal sensors are seen as the next evolutionary advancement in the field of health monitoring [15, 16]. These systems integrate several sensing technologies to measure many parameters at once, on a single instrument [17]. A device comprising optical and mechanical sensors can track both cardiovascular state and physical activities, giving a better picture of the patient's conditions. This chapter describes the different types of hybrid and multimodal systems that positively impact, such as increased data quality, higher patient compliance, and opportunities to create more individual health interventions.

Remote surveillance is most valuable in geriatric care, as keeping patients' chronic diseases under control is always challenging [18]. Diabetes, hypertension, and heart diseases, for instance, call for relentless assessment of body's various parameters in elderly people [19]. Responsive wearables, implanted sensors, and telemetric monitoring tools are intended to answer these needs, providing continuous support and monitoring [20]. The current chapter highlights the deployment of these long-term monitoring solutions with regard to patient care, early health disorder detection, and for the enhancement of the quality of the life of the elderly.

The future of biomedical sensor technologies lies in its ability to offer solutions at a comparative cost and in large volumes [21]. Since such demand for enhancing the capability of health monitoring systems continues to rise, the ability to manufacture the sensors in a large scale, while at the same time keeping the costs affordable is paramount [22]. Chapter 4 of the book is devoted to the issues of manufacturing scalability, material costs, and system integration [23]. It also looks at how strategies for realizing economies of scale, the efficient use of material, and minimizing regulatory and compliance costs are possible [24]. In addressing these factors, this chapter hopefully sheds light on the steps required to bring advanced biomedical sensors to a more affordable price point to allow for their increased use throughout the healthcare systems [25].

2 Introduction to biomedical sensors

2.1 Definition and classification of biomedical sensors

Biomedical sensors are unique devices that are developed to monitor the quantitative and qualitative variables in human bodies and translate them early into electrical signals. These sensors have become an integral part of today's healthcare technologies, mainly in the interrelated monitoring of various bio signals, including heart rate, blood glucose, and oxygen saturation levels. Biomedical sensors have made it possible to incorporate wearable devices that enable timely diagnosis of diseases as well as improving the patient care services.

Biomedical sensors can be categorized by a number of ways, depending on the type of physiological parameter used to measure, the interaction with the body as well as the mechanism of sensing.

Classification based on physiological parameters:

Electrophysiological sensors: These sensors measure electrical activity in the body, including cardiotocograms for heart activity, encephalograms for brain activity, and myograms for muscle activity.

Biochemical sensors: These sensors are built to identify particular chemicals in the body, for example, glucose sensors for the screening of blood sugar levels, lactate sensors for monitoring metabolic rates, and pH sensors to measure the body's acidity level.

Physical sensors: These demonstrate physical characteristics; thermometer for temperature, sphygmomanometer for blood pressure, and accelerometer for activities.

Optical sensors: These sensors apply light to determine parameters like oxygen concentration using pulse oximetry, tissue diagnosis, and blood flow using near-infrared spectroscopy (NIRS).

Mechanical sensors: Such sensors include tension sensors for respiratory activity and piezoelectric for heart rates, among others. The piezoelectric sensors on the human body generate minor electric current, in response to mechanical pressure.

Classification based on the nature of interaction:

Invasive sensors: These kinds can be implanted into the body with the purpose of directly measuring physiologic variables, e.g., glucose monitors under the skin.

Noninvasive sensors: These sensors do not penetrate the skin's surface and include wear and tear heart monitors, pulse oximeters, and thermometers.

Minimally invasive sensors: These sensors do not have to be embedded into the body; they only require contact with the body, like micro-needle glucose monitoring sensors within the body cavities.

Classification based on sensing mechanism:

Electrochemical sensors: These types of sensors use electrochemical process for the determination of determined analytes in body fluids, which are used in glucose and lactate determination.

Optical sensors: Sensors of this type rely on optical methods to measure changes of biological tissues, for example, pulse oximeter and optical coherence tomography.

Piezoelectric sensors: These sensors translate mechanical stress into electric signals, and are widely applied in pressure sensing and motion detection.

Capacitive sensors: These sensors measure changes in capacitance due to movements of the body tissues; they are used in touch switches and respiratory rate measuring devices.

Magnetic sensors: These sensors detect magnetic fields resulting from physiological processes, and are used in magnetoencephalography (MEG) and such other equipment.

This extensive classification demonstrates the available variety of biomedical sensors for constant surveillance, which proves the specificity and specialization of the sensors, aimed at addressing multiple clinical demands. It is therefore important to identify and define the various types and classifications of such sensors in order to advance and improve the implementation of continuous health monitoring systems, thereby enhancing personalized healthcare solutions.

2.2 Historical evolution of biomedical sensors

Below, we look at the history of biomedical sensors to show the steps of evolution and corresponding changes in technology and the approach to health monitoring and diagnosis. Biomedical sensors technology dates back into the early twentieth century, when basic instruments for measurements were developed. The first chapter described the creation of basic tools, including sphygmomanometer for blood pressure and the electrocardiograph for heart activity. These early sensors served as stepping stones to the sophisticated ones that are embraced today by laying down the possibility of devices to furnish important health information.

A better type of sensors that came in the mid twentieth century is a result of technological advancement. Proliferation of solid-state electronics, coupled with integrated miniaturization, enabled the fabrication of devices that could provide continuous monitoring. Portable ECG machines and pulse oximeters were brought into use during this period to make quick evaluations of cardiac and respiratory rates. The combination of the electronic elements helps to be more accurate and as used, improves the engineering, compared to mechanical and analog systems.

The last few decades of the twentieth century and through the twenty-first century, we have entered an era of major gains in biomedical sensor technology because of material science, biotechnology, and information technologies. Wearable sensors like smart watches and fitness trackers are some of the examples of consumer health and monitoring, a common aspect of this era. Such devices, integrated with new and improved sensors for the monitoring of relevant parameters – basal rate, activity, and sleep – substantiate a gradual move toward constant health surveillance and preventive healthcare. The employment of biosensors that incorporate biological components with electronic transducers extend the scope of monitoring physiological variables, including glucose concentration and hormonal changes.

In recent years, biomedical sensors have been integrated with AI and machine learning, which pave way for smart self-learning monitoring systems. The newest development in AI automated sensors include a more detailed report on the analysis of data to help in the early diagnosis of health problems. AI has complemented sensor technology, resulting in the development of smart sensors that are used to produce health enhancement recommendations based on actual biometric data fed into the sensors.

The process of evolution regarding biomedical sensors reflects an incremental innovation process, which is characterized by the development of simple measuring instruments to advanced sensors, based on artificial intelligence. Civilian, military, and commercial SATs' have all improved health monitoring's viability as a field since every stage of development further develops the corresponding technologies when it is, in general, a pattern that describes innovation's evolution.

2.3 Key functionalities of biomedical sensors in health monitoring

In the context of ambient and continuous health monitoring, biomedical sensors play a crucial role in propelling the medical progress by providing additional features that represent a vast improvement over conventional health check. These sensors are intended to measure and record physical variables while acquiring precise physiological signals, thereby delivering a real-time updated general wellness profile of a person. The following key functionalities highlight the critical roles that biomedical sensors play in health monitoring:

2.3.1 Real-time data acquisition and monitoring of data

Most biomedical sensors are capable of continuously monitoring a range of physiological indicators, including but not limited to pulse rate, blood glucose levels, body temperature, and blood oxygen saturation. This real-time data acquisition allows for ongoing updates of health parameters, facilitating rapid detection of deviations from normal values. As illustrated in Figure 1, such sensors provide immediate feedback, enabling healthcare providers to promptly adjust medical interventions and modify treatment plans when necessary. This capability enhances patient care and helps reduce the risk of acute health complications.

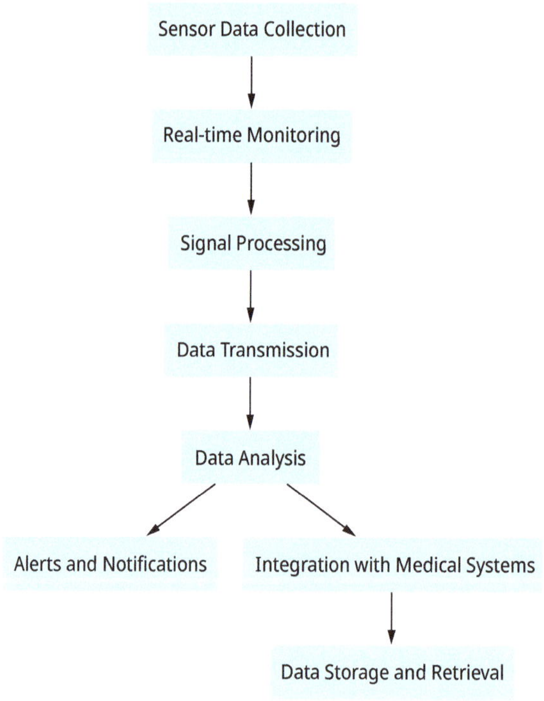

Figure 1: Key functionalities of biomedical sensors in health monitoring.

Another feature of anthropometry as a field of measuring human body is the high precision and accuracy of measurement.

Skill in measurement is indispensable for successful health oversight, as is the ability to maintain finite accuracy in those calculations. The development of biomedical sensors is made in such a way that they have improved precision and minimal variability in the values obtained, in order to promote accurate diagnosis. Development in manufacturing sensors and integrating advanced signal processing algorithms and methods of calibrating the system add up to the high degree of accuracy in these measurements such that their usefulness to healthcare providers for decision-making is not doubted.

2.3.2 Wearable and implantable device compatibility

Biomedical sensors are thus developed to be incorporated in wearable and implantable devices, which makes the choices beyond chronic disease monitoring flexible. Smartwatches and fitness trackers offer product-oriented methods for permanent body surveillance that are nonintrusive. While the wearable sensors provide external and physically tangible measurements of the body, implantable sensors provide slightly more direct measurement over the internal physiological conditions, like intracranial pressure and heart rate. They also make it possible to monitor virtually any condition on a continuum, from simple check-and-balance monitoring of healthy lifestyles for busy people to severe chronic disease management.

2.3.3 Transmission of data and material and person-to-person interaction

Success in the communication to the healthcare providers of the sensor data is crucial for detailing the health status. Biomedical sensors are enhanced with communication functionalities that depend on Bluetooth, Wi-Fi, etc. to send information to remote monitoring systems. This connectivity ensures that the data is transferred on time and securely, thus improving the on-time retrieval of information to make decisions regarding patient care.

2.3.4 Personalization and adaptation

Contemporary biomedical sensors are created in a way that they capture patient's needs in a particular healthcare situation. These sensors can be adjusted based on the various features related to the physiologic attributes and pathophysiologic conditions of the patients. For instance, individual algorithms are used in adjusting the monitoring parameters to the acuteness and specificity of the patient's conditions, increasing the applicability and efficiency of the collected information.

2.3.5 Early warning system and predictive modeling

Biomedical sensors are important in that they help in the diagnosis and prognosis of possible human ailments, through analysis. Through consistent tracking of physiological information during daily activities, these sensors detect new trends that suggest development of illnesses. Applied artificial intelligence, in particular machine learning, enables identification of high-risk patients and thus potential medical events like heart attack, and allow for early intervention.

2.3.6 Navigation and alert interfaces

Biomedical sensors also apply their functionality in user interfaces that are meant for improving patients' interaction and adherence. Most of the sensors come with easy-to -read displays and notification systems that give real-time results of the state of health. The messages and notifications options can be tailored to send alerts whenever any out-of-the-ordinary number is received, in order to help the users manage their health better, and reminding them to follow the prescribed treatment plans.

These functionalities demonstrate the benefits of biomedical sensors in state-of-the-art health monitoring systems and show that such systems can not only deliver but indeed do deliver accurate, up-to-date, and individualized health information. Sensor technology is still in the process of expanding its capabilities, and the future development of this technology only seeks to amplify these traits and advantages toward serving patient needs with regard to their health.

2.4 Role of sensors in preventive healthcare

Biomedical sensors are very crucial in the area of preventive healthcare, where the detected health complications can be treated before they worsen. The major goal of preventive healthcare is to recognize the causes of sickness and prevent their development into chronic diseases, which has a positive impact on the general health and the overtaxing of facilities.

The constant tracking of physiological parameters is one of the basic responsibilities of biomedical sensors when it came to preventive healthcare. Health monitoring sensors intended for use in capturing data at regular intervals, including pulse rate, blood pressure, and glucose levels, are useful in identifying symptoms as well as potential signs of a certain health complication. Through the constant collection and study of such data, sensors make it possible to detect changes in physiological thresholds and, therefore, the existence of pathological states that require treatment.

Also, biomedical sensors help to identify the patients with certain risk factors and known diseases so as to develop individual approaches for their corresponding management. For instance, sensors used in diabetes glucose monitors present an opportunity for the constant monitoring of glucose levels in the body of diabetic patients to help adjust their diet, exercise regime, and the effect of diabetes on the body. In the same respect, wearable sensors that track movement and cardiovascular signs can

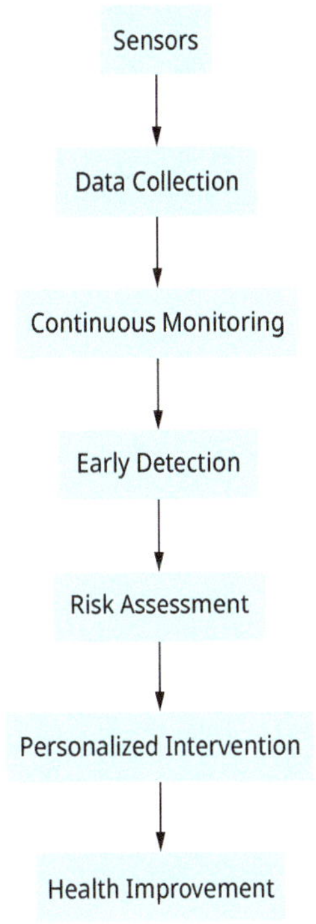

Figure 2: Role of sensors in preventive healthcare.

help improve the lifestyle of a person who has been diagnosed with heart disease and recommend an exercise regimen.

Biomedical sensors, when combined with high-end data analytics and artificial intelligence, play an even bigger part in preventive healthcare. In developing AI, algorithms are able to read the data acquired by the sensors, and subsequently diagnose potential health complications that are likely to occur before the symptoms show up. This predictive capacity proves particularly useful in chronic diseases management and in helping to avoid deterioration, hospitalization, and the resultant negative effects on patients' well-being.

Biomedical sensors support preventive healthcare by enabling remote patient monitoring and treatment through virtual sessions. Portable devices that capture physiological signals and transmit them to caregivers facilitate continuous, virtual supervision particularly valuable for patients in remote areas with limited access to healthcare facilities. As shown in Figure 2, this capability enhances both the accessibility of preventive healthcare services and patients' ability to receive timely responses and recommendations from healthcare providers.

2.5 Current market trends and innovations in biomedical sensors

The biomedical sensors market is also growing at a very high rate and is continuously changing, which is attributed to the increasing incorporation of sophisticated novel technology for health monitoring applications. This part provides an overview of the emerging trends that are prevalent in the most contemporary area of biomedical sensors.

2.5.1 Emergence of wearable sensors

Wearable sensors are used widely nowadays because of the comfort and simultaneity of measurements in comparison to other methods. These are wearable sensors that can be worn in devices like smart watches, fitness trackers, as well as smart cloth; the aim of these sensors is to track parameters such as heart rate, physical activity, and even sleep. New advancements have aimed at improving the precision and wearability of these gadgets and by using kinetic textiles with features of flexibility and elasticity that adapt to the body more closely, with increased performance and reliability. The integration of far more sophisticated biosensors in wearables has also enhanced their ability to monitor parameters such as blood glucose levels, blood pressure, and oxygen saturation, and deliver full-body health updates to the user.

From the present work, the following specific research questions emerged:

Therefore, implantable sensors can be looked at as a major improvement with regards to aggregative, internal monitoring of healthcare. Some of the advancements in this area are biocompatibility and advancement of mini sensors to sizes that can be implanted inside the human body. Recent developments have enhanced the durability and reliability of implantable sensors to include aspects such as glucose monitoring in diabetes patients and intracranial pressure in neurological disease. Subject categorization work has also been done to design wireless, battery-less implantable sensors, which are also operated through energy harvesting techniques to minimize the issues regarding the lifetime of the implantation devices and the safety of patients.

2.5.2 Combination of AI and ML

The development of the biomedical sensor market is influenced by the integration of AI and ML technologies into the product. AI tools are therefore being used as signal processing engines to offers improved accuracy in the machinery sensors, besides minimizing noise and optimizing pattern recognition. Other emerging technologies integrated with data analysis models in real-time are machine learning, which allows one to identify potential treatment, before the matter becomes acute. The integration also improves the performance of individual sensors but also contributes for the improvement of advanced intelligent health monitoring systems.

2.5.3 Multimodal and hybrid sensors, and their development

The integration of different sensors is gaining consideration as solutions to achieve multiple and diverse modalities of health conditioning. These sensors use multiple modalities, including optical, electrical and chemical, for the measurement of a number of physiological parameters at one time. Using multiply sensing technologies allow for enhanced and accurate health monitoring systems. For instance, the hybrid sensors include capabilities for heart rate monitoring, blood oxygen level detection, and physical activity, which give a broader picture of the client's condition. This trend can be explained by the shift to more advanced and complex health monitoring platforms.

2.5.4 New techniques and technologies of noncontact and telemetric recording

Contactless transducers and home monitoring devices are steadily growing popular because they are more convenient to patients. New technologies in optical sensing and bioimpedance sensing are progressively allowing the monitoring of physiological signals without stern procedures. Tele-monitoring technologies that have resulted from developments in wireless communication and cloud computing make it possible to obtain health data from remote and relay updates to physicians and nurses. This trend is in line with the increasing embracement of telemedicine. Heinrichs and Hunt (2014) note that this trend supports the escalating use of telemedicine and remote patient management solutions, especially for chronic illnesses and to areas that lack adequate care resources.

These market trends and innovations demonstrate the constant growth of biomedical sensor technology, the constant attempt to refine health monitoring capacity, and the constant push to improve patient prognosis and meet new healthcare demands. It is believed that with such advancements, new materials, AI analytics, and multimodal sensing would continue to revolutionize biomedical sensors, and advance the future of the field-invasive sensing techniques, such as optical sensors and bioimpedance sensors, which are enabling the measurement of physiological parameters without the need for invasive procedures. Remote monitoring technologies, facilitated by advancements in wireless communication and cloud computing, allow for continuous health data collection and transmission to healthcare providers from remote. This trend supports the growing adoption of telemedicine and remote patient management solutions, particularly in managing chronic diseases and providing care in underserved areas.

These market trends and innovations highlight the dynamic evolution of biomedical sensors, reflecting the ongoing efforts to improve health monitoring capabilities, enhance patient outcomes, and address emerging healthcare challenges. As technology continues to advance, it is expected that the integration of new materials, AI-driven analytics, and multimodal sensing further transform the landscape of biomedical sensors, driving future developments in the field.

3 Types of biomedical sensors

3.1 Wearable sensors for continuous monitoring

Wearable sensors have recently emerged as a vital component of continuous health assessment due to their advantages in real-time and efficient monitoring of an individual's well-being. Designed to be integrated into various wearable devices, these sensors continuously measure key physiological parameters. As illustrated in Figure 3, they are increasingly used in health monitoring, physical activity tracking, and health promotion applications.

Types of wearable sensors

Wearable sensors cover a large range of technologies based on individual health monitoring requirements. Common types include:

Accelerometers and gyroscopes: The first kind of sensors is responsible for the detection of motion and position, for monitoring of the physical activity, for analysis of gait, and for position correction. They are popular in wrist-wearable devices for tracking exercise and the overall activity during exercise sessions.

Photoplethysmograms (PPG): PPG sensors optically measure motion, reflecting changes in blood flow and offering details of heart rate, blood oxygen, and stress. They are normally used in smartwatches and fitness trackers.

ECG: ECG sensors are used to record the electrical impulses of the heart; they can record complications in heart rates and health. Technology-enhanced wearables today have features such as ECG scanners in watches for continuous monitoring of heart and any abnormality.

Thermistors and thermocouples: These sensors determine the temperature within the body and across the skin to help determine fever, hypothermia, and such other heat-related disorders.

Technological innovation wearable sensors

Wearable sensors technology has recorded major advancements in recent years in terms of effectiveness and capability. Key advancements include:

Flexible and stretchable electronics: Wearable sensors themselves have benefited from the advancement of flexible and stretchable technologies. These materials fit the

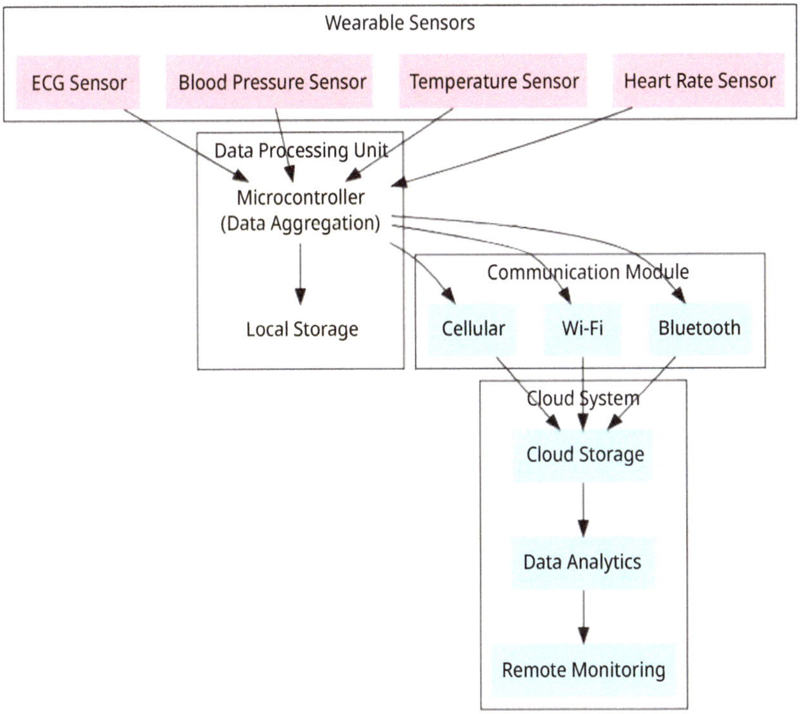

Figure 3: Wearable sensors for continuous monitoring.

configuration of the body, improving the precision of measurements as well as comfort for the user.

Advanced integration technologies: MEMS and printed electronics offer a way of developing a range of sensing systems within a single, small form factor. They also enable us to monitor more in a single wearable device, while integrating data from various sensors.

Enhanced connectivity and data transmission: Wearable sensors have become more connected with Bluetooth, Wi-Fi, and cellular technologies to send the collected data to smartphones and cloud for real-time monitoring and analysis.

Wearable sensors practical use

Wearable sensors serve a broad range of applications in health monitoring and wellness:

Chronic disease management: Wearable sensors are used to track and manage chronic ailments like diabetes, hypertension, and cardiovascular illness. For instance, CGMs offer glucose level status in the continuous process while blood pressure monitor only shows the status of blood pressure.

Fitness and physical activity tracking: Fitness trackers, such as smartwatches, have accelerometers and PPG for monitoring the user's movement, including steps, calories burned, and exercise. The pros of these devices aim at creating and encouraging the achievement of goals toward better health.

Health and wellness monitoring: Wearable sensors can track a number of health parameters, including sleeping cycles, stress levels, body temperature, and more. The information derived can help the user consider patterns, recognize signs of health risks, and make suggestions for a better quality of life.

This paper focuses on some of the difficulties faced in the design of wearable sensors.

Despite their benefits, several challenges persist in the development and deployment of wearable sensors.

Accuracy and calibration: Maintaining high accuracy and proper calibration of the measurements are still issues, particularly when the measured environment conditions change and for measurements in different physiological states.

Battery life and power management: Battery longevity is an important factor to ensure constant surveillance, while simultaneously supporting the features needed to run a smart home, thus keeping it energy-efficient is more of a problem.

Data privacy and security: The relentless collection and forwarding of perishable health information requires application layers of security to maintain privacy and data preservation.

Future trends and directions in wearable sensor applications

Future advancements in wearable sensor technology are anticipated to focus on several areas:

Enhanced sensor capabilities: Progress in multimodal sensors bring in more physiological parameters that can be monitored for better health information.

Integration with AI and Machine learning: Coupling the wearable sensors with Artificial Intelligence and Machine learning algorithms improves their capacity in understanding complicated signal data, capability in offering forecasts, as well as in prescription of health information.

Improved user experience: Both materials and design are being developed to provide increased levels of comfort and beauty, as well as greater interaction with the user so as to make it even easier to introduce wearable sensors into one's everyday existence.

Imminent changes in device design are expected due to the availability of wearable sensors, enabling invaluable continence monitoring and positive contribution to healthcare management. It is for this reason that there is an expectation that, with development of technology, portable sensors are going to broaden their functions, which are going to improve the impact in the area of health and wellness.

3.2 Implantable sensors for internal monitoring

Wearable and implantable sensors: Implantable sensors are the next generation devices for continuous monitoring of health status for long-term, real-time data assessment of the body's physiological parameters on internal data. They are intended to be implanted within the human body to continually measure a number of internal conditions necessary for the practitioner to make key decisions in the management of the patient.

Materials' biocompatibility and product design.

Implantable sensors' design considers the organism's compatibility due to the possibility of the formation of complications within a human body. Recent advancements in material engineering have provided biocompatible polymers, metals, and ceramics that can last in the biological environment without stimulating the formation of an inflammatory response. Further, modern technologies have allowed the development of ever smaller constituents of sensors and their implantation with less invasive actions. In addition, the increased use of flexible and conformal materials to integrate with and conform to the body's internal tissues has increased the comfort and efficiency of such devices.

Classification and uses of implantable sensors

Implantable sensors are designed to monitor a range of physiological parameters, each serving distinct medical purposes:

Glucose monitoring sensors: Diabetes management is closely associated with such sensors, which continuously measure the levels of glucose to ensure the proper regulation of blood sugar concentration. Electrochemical sensors use sharp and long-life elements, providing high sensitivity in the measurements, and seldom requiring calibration.

Intracranial pressure sensors: Basically, found in neurological applications, these sensors monitor intracranial pressure, which is crucial in several diseases like TBI and hydrocephalus. Among the recent developments are wireless sensors that do not require external connectors and, therefore, are not inflamed.

Cardiac sensors: ICCAs measure HR, HRV, and other data from the patients' heart and help in the modality of detection of arrhythmias and heart failure. Progress in this area has seen the fabrication of shoe box-sized devices that can be installed within the chest cavity surrounding the heart in order to provide accurate and constant observation.

Pressure sensors: These sensors calculate pressure within the different body organs such as the bladder and the gastrointestinal system. These are bendable pressure sensors that can mimic the form of internal organs and offer almost painless measurements.

Temperature sensors: Febrile thermal condition recording require implantable temperature sensors. Recent advancements advance the precision and speed of these sensors in the clinical setting.

Suppliers and energy regulation

One of the major concerns in constructing the implantable sensors is related directly to power delivery mechanisms. The existing battery-based sensors pose a problem in terms of lifetime and must be surgically inserted. New technologies incorporate support energy management, where sensors, for instance, can be powered through wireless energy transfer. Application of energy-harvesting technologies, including thermo active and kinetic, for example, is also under consideration to help prolong the functionality of implantable devices and thus minimize the cases when these devices need to be replaced.

Transmitting the data and the integration of the data

All implantable sensors send the collected data to other devices for further analysis. Another area that has developed fantastic progress is the wireless communication technology, where RF and Bluetooth can support the data transition of sensors (fm2,046). The compatibility of these sensors with external monitoring devices like smartphones and health monitoring platforms enables real-time data access as well as remote patient monitoring. This connectivity is crucial for the early identification of diseases and for individualized approaches to treatment.

In some of the guidelines, the concerns that have arisen are to do with aspects such as instruction language, evaluation criteria, knowledge incorporation and learning resources accessibility, and assessment method uniqueness and validity, with a view to giving direction toward their future improvements.

The application of implantable sensors faces several problems: biocompatibility limits the duration of implantation; the reliability of data collection is questionable; and data security remains an important concern. It is important that these sensors are functional and maintain their reliability in the body for long periods of time, to be effective. The issues of patient data privacy and data security that address both problems of protecting patient data from unauthorized access and providing integrity to patient data also become more apparent as the technology progresses. Further research and development probably inevitably continue to be centered on the optimization of the function, security, and bio sustainability of the implantable sensors, and the novel materials and approaches in the realization of these devices.

Personal, implantable monitors for internal health conditions are one of the most significant innovations in the sphere of medical equipment's constant, comprehensive, and sensitive measurements of the internal environment. The continued advances of these sensors make it easier to monitor chronic conditions and advance the clinical process of diagnosing, treating, and managing illnesses, with the overall potential of favoring individualized medicine with accurate feedback.

3.3 Noninvasive sensors

As a subfield within the broader domain of biosensors, noninvasive sensors have gained prominence due to their ability to collect multiple physiological parameters without penetrating the human body. This approach enhances patient comfort while minimizing the risk of complications typically associated with invasive procedures. As shown in Figure 4, this section provides a concise overview of the main types of noninvasive sensors, their operational principles, and their applications in continuous health monitoring.

3.3.1 Optical sensors

Optical sensors employ principles of operation that are founded on the use of light for the measurement of physiological characteristics. One of the most widely known optical sensors is the pulse oximeter which measures blood oxygen saturation levels and pulse rates using light. This is done using the principle of a photoplethysmography, where the amount of light absorbed by the oxygenated and deoxygenated blood hemoglobin clearly differs. Other optical sensors include NIRS sensors, used in tissue oxygenation and blood flow measurements and reflectance, as well as transmittance sensors, utilized in glucose monitoring of diabetic patients. Optical sensors can be used in a quasi-continuous and real-time mode because they are noninvasive in most cases.

3.3.2 Electrochemical sensors

Electrochemical sensors monitor physical characteristics by ascertaining the electrical activity of a chemical process. An example that one uses often is the without-holes sweat sensing, which consists of determining the composition of sweat so as to check the position of electrolytes. These sensors routinely use microelectrode arrays that help to record and measure the electrochemical activity with metabolic and hydration indicators. New technologies in electrochemical sensor system enable wearable devices for the determinations of biomarkers in sweat instead of the regular invasive blood sampling.

3.3.3 Acoustic sensors

Ultrasound and acoustic wave sensors are used to monitor a number of physiological variables using sound waves. Ultrasound sensors for instance are used in imaging systems for examining internal organs, blood flow, and any pathology. Other biomechan-

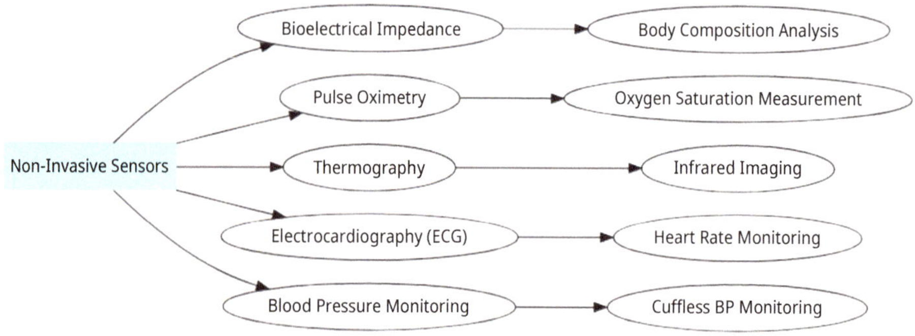

Figure 4: Noninvasive sensors.

ical properties that can be measured with high-frequency acoustic sensors include tissue stiffness. These sensors offer highly specific knowledge on the inner state of the body without the interference of surgeries and are used both for diagnostics and prognosis.

3.3.4 Bioimpedance sensors

Bioimpedance sensors identify how much tissue resist an electrical current by using of an electric current that passes through the body. These sensors can measure changes in impedance and the data that is fed back can reflect tissue composition, fluid status, and the general health state. In body composition analysis, bioimpedance techniques are applied in applications such as body composition determination, cardiac recovery, and fluid status determination. Optical and electrical bioimpedance measurements can be noninvasive and are ideal for constant supervision of changes in the body's physiological conditions, and can be incorporated into wearables health monitors.

3.3.5 Thermal sensors

Thermal sensors record changes in skin temperature of the body. Examples of thermal sensors include infrared thermometers and temperature-sensitive patches, by which temperatures are measured from the body without touching the patient. These sensors are suitable for use in detecting fever, observing the body temperature while people are physically active, and for evaluating thermal balance. Recent advancements in thermal sensors have made it possible to produce wearable devices that track temperatures consistently, and provide useful information for diseases like fever and hypothermia.

Vibration sensors

Vibration sensors encompass mechanical vibrations and movements in the body with certain physiological conditions and activities. Vibration sensors for physical activity, posture, and movements include accelerometers and gyroscope found in most fitness trackers and smartwatches. These sensors help to determine and quantify the amount of physical activity in daily life, characterize gait, and perform fall detection, which improves the accessibility of musculoskeletal and neurological condition supervision.

PPG sensors

Two PPG sensors operating at different wavelengths measure light absorption by tissues and blood in different ways to image the microvascular bed of the skin. Due to the alterations in light absorbed by the pulsatile blood flow, PPG sensors can inform estimates of the HR, BP, and vascular profiles. PPG technology has been often implemented in wearables, such as fitness trackers and smartwatches, which allow detecting various cardiovascular indicators in a noncontact way.

3.4 Biosensors

Biosensors are defined as a type of biomedical sensors, specifically developed to recognize biological substances and transform the interaction between these substances into an electric signal. They play an essential part in disease detection as well as treatment by giving accurate, selective, and rapid quantitative analysis of multiplex biological markers. This section aims at discussing the different categories of biosensors, the operation mechanisms of biosensors, and their practical utilization in the health sector.

Principles of biosensor operation

The working of biosensors can be explained by means of a biological element interacting with a transducer element. The biological component has to bind with a target analyte, which is glucose. The transducer transforms this interaction into an electrical signal subsequently. Electrochemical, optical, and piezoelectric transducers can be classified as the commonly used transducers of biosensors, in which each type has its own advantage based on the situation. For instance, electrochemical biosensors are relatively preferred for glucose sensing because of their high sensitivity and user friendliness, while optical biosensors are preferred because they are label-free and facilitate biolayer sensing.

Electrochemical biosensors

Electrochemical biosensors represent the largest and most popular biosensors because they are sensitive, easy to use, and inexpensive. These sensors quantify the variation in electrical current that accompanies the biochemical interaction of the bioreceptor with the target analyte. Some examples are blood glucose meter for diabetes, regulated by glucose oxidase as the bio receptor and electrochemical transducers, which reduce the current change. Finally, electrochemical biosensors are employed for the identification of biomarkers that relate to different diseases such as cancer and cardiovascular disorders, for their early screening and tailored therapy.

Optical biosensors

Optical biosensors work by using light to identify any variation in the biological processes. These sensors generally use surface plasmon resonance (SPR) and fluorescence instrumental methods for evaluating the bioreceptor's analyte binding events. Some features of optical biosensors include high sensitivity, so there is no need for a label and thus the possibility to monitor the process in real-time, for example, in the case of pathogen detection, where changes in the optical property show the presence of certain bacteria. Optical biosensors are widely used in genomics and proteomics for the identification of targeted nucleic acid in a heterogeneous background.

Piezoelectric biosensors

Piezoelectric biosensors utilize the piezoelectric effect, which occurs when an electrical charge is generated in a piezoelectric material due to mechanical strain. In these biosensors, the binding of a biological analyte to the bioreceptor causes a change in mass on the surface of the piezoelectric material, leading to a shift in its resonant frequency. This type of biosensor allows biomolecular interactions to be monitored in real time with high sensitivity. For example, it can be used to identify biomolecules such as proteins, where the frequency shift is proportional to the analyte concentration, enabling quantitative measurement of biological fluids.

Biosensors in healthcare

The demand for biosensors in healthcare remains immense because they allow for accurate, fast, and highly localized measurement. They are used in point-of-care diagnostics as a medical instrument for regular and at-home screening, including glucose monitoring for diabetes, cholesterol tests, and some screening for infections. Biosensors are also used in the field of personalized medicine, so that enablers can control the particular biomarkers to address targeted patients. Besides, biosensors are used in drug development and clinical research to give information on drug reactions and drug effectiveness. The continuous innovation in biosensor technology such as advances in multi-biosensor and biosensors integrated with digital health technologies is anticipated to continue to increase their use in the health sector.

Biosensors are a promising and developing category of biomedical sensors, where improvements in the sensors' performance and potential uses are continually being pursued.

3.5 Optical, chemical, and mechanical sensors

There are many ways that biomedical sensors are categorized, depending on the sensing principles as well as their application. Integrated optical and chemical, and mechanical sensors form the core three, all of which are essential in the constant health check of the system. This section discusses these kinds of sensors, the manner in which they work, and their uses in healthcare.

Optical sensors

Optical sensors operate by detecting the interaction between light and biological tissues. These sensors utilize various optical techniques such as absorption, fluorescence, and reflection to measure physiological parameters. As illustrated in **Figure 5**, each method enables the noninvasive monitoring of specific biological signals with high precision.

Pulse o: Probably the most widespread example of optical sensors is pulse oximetry that assesses blood oxygen saturation and pulse frequency. In most of the cases, this system employs red and infrared light to contrast between both forms of hemoglobin within the bloodstream.

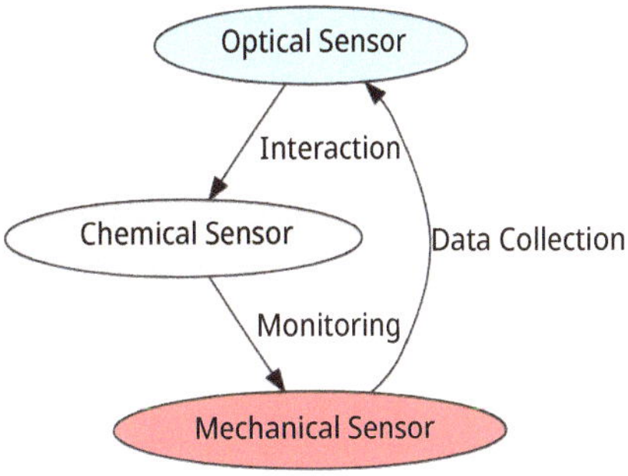

Figure 5: Optical, chemical, and mechanical sensors.

OCT: OCT is used as an imaging modality, which does not involve making contact with the tissue sample and offers quality tomographic imaging of higher resolution of the biological tissue. It is widely used in ophthalmology for imaging of retina and in cardiology for imaging the coronary arteries.

Fluorescence sensors: These sensors work by identifying particular biomarkers through the measure of the fluorescence of a sample, which was earlier exposed to a particular wavelength. In diagnostics, they are employed in assays to detect tumor markers in the patient's body.

NIRS: NIRS employs near-infrared light to photograph the tissues to assess the concentrations of tissue oxygenation and hemoglobin. It is used in tracking brain and muscle oxygenation, especially during exercise.

Optical biosensors: These sensors integrate optical structures with biological receptors that selectively recognize target biological entities, for example, for glucose measuring in diabetes mellitus-given analytes in body fluids.

Chemical sensors

Chemical sensors are used to measure the concentration of specific chemical species in living organisms. Most importantly, they are used to measure biomarkers and other chemical indices that are concerning to health.

Electrochemical sensors: Electrochemical sensors quantitatively determine the concentration of molecules from the changes of the underlying electrical characteristics, including current. Some of the applications include glucose for diabetes monitoring and lactate sensor for metabolic monitoring.

Enzyme-based sensors: These sensors employ certain enzymes to catalyze reactions with the target analytes to form changes in signals that can be measured. Aerometric glucose sensors employing enzymes to catalyze the oxidation of glucose are well suited for ambulatory monitoring of blood glucose in diabetic patients.

Chemical implants: Devices like intracranial pressure sensor are examples of implanted sensors that rely on chemical sensing for real-time monitoring within the body. These sensors utilize free-running electrochemical sensors.

Colorimetric sensors: Qualitative sensors identify an analyte using color changes as a result of reactions between the analyte and the indicator. They are applied in different analyses nowadays, the well-known are in urine analysis and infection detection.

Chemical array sensors: These sensors employ diverse chemical sensing components to quantify and identify multiple analytes at once. Some of these are diagnostics by sensing the breath for diseases and monitoring the condition of the environment, among others.

Mechanical sensors

Physical sensors record changes within biological systems. Previous to analysis for advanced pathophysiology, they were utilized to check parameters that are connected to physical health and mechanical operations of the human body.

Strain gauges: Strain gauges are used to measure mechanical deformation by forces applied to a specimen. In biomedical applications, they are incorporated in wearable gadgets in order to track movements of joint as well as muscle activity.

Pressure sensors: These sensors determine the pressure of liquids as well as gases in the human body. Most frequently, their reliance can be identified in cases of tracking blood pressure, intracranial pressure, and respiratory pressures.

Force sensors: Strain gages measure the level of force applied on any biological part. They are applied in prosthetic and orthotic practice to evaluate the effectiveness and make adjustment of mechanical limb substitutes and orthoses.

Accelerometers: Accelerometers record the velocity and acceleration; the information involves a device's motion and position. Along with health monitoring, they are incorporated into wearable gadgets to track movement, gait, and falls of old people.

Tactile sensors: Tactile sensors capture the force being applied on a certain plane. For instance, they are used in surgery using robots to control prosthetic limbs to get force feedback and information on how tight a grip is.

All types of sensors, optical, chemical, and mechanical, are valuable within the biomedical monitoring discipline since they generate important information to diagnose, control, and enhance the health of the patients. These sensors function in such a way that they give a holistic view of healthcare monitoring and physiology of the human body.

4 Technological advancements in sensor design

4.1 Flexible and stretchable electronics in biomedical sensors

Biomedical sensors have come under dramatic improvement in their design with the advent of flexible and stretchable electronics. These innovations overcome the defects of conventional brittle electronics to improve sensor efficiency, as well as broaden the scope of smooth health check.

4.1.1 Flexible and stretchable electronics: An overview

Wearable and conformable electronics are electronic devices designed to adapt to various shapes and movements. They use materials such as organic semiconductors, conductive polymers, and flexible substrates to create fully functional electronics that can bend, twist, and fold. Integrating these materials into biomedical sensors enables the development of wearable and implantable devices that are more comfortable, versatile, and equipped with enhanced functionality.

4.1.2 Advantages of flexible electronics in biomedical application of sensors

Enhanced comfort and wearability: Conformal sensors are made possible through flexible electronics, eliminating irritation that hampers patient adherence with most monitors today. This is crucial in the case of wearable sensors, especially those that require the subjects to have them on for protracted durations.

Improved sensor integration: First of all, the characteristics of stretchable and flexible sensors enable end-users to incorporate them into numerous garments and accessories. By integrating these features, the user can constantly be monitored without having to wear any equipment.

Durability and longevity: Compared with a stiff sensor, flexible electronics offer more adaptability and are less easily affected by mechanical stress. The increased durability is a plus for longevity of the sensors showing potential for good heath monitoring for the long term.

Adaptability to dynamic environments: Bendable sensors can continue to be useful while conforming to the shape changes in the human body, for example, when stretching. Such flexibility is important in achieving the correct measurement of the parameters like heart rate and muscle activity during exercises.

In the following sections, key technologies in flexible electronics are described as follows:

Conductive polymers: Conductive polymers, including polyaniline and PEDOT, are employed to design and fabricate stretchable and bendable electronic devices with further enhanced electric conductivity and mechanical compliance. These materials are essential for the realization of pliable sensor electrodes and interconnects.

Organic semiconductors: For similar reasons of flexibility and low weight, organic semiconductors like OLEDs and organic photovoltaic are used in flexible electronic devices. They support the fabrication of reconfigurable multipurpose sensors with electrical and optical properties.

Flexible substrates: These material types include polyimide, polyethylene terephthalate, and PDMS that form the platform for flexible electronics substrates. Some of these materials are chosen for their mechanical flexibility, some for chemical stability, and some for compatibility with electronic devices.

Hybrid materials: Composites based on inorganic and organic phases are employed in optimizing the characteristics of flexible sensors. These materials exploit the beneficial features of both types to design sensors with enhanced electric/magnetic and mechanical response characteristics.

4.1.3 Uses of flexible electronics for biomedical sensors

Wearable health monitoring devices: Wearable electronics include flexible smartwatches, fitness trackers, and flexible health-monitoring patches that can be worn on the skin, employing physiological parameters, include heart rate, body temperature, and sweat rates.

Flexible biosensors: A biosensor that comprises an ultralow power flex circuit can be adopted for biomarker and pollutant monitoring. Examples include glucose and lactate sensors, and electrochemical sensors for the detection of different analytes.

Implantable devices: Flexible electronics can be applied to implantable sensors in monitoring glucose level of diabetic patients and intracranial pressure of neurological disorder patients. Due to portability, these devices reduce tissue injury of the patient, which in turn improves their level of comfort.

Smart textiles: Conductive ink and flexible electronics enable creating intelligent clothing by integrating textile electronics that can measure key physiological signs and give signals to the wearers. Some of them are outfits that include athletic apparel, which contain sensors for performance tracking alongside health checks.

4.1.4 The future of challenges

Material performance and stability: It still remains difficult to achieve high long-term stability and performance of flexible substrates under various environmental demands. Research should aim at enhancing the durability of these materials so as to increase the useful lifespan of the flexible sensors.

Scalability and manufacturing: The practical utility of flexible electronics is highly dependent on manufacturing processes that are efficient, and more importantly cost affordable at larger scales. These challenges are being addressed by existing and newly developing printing technologies as well as roll-to-roll processing.

Integration with advanced technologies: Several potential future evolutions of flexible electronics include wireless technology, energy conversion, and even big data analysis to dedicate possibilities of biomedical sensors.

Flex electronics and stretchable electronics are on the brink of providing drastic improvements in the biomedical sensors arena in terms of comfort, adaptability, and conventional performance. These technologies have been stated to improve more innovations in health monitoring as research and development go on.

4.2 Nanotechnology-enhanced sensors for high sensitivity

In biomedical sensors, nanotechnology has come in as a powerful tool that can provide accuracy that is unattainable at the conventional coarser scale of material. Nanotechnology has shifted the sensor industry into higher level of sophistication by facilitating the sensors to identify biological markers as well other changes in the human body with great accuracy.

Some of the solutions include the use of nanomaterials in order to improve sensitivity.

Biomedical sensors, such as nanoparticles, nanowires, and nanotubes, have unique physical and chemical characteristics, which make them more sensitive. These materials have a large surface area-to-volume ratio, thus enhancing the interaction between the sensor and the target analyte, and hence a higher sensitivity is achieved. For instance, gold nanoparticles find application in biosensors because of the optical properties and signal amplification effect via LSPR.

4.2.1 Quantum dots in biosensing

Quantum dots (QDs) are submicron semiconductor particles known for their size-dependent fluorescence properties, which make them highly suitable for integration into biosensing technologies. Their exceptional photostability and brightness enhance the sensitivity of fluorescence-based sensors. As shown in Figure 6, QDs are widely used in applications requiring the detection of scarce signaling molecules and cellular receptors with high sensitivity. Additionally, their tunable emission spectra enable the simultaneous detection of multiple targets using a single set of fluorophores.

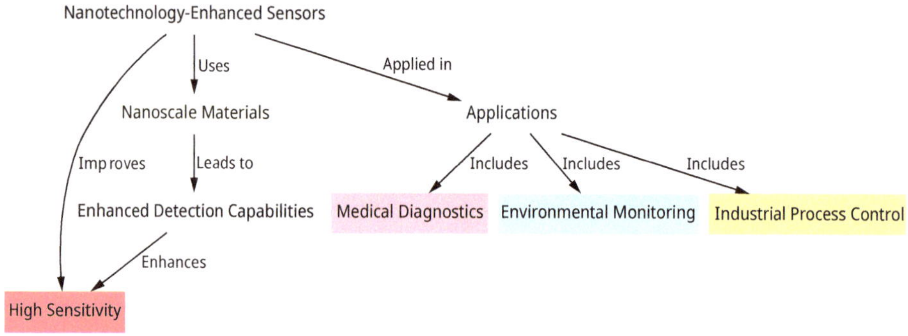

Figure 6: Nanotechnology-enhanced sensors for high sensitivity.

4.2.2 Carbon nanotube and graphene in sensors

Carbon nanotubes (CNTs) and graphene are two carbon allotropes distinct in electronic, mechanical, and thermal characteristics, which have shown tremendous catalytic impact on the sensor research field. CNTs, due to their high electrical conductivity and large surface area, are used in electrochemical sensors, whereby variations in current, reflective of the presence of target analytes, are measured. Graphene interacts as a high surface area conductor in the amplification process of different sensor technologies, including biochemical sensors and chemical sensors.

4.2.3 Nano coatings for high performance

Nanostructured coatings involved the integration of nanomaterial composites into sensor surfaces to encourage enhancements in performance. These coatings can change the physical and chemical characteristics of the sensor's surface to make it more attractive to an analyte molecule, hence improving both sensitivity and selectivity. For example, based on nanostructured coatings, the binding capacity of biosensors can be enhanced to increase the surface area for reaction with target analytes, ensuring higher sensitivity.

4.2.4 Afflation of combining nanotechnology with microfluidics

A marriage between nanotechnology and micro fluidics has taken the sensor design to a new level. Microfluidics deals with the movement of small volumes of fluids through microchannels and that when integrated with nanotechnology, produces compact, sensitive, and fast sensors. There is the employment of nanomaterials inside microfluidic systems for biosensing and improvement of biomarker detection as well as for integrated sample preparation and for on-chip analysis. Combined, these result in higher sensitivity and shorter times for detection, which makes the process ideal for use in the diagnosis at the point of care.

4.2.5 Novelty in nanotechnology: nanostructured sensing platforms

The inclusion of nanomaterials by current advances in nanostructured sensing platforms results in improved accord sensitivity and multiscale function. These include optical sensors using nanoparticle with SERS and field-effect transistors using nanowire for the detection of biomolecules, without the need for label. These platforms take advantage of the unconventional characteristics of nanomaterials to provide very accurate measurements of low concentrations of analytes.

Nanotechnology has brought broader possibilities to biomedical sensing through refinement of technology to produce some of the most sensitive sensors for the identification of physiological and biochemical parameters. It is also anticipated that as research advances at the nanoscale levels, nanotechnology would open new horizons for improvement in sensor development and application in health assessment, prognosis, and disease prevention.

4.3 Microelectromechanical systems (MEMS) in biomedical applications

MEMS technology reflects a brand-new stage for sensor technologies, with their unique feature being that combines mechanical and electronic elements on a micro scale. MEMS technology, when applied to biomedical fields, has improved the accuracy, functionality, and flexibility of sensors used in chronic health monitoring systems. This section explores the use of MEMS in biomedical engineering with focus on its main interventions in the health sector.

4.3.1 Overview of MEMS technology

MEMS technology deals with the integration of micro mechanical structures, together with microelectronics on micro scales, in a single base chip. These devices are generally produced using semiconductor processing technologies that provide preferred accuracy and repeatability. MEMS sensors work through a change in physical variables, including pressure, acceleration, and displacement, in the form of mechanical displacement that is converted to electrical signal.

4.3.2 MEMS sensors used in physiological monitoring

MEMS sensors are useful in physiological monitoring, particularly for being small and highly sensitive. Key applications include:

Pressure sensors: MEMS pressure sensors are used in BP monitoring, ICP, and IOP monitoring. Being more accurate and highly reliable, it is easier to determine certain pressure changes that are so vital in issues like hypertension.

Accelerometers: MEMS accelerometers are applied to communicate with acceleration and dynamics, while physical movement data enables the observation of mobility and falls. Part of wearable devices that capture the movement and body position assist in related therapy and anti-fall measures.

Microfluidic devices: MEMS facilitates fabrication of micro fluidic systems that control as well as analyze micro volumes of fluids. These devices are employed in point of care diagnostics and lab-on-chip systems, where biological samples are analyzed with great speed and high sensitivity.

New developments in micro and nano electromechanical system-based biosensors

Recent advancements in MEMS-based biosensors have expanded their applications and capabilities in biomedical fields:

Lab-on-a-chip systems: MEMS technology has enabled miniaturization of analytical functions and the development of high-density lab-on-a-chip microsystems. These systems also allow for the simultaneous analysis of multiple biomarkers from a small volume of blood sample, and improve diagnostics.

Implantable MEMS devices: Advancements in MEMS technology enable the realization of new implantable sensors that monitor internal physiological status in real-time. They are developed from a material that is compatible with body tissues and using miniaturization technology. They can be implanted with the least discomfort for a long period.

High-resolution imaging: MEMS has proved to have enhanced the operations of imaging systems for micro-optical and micro-endoscopes. These systems provide better image quality with less size and cost; thus, they are best suited for interventions that require minimal invasions and diagnosis.

In broadband telecommunications, MEMS has the following potential:

MEMS technology offers several advantages in biomedical applications.

Miniaturization: MEMS sensors are different from traditional sensors in that they are much smaller, enabling wearable / implantable applications. This miniaturization improves patients' comfort and allows overall supervision without interrupting their normal lives.

High sensitivity and accuracy: MEMS sensors are highly sensitive and accurate in measuring physiological parameters. Since wireless MEMS technology for acquiring signals is extremely accurate, it provides accurate data needed for diagnosing and managing illnesses.

Low power consumption: MEMS devices are extremely low-power devices, thus they are well suitable in battery-powered wearable devices. This low power consumption has the effect of extending the useful life of equipment and lowering the required maintenance frequency.

Possible future development trends of MEMS

The future of MEMS technology in biomedical applications is poised for further innovation:

Integration with AI and Machine learning: AI and machine learning are expected to improve MEMS sensor data analysis and prediction, with the integration of MEMS sensors with the technologies. This integration has led to intelligent as well as adaptive health monitoring systems.

Advanced materials and manufacturing techniques: Current research on microsystems includes study on the technology of MEMS with enhanced parameters such as performance, biocompatibility, and durability of the devices. Development in such areas increases the variety of uses and improves the performance characteristics of the MEMS-based sensors.

Expansion into personalized medicine: MEMS technology is expected to be included in the field of personal medicine; thus, health-monitoring services could be provided according to the patients' profile and requirements.

MEMS have ended the conventional approaches of biomedical sensors, which have enabled the development of compact, accurate, multifunctional modalities for monitoring health. MEMS technology is likely to evolve and progress in the near future, which open new options for biomedical sensors as instruments for patients' treatment and helpers in medical experiment.

4.4 Advances in wireless and battery-free sensor technologies

Modern developments in wireless and battery-less sensor technologies are expanding to new technological frontiers in biomedical monitoring by extending the potential for conveying a steady flow of information about people's health, while avoiding the problems of inconvenience and limitations associated with batteries. These innovations help to solve some issues that include power constraints, integration of small devices, and handling of real-time data.

4.4.1 Wireless communication technologies

The lack of cables is another important characteristic of today's biomedical sensors since they all rely on wireless schemes. Several advancements have significantly improved the performance and reliability of wireless sensor systems.

Bluetooth low energy (BLE): Based on the current literature, BLE technology is accepted as a standard for low-energy and short-range communication in wearables sensors. This makes it suitable in devices with high frequency data update requirement, whilst supporting long battery life.

Near-field communication (NFC): NFC technology enables short-range high-frequency wireless communication between various sensors and other devices. It is used in such areas as body glucose monitoring without physical contacts and identification of patients, where data transfer has to be fast and secure.

Zigbee and LoRaWAN: These protocols are used in sensor networks that entail high-range transmission and minimal power usage. Zigbee is ideal for devices that need multiple connections with sensors within the same personal area network, while LoRaWAN is ideal for knowledge transfer across expansive areas, as required in remote health monitoring.

5 G connectivity: The emergence of 5 G technology improves the data channeling process by providing increased speed of transmitting data with low latency, making the wireless sensors more effective for conveying real-time health information. This advancement aids the growth of complex health monitoring systems that need processing and feedback with little delay.

Wireless power transfer (WPT): The advancement in WPT technology can allow the charging of sensors wirelessly such that they do not require battery replacement. This is especially helpful in implantable sensors and in long-term wearable devices.

4.4.2 Battery-less sensor technologies

Battery-less sensors rely on alternative energy sources to operate without the need for conventional batteries. These technologies overcome the limitations of battery life and significantly reduce the need for frequent maintenance. As illustrated in Figure 7, one key approach is energy harvesting, where ambient energy from the environment is captured and converted into electrical power for sensor operation. Emerging energy scavenging techniques include piezoelectric energy harvesting, which generates power from mechanical vibrations, and thermoelectric energy harvesting, which converts heat gradients into electrical energy.

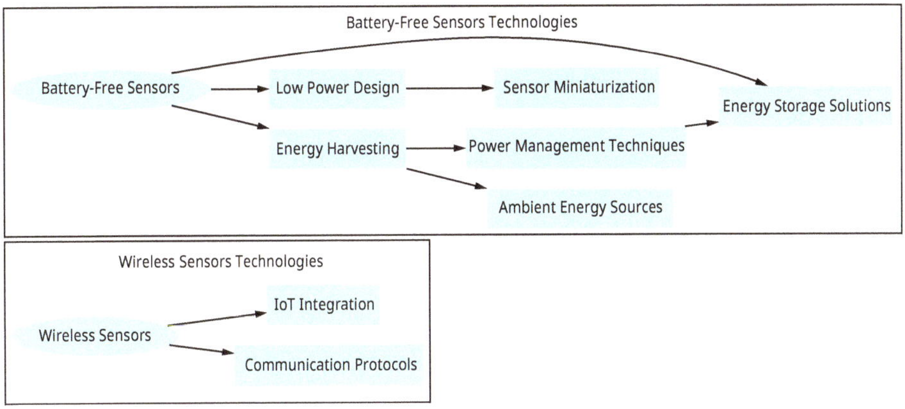

Figure 7: Advances in wireless and battery-free sensor technologies.

RFID technology: Radio Frequency Identification (RFID) sensors work when the devices receive energy from an external radio frequency field. These sensors are commonly integrated in low-power use applications like monitoring medical implants and intelligent IDs for patients.

Solar-powered sensors: Wearable sensors with incorporated solar cells can use the available light to produce energy. This technique is particularly beneficial for sensors situated in environments of natural light, decreasing the usage of batteries and increasing the equipment's lifespan.

Electromagnetic induction: In this technique, a current is created in the components of the sensor by applying electromagnetic fields, where the components do not have their own power source. It is useful in diverse sensor structures such as those that are applied in health-focused devices, including health monitors.

Biofuel cells: Biofuel cells produce electricity through bio redox reactions, for example, glucose, and can be employed to supply electrical energy to implantable biosensors. This technology is based on sustainable energy delivery system, completely mimicking body's own chemistry for charging batteries that are used for powering up the metabolic parameter sensors.

In the same way that miniaturization and integration became keywords in the field of microelectronics, where computer systems and their peripheral equipment appear to be inextricably merged, an analogous combination of merging elements at a more symbolic level seems to occur with the two-visual semiotics.

Advancements in miniaturization and integration technologies are enhancing the capabilities of wireless and battery-free sensors:

MEMS: MEMS allows making small sensor/actuator elements with idiopathic wireless communication possibilities. The capacitive and resonance of MEMS-based sensors are utilized due to their efficient functioning and higher tolerance for miniaturization for wearable and implantable medical applications.

Flexible electronics: The advancement of flexible and stretchable electro material enables the formation of skin-like sensors that can easily conform to different body surfaces of a patient. These allow designing of flexible sensors that people could wear for long-term use without discomfort and which can move with the body.

System-on-chip (SoC) integration: The system-on-chip technology applies many functionalities and communications of the sensor and its components in a single chip, cutting the size of complex sensor systems. This integration optimizes the usage of such wireless as well as battery-less sensors and enhances their efficacy in real-life applications.

Improvement of the product's durability and reliability

Technological advancements have also focused on improving the durability and reliability of wireless and battery-free sensors:

Robust packaging materials: Innovations in the packaging of the sensors prevents them from the effects of moisture as well as mechanical stress, improving the durability of the sensors.

Self-healing materials: Integrating concepts of self-healing materials in a sensor help to increase the durability of devices and make necessary repairs to minor damages to sensor.

Temperature and humidity tolerance: The development of different material of sensors and their structure has enhanced the ability of holding different climates and temperature levels, which widely enhance using the sensors in different types of environments.

These developments in wireless and battery-less sensor platforms are opening up new opportunities for better, longer lasting and more diverse biomedical monitoring applications. These solutions help address problems in terms of power infrastructure, scale, and data transfer, and thus improve the performance of current health monitoring systems.

4.5 Hybrid and multimodal sensor systems for comprehensive monitoring

Hybrid and multimodal sensors are one level up in biomedical monitoring, providing greater capabilities due to the ability to evaluate the patient's condition more thoroughly. These systems incorporate many sensing subsystems to monitor and/or analyze the parameters of human body to overcome the drawbacks of single-mode sensor.

4.5.1 Multiplexed multiple sensing

Hybrid sensor systems integrate two sensing modalities, which include optical, chemical, or mechanical sensors, in a single system. It also means that there can be measurement of more than one parameter at a given time, which makes for a proper assessment of a patient. For instance, a single sensor combines electrochemical and optical principles and can measure glucose and blood oxygen saturation, which can be of definite use for diabetics.

Increased efficiency, data accuracy, and reliability

Hybrid systems are developed with the use of multiple sensing technologies and helps deliver more accurate and reliable results in terms of health. One modality offsets the weaknesses of the other to eliminate areas of vulnerability and improve the capabilities of each sensor. For example, integrating optical and mechanical sensors would ensure that wherever there are problems like motion artifacts where wearable devices are used, there be better approaches to data acquisition.

Pages 49: The emergence of instruments for multimodal health monitoring

Wireless vital signs are novel health monitoring gadgets that are created to have multiple parameters that are sensed by several sensors. These devices can track the number of beats per minute, the number of breaths per minute, activity, and even temperature, all at once. Illustrations include state-of-the-art wearable devices that incorporate accelerometers, gyroscopes, and optoelectronic sensors to obtain detailed outlines of the physical activity as well as physiological indices. Such multimodal systems can allow for constant and highly granular tracking of health and are especially useful when it comes to controlling various chronic diseases and providing treatment plans.

4.5.2 Patient compliance and comfort can be easily enhanced

Combining several sensors into one monitoring device also improves patient compliance and comfort. Multimodal sensors, with integrated multiple sensors, are typically developed to be small in size, lighter in weight, and less invasive to patients, compared with the use of multiple devices. For example, a single wearable device that has both the heart rate monitor and activity tracker replaces another device and thus enhances the overall compliance of patients to adhere to monitoring, improving user experience.

4.5.3 Data fusion by applying ensemble learning and data mining

Heterogeneous and distributed sensor systems create a large amount of data from multiple sources, hence requiring sophisticated data fusion and analysis procedures.

These techniques amalgamate all data from various sensors, giving a clear and concise picture of the health state of the patient. Artificial intelligence and machine learning are used to process and analyze the large volume of data obtained from the multisensor modalities in order to make immediate and personal health decisions. For example, data from heart rate monitors, glucose sensors, as well as activity trackers could give comprehensive information about a patient's cardiovascular health, metabolism, and activity, in comparison with traditional checkups that are done once a week, and therefore the intervention could be more efficient.

4.5.4 Some applications in personalized and preventive medicine

Hybrid and multimodal sensor systems are of great importance in terms of personalized and preventive health interventions for individual health status. These systems allow for specific approaches to health monitoring, derived from a broad understanding of the patient's physical condition. For instance, multimodal sensors used in continuous glucose monitoring can give information on how diet, exercise, and stress influence the levels of glucose in the blood, to aid develop suitable strategies in controlling the diabetes. Further, HS can play an important role in support of the preventive measures by detecting early symptoms of potential health problems and thus calling for timely actions against diseases.

Hybrid and multimodal sensor systems are a significant innovation in biomedical monitoring, which extends all the crucial abilities required in the total health overview and management. Due to the drawbacks of single-mode sensors, these systems utilize a number of the abovementioned technologies to enhance the accuracy of data collected, patient compliance, and overall healthcare delivery results. Even so, devices with hybrid and multiple modal sensors are bound to become crucial as multifaceted technology expands in the future toward fulfilling personalized and proactive healthcare goals.

5 Biomedical sensors in chronic disease management

5.1 Sensors for cardiovascular health monitoring

Cardiovascular health plays a crucial role in identifying potential health complications. The advancement of wireless sensors has significantly expanded the range of cardiovascular parameters that can be monitored, while also enhancing the quality and convenience of data collection. As shown in Figure 8, this section explores the types of sensors employed in cardiovascular health monitoring, their applications, underlying technologies, and their impact on patient management.

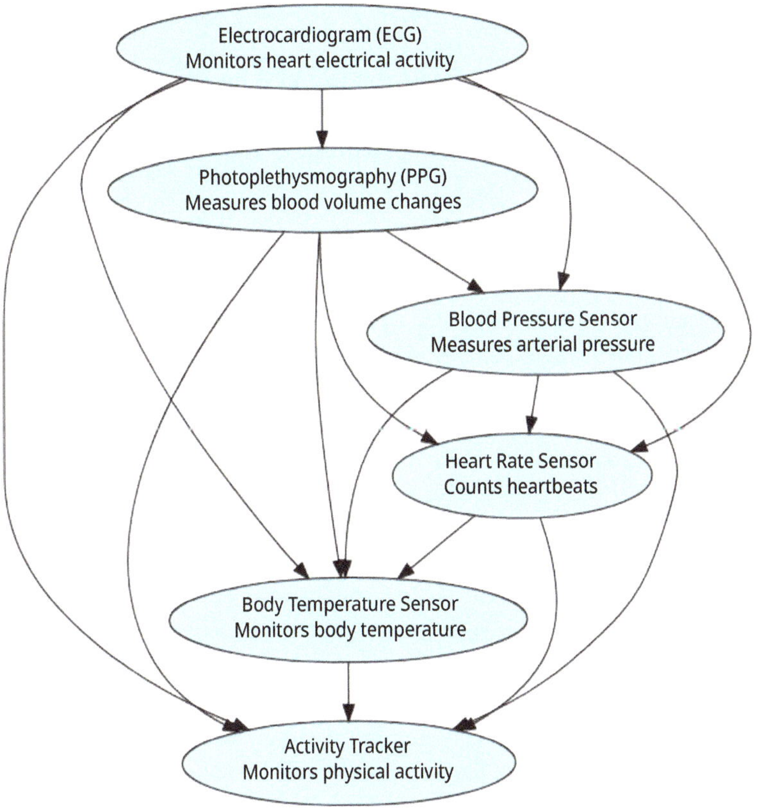

Figure 8: Sensors for cardiovascular health monitoring.

5.1.1 ECG sensors

ECG sensors are used for capturing electrical signals that cause the heart to pump blood. These sensors capture the electrical signals that provoke cardiac contractions, and afford critical data on cardiac pacing and efficiency.

Wearable ECG monitors: Currently, modern wearable ECG monitors track the heart rate and rhythm control. These devices are intended for assessing and monitoring arrhythmias, for instance, atrial fibrillation, and allows the user to be notified immediately.

Holter monitors: Holter monitors are less bulky ECG instruments and can be used for 24–48 h recordings in order to offer detailed information on the activities of the heart over a longer period. These monitors are utilized in monitoring repeated and periodic arrhythmias, and evaluate the success of treatments targeting the heart.

Implantable loop recorders: Implantable loop recorders are small devices that are buried under the skin to record the heart rhythm for months and even years.

These are especially good for making the diagnosis of paroxysmal tachycardia and syncope of unknown etiology.

Patch ECG monitors: Adhesive ECG patches are intended for short-term use, thus offering real-time heart rate information discreetly and comfortably. Outpatient, event monitoring is done and addressed through these patches.

5.1.2 Blood pressure sensors

Blood pressure sensors track the force of blood pushing on the arteries' walls, which is indispensable for diagnosing/high blood pressure and managing cardiovascular diseases.

Wearable blood pressure monitors: This type of blood pressure monitoring device is wearable; most are wristbands to monitor the patient's blood pressure. These devices operate on oscillometric principles to give reliable pressures.

Ambulatory blood pressure monitors: Emerging home and outpatient devices include ambulatory arm blood pressure monitors, in which patients wear these portable monitors; these can be worn for 24h. They give a detailed picture of the shifts in blood pressure during the course of normal day, helping in establishing hypertension diagnosis and in monitoring the effectiveness of treatments.

Invasive blood pressure sensors: Some routinely invasive blood pressure monitors include catheter-based, and this is common in the setting of the hospitals. These sensors give consecutive and accurate blood pressure information that is crucial for intensive care as well as for operations.

Noninvasive continuous blood pressure monitors: Modern continuous monitors include PTT that allow reading BP without constrictive bands.

5.1.3 Pulse oximeters

Pulse oximeters are used to measure the level of oxygen in the blood and the pulse rate to help understand respiratory and cardiovascular conditions.

Finger and Wrist pulse oximeters: Finger and wrist pulse oximeters are handy and convenient in use, and perform constant measurements of blood oxygen and pulse rate. They are most beneficial for sufferers of respiratory diseases.

Pulse oximeter patches: Noninvasive cuffless and adhesive pulse oximeter patches are easy to use and can be worn on the skin for all types of monitoring. They are more cumbersome and time-consuming as compared to traditional pulse oximeters, but can be used for short-term monitoring of blood oxygen levels. These patches are designed to allow for long-period monitoring in different environments such as home care and in hospitals.

Wearable pulse oximeters: They are like fitness trackers attached to clothing as smartwatches are like to rings, allowing users to track their blood oxygen levels and heart rate during exercise.

Pulse oximeters for critical care: Offering features of continuous monitoring and alarms for low oxygen saturation to different patients with various conditions, the use of advanced pulse oximeters are recommended in critical care areas for patients with serious respiratory complications.

5.1.4 Heart rate variability (HRV) sensors

HRV sensors calculate the difference in the time intervals between successive heart beats, giving information on the function of ANS and the state of the cardiovascular system.

Wearable HRV monitors: Chest straps and smart watches are used as wearable devices to record constant data of HRV and for autonomic regulation of stress levels. The cardiac autonomic function is evaluated using high-frequent variability (HRV).

ECG-based HRV measurement: Contactless ECG-based HRV sensors present high accuracy of heart rate variability by processing the ECG signal. These sensors offer advance information regarding cardiac autonomic control and can be applied both in clinical practice and in research.

HRV monitoring in fitness devices: Most fitness devices use HRV sensors to allow users to monitor their recovery and training volume.

HRV sensors for clinical assessment: Clinical applications of HRV sensors in devices for patients with cardiovascular diseases such as heart failure and ischemic heart disease provide information on functional systems and disease progress.

The latest technology that we are going to talk about today is the integrated cardiovascular monitoring system or the ICMS.

Current sophisticated multifunctional CVMS incorporates various methods of sensing in one instrument for a systematic approach to the assessment of cardiovascular status.

Multiparameter cardiovascular monitors: Technological advancement has seen these systems use ECG, blood pressure, and pulse oximetry in a single unit, giving an overall picture of the cardiovascular status. They are utilized in both outpatients as well as for inpatient care.

Smart cardiovascular patches: Specific intelligent patches containing several sensors allow for ECG, heart rate, and blood pressure to be measured constantly without any inconvenience. These patches are meant to be worn for the purpose of continuous and remote supervision of patients.

Cardiac event monitors: Cardiac event monitors use several electrodes to acquire and register potential major cardiac events, which include arrhythmias episodes.

Wearable health platforms: Mobile health platforms combine all cardiovascular sensors and analyze the data to provide cardiovascular status updates and alarms for proactive cardiovascular management and treatment.

These sensors, used for cardiovascular disease risk assessment, are very essential for the diagnosis, control, and handling of any condition that affects the heart. By providing healthcare information that are accurate and updated to the physiological and pathological status of patients, they support healthcare givers in decision-making for good patient management and consequently better quality of patient life.

5.2 Glucose monitoring sensors for diabetes management

Self-monitoring glucose sensors are vital in diabetes because they produce essential information that allows gunners to regulate glycemic control. These sensors provide the current daily indication of blood glucose concentrations, which enables individuals to make important lifestyle and/or therapeutic modifications, as necessary. This section looks at the advancement of glucose monitoring sensors with regard to the function and usefulness in diabetes treatment.

5.2.1 General information on glucose monitoring devices

Several technologies are employed in glucose monitoring, each with its distinct mechanisms and applications.

Fingerstick glucometers: The oldest form of portable devices, fingerstick glucometers, uses a small amount of blood from a finger prick to determine blood sugar levels. This method also produces results in real-time and is employed on account of it versatility and inexpensive nature.

Continuous glucose monitoring (CGM) systems: CGM systems enable patients to monitor their glucose levels at any one time with the use of a sensor placed in the subcutaneous tissue. These sensors monitor glucose levels in the interstitial fluid and relay the information to a receiver in a smartphone where one is able to get an overview of the pattern and occurrences of hyperglycemia.

Flash glucose monitoring systems: Flash glucose monitoring systems encompass a sensor placed on the skin to check on glucose in the form of interstitial fluid. Compared with CGMs, the flash systems are more discreet since users need to swipe the sensor with a reader device for glucose measurement, but at the same time offer continuous monitoring.

5.2.2 The expectations to improvements in glucose sensors via technology

Recent advancements in glucose sensor technology have significantly improved their accuracy, comfort, and convenience.

Enhanced sensor accuracy and stability: New material for the sensing layer and innovative design of the devices have brought more accurate and stable readings of glucose level. New trends are being witnessed in both baseline modification and signal processing aspects, such as improved enzyme coatings and algorithms, thus reducing errors and the shelf lifetime of the biosensors.

Wearable and noninvasive sensors: Modern extra-invasive glucose control transferring techniques, for example near-infrared, objectives to do away with blood tests. While currently under development, new technologies have the potential of affording users more comfort and convenience.

Integration with smart devices: Presently, advanced glucose sensors are fused with smartphones and other smart devises. This integration allows people to monitor glucose level trends, monitor their data, and even receive alerts via their mobile devices. Increased connection also means sharing of information with healthcare providers in the management process.

A variety of benefits are affiliated with continuous glucose monitoring.

Continuous glucose monitoring systems offer several advantages over traditional fingerstick methods.

Real-time glucose tracking: All CGM systems offer the subject a constant stream of glucose concentration readings, enabling one to understand changes in their levels during the day. This immediate monitoring assists better management of the disease than was possible with normal periodic monitoring, and cuts the risk of experiencing complications due to glucose irregularities.

Predictive alerts: Most CGM systems provide alarms that are used to indicate that the user's blood glucose levels are going high and soon reach alert limits. These alerts enable avoidance of such periods of hyperglycemia and thus enhance general glycemic management.

Improved glycemic control: Because of more frequent intervention and constant feedback, there are better titrated insulin doses and carbohydrate intake, which results in overall improved glycemic control and subsequently, lower HbA1c levels.

5.2.3 Compatibility with insulin delivery systems

Integration of glucose sensors with insulin delivery systems, such as insulin pumps and automated insulin delivery (AID) systems, represents a significant advancement in diabetes management.

Closed loop systems: The closed loop systems, referred to as artificial pancreas systems, incorporate CGM data, in conjunction with insulin delivery, which alter insulin doses in response to glucose measurements. This integration is useful for maintaining blood sugar within a desired range with little input from the user.

Smart insulin pumps: The current insulin pumps are equipped with a state-of-art sensor that regulates insulin dose according to CGM results. Basal and bolus rates of insulin can be achieved, coupled with precise glucose patterns, making it easier to avoid glycemic oscillations.

5.2.4 Issues and innovations

Despite the advancements, several challenges remain in glucose monitoring.

Cost and accessibility: Unfortunately, some of the most modern glucometers and other CGM systems that provide significant benefits to diabetic patients are too costly to be afforded by many patients. Continual work is being done on this front to try and bring down the costs of these resources.

Sensor accuracy and calibration: Although innovation has enhanced the performance of the sensors, frequent calibration and maintenance are necessary to affirm the quality of the data received. Some of the research activities are centered on designing sensors that need less calibration but offer higher accuracy.

User comfort and compliance: Maintaining user comfort and promoting adherence to continuous glucose monitoring remain two key issues. Constant advancements made in relation to this area are directed toward increasing sensors' comfort for their users.

It is also important for glucose monitoring sensors because such sensors serve as an invaluable data source that would support improvements in the general management of diabetes, glucose control, and other related health outcomes. The optimization of technology and its connection with other diabetes care also aids to strengthen the glucose mapping systems that eventually result in good results for diabetes patients.

5.3 Respiratory monitoring sensors for pulmonary diseases

Respiratory monitoring sensors are of great importance when it comes to management of pulmonary diseases as they offer real-time monitoring of the respiratory function. These sensors are useful in monitoring the course of chronic illnesses like asthma, COPD, and ILD in order to make management of the condition better and result in better results for the patients.

Based on the concept of respiratory monitoring, there are a wide range of sensors as described below.

Pulse oximeters: Pulse oximeters are used to monitor blood oxygen level (SpO2) and the pulse rate using optical sensors. By fluorescing light through a part of the body, such as through the finger top, the device is capable of detecting the amount of light absorbed by the oxygenated and deoxygenated hemoglobin – information that is very vital as regards the levels of oxygen in the blood. Any patient with a chronic respiratory disease should use pulse oximeters to determine if their oxygen levels are adequate, for management decisions.

Spirometers: Spirometers are used to determine the amount and speed of air inhaled into, and expired out of the lungs. These devices are used for the identification of the severity of diseases like asthma, depending on the FEV1/FVC values. Contemporary spirometers allow using electronic sensors and data analysis as well as tracking changes in a patient's lung function within a time period.

Capnometers: Capnometers are devices that quantify the level of CO_2 in the air reflected by the patient's breath. These CO_2 sensors help in giving information on the efficiency of the ventilator and the gases in the lungs, from the level of CO_2 in the air. The use of capnometers is especially convenient because patients with respiratory diseases using ventilation, such as severe asthma exacerbation, can benefit from it.

Impedance pneumography sensors: Impedance pneumography assesses the variation of thoracic impedance, resulting from respiratory activity and consequently the movement of air into and out of the lungs. Using electrical currents that range from 5 to 20 mA across the chest and measuring the change in impedance, they can also estimate the respiratory rate and the tidal volume. This noninvasive method proves to be valuable in evaluating respiratory function in patients with chronic pulmonary disorders.

Wearable respiratory sensors: Over the years, mainly due to the availability of modern nondisposable electronics, wearable devices have been developed that allow continuous measurement of respiratory parameters. Such sensors are used with smartwatches. These sensors can monitor respiratory rate, oxygen saturation, and other related parameters in real-time. The use of wearable respiratory sensors is also noninvasive and longitudinal; this less intrusive type of monitoring is likely to enhance compliance and generate considerable data in chronic respiratory diseases.

5.3.1 Uses in managing diseases

Asthma management: Respiratory monitoring sensors are of great utility in asthma management, in that they measure changes in lung function, oxygen saturation, and respiratory rate. Data from spirometry and pulse oximetry can then be utilized by patients and healthcare providers to recognize signs of deterioration, alter medication dosages, and adopt lifestyle changes to manage the disease.

COPD monitoring: In patients with COPD, respiratory sensors yield demographic data and the state of the lungs, which are useful in controlling the symptoms and preventing the development of COPD exacerbations. Wearable sensors and capnometers, used as constant surveillance equipment, assist in the modification of methods, regulation of oxygen therapy, and for assessing the efficacy of implementations.

Assessment of disease progression: While chronic pulmonary diseases require constant observation of additional respiratory sensors to evaluate the disease progression and effectiveness of the treatment, such factors ensure the identification of the changes in the respiratory parameters over time, hence enabling the healthcare providers to come up with the correct decision when it comes to the management of treatment services and interventions.

Emergency response and alerts: Modern respiratory monitoring sensors are capable of being set to generate alarms in case certain abnormal respiratory rates are detected. This feature is especially useful in patients with critical diseases of the lungs and thus could prevent severe complications.

5.3.2 Challenges and considerations

Accuracy and reliability: The sensor technology used in respiratory monitoring has been considered vulnerable to factors such as position of sensors, movement of the patient, and environmental factors. There is a need to standardize the sensors to provide dependable results in disease control and security of the patients.

Data integration and interpretation: Real-time respiratory rates analysis and monitoring based on the data from different sensors is challenging. Ordering, integration, and analysis of patient's multimodal data are needed, especially for the purpose of decision support.

Patient compliance and comfort: In particular, compliance with respiratory monitoring is critical when managing diseases, as evidenced by patient's noncompliance with wearing a face mask emanating from virus breakage. Comfort and easy to use of these respiratory sensors is another factor that determines its usage, where a poorly designed sensor remains a collector's item despite its high-tech functions.

Pulmonary disease management can significantly benefit from using respiratory monitoring sensors as diagnostic tools, especially in an outpatient setting, and as adjuncts to the treatment of chronic respiratory conditions. Future advancements in technology predict that advanced and accurate, easy-to-use and integrated respiratory sensors would further improve disease management, hence optimizing patient outcomes.

5.4 Neurological monitoring

Neural check is one of the most effective approaches to handling neurological conditions to assess brain and neuronal communications and total neurological well-being. Recent advancements in biomedical sensors have greatly improved the diagnosis and treatment of many neurological disorders, including epilepsy and neurodegenerative diseases, among others.

5.4.1 EEG sensors

EEG electrodes are essential components in neurological assessments, used to detect the brain's physiological activity by capturing electrical signals from neurons. EEG sensors are commonly applied in the diagnosis and monitoring of conditions such as epilepsy, sleep disorders, and brain tumors. As illustrated in Figure 9, recent advancements have led to the development of multifunctional EEG devices that are more affordable, compact, and suitable for use in natural environments eliminating the need for bulky, complex equipment. These innovations enhance mobility and allow for continuous monitoring, facilitating the analysis of seizure activity and the evaluation of cognitive functions in the management of neurological disorders.

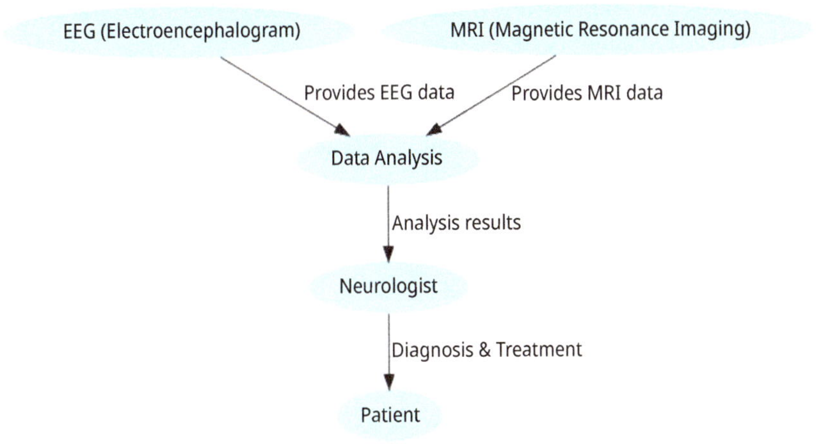

Figure 9: Neurological monitoring.

The study uses Functional Near-Infrared Spectroscopy (fNIRS).

fNIRS is an optical technique for evaluating the brain activity without any need for touching the head, by measuring changes in the oxygenated blood volume. This technique is used by means of fNIRS sensors to evaluate the cerebral blood flow and brain function during cognitive task's incidents. It was applied in clinical trials and received great benefits in the diagnosis of brain dysfunction in patients with stroke, TBI, and neural disorders. Advances in the field fNIRS are the introduction of portable fNIRS system in studying cortex oxygenation and cerebral blood flow continuously in real-time.

5.4.2 ICP monitors

ICP sensors are essential for patients developing certain conditions, such as traumatic brain injury (TBI). These sensors keep a check on the pressure that is exerted inside the cranial cavity and contain helpful information about the condition of the brain. Today's ICP monitors are made with sophisticated materials and technologies to provide a high degree of precision and durability. Sometimes, they have wireless features and connectivity with other monitoring systems, which enable intracranial pressure to be monitored continually and from a distance.

5.4.3 Brain–computer interface (BCI)

BCIs are a new form of neurological monitoring that enable the brain to interface with other devices. BCIs rely on different modes of sensing, including electroencephalogram and implantable sensors, which help in identifying the brain signals. These systems are employed to embrace individuals with quadriplegic disorders and others who are not capable of voluntary movement; the technique allows manual control of assistive technology, solely through brain exercises. Current innovations in BCI concern itself with how the signals are made finer as well as how the hardware maybe rendered more comfortable and how the new generation of prosthetic devices as well as communication appliances are incorporated into the system.

5.4.4 Portable electrical stimulation apparatus

Portable neurostimulation devices also provide a unique paradigm to neurological monitoring and treatment. Treatment tools employ electrical currents to adjust the putative neuronal activity to help patients with conditions such as chronic pain, depression, and Parkinson's disease. New trends in wearable neurostimulation are emerging; these include but are not limited to miniature, easily wearable devices, which can be used for continuous stimulation. Many of these devices also include some sort of apparatus for measuring physical responses to stimulation and adjust the therapy regimen according to the feedback received.

5.4.5 Intelligent prostheses and adaptive neuroprosthetics

Smart implantation and responsive neuro technology can therefore be viewed as sophisticated applications of constant neurological monitoring and treatment options. These implants, which usually come with sensors and stimulation systems, are intended to perform a dynamic interfacing with neural activity. Some examples include deep brain stimulators that can have programmable modes that change stimulation parameters according to real-time neural recordings. They are employed for diseases such as Parkinson's to deliver applicable clinical treatments as per the neurological state of the patient.

Neurological monitoring with biomedical sensors has improved the diagnosis, continual tracking of data in real-time, and patient-tailored therapy for neurological disorders. These technologies enhance the scope of an easily integrated approach in the neurological health of patients, while enhancing the quality of the years of research done in neuroengineering and neurology. More precious developments are assumed to increase the accuracy, usability, and efficiency of neurology monitoring techniques in future years.

5.5 Long-term monitoring solutions for geriatric care

Intermittent monitoring systems are important in geriatric care because elderly patients often suffer from multiple chronic illnesses requiring comprehensive management. The use of long-term remote biomedical sensors offers significant advantages by improving the quality of care, enhancing health outcomes, and enabling timely interventions for older adults.

5.5.1 Wearable health monitoring devices

According to findings, wearable health monitoring devices are widely used in the management of chronic diseases in the geriatric population. These devices are easy to wear, avoiding any inconvenience to the patient, and are intended for constant monitoring of different vital parameters, including pulse, blood pressure, and blood sugar level. Some of the enhanced wearables also target aspects such as falls, physical activity, and sleep, which are essential for elderly users. For example, Smartwatches with cardiovascular features can help caregivers if a child is experiencing cardiac arrhythmias blood pressure.

5.5.2 Remote health monitoring systems

Tele-care technologies enable continuous acquisition and transmission of health information from the elderly to the healthcare givers without having to make frequent home visits. Such systems often include measurements by sensors of indices like glucose, pressure, and saturation of oxygen. Information is sent over secured networks to the healthcare givers, who in turn can work on the information to make informed decisions on the changes in the treatment plan. Remote monitoring systems assist in controlling chronic diseases like diabetes and hypertension and coming up with a plan without going through regular hospital appointments for the elderly, who face mobility problems.

5.5.3 Wearable sensors for internal recording

As for implantable sensors, it is possible to monitor internal processes of chronic conditions, for which accurate and long-term surveillance is needed. These are implantable sensors that can monitor parameters such as intracranial pressure, glucose, and heart function. Some of the conditions that are best monitored through implantable sensors are diabetes and heart diseases because they give real-time information that is very vital in the management of the diseases. Recent developments in biocompatibility of materials as well as the above-mentioned aspects of miniaturization have enhanced the safety and comfort of these devices, making them potential candidates for use in elderly patients.

5.5.4 Integrated health monitoring systems

Central monitoring systems are developed from a single patient-health-monitoring unit and include more sensors and technologies to produce an overall and detailed information of the patients' state. Such systems comprise wearable devices, home-based sensors, and other telemonitoring methods for multiple health parameters' monitoring. For instance, an integrated system consolidates sensors for heart rate, blood glucose, and fall to ensure that it covers all aspects. Such systems facilitate team work, in that they make it possible for several caregivers to view several facets of a patient's condition and make decisions based on the information presented.

5.5.5 Fall-risk identification and mitigation systems

Many fall-detection and prevention systems are essential components of monitoring in geriatric settings for long-term care. In elderly people, fall-risks are prevalent and often result in severe consequences. Present-day automated fall-prevention devices incorporate accelerometers, gyroscopes, and pressure sensors to detect a fall and notify the caregiver services instantaneously. Further, certain systems offer features for assessing potential falls by analyzing successfully whether changes in gait can be made before a physical fall takes place. Such systems improve the level of safety and the possible level of autonomy for elderly people by offering the ability to respond to falls as soon as they happen.

Environmental monitoring and integration of smart home

Environmental assessment and intelligent homes improve geriatric care by specifically reducing environmental hazards for the elderly. The home environment can be sensed through analyzers placed within a home to measure a variety of values such as room temperature, fresh air levels, and presence of hazardous items like gas moisture. Health monitoring devices can be connected to home systems for smart home systems to deliver comprehensive care by adjusting the environment and notifying the caregivers about the problems. For instance, automated systems can control heating and cooling, depending on the health input and notify caretakers if an elderly person is inactive from the usual rhythm.

5.5.6 Challenges and considerations

Several challenges have to be considered when aiming for adopting long-term monitoring solutions in geriatric care – these include the ability of patients, their geriatric age, their comfort to use the devices, proper data privacy and security issues, and correct incorporation into the existing geriatric care practice. Also important are aspects such as possible psychological effects of constant supervision on the elderly and the necessity of caregivers' training for the proper interpretation and utilization of the data received from the sensors.

Telemonitoring technologies for the elderly care have significant roles in the long-term management of illnesses in the elderly population. Specifically, these solutions offer enhanced, real-time, thus more effective, ongoing information and intervention, hence a positive impact on the specific group's health.

6 Challenges in biomedical sensor development

6.1 Sensor accuracy and calibration issues

Peculiarities of biomedical sensors require close attention to issues of sensor accuracy and calibration as far as they determine the viability and efficiency of health monitoring systems. For getting more accurate health results that are credible for correct diagnosis, medical treatment, and subsequent handling of patients, it is crucial to provide high accuracy and correct calibration.

6.1.1 Role of precision on biomedical sensors

In this case, accuracy means how effectively the sensor has the capability to assess the physiological parameters without significant variation. Faulty sensors produce wrong information, which can contribute to wrong diagnoses and wrong treatments of health problems. For example, faulty glucose sensors cause wrong insulin measurements for diabetic patients, and so with wrong blood pressure measurements for cardiovascular disease patients. Hence, high accuracy and steady value outputs are core to the utility of biomedical sensors in clinical and personal healthcare applications.

6.1.2 Calibration procedures

Calibration refers to the process of aligning a sensor's output with a precise, measurable reference value. This involves using reference devices to compare and adjust the sensor's readings, ensuring continued accuracy over time. As shown in Figure 10, periodic calibration is essential to maintain sensor performance, especially in the face of environmental changes, component wear, and signal drift from original settings. The primary goal is to guarantee accurate data output, and when discrepancies occur, adjustments are made based on standard reference values. Regular calibration enhances the reliability of biomedical sensors and underscores the importance of robust, validated calibration procedures.

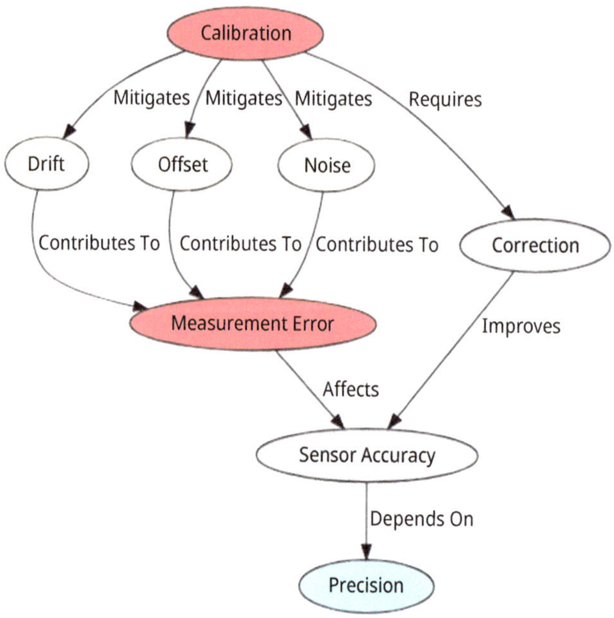

Figure 10: Sensor accuracy and calibration issues.

6.1.3 Sources of accuracy issues

There are various factors that can attest to accuracy problems in biomedical sensors. These include:

Environmental factors: Fluctuations in temperature, humidity, and even electromagnetic interference affect the sensor's abilities. Biomedical sensors should be able to function properly, irrespective of the prevailing environmental characteristics, hence the need to design portable biomedical sensors.

Biological variability: The sensor is influenced by the difference in politics across individuals, be it the skin color of the tissues. There is biological variability of the human body that sensors should capture, and hence, sensors must be standardized.

Sensor degradation: Sensors, as any other equipment, can also become damaged over time from exposure to bodily fluids and physical abrasion. This is primarily due to problems associated with ageing of the sensors and the need to recalibrate the system to reflect the changes as they occur.

Interference from other substances: Chemical-type sensors respond not only to the quantity of interest but also to any chemical interferences in the sample. For instance, glucose sensors are affected by presence of other sugars in the blood stream. Such interference poses some challenges and it necessitates an effective calibration and correction algorithms.

6.1.4 The main challenges in calibration techniques

Calibration techniques for biomedical sensors can face several challenges, including:

Complexity of calibration processes: Of course, there are sensors that need multiple-point calibration with certain equipment. These processes often tend to become complex and making them simple without compromising on the accuracy is difficult.

Standardization issues: In order to manage heterogeneity, there is a need to agree on calibration procedures and the use of calibration metrics. A high level of variability means that sensors cannot be standardized, and this brings about differences in their operation and the interpretation of results supplied by these gadgets.

Real-time calibration: However, in dynamic situations where reconfiguration is necessary, as in the cases of wearable sensors, calibration in real-time is a problem. Work is still underway to come up with ways of making real-time calibration without affecting the sensor functionality.

6.1.5 Modern technologies related to calibration

Coping with inaccuracy is one of the emerging problems related to sensors, and new improvements in calibration technologies are trying to solve some of them. These include:

Self-calibration techniques: Some sensors contain a self-biasing circuit to adjust the output in proportion to the internal measurements of the sensor. It helps to avoid frequent setting by an operator, achieving and maintaining stability of the calibrator.

Adaptive calibration algorithms: Increased accuracy of calibration is achieved by using complex algorithms that are capable of learning from changes and individual differences. Adaptive calibration methods are in the field of AI and therefore machine learning approaches are being investigated.

Integrated reference systems: Depending on whether the sensors have incorporated reference systems of built-in calibration standards, the use of validations requires less dependence on other calibration methods.

To guarantee the highest level of sensor performance and solve calibration problems, it is important to consider the basic requirements for the creation and application of biomedical sensors. For reducing and controlling the sources of errors affecting the accuracy of biomedical sensors, key technologies and sound calibration methods make it achievable to obtain higher reliability and performance of the sensors, which help to improve healthcare management and the patients' condition.

6.2 Power consumption and energy harvesting solutions

Energy intake and utilization are probably the major constraints in the creation of biomedical sensors, especially in monitoring systems where power supply is not perennial. Control of power usage and the integration of sustainable energy measures are desirable for the operation, durability, and feasibility of these devices.

6.2.1 Power consumption challenges

Biomedical sensors typically function in areas where power conservation is an issue. Stepped-up energy usage also shortens sensor life, a major issue for wearables and implantables, where frequent battery replacements can be disruptive. Power consumption issues are further exacerbated, especially because of the need to develop very elaborate sensor and data processing systems. One of the major challenges that arose in the development of these sensors was the need to properly distribute the power needs of the components with the need for processing accuracy and the ability to generate real-time monitoring information.

6.2.2 Low-power design techniques

Regarding power consumption problems, designers apply a number of low-power strategies. Others are selecting sensors that consume low power on their circuit boards, integrating low-power microcontrollers and signal processing units, and incorporating power-efficient communication schemes. Further, sleep and wake modes are usually adopted to cut down energy power during inactive operations. Because of the proper control of these aspects, it has become feasible to increase the usable time of biomedical sensors and decrease the frequency of battery recharge.

Energy-harvesting technologies came into existence when researchers started looking at materials and systems that can convert ambient energy into usable electrical energy.

Energy scavenging solutions present unique prospects for addressing power consumption difficulties through AMB by converting environmental energy into electrical power. Various energy-harvesting methods are being explored for biomedical sensors, including:

Photovoltaic energy harvesting: Solar cells can turn light falling on them into electricity. This method is especially valuable when applied to wearable sensors that are exposed to light conditions. However, it is not efficient when used indoors with low lighting.

Thermoelectric energy harvesting: Thermoelectric generators use variation in temperature between the body and its surroundings to produce electrical power. It is applicable to body-worn sensors in conditions where temperature gradients are usable.

Piezoelectric energy harvesting: Piezoelectric materials produce electric current in the form of voltage when mechanically deformed systems vibrate. This method can be used in the wearable sensors that are affected by motion-like sensors used in physical activity tracking.

Electromagnetic energy harvesting: This technique involves receiving electromagnetic waves with the surrounding area like radiofrequency RF signals. It can be used where there are RF signals present so that these signals can be translated into usable power for the sensors in use.

6.2.3 Energy harvesting with integration of sensors

Design of the sensing device introduces complications for energy-harvesting technologies as both must be considered, in relation to energy needed. Energy conversion and storage are fundamental for the harvested energy to be capable of providing sufficient power to support the functioning of the sensor. It is also necessary to consider the dimensions as well as the form of the energy harvesting components in order to fit them within the form factor of the sensor device without a negative impact on the form factor of the device.

6.2.4 New battery technologies

Another solution is energy harvesting and there is no doubt that battery technologies are integral in solving the power consumption challenges. Improvements in battery chemistry and the consequent design, for instance, flexible and high-capacity batteries, improve the efficiency and durability of biomedical sensors. New developments with micro batteries and thin-film batteries present advantages to miniature and implantable sensors due to the ability to store greater amounts of energy within the same amount of space.

6.2.5 Hybrid power solutions

Intelligent battery power systems, where energy harvesting is integrated with the existing battery types, is a realistic solution to power intake problems. Since both energy harvesting and rechargeable batteries are incorporated in sensors, the harvested energy can be used to prolong the battery lifetime and minimize its replacement. Hybrid systems also allow for the systems to run for long without requiring a source of power, other than when conditions for energy harvesting are inappropriate.

6.2.6 Annotated bibliography of future directions and research

One can only expect that on-going work on the improvement of power consumption and energy harvesting systems continue to deliver better solutions to biomedical sensors. New work would include better materials for energy harvesting, better algorithms to use the harvested energy, and new ways of using energy harvesting along with sensors. These developments are anticipated to power the next generation of biomedical sensors, making it possible for far better longitudinally and dependable assessment of health.

Energy dissipation and the ability to design and integrate efficient methods for energy harvesting are critical to the development of biomedical sensors. Through a system-level approach that targets low-power design and by consideration of various energy-harvesting technologies as well as enhancement in battery systems, it is possible to design and implement practical sensors for long-term health monitoring.

6.3 Data privacy and security in sensor networks

Privacy and security are major concerns in the design of biomedical sensor networks, as these systems continuously collect, transmit, and analyze highly sensitive patient data. Safeguarding this information ensuring confidentiality, integrity, and adherence to patients' rights of access is critical, as any breach may result in serious privacy violations. As illustrated in Figure 11, this section examines the key challenges, emerging opportunities, and strategic approaches for ensuring data privacy and security in biomedical sensor networks.

6.3.1 Protecting health information that can be identified

Biomedical sensors produce and transmit identified health information, lifestyle records and disease histories, biometrics, and immediately observed health conditions. It is important to maintain anonymity of this data to uphold patient rights and, specifically, their right to privacy. Security measures are used to protect information at rest and in transit so that anyone who is not supposed to see the content should not be able to gain access to it. End-user encryption methods and progressive algorithms as well as key management security measures must be implemented at each stage of the data life cycle.

6.3.2 Protection of the communication media

An issue is the ability to introduce new physical layers for the transmission of data between the biomedical sensors and the healthcare systems, which are inherently prone to interception and unauthorized access. Booking these channels requires effective communication measures and policies, which are critical in guaranteeing adequate security measures for such a channel. Security measures such as the Transport Layer Security (TLS) and Secure Socket Layer (SSL) are used to ensure that the data transmitted across the networks is encrypted to prevent third parties from accessing the data. Further, authorization controls like two-factor authentication, there is MFA of users, and the use of digital certificates ensure that only specifically permitted devices access the data.

6.3.3 Overcoming some of the concerns over privacy with wearable devices

Integration of sensors as wearable biomedical devices poses major data privacy concerns since they are designed to interminably assess and emit health information. The collection of personal health information on a continual basis also leads to questions

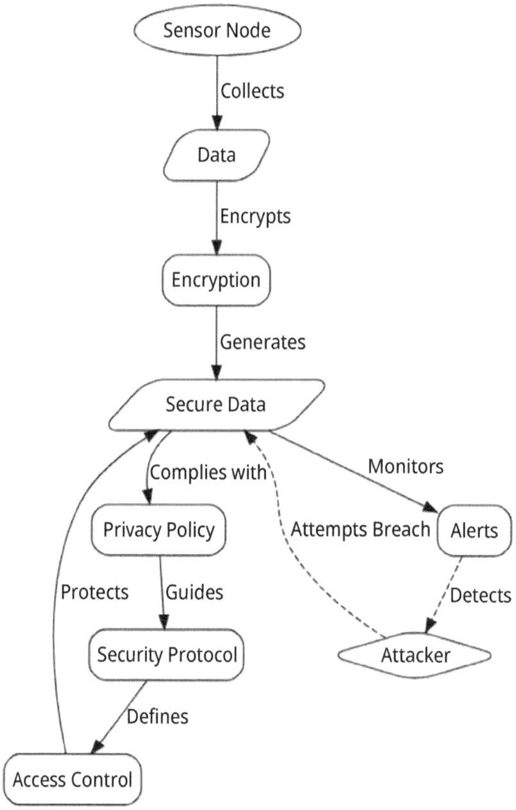

Figure 11: Data privacy and security in sensor networks.

being asked on the use and disclosure of such information. These issues can only be resolved with the right privacy policies that state and policies that explain to the user how their information is used. In addition, they can prevent overly sensitive personal data information from being shared, while at the same time maintaining invaluable health data.

6.3.4 Maintaining the directory of information files

Such sensor networks in biomedical applications require storing huge amounts of health information in central architectures. The use of password control and other security measures depends on security measures and the advantages of data security. Storing of data should require limited forms of access, particularly to personnel, through the use of role-based access control and the need to analyze the logs of access frequently. Also, data should always be encrypted when stored and when transmitted, to minimize the risks of theft and breach.

6.3.5 Acceptance of compliance with regulatory standards

It is envisioned that adherence to regulatory paradigms within the biomedical sensor networks must be complied to ensure that data privacy and security are maintained. HIPAA in USA and the GDPR regulation in EU countries define exacting standards for confidentiality and security of health information. Satisfying these regulations encompasses exercising measures to protect personally identifiable information, conducting security reviews periodically, and ensuring that the dealing with data complies with legal as well as ethical procedures.

6.3.6 Security threats and responding to vulnerabilities

Biomedical sensor networks are not immune to diverse forms of threats such as cyber-attacks, loss of data, and malicious software. Constant risk and vulnerability scans must be accomplished to identify and eliminate possible security threats. EDS involve the use of software known as intrusion detection systems by having frequent security patches and updates in place, which can prevent threats. Further, it is possible to improve the protection of the sensor networks with reliable security characteristics that include secure boot mechanisms, hardware security modules, and others.

6.3.7 Security versus ease of use

Their wants: one of the toughest obstacles in realizing biomedical sensors is mainly the concern of data security versus operational convenience. This paper establishes that acquiring and implementing high levels of protection degrades the efficiency and utility of sensory systems. There is therefore the need to develop security solutions that, in one way, do not inhibit the performance of biomedical sensors. Smooth externally applied authentication methods like facial recognition, password, and replacing it with something like passkey can bring about a positive change in security, while not compromising too much on convenience.

Protection and security of data are major issues that are central to the integration and functioning of biomedical sensor networks. These challenges can only be met through a multifaceted solution, which involves practicing rigorous encryption, controlling data transfer and storage, sustaining with the regulatory principles, and achieving a tightly knit security. When these factors are addressed, the developers of SMART technologies, healthcare practitioners, and patients can be confident that personally identifiable health information of patients is kept secure, and the promise of biomedical sensor innovations in patient care remains intact.

6.4 Wearability and patient compliance

Comfort and user compliance are two of the biggest issues concerning the design of biomedical sensors, more so if these devices are to be used for long-time monitoring across various population groups. It is thus critical to overcome these challenges to enable biomedical sensors play the right role in health monitoring and management.

6.4.1 Environmental comfort and design implications

This is why biomedical sensors that have a great design ensure that they do not cause any discomfort to the patients who use them. Measuring devices that cause discomfort, also result in low compliance from the patient and thus poor measurements. The improved flexibility in the design of wearable sensors is attributable to emerging features in material science such as the possibility to incorporate flexible and lightweight wearable material with enhanced breathability. New breakthroughs like stretchable electronics and skin-like adhesives guarantee that the sensors that wrap around the body's curves are nonirritant. Ergonomic designs that also do not exert much bulk and weight on the user equally enhance the comfort.

6.4.2 Durability and longevity

The above-discussed key performance indicators indicate that high measurements of durability and longevity are vital and more preferable in biomedical sensors, particularly in the long-term monitoring endurance of the patient. Unfortunately, wearables and sensors must be ready to face the day-to-day use and can be subject to moisture and sweat abrasion. These prognosticators become more durable as a result of the progress of material sciences and protective coats used in these devices, hence they do not need frequent replacements. For instance, coatings such as water protection and sweat proofing of sensors minimize the likelihood of their failure under some circumstances.

6.4.3 Ease of use and integration

The do-culture and the ability to wear biomedical sensors seamlessly in day-to-day life radically affect compliance. Sensors should be easy to use, with little training needed to work with them. Clear and uncluttered designs, as well as clear control elements, can make using the developed apps easy-to-use and thus enhance patient compliance with existing health management practices. For example, wearable sensors that sync well with the user's smartphone electronic health record allow them and their doctors to track health information easily. Their less emphasis of involving complex set up gives better user experience.

6.4.4 Beauty and customization

On patients' acceptance and compliance, the looks of products matter. Sensors that patients find unattractive are not likely to be worn regularly. Acceptance of sensors can be improved by the advancement of sensors that are more complementary to the clothing. Key elements for self-customization, including designs and colors, enables patients to choose devices that suit them best in relation their preferences, thus enhancing wearability.

6.4.5 Effects of psychological and patient involvement

The discussed psychological effect of biomedical sensor affects the compliance of the patient. This is mainly because there are devices that are designed to measure health indicators over some period continually. Constant evaluation can cause anxiety in the patient, especially if the patient feels he/she is at risk of developing some health condition. To this end, the promotion of the virtues of constant surveillance and the assistive nature of sensors in healthcare is likely to increase patient activation. Information concerning how the data from the sensors is handled, and the extent to which patients feel in control of their health would remove these concerns and improve compliance.

6.4.6 Barriers to wearability

Possible factors that limit the wearability of biomedical sensors include social economic status, cultural endowment, and health literacy. Some types of sensors are costly, making them difficult to acquire, and may, therefore, hamper proper usage and usability. The use of cost-efficient solutions that are also easily paid for through insurance is also core to the affordability objectives to make their use wide spread. Also, culturally appropriate designs of products and services as well as educational programmers can help patients overcome health literacy learnt barriers to use.

6.4.7 Data privacy and security

Patients are reluctant to wear these devices due to their discomfort with an organization's ability to track their identification, violating their data privacy and security. Privacy is another important aspect that forms the design of communication of sensor data and the abilities of patients to control their information. It is possible to overcome concerns and motivate patients to wear sensors more frequently by employing reliable encryptions, defining clear policy on data handling and management, and making privacy mechanisms easily understandable.

First, on the issue of wearability and patient compliance, it is axiomatic to the design and deployment of biomedical sensors. Understanding the comfort factor, product durability, easy usability, attractiveness, psychological effects, as well as breaking the barriers help developers to enhance patient compliance, the usage of which optimizes the efficiency of continuous health monitoring systems.

6.5 Scalability and cost-effectiveness of sensor technologies

Scalability and cost are clearly pivotal factors influencing the adoption and widespread implementation of biomedical sensors. These considerations directly impact the availability of sensor technologies across diverse healthcare settings. As shown in Figure 12, addressing these challenges is essential for advancing sensor development and ensuring broader accessibility in both clinical and remote healthcare applications.

6.5.1 Manufacturing scalability

Manufacturing scalability involves the dimension of the extent to which biomedical sensors can be produced, while maintaining the quality of the product. Since scalability is a key factor, sensor designs must be made suitable for large-scale production in terms of affordable materials and processes. It is found that strategies, including microfabrication and using reusable components, are employed for mass production. Roll-to-roll processing and automatic assembly for example, are manufacturing methods that can increase the quality and lower the cost of production. Though, attaining scalability introduces issues associated with standardization and stability in large production lots.

6.5.2 Materials used and availability

For biomedical sensors, issues relating to the type and price of materials affect the cost of manufacturing. This is particularly true with specific types of high-performance sensors that need unique materials, including developed polymers, nano-materials, and biocompatibility materials, which are costly and hard to come by. Exploring improved paste materials with performance characteristics at an equivalent level, but at a cheaper price, is one of the major working directions. In the same way, the proper management of the use of materials during manufacturing help in minimizing wastes, thus helping in the control of costs. Another criterion is the requirement for a stable procurement of key materials so that no disruptions in supply to production occur.

6.5.3 Economies of scale

Thus, specialization has great importance, where costs per unit of biomedical sensors products also decline with increase in volumes, as indicated by economies of scale. Manufacturing is about economies of scale because fixed costs are apportioned over a larger number of products, thus enabling the lowering of costs per unit. To take advantage of scale economies, there is a need to focus on setting up efficient production facilities, high level of automation, and flow of supplies. Extended cooperation with other industries and making use of existing distribution platforms also help to attain cost efficiencies as well as general cost reduction in the sensor technologies.

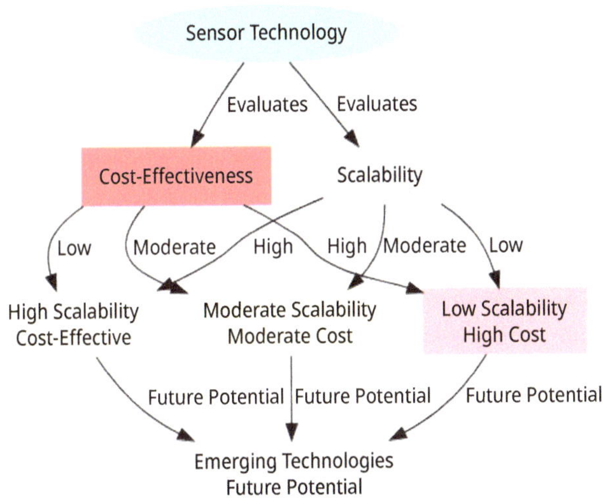

Figure 12: Scalability and cost-effectiveness of sensor technologies.

6.5.4 Areas of integration and system-level cost aspects

Biomedical sensors are neither solely evaluated in terms of cost nor in terms of an individual sensor's lifespan, but rather on the sensors' efficiency and cost as part of a greater system within the healthcare industry. This means that the total cost related to a system involves issues to do with data handling, interfacing with other systems, and maintenance of the system. For instance, sensors that need further data processing tools to analyze the collected information cost extra amounts of money apart from the initial cost. When the designed sensors include signal processing in their functionality with existing healthcare systems, then system's costs lowered. Also, the development of sensors that do not require maintenance/calibrations frequently make a long-term positive impact on cost.

6.5.5 Market presentation and price-setting directions

Market demand is an essential determinant of the price strategies of biomedical sensors. The peculiarities of targeting and the competitive struggle in the highly demanding markets call for pricing strategies that contribute to production with acceptable at price levels and at the same time with high profits. One way to make higher prices acceptable is to create novel sensors with capabilities that the existing models do not satisfy. On the other hand, for sensors that are meant to be used in high volumes for different industries markets, cost-containment methodologies need to be put in place. There are very many pricing strategies that manufacturers of sensors can adopt so as to lower the prices, yet make very attractive profits; pricing models such as tiered pricing models can be used to meet this end.

Due to the risk associated with operating within the above legal environments, companies face regulatory and compliance costs.

It is seen that compliance and regulatory demand also form another supplementary layer of cost to the biomedical sensors. Some of the areas that elicit testing, certification, and documentation to ensure that the sensors have met the safety and performance requirements of different overarching regulatory bodies include: These requirements can enhance relatively high development and production costs. Possible measures that can be taken to deal with the issue include reconciling the time taken to seek regulatory approval and cutting costs by setting standard processes testing. Also, early involvement of regulatory agencies and compliance to regulatory requirements in design and documentation reduce chances of hold up.

This is the case, given that scalability and cost remain major challenges that need to be overcome in this field in order to extend the use of biomedical sensors, especially to the larger population. By using such approaches to improve manufacturing methods, minimize the consumption of consumable materials, and enhance system integration, in association with carefully managed market demand and regulatory costs, there is a major opportunity to advance the application of cost-effective and scalable sensor technologies to promote innovation in healthcare.

7 Conclusion

The implantation of new types of biomedical sensors in healthcare systems is a significant progress in the treatment of chronic diseases and in the possibility of effective health control. This chapter discussed the nature of biomedical sensors for artificial intelligence purposes, the process of creating such sensors, as well as the trends that the progress of such sensors follows. In considering the nature and implementation of several forms of sensors, including the optical, chemical and mechanical sensors, the chapter has shed light on how these new technologies are likely to revolutionize and enhance the capability of reporting current and accurate reports on the state of health.

Optical sensors that capture key biomarker values nondestructively and chemical sensors that provide high selectivity to the corresponding substance are essential parts of the current health monitoring devices. Mechanical sensors help to gather data about the physical activity and physiological transformations, which also brings producers more prospective data for the evaluation of health. Integration of these sensor technologies improves the holistic and accurate way of monitoring the person's health, making the multi and hybrid systems to capture the real health status of the person effectively.

In this chapter, sensors are recommended in long-term monitoring solutions, especially when caring for geriatric patients. As is evident, the elderly are usually dealing with more than one chronic disease and thus prefer continuous and even real-time, monitoring. Wearable devices, implantable sensors, and remote sensors have been described as essential interventions that are capable of providing optimal chronic disease care and improving the standard of elderly patients' lives. These solutions help experts identify the disease in its initial stage and begin the treatment process immediately plus they assist in designing individual treatment plans.

Scalability and cost considerations continue to be particularly important regarding the more general adoption of biomedical sensors. Sustainability in the creation of sensors is critical due to the requirement for mass production at an affordable price point to achieve efficiency for a widely applicable range of contexts in the healthcare industry. This chapter has explored ways that firms can seek for scale efficiencies and improve production, material consumption, and system interfaces, ultimately decreasing costs. These are factors that are crucial in ensuring that the newer technologies in sensors are put to greater use by more people, ensuring improved use of sensor technologies in the healthcare system.

References

[1] Andreu-Perez, J., Leff, D. R., Ip, H. M., & Yang, G. Z. (2015). From wearable sensors to smart implants-toward pervasive and personalized healthcare. IEEE Transactions on Biomedical Engineering, 62(12), 2750–2762.

[2] De Fazio, R., De Vittorio, M., & Visconti, P. (2021). Innovative IoT solutions and wearable sensing systems for monitoring human biophysical parameters: A review. Electronics, 10(14), 1660.

[3] Soh, P. J., Vandenbosch, G. A., Mercuri, M., & Schreurs, D. M. P. (2015). Wearable wireless health monitoring: Current developments, challenges, and future trends. IEEE Microwave Magazine, 16(4), 55–70.

[4] Kazanskiy, N. L., Khonina, S. N., & Butt, M. A. (2024). A review on flexible wearables-Recent developments in non-invasive continuous health monitoring. Sensors and Actuators A: Physical, 114993.

[5] Pantelopoulos, A., & Bourbakis, N. G. (2009). A survey on wearable sensor-based systems for health monitoring and prognosis. IEEE Transactions on Systems, Man, and Cybernetics, Part C (Applications and Reviews), 40(1), 1–12.

[6] Gupta, P., & Pandey, M. K. Role of AI for smart health diagnosis and treatment. In Smart medical imaging for diagnosis and treatment planning, (pp. 23–45). Chapman and Hall/CRC.

[7] Prabhakar, P. K. (Ed.). (2024). Biomedical research developments for improved healthcare. IGI Global.

[8] Alshamrani, M. (2022). IoT and artificial intelligence implementations for remote healthcare monitoring systems: A survey. Journal of King Saud University-Computer and Information Sciences, 34(8), 4687–4701.

[9] Nasiri, S., & Khosravani, M. R. (2020). Progress and challenges in fabrication of wearable sensors for health monitoring. Sensors and Actuators A: Physical, 312(112105).

[10] Sempionatto, J. R., Lasalde-Ramírez, J. A., Mahato, K., Wang, J., & Gao, W. (2022). Wearable chemical sensors for biomarker discovery in the omics era. Nature Reviews Chemistry, 6(12), 899–915.

[11] Cammann, K., Lemke, U., Rohen, A., Sander, J., Wilken, H., & Winter, B. (1991). Chemical sensors and biosensors – Principles and applications. Angewandte Chemie International Edition in English, 30(5), 516–539.

[12] D'Orazio, P., & Meyerhoff, M. E. (2014). Electrochemistry and chemical sensors, (pp. 151). St. Louis: Elsevier.

[13] Homayounfar, S. Z., & Andrew, T. L. (2020). Wearable sensors for monitoring human motion: A review on mechanisms, materials, and challenges. Slas Technology: Translating Life Sciences Innovation, 25(1), 9–24.

[14] Wilson, J. S. (2004). Sensor technology handbook. Elsevier.

[15] Tang, Q., Liang, J., & Zhu, F. (2023). A comparative review on multi-modal sensors fusion based on deep learning. Signal Processing, 109165.

[16] Muhammad, G., Alshehri, F., Karray, F., El Saddik, A., Alsulaiman, M., & Falk, T. H. (2021). A comprehensive survey on multimodal medical signals fusion for smart healthcare systems. Information Fusion, 76, 355–375.

[17] Ates, H. C., Nguyen, P. Q., Gonzalez-Macia, L., Morales-Narváez, E., Güder, F., Collins, J. J., & Dincer, C. (2022). End-to-end design of wearable sensors. Nature Reviews Materials, 7(11), 887–907.

[18] Young, H. M. (2003). Challenges and solutions for care of frail older adults. Online Journal of Issues in Nursing, 8(2).

[19] Kim, B. Y., & Lee, J. (2017). Smart devices for older adults managing chronic disease: A scoping review. JMIR mHealth and uHealth, 5(5), e7141.

[20] Majumder, S., Mondal, T., & Deen, M. J. (2017). Wearable sensors for remote health monitoring. Sensors, 17(1), 130.

[21] Raparthi, M. (2022). Quantum Sensing Technologies for Biomedical Applications: Investigating the Advancements and Challenges. Journal of Computational Intelligence and Robotics, 2(1), 21–32.

[22] Pramanik, P. K. D., Upadhyaya, B. K., Pal, S., & Pal, T. (2019). Internet of things, smart sensors, and pervasive systems: Enabling connected and pervasive healthcare. In Healthcare data analytics and management, (pp. 1–58). Academic Press.

[23] Prabhu, V., Kumara, S., & Kamath, M. (Eds.). (2003). Scalable enterprise systems: An introduction to recent advances, (vol. 3). Springer Science & Business Media.

[24] Esty, D. C., & Porter, M. E. (1998). Industrial ecology and competitiveness: Strategic implications for the firm. Journal of Industrial Ecology, 2(1), 35–43.

[25] Pramanik, P. K. D., Upadhyaya, B. K., Pal, S., & Pal, T. (2019). Internet of things, smart sensors, and pervasive systems: Enabling connected and pervasive healthcare. In Healthcare data analytics and management, (pp. 1–58). Academic Press.

B. Yamini, R.M. Dilip Charaan, Janani G., S. Sree Dharinya, G. Parthasarathy

Machine learning-driven personalized medicine: customized drug delivery systems and patient-specific material applications

Abstract: Due to the modern developments in genomic studies and technologies in computing, the concept of a new form of treatment called the personalized medicine has emerged, which can be translated as one's unique treatment plan considering genetic makeup. This chapter of the book focuses on the principles of profile-specific pharmacotherapy and the modern technologies it is based on to describe the powerful potential of personalized therapy in preventive healthcare and early diagnosis. Integrating machine learning techniques with the field of bioinformatics simplifies the analysis of what used to be complex biological data, which aids in the recognition of such genetic biomarkers and the development of smart medication delivery systems. Nonetheless, there are challenges such as access to resources inequality, very expensive, and moral issues; these need to be addressed to ensure equal use within different groups. This chapter also examines how personalized medicine complements conventional care in order to clear barriers and enhance opportunities for tailored care. It speaks strongly for intercultural genetic research and context-sensitive teaching. This chapter aims to help the readers gain a clearer understanding of how the concept of personalized medicine can revolutionize healthcare systems and improve the quality of patient lives worldwide, through analysis.

Keywords: personalized medicine, genomic research, bioinformatics, preventive healthcare, individualized treatment, equitable access

B. Yamini, Department of Networking and Communications, School of Computing, College of Engineering and Technology, SRM Institute of Science and Technology (SRMIST), SRM Nagar, Kattankulathur 603203, Chengalpattu District, Tamil Nadu, India, e-mail: yamini.subagani@gmail.com
R.M. Dilip Charaan, Department of Computer Science and Engineering, Vel Tech Rangarajan Dr. Sagunthala R&D Institute of Science and Technology, Chennai 600062, Tamil Nadu, India, e-mail: dilip_charraan@yahoo.co.in
Janani G., Department of Information Technology, RMD Engineering College, Chennai, Tamil Nadu, India, e-mail: janani.it@rmd.ac.in
S. Sree Dharinya, School of Computer Science Engineering and Information Systems, VIT University, Vellore, Tamil Nadu, India, e-mail: dharinyasathish@gmail.com
G. Parthasarathy, School of Computing and Information Technology, REVA University, Bangalore, Karnataka, India, e-mail: parthasarathy.g@reva.edu.in

https://doi.org/10.1515/9783111503202-005

1 Introduction

The concept of personalized medicine is arguably the next generation of medical practice based on recent studies in genomics and computational technologies [1]. Personalized medicine is focused on making adjustments in the course of managing patients' medical conditions on the basis of patients' unique genetic makeup, environment, and other aspects of their lives as opposed to conventional interventions that many a times rely on a protocol that has been tried and tested [2, 3]. This shift towards customized therapy is supposed to improve the efficacy of the treatments and health outcomes, and decrease the incidence of adverse drug reactions [4]. Personalized medicine aims at enhancing tailored therapeutic intervention options through integrating genetic data with clinical practice [5, 6].

Computer tools and bioinformatics that have revolutionized the analysis of genetic data are the basis of personalized medicine [7]. The upsurge in large-scale sequencing data is realized by high-throughput sequencing technology and, as such, requires sophisticated approaches for data analysis [8]. Actual treatment plans, those employing variations of the gene to medication reaction in addition to the disease predisposition, are established by the help of bioinformatics [9]. Several therapeutic approaches can be personalized, as well as can predict the patient's reactions to the medications by using machinery learning techniques [10].

It appears that individualized medicine has helped shape early diagnosis and prevention of diseases [11]. Healthcare professionals can find patients who are predisposed genetically and environmentally to certain diseases long before they exhibit symptoms [12]. Biomarkers such as genomic screening and diseases like cancer and cardiovascular problems become detected at an early and manageable stage, after which prompt treatment, which has a very significant chance of halting the progress of the disease, is administered [13]. This proactive strategy of healthcare focuses on early identification and unique appropriate control measures as compared to the harm control processes [14].

The concept of personalized medicine has been claimed to be challenging to be implemented in the following ways [15]. Lack of equal access to individualized healthcare remains an issue due to social, spatial, and systemic differences in obtaining individually targeted solutions through access to high technologies in diagnostics and treatment [16]. Secondly, with the current genomic research databases mainly characterized by White European origin, precision medicine application appears restricted to only those subpopulations [17, 18]. To make sure that the benefits of customized treatment are as widespread as possible, it is critical to address these imbalances [19].

Personalized medicine as a concept speaks of the possibility and the challenges as the conventional medical model is enriched by incorporation of this concept [20]. The operational precision and effectiveness of individualized approaches forms the base and requires the use of traditional management plans for the best results [21, 22]. The difficulty here is to ensure the use of both individual and systemic approaches that would be integrated and efficient [23, 24]. Research studies show that if these integration challenges are solved and the research and technologies of personalized medicine are further developed, there are current possibilities to revolutionize healthcare and to strongly enhance the therapy outcomes [25].

2 Historical background and evolution of personalized medicine

2.1 Origins of personalized medicine

The concept of pharmacogenomics that brought some ideas of individualized medicine has early roots in the notion of splitting medicine usage according to an individual's genetic data. The field of pharmacogenomics, which emerged in the middle of the twentieth century aimed to understand how genes explain how drugs and other agents operate in the body. This means that by addressing the observation in patients that pharmacological response both in terms of effectiveness and safety varies, this strategy intends to show the way for more precise and effective treatment.

The major scientific advancement that set the ball rolling on the concept of personalized medicine was the discovery of the double helix DNA by Watson and Crick in 1953. The advancement means the scientific world now has a clearer idea of the means by which the genetic code holds sway over health and illness. The FAIR act was a basis for the Human Genome Project, which began in 1990 and came to completion in 2003. As a project that strived to map the whole human genome the initiative sought to identify and sequence the 20,000 to 25,000 genes that define man's biology. The extreme success of the Human Genome Project gave a new plot to genetics and created many chances for individualized therapy.

Genomics initiated in the early 2000s lead to identification of several genetic variations associated with different diseases and patient responses to treatment. The knowledge of specific indicators that determine the rate of a given person's susceptibility to specific diseases and reaction to various medications also saw major developments during this period. P4 Medicine adopted a principle that was popularized by pharmacogenomics – right drug, right dose, right patient.

Over this period, pharmacogenomics evolved to embrace and accommodate other influences that have an impact on human health in addition to genetic influences such as lifestyle and environmental factors. Management and implementation of these variables to tactics in the strategy for individualized treatment marked a shift to better understanding of patient-centered care. It also revealed the fact that the dependence of various diseases on heredity factors should not rule out the other necessary environmental factors.

In the perspective of biomarker validation, personalized medicine has progressed because of high-throughput technologies such as bioinformatics and next generation sequencing. These technologies therefore make it cheaper and faster to analyze large genomic data sets, and so open up the possibility of identifying new biomarkers and potential therapeutic targets. As a result, the wing of individualized medicine analysis is much broader than that of pharmacogenomics, expanding to how individualized medications are developed, how preventative techniques are fit to an individual, and checking for disease susceptibilities.

Indeed, cancer care has been the forerunner of thoughtful, intense intentions toward hopeful therapeutics since genetic discoveries were effectively applied into tendency of oncology. Molecular alterations that are directed toward the cancers being treated are also a good example of personalized medicine in cancer therapy, for instance, trastuzumab in HER2-positive breast cancer and imatinib in chronic myeloid leukemia.

2.2 Milestones in the development of personalized healthcare

Personalized medicine has progressed through several evolutionary milestones relevant to the given field, which has made great leaps. These achievements describe the development from fundamental conceptual models to advanced statistical models to considerate customized medical treatment management.

Pharmacogenomics: historical perspective and development (1950–1960)

The idea of pharmacogenomics started to emerge as soon as scientists started to find the genetic differences affecting metabolite pathways of some drugs. Some of the earliest pieces of work clearly identified differences in the activity of enzymes, which can cause variations in drug metabolism among people. These years provided the basis for comprehending the issues connected with the genetic effects of drug effectiveness and side effects.

The Human Genome Project announced in February 1988 was formally launched in October 1990.

An initiative with scientific approach from all over the world known as Human Genome Project (HGP) focused on mapping the whole human genome. The astonishing feat completed in 2003 furnished anatomic details, which are part of personalized treatment, for instance, detailed genetic blueprint. It helped in increasing appreciation of inheritance and tailor-made solutions by offering an understanding of the varieties of genetic makeup the human body has concerning various diseases.

Advancements of technology in genomic sequencing in the 2000s

Next-generation sequencing, the advancements of which formed the basis for the achievements of the HGP, began in the early 2000. Technological advancements such as next-generation sequencing (NGS) enhanced genomic sequencing in terms of cost and time thus embracing its use in clinical diagnosis. This development led to a criterion that allows genetic data to be used in the customized medicine system.

The latter was inaugurated by the arrival of targeted therapies in the course of the decade 2000–2010.

One of the key shifts in customized medicine was the development of targeted medicines. Imatinib for treating CML and trastuzumab for treating HER2 + BC were among examples of genetic breakthroughs that were used in creating drugs, which acted specifically once certain biological markers were identified. These treatments demonstrated more about steps that are used to plan and execute interventions that improve patient outcomes more.

Big Data integration and AI (2010–present)

There has been a shift in the customized medicine subsector through the integration of AI and big data analysis. Genetic, clinical, and environmental data are gathered for a large number of patients and then preprocessed to look for tendencies and predict the responses by means of large and complex formulas and machine learning. Therapeutic and diagnostic treatments get enhanced by the use of AI solutions and help in designing individual therapy plans.

Personalized medicine initiative development and partnerships (2010s onwards)

Several efforts and partnerships have been created to advocate the adoption and innovation of personal medicine. The most recent example was the Precision Medicines Initiative in the United States and similar initiatives in other countries. These programs aim at facilitating collaboration between academics and/or physicians on the one hand and patients on the other, and fostering research and exchange of data in the domain of personalized medicine with the vision of enhancing the latter's implementation in clinical work.

Individualized therapy in mainstream ambulatory care setting: current and future

The latest achievement is the further entrenchment of personalized medicine into a myriad of typical clinical operations. Genome sequencing and Big Data utilization in medicine and precision medicine are making their way into routine practice more and more often. These trends have shifted towards more personalized care designated as an effort to enhance patients' outcomes and interventions and prevention processes.

2.3 Evolution of treatment approaches

One of the major revolutions in the flow of treatment strategies in medicine is a shift from the generalized approach to treatment to the individualized approach to treatment: a shift that personifies future-focused and precision medicine. An analysis of the evolution of treatment approaches suggests that while, initially, strategies are very global, addressing the context generally, specialist approaches have been developed to consider different features of different patients.

It is important to appreciate the ability of the theory to impose generalized treatments in hierarchical structures.

At first, there were very few individual and varied medical cures with most methods being rather crude and pragmatic, not indicative of the actual condition of the patient. The usage of these treatments presumed that the patients of different categories of diversity could be treated uniformly. Their medication included broad-spectrum antibiotics and standardized drug treatments that target multiple conditions while not taking into account the patient's genetic makeup to metabolize drugs.

After understanding the cause of those related diseases, began the discovery of what are today known as targeted therapies.

The mid-twentieth century marked a pivotal shift toward more precise and personalized treatments, largely driven by advances in medical research and technology. Increased understanding of specific disease mechanisms and the identification of molecular targets enabled the development of targeted drug delivery systems. As illustrated in Figure 1, this era saw the emergence of therapies directed at particular biological pathways, allowing for more tailored treatment approaches. Notably, the advent of targeted cancer therapies aimed at defined genetic markers in tumors represented a significant advancement over traditional, non-specific chemotherapy.

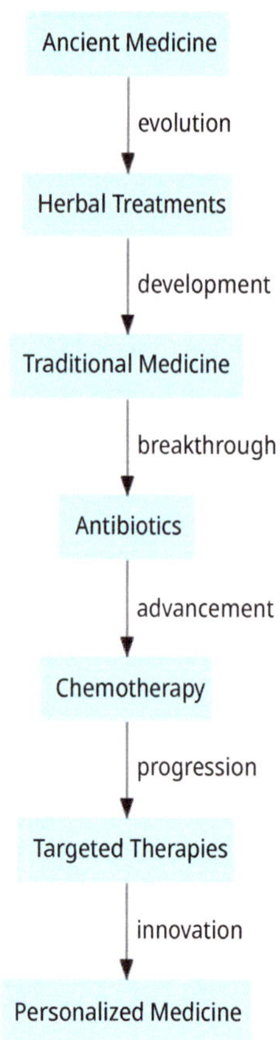

Figure 1: Evolution of treatment approaches.

2.3.1 Multiplatform and personalized medicine

Advancements in the past few years have also made it possible to combine various data that are multidimensional including proteomics, metabolomics, and genomes in improving treatment methodologies. These broad databases are applied by precision medicine to build up intensely individualistic treatment therapies that involve the complexity of biological systems. Several omics layers can be integrated, and the healthcare practitioners are capable of enhancing the accuracy of the models involving illness development, as well as therapy response.

2.3.2 Evolution of possible ways of treatment

It was expected that with advances in ML and AI, future therapeutic methods would become more personalized and data-oriented. Because these technologies can analyze vast amounts of data and discover new patterns and correlations, they have the potential to revolutionize customized medicine. The dynamic adjustment of therapies based on patient data could be enhanced by integrating AI and ML with real-time wearable devices. This evolution marks a transformation of healthcare from a reactive system to one that continuously adapts to meet patient needs.

2.4 Key figures and contributions in the history of personalized medicine

Several key pioneers have made significant contributions to the advancement of personalized medicine through their innovative research and strategic approaches. As illustrated in Figure 2, these individuals laid the foundational groundwork for today's trends in individualized and personalized medicine, helping to transform the field into a practical and accessible clinical discipline.

2.4.1 Sir Francis Crick and James D. Watson

The Nobel Prize was awarded to Sir Francis Crick and James D. Watson for their discovery of the DNA double helix structure in 1953. This breakthrough provided critical insights into the molecular structure of genes and heredity, forming the foundation for an individualized approach to medical treatment. What they found helped explain how genetic information is stored and transmitted from one generation to the next, paving the way for the development of genomics and personalized medicine.

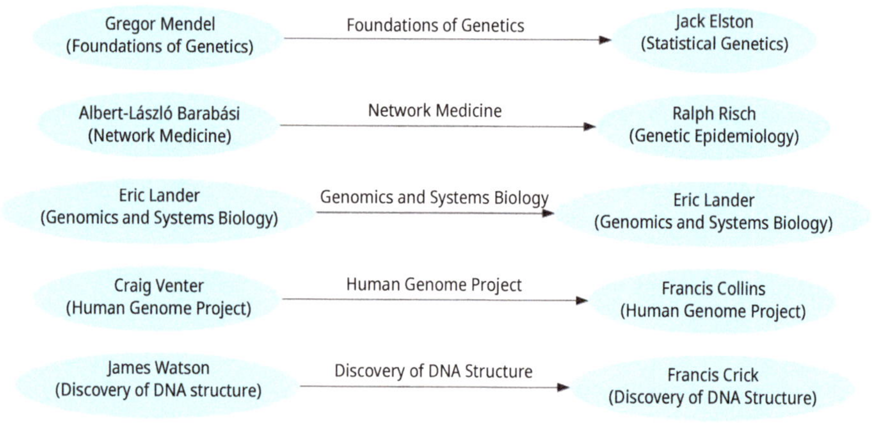

Figure 2: Key figures and contributions to the history of personalized medicine.

2.4.2 Linus Pauling

Molecular biology and genetics were enriched by two-time Nobel price winner Linus Pauling. Through his work about nature of chemical bonds and structure of proteins, particularly the investigations of sickle cell anemia, he explained how mutations caused diseases. Pauling's work has given impetus to molecular medicine, which was at the basis of current therapies adapted to individual geno-molecular characteristics.

2.4.3 Baruch S. Blumberg and Robert L. Wilner

The presented paper was devoted to the substantial contributions of Robert L. Wilner and Baruch S. Blumberg for the enhancement of knowledge regarding the variations at genetic level and their correlation with diseases. The efficacy of the discovery made by Blumberg concerning hepatitis B virus and liver cancer showed that genetic and viral risks have to be considered when conducting disease risk assessment. Their work emphasized that the basis for prevention and treatment of the disease can be identified precisely with the help of the concept of the personal genome and the environment.

2.4.4 Leroy Hood

Leroy Hood was a systems biology scholar whose achievements relate to the progress of high-throughput technologies as well as the adaptation of the technologies for individualized medicine. His contribution was in the Human Genome Project and development of new tools such as automatic DNA sequencers and microarrays. Due to Hood, large-scale genetic analysis became possible and such large-scale processes are required to study individual differences, which can then be used for personalizing medicine.

2.4.5 Eric Lander

Eric Lander is a well-known name in genomics and the Human Genome Project. The possibility of developing complex diseases and gene-based therapies that are unique to an individual are presented in Lander's research as some of the new insights into the hereditary components of complex diseases. He has played an essential role in the process of translating genotype information into phenotype, relevant in personalized medicine.

2.4.6 Francis Collins

Among those calling for personalized medicine most loudly was Francis Collins, the director of the National Institutes of Health (NIH) and head of the Human Genome Project. Framing of genetic diseases and development of personal treatment modalities have come to be enhanced immensely due to leadership in sequencing human genome by Collins and his research in genetics. To enhance treatment delivery, his work has revealed how essential it is to integrate genetic data into the practice.

2.4.7 George Church

George Church was also involved in various projects, which involve genomics and synthetic biology; he was a professor in the department of genetics in Harvard Medical School. Church has done significant work on genome sequencing, gene editing technologies, and personal genomics come in handy in shaping the future for personalized medicine. His work has been devoted to leveraging these technologies to build individualized steady-state therapies and enhance disease control methods.

2.4.8 James P. Allison, Elon Musk, and Steve Jobs

For the development of cancer therapy by means of an enhanced ability of the immune system to find cancer cells, James P. Allison and Tasuku Honjo received the Nobel Prize in Physiology. They identified immune checkpoint inhibitors that harness the body's immune system to attack tumor cells, changing the treatment of a number of forms of cancer. This strategy was a good example of how personalized medicine concepts are applied in developing individualized therapies for cancer patients.

All prominent scientists engaged in the development of personalized medicine and all have provided original contributions and discoveries in the respective fields. Altogether, their contributions advanced PM from a conceptual idea towards a variety of actual practices that shape progress in healthcare and make the targeted, patient-specific approaches achievable.

2.5 The impact of genomic research on personalized treatment strategies

The emergence and integration of advanced genetic information into clinical practice have significantly propelled the development of personalized medicine. As shown in Figure 3, genomic research particularly through high-throughput sequencing technologies has delivered unprecedented insights into the genetic underpinnings of numerous diseases. This progress has enabled a highly individualized and targeted approach to treatment, marking a major shift in modern healthcare.

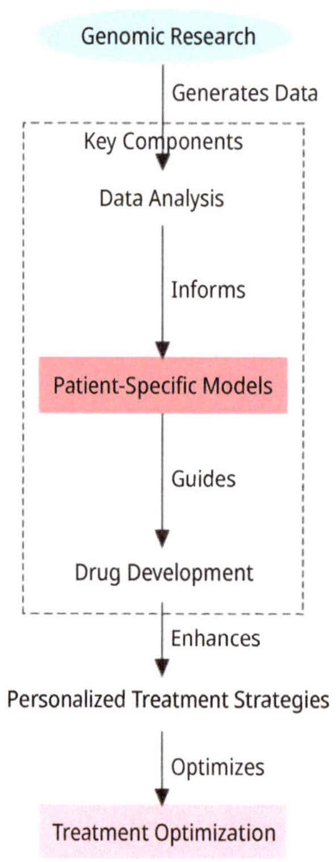

Figure 3: The impact of genomic research on personalized treatment strategies.

One of the major impacts of genomic research was the identification and description of genetic policies associated with disease risk, protection, and treatment response. Thus, genomic research has now opened the possibility of constructing precise medicine techniques based on the enumeration of the genetic polymorphism implicated in an individual's pharmacogenomic response. For instance, genetic changes in malignancies like BRCA1 and BRCA2 in breast cancers have led to the development of treatments that address the mutations ailing the cancers exclusively, increasing the quality of treatments without straining the side effects.

Genomics has facilitated the customization of medication regimens individualized on the gene makeup in people. In the brigade of genomic investigation, pharmacogenomics concerns the effects of genotypes on the disposition and reaction to drugs. This knowledge minimizes adverse drug responses and enhances therapy outcomes due to efficient drug dose and selection. For example, it is acceptable to use genetic testing in order to ascertain the dose of warfarin, anticoagulants, and thereby reduce the possibility of disorders linked with wrong medication.

Besides pharmacogenomics and cancer, the study of genomes has advanced complex diseases including neurological disorders and cardiovascular disorders. Scientists have identified gene patterns that lead to higher rates and progression of particular diseases through studying the relationships within human genome. Thanks to this information, patients with such diseases as, for instance, familial hypercholesterolemia disease were provided with more efficient management opportunities due to the elaboration of individual treatment programs that are based on the therapy of certain genetic hubs.

Preventive medicine has been incorporated into the clinical application of genomics. Some individuals are born with a high-risk factor for specific genetic diseases; thanks to genomic investigation, it can now be addressed through early assessment and prophylactic measures. For instance, active monitoring and primary prevention stem from the genetic screening for hereditary cancer syndromes, reducing the risk of cancer significantly.

A number of issues are still pertinent regarding the application of genomic research to treatment of individual patients. Challenges concerning the multiplicity of interpretation of genetic data, the availability of large and diverse genomic datasets, and the contentious question of genetic information must be solved. There is always new research, and, particularly, new technologies developed in the field of genetics, which must help avoid such complications and properly apply the results of the work done in this area to create individual treatment plans.

Taking everything into consideration, genomic work has profound implications on individualized therapeutic programs and suggests the potential for better, quicker, and individualized medicinal intervention. It is likely that as the concept unfolds, the integration of genetic data into the clinical work environment stimulates more innovative contributions to personalized medicine and subsequently improves patients' outcomes with modern treatment.

3 Fundamental concepts of personalized medicine

3.1 Defining personalized medicine

Precision medicine, defined as the state of the art in both the delivery of healthcare and a novel and innovative model of healthcare, is the way to develop prevention and therapeutic strategies that are tailored to meet each individual patient's needs. Personalized medicine focuses on the individualization of care based on constant understanding of the patients' specific traits in contrast with the regular treatment strategy where the same treatment was given to patients regardless the group they belong to.

In essence, personalized medicine integrates a patient's genetic profile with environmental and clinical factors to guide medical decision-making. As illustrated in Figure 4, this approach leverages genome-specific biomarkers and genetic signatures that reveal molecular alterations and predispositions influencing disease susceptibility and treatment response. By accounting for biological disorders that affect key mechanisms within target tissues or organs, personalized medicine aims to optimize the efficacy of therapeutic interventions while minimizing adverse effects.

The definition of personalized medicine encompasses several key components:

Genetic profiling: Generally, the use of biomarkers is critical to personalized medicine so that scientists can obtain the required genetic data about a patient's genetic makeup. This includes the determination of risks for inherited diseases, the determination of how gene defects influence pharmacokinetics, and the search for new drug targets.

Integration of multiomics data: Precision medicine often times includes multiomics data; these are genomics, transcriptomics, metabolomics, and proteomics. This holistic approach provides information on all aspects of the patient's biological status, which in turn facilitates a higher-order understanding of why certain diseases exist and the impact of certain treatments.

Tailored therapeutic strategies: This is why developing and providing individual therapies is the aim of personalized medicine. This means selection of medications and treatments most likely to benefit the patient because of their physiological and genetic characteristics and adjustment of dosages to control adverse effects.

Preventive and predictive measures: Many times, personalized medicine is primarily confined to treatment plans; it has aspects of prevention as well as prediction. Personalized medicine is the practice of diagnosing high-risk individuals for certain conditions through screening and genetic risk profiling for preventive strategies that minimize occurrence and effects of diseases.

Patient-centric approach: One of the tenets of truly patient-centered medicine is the emphasis on the patient. Engaging more patients in a decision-making process and utilizing detailed personal data as the basis for proper treatment and management of one's health are its key features.

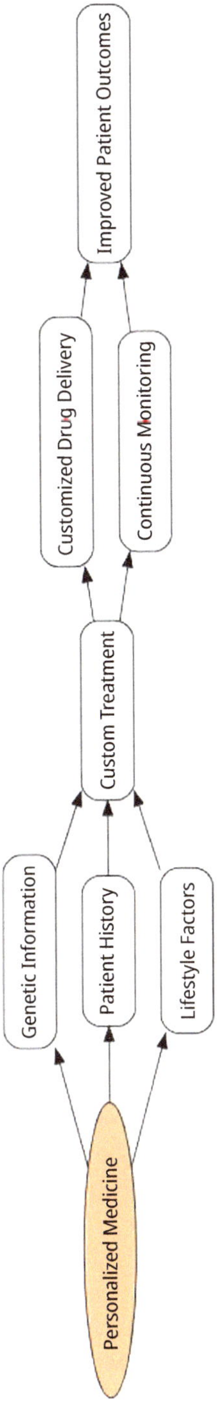

Figure 4: Defining personalized medicine.

As a result of new advancements in genomes, data science, and technology, personalized medicine cannot be farther from traditional approaches. As it moves closer to more individual and effective treatments it can revolutionize healthcare and in due time achieve superior patient outcomes and a more efficient healthcare system. The meanings and application of personalized medicine are further elaborated with the advancement of knowledge by research and technological improvements. This creates the ability for the discipline to address more medical disorders and specific patient needs.

3.2 The role of genetics and genomics in personalized treatment

Genetics and genomics remain powerful players in the field of personalized medicine as principal building blocks in targeting treatments specifically to the patient's individuality. These categories mean therapy strategies could be tailored to get the best benefits without side effects since one can know full details of genetic and genomic make up.

Traditional molecular biology having focused on individual genes and their function in the inheritance process has traditionally been at the core of disease etiology and variation in drug efficacy. By using the genetic methods, researchers have compared the relevant gene markers and mutations, which are connected with certain diseases. These genetic markers define illness risks as well as help in diagnosing hereditary disorders. For example, variations in the BRCA1 and BRCA2 genes are associated with an increased risk factor concerning ovarian and breast cancers. Patients' healthcare organizations including physicians get to offer directions and specific prevention by using genetic tests to identify these mutations to fashion out specific treatment for every individual depending on that genetic risk.

A broader assessment of the genetic framework of health and illness made using the more encompassing discipline of genomics, which is a descendent of genetics, includes an analysis of an entire genome rather than just a selection of gene loci. Any exploration of genetic factors and environments with the understanding of gene functions and impact of variants on disease processes, and therapeutic effects are considered genomic research. Novel technologies mean one can pinpoint deviations anywhere across the whole genome rather than specific areas as in previous techniques used for genetic analysis, making it possible to gain a much better insight into an individual's genotype. Some of the instances of such technologies are known as exome and whole-genome sequencing technologies. This data is needed for devising personalized treatments based on the genetic profile the patient carries as a unique identifier.

Genetic and genomic knowledge also help practices in the prevention of diseases through high-risk stratification of genotypes of persons susceptible to diseases. For instance, people who inherit genes that make them susceptible to cardiovascular diseases can enroll in prevention measures, which include changing their lifestyles and taking preventive measures before the onset of diseases due to their genetic structure gets activated.

Along with the great changes genetic and genomic science bring there are several challenges that have to be overcame. These are challenges such as the inability to analyze vast amount of genomic data, absence of guidelines for implementing genomics into clinical practice, as well as ethical questions touching on genetic privacy as well as utilization of genomic data. Overcoming these barriers in order to implement the ideas of personalized medicine at their best, it is necessary to advance research activities and the creation of new genetic technologies.

3.3 Understanding the patient's genetic, environmental, and lifestyle influences

Personalized medicine is grounded in the understanding of gene–environment interactions, and Phenome-Wide Association Studies (PLE) serve as a powerful tool to visualize and analyze these complex relationships. As shown in Figure 5, this holistic approach recognizes that health cannot be defined solely by genetic factors; rather, it is shaped by a combination of genetic, environmental, and lifestyle influences. The integration of these elements is central to the practice of personalized medicine.

3.3.1 Genetic influences

The MGT 507–Genetics and Genomics shows how genetics was central to understanding human health and vulnerability to diseases. Every individual has some kind of gene polymorphism that affects the disease predisposition, pharmacogenetics, and effects of treatment. Due to advancements made in the field of genomics, it has become relatively easy to find out specific genes associated with quite literally almost any kind of disorder, right from the rare genetic disorders to common

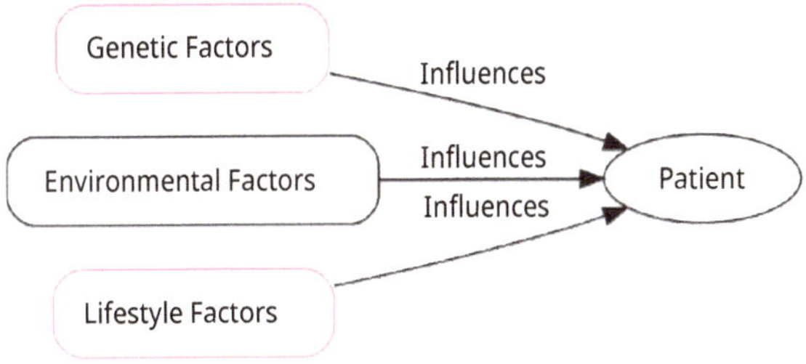

Figure 5: Understanding the patient's genetic, environmental, and lifestyle influences.

scourges like diabetes and heart disease. These genetic factors common to a community also made the prevention measures and the provision of tailored treatment programs dependent on an individual's genetic makeup possible. For instance, consciousness of genetic vulnerabilities to definite cancers determines interventions and testing modalities.

3.3.2 Environmental influences

Environmental factors can be almost anything from geographical location, exposure to pollutants, job dangers, and dietary habits that one can think of. Such outside variables can have the capacity to greatly influence a person's health experience, change the risk of disease, and interpersonal relationships with genes. For instance, the genetic susceptibility can become worse due to exposure to hazardous substances such as heavy metal and lead to disorders such as cancer. Wikipedia Integration generates an enhanced understanding of the developmental origin of illness and the construction of novel individualized treatments that incorporate both genetic and environmental data.

3.3.3 Lifestyle influences

Concepts and processes related to determinants on the health impact include social support, and regime changes, lifestyle, and diet, physical movement, and substance abuse including smoking and alcohol end use. All these aspects can determine the genesis and evolution of diseases, intervene in genetically determined processes, and modulate the outcome of therapies. Personalized medicine involves lifestyle profiling to obtain information about a person's health status. For instance, diet and exercise plans can be adjusted based on individual lifestyle factors to address metabolic disorders such as obesity, thereby improving overall wellness and enhancing therapeutic outcomes.

3.3.4 Integration of factors

Combining the genetic, environmental, and lifestyle factors in the development of different aspects of an individual is fundamental for complete personalization of the treatment. It leads to development of a specific treatment plan for a patient that would be based on all the relevant aspects of that patient's life. For example, patients with the hereditary propensity for cardiovascular disease are recommended changes in food to eat and their physical activities, ways to minimize environmental risks, and drugs to take.

3.4 Differences between precision medicine and personalized medicine

Precision medicine and personalized medicine, though often used interchangeably, convey distinct meanings and concepts that are crucial to healthcare. It is important to understand these differences to appreciate how each approach contributes to the development of medical treatments and patient care.

Precision medicine is a large concept in therapy and makes use of different types of data – genetic, environmental, and lifestyles, to provide more accurate treatment frameworks for groups of patients who are alike in some ways. Precision medicine seeks to modify interventions based on patterns and shared characteristics by patients and risk factors, in an effort to target the right therapy. For instance, precision medicine requires categorizing cancer types through genomic markers in order to subdivide them according to unique genetic changes in order to design therapies that effectively counter the mutations.

The translation of pharmacogenomics into clinical practice is a process of extending a number of medications and the dosages of the chosen ones for each patient. Personalized medicine implies the adjustment of the medical care to the patient's individual features. This approach goes further than the patterns common in precision medicine that has adopted an individualized approach to the disease dependent on the patient's genes, previous medical records, and preferences. It is a microbots healthcare concept that involves entering tailored treatment options for an individual patient for improved efficiency and least harm. For example, personalized medicines include prescribing a drug taking into consideration the genetic structure of the patient to achieve the optimum amount of medication that will not cause an allergic reaction.

The major contrast between the two approaches is the degree of coverage. It is mostly about classifying diseases and deriving treatments for different clusters of patients having common genetics; that is why it deals with patterns inside communities. On the other hand, personalized medicine is more patient-centered, which encompasses including an array of specific characteristics of a given patient in order to provide care that is shaped to fit exact specifications.

This analysis also highlighted another difference between the two sets of approaches: their use. While precision medicine uses large-scale information about populations to understand and modify the treatment patterns of diseases, the concept of personalized medicine is more focused on using patient's data and preferences when making decisions. Personalized medicine intends to tailor preventive measures and therapeutic interventions to specific persons and their diseases, while precision medicine intends to deliver the most effective preventive measures and therapeutic intercessions to given populations of patients with the same disease.

Precision is a dramatic improvement in medical treatment; precision medicine has given the tools to find those treatments for subpopulations. Personalized medicine makes the treatments tailored to the patient. As these fields advance, it is believed that integrating the two provides a better approach through improving healthcare with the end result of improving health systems for the better.

3.5 The interplay between personalized medicine and conventional healthcare approaches

Personalized medicine can be regarded as a great leap beyond the traditional systems of delivering an individual treatment, as it shifts to the core of creating and applying medical treatments. Whereas conventional medicine has always focused on standardized solutions applicable to as many cases as possible due to their universality, personalized medicine is built around the idea that has never been so much as possible due to holistic approaches combining genetic, environmental, and individual life characteristics of the patient.

Traditional medical management techniques are notable for their tendency toward uniformity and usage of general norms and average standards. Standard methods of treating an illness entail prescribing identical treatment procedures to all patients with the same disease, according to the averaged results derived from research studies. Although this treatment approach has worked well in many patients, its weakness is that it does not factor issues such as divergent disease presentation, differences in the rate of drug metabolism, and differences in drug sensitivity among patients leading to variation in the outcomes.

Personalized medicine overcomes many traditional constraints by aiming to deliver tailored healthcare solutions that align with each patient's unique needs. As illustrated in Figure 6, recent advances in genomics have enabled detailed mapping of individual genetic profiles and their associations with disease susceptibility and treatment outcomes. When combined with data on lifestyle and environmental factors, personalized medicine facilitates the development of treatment regimens that are precisely aligned with each individual's genetic makeup.

The complementarity of precision medicine with conventional treatments can be illustrated in several areas. For example, genetic and molecular forward practice improves the traditional practice through focusing on the patient individual characteristics. In oncology, for instance, while routine approaches to management involve conventional therapeutic protocols, individualized medicine entails molecular characterization of tumors to determine unique molecular changes and then prescribe drugs accordingly. Combination of individual approaches with standard treatment procedures elevates the effectiveness of interventions, besides making them specific, which in turn enhances the overall treatment processes.

In addition, personalized medicine works hand-in-hand with conventional systems and is underpinned by prediction and prevention. Intensive care usually includes a cure after a virus has already manifested itself in the patient. Personalized medicine however uses genetic and environmental information about a patient so as

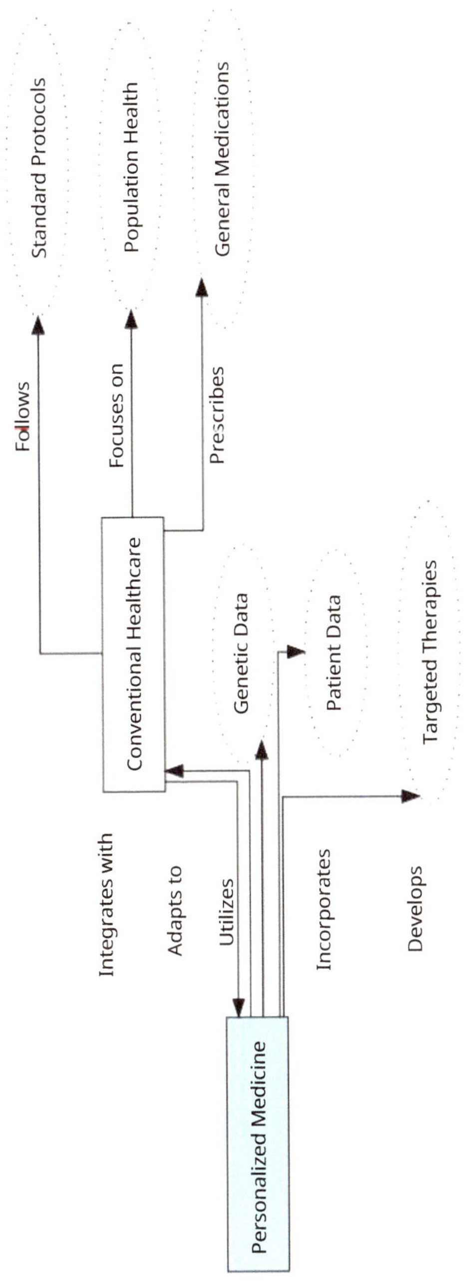

Figure 6: The interplay between personalized medicine and conventional healthcare approaches.

to prevent diseases and its effects before manifesting themselves. Taking charge of one's own healthcare could prove to be beneficial to the society as it helps one to seek safe and effective treatment of chronic diseases before deterioration that may need further complicated and risky treatment.

4 Technological advancements enabling personalized medicine

4.1 The role of next-generation sequencing in personalized healthcare

Next-generation sequencing (NGS) represents a major technological advancement in personalized healthcare, offering high-throughput and in-depth characterization of an individual's genetic makeup. As illustrated in Figure 7, NGS enables accurate, large-scale genomic analysis, supporting a wide range of applications that are fundamental to the advancement of personalized medicine. By allowing for the rapid and cost-effective sequencing of entire genomes, NGS has revolutionized both genomics research and clinical practice. Unlike earlier sequencing methods, which were time-consuming and expensive, modern NGS platforms can simultaneously process millions of DNA fragments. This capability has enabled the creation of highly detailed genetic maps for individual patients, leading to more precise diagnoses and improved treatment strategies.

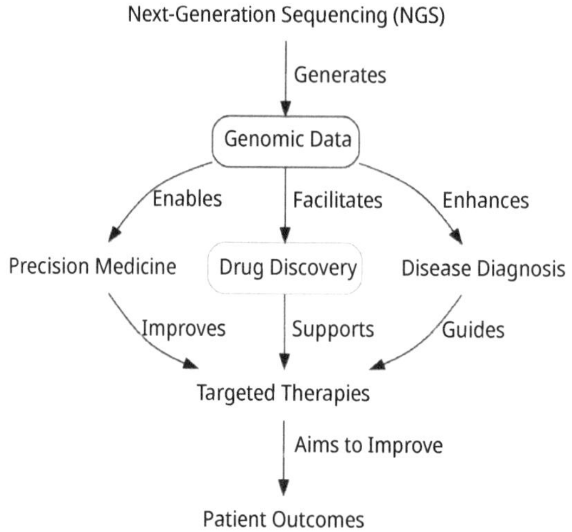

Figure 7: The Role of next-generation sequencing in personalized healthcare.

Genetic factors associated with medication response as well as disease susceptibility are another area of use of NGS in SHP. Using genome sequencing, experts and healthcare practitioners define specific changes to the DNA and differences in a person's susceptibility to various diseases such as cancer, heart diseases, and rare inherited diseases. Such information is important for initiation of therapy that is based on the genes causing diseases, in this way being more effective and individualized.

In the case of oncology, NGS is critical to the tumor profiling, specifically studying the alterations that result from the development of cancer cells in a patient. When using such a wide set of genetic information, it is possible to define necessary mutations that can be targeted by certain medicines, maximizing the efficiency of the therapy, as well as minimizing the number of side effects. For example, mutations that occur in non-small cell lung cancer, due to changes in genes such as EGFR have been identified to have specific medications that work to precisely eliminate the cancers that are caused by these abnormal genes, thus enhancing medical outcomes.

Thus, NGS extends the pharmacogenomics field by giving an understanding of how these genetic differences influence the reaction to the drugs and their processing. With this kind of data, personalized medicine decreases unwanted reactions to medications as well as increases effectiveness of therapeutic intervention by individually selecting the right drug, the right dose, and combination of drugs based on the individual's genetic makeup. For making therapies person-oriented and for individual gene makeup of the person, genetic tests are helpful in selecting and prescribing drugs such as antidepressants and warfarin for which the CYP450 enzyme family needs to be tested.

Health informatics is therefore an area where personalized healthcare is likely to offer the greatest benefit of applying NGS; but there are also disadvantages. The era of Big Data means that an enormous volume of data is produced in clinical practice, and their evaluation and implementation require special knowledge of bioinformatics and developed computing facilities. To achieve the goals of the NGS in personalized medicine additional challenges are described: data privacy issue, ethical dilemmas, and the emergence of specific criteria for genetic testing.

4.2 Integration of Big Data and artificial intelligence in personalized medicine

Smart treatment, which expands the ability to individualize medical treatments with reference to certain patients' characteristics, is one of the most significant innovations in healthcare thanks to the use of Big Data and AI. All these technologies provide positive tools for analyzing big and complex data, identifying tracks and advising unique treatment strategies that boost patient outcomes.

Big Data refers to large and complex datasets in terms of the volume, velocity, and variety, which stems from various forms of information such as genomics, electronic health records, clinical trials, and patient-reported information. This data itself is enormous, and it is evident that there are both advantages and limitations when it comes to integration to personalized medicine. These large datasets call for sophisticated analysis to produce meaningful patterns and correlations that guide treatment.

Big Data research immensely relied on applying AI especially ML and DL to extract information from the large datasets. The spectrum of such AI tools is aimed at handling and analyzing vast and complex datasets, wherein conditions and trends sometime overlooked by conventional analytical approaches are detected and defined. For example, ML algorithms analyze large swaths of genomics data to look for genetic markers associated with specific diseases. In this regard, it is possible to develop prediction models that individuals can use to predict the risk for an illness, handling of the illness, and therapy needed for the same based on machine learning skills.

In personalized medicine, Big Data and AI are enhancers in several ways: First, AI advanced analysis improves the overall accuracy of diagnostic procedures considering multiple data inputs, such as images, genetics, and health records. For instance, forms of AI can help examine medical imaging data to identify early signs of diseases, which would be missed by other approaches.

Second, establishing personalized treatment programs has also benefited from the AI and Big Data work integration. AI can predict a patient's reaction to a specific drug because it analyzes the results of the other patients with similar genetic and medical histories. By pointing towards fewer trial error instances and less side effects, this precise predictive force allows doctors to select the most successful medications with far greater accuracy.

Third, new biomarkers and treatment targets are identified through the utilization of both AI and Big Data. Big Data analysis can be used to discover new targets for drugs, as common genetic patterns can be associated with the illness prognosis that was not known earlier. This technique benefits the development of biopsy drugs for specific patients by utilizing molecular qualities of target patients.

There are still challenges to the incorporation of Big Data and AI toward personalized medicine, albeit the gains are obvious. There are such critical issues as the protection of data – own and others', formatting of the data, and the outcome of the AI and its analysis. Collaboration is needed across disciplines to ensure that the AI applications are designed and implemented in clinically appropriate and ethically and medically sound manner.

Taken together they could be symbolic of a revolution in not only the field of medicine but also a revolution in ensuring accurate, successful treatments and even medical cures that have been tailored to each patient's particular needs. Improved patients' health status and increased specificity in healthcare are expected as a consequence of integration of these technological advancements in the personalized medicine field.

4.3 Innovations in biomarker discovery and their impact on treatment customization

Emerging trends in biomarker research have pioneered the advancement of the field of targeted therapies for patients. Biosignature biomarkers are biological molecules that offer clues to a disease condition on a certain therapy and are essential aspects of pharmacogenomics, which involves building drugs tailored to the peculiarities of each patient. The technological improvements of the recent past have gone a long way to increase the discovery and utilization of biomarkers, a major revolution in the customization of treatment.

Some of the major breakthroughs in biomarker discovery have been driven by advancements in high-throughput screening techniques. As shown in Figure 8, these technologies enable the simultaneous testing of hundreds of biological samples, significantly accelerating the identification of novel biomarkers. Methods such as mass spectrometry and next-generation sequencing (NGS) are essential for detecting biomarkers linked to a wide range of diseases, including cancer, cardiovascular conditions, and neurological disorders. These advanced tools are capable of identifying subtle molecular changes that may go undetected by conventional methods, thereby improving diagnostic precision and therapeutic targeting.

New trends in biomarker discovery, driven by omics technologies such as genomics, proteomics, and metabolomics are transforming biomedical research. Molecular genetics reveal how polymorphisms influence disease risk and drug response (pharmacogenomics), while molecular profiling and metabotyping identify relevant proteins and metabolic states, respectively. Thus, the integration of information from these various levels of omics platforms enables researchers to define biomarkers that contribute to understanding disease pathogenesis and identifying potential therapeutic targets.

These changes have left none on the treatment customization untouched. Biomarkers help in partitioning of patients in terms of molecular character and allow specific drug and therapeutic interventions to be designed to address the corresponding disease. For instance, the identification of genetic mutation in a tumor has culminated into development of drugs that specifically block the proteins by means of which these mutations manifest themselves, thus leading to development of efficient and less toxic effectation therapies. Like the former, biomarkers that reflect drug metabolism can act as surrogate markers directing dose adjustments depending on patients' benefits and risks.

Biomarkers are applied not only for therapeutic purposes, but for diagnosis as well as for the prediction of the disease. Prognostic biomarkers are used in determining the disease progression and likely outcomes, which in turn guide treatment regimens that would best fit the disease progression prognosis. Another advantage of the use of biomarkers was that they can produce a daily reflection of treatment outcomes in response to the action taken.

Several issues are still outstanding after these developments in biomarker discovery research. There are multiple challenges such as the system and network of biomarkers,

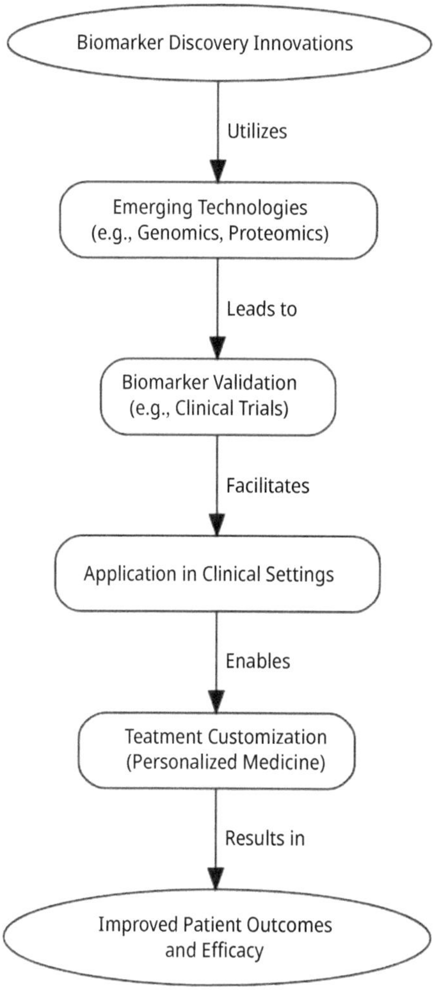

Figure 8: Innovations in biomarkers discovery and their impact on treatment customization.

including the requirements for biomarkers to be replicated in different population cohorts, as well as the issue of implementing biomarker information into routine practice. These difficulties indicate that much more research and work needs to be undertaken as also improved biomarker identification, validation, and application protocols.

4.4 Role of CRISPR and gene editing technologies in tailoring therapies

CRISPR and gene editing technologies have emerged as some of the strongest forces that have boosted the field of personalized medicine through enabling one to make specific genetic changes. These developments have led to the formulation of disease-specific treatments that directly target the foundation of genetic diseases, providing new treatment possibilities that are not possible through usual therapeutic techniques.

Among the most transformative advancements in recent biotechnology, CRISPR-Cas9 stands out as a revolutionary tool in the field of personalized medicine. As depicted in Figure 9, this gene-editing technology enables precise modifications to specific regions of the genome, making it possible to correct mutations responsible for various genetic disorders. For patients with inherited conditions, CRISPR allows scientists to edit or replace defective DNA sequences, paving the way for the development of individualized, genetically tailored therapies that directly address each patient's unique genetic profile.

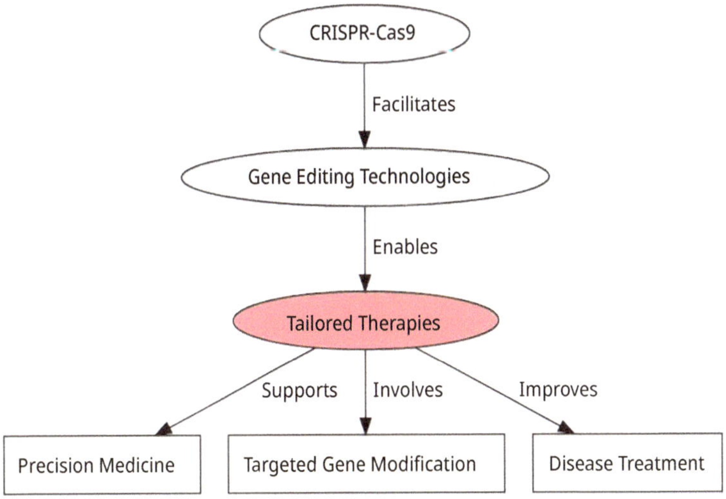

Figure 9: Role of CRISPR and gene-editing technologies in tailoring therapies.

Personalized medicine is one of the main applications of CRISPR, and one of its main areas of use is in treating genetic diseases. The usefulness of such techniques includes being able to cure diseases such as sickle cell anemia and cystic fibrosis, among others that are as a result of single gene defects. CRISPR technology offers the opportunity to work on cure rather than symptoms – this is in contrast to treatments based on genetic prescription built to address the specific genetic abnormalities in every individual. Current scientific and clinical investigations are underway to determine the possibility and efficacy of these gene modification techniques in the management of different inherited diseases.

CRISPR and other gene-editing technologies are being used not only in individualized cancer therapy but also heredity issues. Cancer develops because of the specific genetic changes that are inherent in most cancerous tissues. Scientists can develop analytic drugs that have an effect exclusively on cancer cells and do not harm sound cells as that can reverse these modifications. Relative to conventional therapy, this procedure enhances the precision of cancer therapies potentially enhancing their effectiveness while reducing side effects.

Also, the breakthrough of immunotherapy itself could not have occurred without developing a technique in gene editing. Working with separate cells, for instance, T cells, scientists can enhance immune cells' ability to discern and treat cancer cells. For specific cancer antigens, cells are engineered to express chimeric antigen receptor (CAR), as in CAR T-cell therapy. This personalized approach shows rather promising results for the treatment of some types of blood cancer and finds further application in the sphere of oncology.

It is however crucial to note that there are in reality a number of challenges and considerations with regard to CRISPR and the gene editing technologies. As these technologies move towards the preclinical and clinical applications of the technology some of the concerns such as off-target effects, ethical concerns, and long-term effects need to be analyzed in detail. For this reason, constant research and regulatory supervision are crucial for gene-editing treatment to be safe and efficient for the patient.

With the help of CRISPR and other tools of gene editing in personalized medicine the ability to produce drugs-matched individual genetic codes has become possible. These technologies have drastic capabilities to treat genetic diseases, cancer, and other diseases through targeted therapies. The role of gene editing in personalized medicine is believed to grow with both research activities and application, which would open new possibilities for introducing effective individualized treatment of patients.

4.5 The contribution of bioinformatics and computational tools to personalized medicine

The use of computational approaches and bioinformatics has directly improved the possibility of targeted treatments – the heart of the concept of individual medicine, enhancing the distinct possibility of applying medical breakthrough in a manner consistent with a specific patient's needs and constituencies. These technologies close the abstraction level gap between complex genomics and practical applications of genetic data by enabling efficient analysis, comprehension, and utilization of incredibly large amounts of biological information.

Bioinformatics can be defined as the science of analyzing bio sequences through computational methods, concentrating on genetic sequences. Because the amount and complexity of biological data as well as progress in high-throughput sequencing make it challenging to obtain valuable information from these data, bioinformatics is now mandatory. As a result of creating modern program techniques in bioinformatics, the functions of the genetic markers that are connected with diseases are determined, which allows finding new diagnosis indicators and targets for treatments.

Since the multiomics data include proteomics, metabolomics, and genomics, the analysis and integration of the given data require the use of computational approaches.

These methods allow for considering relationships between genetic differences and disease manifestations since designed to address the problems of increased dimensionality and data heterogeneity in multiomics studies. Molecular methods provide great insight into etiology of certain health disorders by integrating the information obtained from multiple sources with the help of computation. This makes it possible to develop customized programs of treatment that incorporate all aspects of the patient's biology.

Bioinformatics is again able to develop models that would help predict treatment plans of patients. Applying statistical models and machine learning algorithm on the patients' data the efficacy of medicine in a given patient is predicted. Using these predictive models according to the genetic and clinical data contributes to the calculation of the optimal therapeutic process and minimizes the possibilities of using ineffective treatment options in the further analysis.

Further, bioinformatics tools can be useful for designing and fine-tuning of targeted drug delivery systems. Computer-aided tools are used to estimate the behavior of molecules and their relationship with a biological target and how they would be processed in different persons. This capability enables treatment through formulation of drugs and methods to reach the intended patient so that therapeutic outcomes improve while side effects reduce.

Bioinformatics in combination with clinical practice has made the usage of precision medicine easier in routine of a healthcare system. Clinical decision support systems and electronic health records are getting supplemented step-by-step with bioinformatics tools to provide implementable information at the point of care. The results of these instances are to enable healthcare practitioners to make informed decisions in relation to patient care and treatment by integrating the genetic information with clinical details – which is the aim of the personalized medicine.

5 Clinical applications of personalized medicine

5.1 Personalized approaches in oncology

The molecular and genetic stratification of breast cancer patients marks a revolutionary shift in cancer treatment, enabling the use of targeted therapeutic interventions based on individual genomic characteristics. As illustrated in Figure 10, this approach moves away from a one-size-fits-all treatment model and toward precision-based strategies that consider the unique molecular profiles of each patient. This transition has significantly improved the accuracy and effectiveness of breast cancer management, leading to better clinical outcomes and reduced adverse effects.

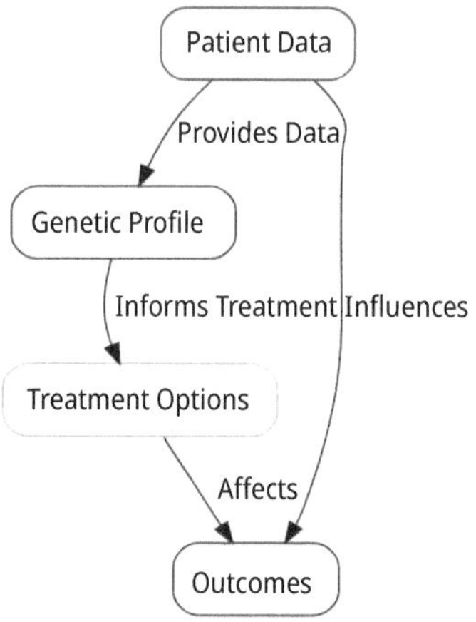

Figure 10: Personalized approaches in oncology.

Genomic profiling to identify specific genetic dysregulation and alterations within cancer cells is the primary approach of personal oncology. As a result of such approaches as NGS, one can identify the set of mutations that drives tumor growth and hence cancer. Oncologists can elect to use a treatment type that would selectively attack the genetic profile present in each patient's tumor through identification of these molecular signatures. New drugs that are more effective, with fewer side effects than ordinary chemotherapy, have been developed to counter some genetic flaws. For instance, tyrosine kinase inhibitors exist for EGFR gene mutation and PARP inhibitors for BRCA1/2 gene mutations.

Individualized oncology methods are immunotherapy and targeted agents, which employ the immune response to deal with malignancies. Knowledge of the types of environments in which tumors grow and how cancers evade immune recognition has given rise to protein regulators of immune response called immune check point inhibitors, which include anti-PD-1 and anti-CTLA-4 antibodies. More examples of what constitutes a personalized treatment plan being used to enhance the immune system's ability to detect and destroy cancer cells are adoptive cell therapies such as CAR-T cell therapy as well as personalized cancer vaccines.

A second key area of individualized oncology care is the use of tumor biomarker testing to inform therapy. For instance, tumor mutational burden (TMB), microsatellite instability (MSI), and selected protein expression are associated with tumor sensitivity to specific treatments. High TMB and MSI are known to correlate with higher rates of response to immune checkpoint inhibitors. Emphasizing these biomarker tests in the clinical arsenal, oncologists are able to identify which treating strategy works for particular patients and provide the best possible results.

The matters demonstrated by constructive oncology indicate how important it is to monitor patients' outcomes and adapt the therapy regimen, accordingly. Liquid biopsies in blood samples, which involve the exam of ctDNA, enable the evaluation of alteration in tumor mutation profiles. Regarding this, treatment management can be adjusted depending on the development of mutations and suitability of the existing regimen of treatment.

There are a few limitations and issues in personalized oncology even today. Issues that remain challenges include the ability to incorporate genomic information, costs of the related new diagnostic and therapeutic technologies, and the use of large patient datasets. Also, the implications of the ethical issues of genetic testing and differential distribution of treatment options to patients have to be discussed to provide the principles of fair in personalized cancer treatment.

5.2 Role of personalized medicine in managing cardiovascular diseases

Precision medicine has gradually expanded the approach of CVD care, paying considerable attention to patient-specific preventive, diagnostic, and therapeutic interventions. It builds upon genetic, environmental, and lifestyle data to optimize the accuracy and usefulness of cardiovascular care, a problem with standard stereotyped treatments.

One of the biggest advances of pharmacogenomics in managing cardiovascular disease has been in the identification and utilization of genetic factors associated with cardiovascular risk. Familial hypercholesterolemia, hypertension, and coronary artery disease are among the many cardiovascular diseases that many are genetically inclined to get. In the fields of personalized medicine, genetic tests are made to distinguish high-risk patients by genetic makeup to prevent them from having serious health issues and to address the problem at its early stage. For better targeting of therapy, for instance, some gene variations that are acted upon by the therapeutic agents have been identified such as genes involved in lipid metabolism and this can be used to justify the use of statins and other cholesterol-reducing drugs.

Equivalent to the traditional personalization concept, the concept of personalized medicine incorporates data including the imaging data and lifestyle parameters along with genetic tests, for informing treatment decisions. Some of the current techniques in diagnosis of structure and function of CHD include cardiac MRIs and CT scans. When such imaging reports are integrated with other patient specific details such as nutritional status and exercise profiles, which are used by personalized medicine, differential treatment plans are developed for each patient.

Cardiovascular diseases are one of the therapeutic areas where pharmacogenomics, a sub-branch of personalized medicine, is invaluable in ensuring appropriate drug therapy. Cardiovascular medicines have different effects on patients depending on these genetic differences in a patients' metabolism. Pharmacogenomic data can also help clinicians modify not only the kinds of medicines given to the patient but also the doses necessary to avoid negative effects while achieving the best potential improvement. For instance, pharmacogenetic differences associated with drug-metabolizing enzymes can enhance the dosing recommendations of anticoagulant and antihypertensive medications and ultimately prevent adverse effects of medication.

In addition, personalized medicine assists in the exploration of advanced technique of treatment where treatments are performed depending on the patient's unique genetic makeup. Technological developments have greatly enhanced the capacity to find new biomarkers for cardiovascular diseases through genomics and proteomics. Individualization allows the development of treatments targeting these aims since non-targeted interventions would be more toxic and less effective than the current therapies. For instance, targeted drug delivery systems and gene therapies are becoming potential solutions to enhanced and sophisticated cardiovascular diseases, which include heart failure and atherosclerosis diseases.

Mobile health, digital health, and wearable technology also help support the concept of personalized medicine in cardiovascular care. Pulse oximeters, blood pressure, temperature, and other health indicators help to gain on-the-spot information about changes in giving targeted interventions and to assess results. Personalized medicine also adapts these data into the management, which means alteration in the routine management plan according the feedback and overall condition of the patient.

5.3 Customizing treatment strategies for neurological disorders

Personalized medicine is an advanced treatment approach that tailors therapy for neurological disorders based on the individual characteristics of each patient. As shown in Figure 11, this strategy leverages emerging technologies and discoveries in genomics, neuroimaging, and biomarker identification to enhance the precision and effectiveness of diagnosing and managing complex neurological conditions, including Alzheimer's disease, Parkinson's disease, epilepsy, and multiple sclerosis.

Lack of simplicity and homogeneity are the characteristic features of most neurological sickness where multiple genetic, biochemical, and environmental factors determine the natural history and response to treatment. It is therefore of highly heterogeneous structure, and traditional treatment methods often involve applying an approach that may not be useful to many people. It is important to understand that, for example, certain diseases in certain patients for which medicines have not been developed yet can be a problem in general for personalized medicine and work is needed to correct these things and offer better treatment.

Among the main developments that enabled enhancing the approaches to managing neurological disorders, the development of genetic and molecular biomarkers is noteworthy. In the genetics context, a lot of genetic variations have been discovered to be linked with neurological disorders mostly through the studies of their causes and possible treatments. For instance, mutations within genes encoding amyloid precursor protein (APP) and presenilins have been identified to cause familial Alzheimer's disease resulting in therapies that seek to influence amyloid-beta synthesis and deposition.

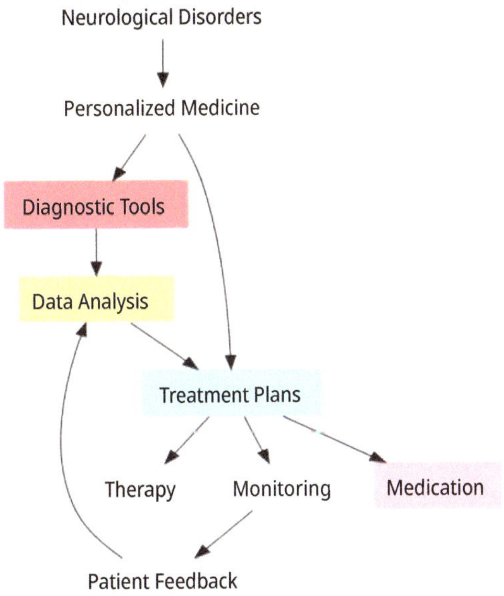

Figure 11: Customizing treatment strategies for neurological disorders.

Biomarkers are also useful in prognosis and response to treatment in neurological disorders as well. Neuroimaging biomarkers include MRI and PET that give details of the anatomical and functional abnormality of the brain thus helping in disease diagnosis at early stages and to triage treatment response. Genetic and proteomic researches in biomarkers from cerebrospinal fluid and blood adjustment biochemistry in brains are biopsy-free diagnostic and prognostic instruments.

Personalized medicine also makes use of the environmental factors that affect the patient especially on the care that is being given. Low-complexity interventions including diet and exercise can be used alongside drugs to enhance the treatment of neurological ailments. Such factors are incorporated in individualized treatment interventions, which help improve the general outcome of therapeutic approaches.

Several challenges have to be addressed in order to get to the potential of personalized medicine for neurology. Some of them are the need for vast research that enables identification of relationships between genetics and diseases, for the development of clinical decision support systems, and for integration of personalization techniques into clinical practice. Ethical concerns linked with genetics testing of patient and privacy also have to be managed adequately to retain the trust and engagement in the process of the patient.

5.4 Applications of personalized medicine in rare and genetic diseases

Molecular medicine has been defined as personalized medicine as it was accepted as a novel therapeutic modality in the management of rare and genetic disorders and carries remarkable promises to the patients who have had very few therapeutic choices in the previous treatment regimens. Personalized medicine in that respect entails matching care interventions with genetic characteristics of the individual client to enhance diagnosis and treatment plans.

In choosing comparison targets, it is therefore understood that rare and genetic diseases pose the most formidable challenge because of their nature and diversity. Conventional diagnostic techniques do not have the required efficiency, often causing the doctor to miss the correct diagnosis. These issues are solved by personalized medicine using genomics-sequencing technologies to identify such genes accountable for these diseases. Whole-genome and whole-exome sequencing are coordinated processes used in the sequencing of an individual's genomic DNA, and the former helps identify carriers of pathogenic variants that would be ignored in standard sequencing. It improves the outcomes of diagnosing because it is accurate and allows clinicians to offer early, appropriate measures.

While being used in the diagnosis of rare and hereditary disorders, the most significant application of personalized medicine is in the development of distinct drugs. Since what genetic basis such ailments have is obvious, scientists can develop drugs that would specifically address these genetic changes in diseases. This method has been particularly useful in diseases such as cystic fibrosis, for which therapies developed to address the CFTR gene mutation have made impressive performance. In the same way, developing more individualistic medication has come up with PARP inhibitors that selectively stick to cancer cells harbored by BRCA mutations raising the treatment positives in hereditary diseases like ovarian and breast cancer associated with BRCA mutations.

Another manifestation of the idea of personalized medicine that has application for rare and genetic diseases is gene therapy. Compared to symptom-focused treatments that have existing gene-based approaches, gene therapy intends to fix the source of a wide range of genetic diseases by inserting genes in a patient's cells. New developments in modern technologies like CRIPPR CAS 9 have given relatively greenish solutions to many genetic disorders including SMA (spinal muscular atrophy) and various forms of genetic blindness.

Personalized medicine also has significance in untangling complicated genetic disorder through early diagnosis and tailored medical approach. For example, in diseases like familial hypercholesterolemia, it is possible to test for genes; therefore once individuals are found to possess those genes, lifestyle changes and the use of drugs can be administered early enough to prevent the situation from escalating. Hugo-Kaplan, A, et al., report that for conditions such as phenylketonuria, patients can develop their meal plan in consideration with metabolic tests, thus avoiding severe structural backups.

There is still a great potential for the development of molecular pathologies that characterize rare and genetic diseases and the introduction of modern approaches in personalized medicine. These diseases are relatively rare, which presents a challenge when obtaining the large data sets that are needed for effective treatment and modelling. Price tags of genomic testing and targeted therapies make them expensive and therefore hard to access. Measures applied to these challenges include multisectoral collaboration toward developing international rare disease genomics databases and international policies enhancing the access to innovative diagnostic and therapeutic resources.

5.5 The impact of personalized medicine on preventive healthcare and early diagnosis

The white paper shows that proactive approach to disease prevention and treatment has been supported by the personalized medicine concept in the sphere of preventive healthcare and early diagnostics. This is because through individual genetic, environmental, and lifestyle data, personalized medicine aids healthcare workers in diagnosing diseases and preventing them for which efforts and resources are boosted in the healthcare delivery system with the overall aim of improving firm outcomes.

This underscores the importance of patient pre-diagnosis, where individuals at high risk of developing certain diseases can be identified through personalized medicine and treated proactively. As illustrated in Figure 12, genetic diagnostics and genomic profiling enable predictive testing for inherited conditions such as diabetes, cancer, and cardiovascular diseases. For instance, the detection of BRCA1 gene mutations allows for the early implementation of monitoring and preventive strategies for breast and ovarian cancers, significantly lowering the risk of disease onset.

Besides genetic screening, biomarkers are also used in the application of personalized medicine to allow for early diagnosis of a disease. Biomarkers are tangible signs of biological events, and can, therefore, be detected to draw blood disease onset. The fact that biomarkers are connected with specific diseases permits the creation of diagnostic means that identify diseases in their early states at which the conditions are most favorable for therapy. This early detection is especially important in oncology, where the earlier the intervention the better the patient's survival and chances of recovery.

It also provides behavioral changes and preventions as a result of patients' specific risk factors for those diseases, which fall under the concepts of personalization of medicine. Drawn from genetic and environmental databases, doctors can suggest certain patterns of behavior like dieting and exercises that could help decrease disease risks. Other components of individual prevention regimens can also comprise targeted pharmacotherapy depending on genetic predispositions, thereby enhancing both effectiveness and side-effect profiles of prophylactic medications.

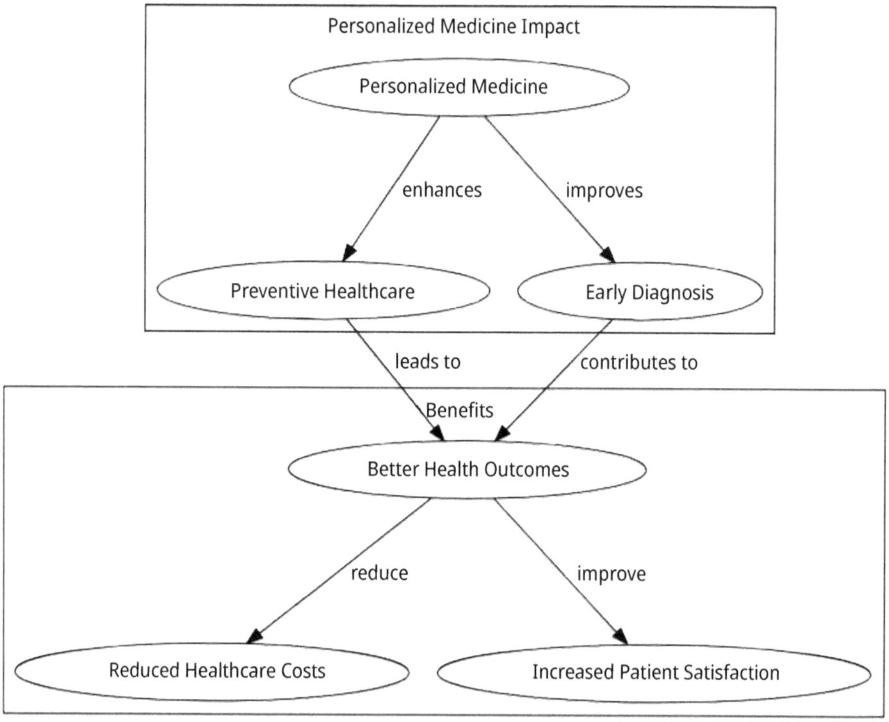

Figure 12: The impact of personalized medicine on preventive healthcare and early diagnosis.

It also means embracing the role of personalized medicine in the broadly conceived preventive framework including infectious diseases. The identification and determination of pathogens genetic map work aids in evolving localized vaccines and an outbreak treatment plan that considers pathogen and host polymorphisms. The application of this approach increases the efficacy of immunization campaigns as well as the lack of ability to control the spread and occurrence of the disease.

Personalized medicine also complements the prevention of diseases, thanks to development in imaging and diagnostic techniques. AI and machine learning are applied in tumor diagnosis by analyzing structural changes and faint features in medical imaging information that depict early disease. Such technologies enhance the specificity of diagnosis and facilitate early treatment that halts disease advancement and multifactorial manifestations.

There are several barriers highlighted from the attempt to adopt personalized medicine in preventive health and early-stage diagnosis, these include; concerns in accessibility, data protected act, and lack of standard protocols. Solving the problem of availability and accessibility of individualized healthcare and solving organizational ethical issues concerning genetics are significant for making use of these approaches more common.

6 Challenges and barriers in implementing personalized medicine

6.1 Ethical and privacy concerns in the era of personalized healthcare

The proper and fair use of genetic information is essential, especially as personalized healthcare continues to evolve. As depicted in Figure 13, the growing reliance on genetic, lifelog, and environmental data in the management of individualized treatments has raised significant ethical and privacy concerns. These include issues related to data protection, informed consent, voluntariness, and equitable access to care challenges that must be addressed to ensure ethical implementation of customized medicine.

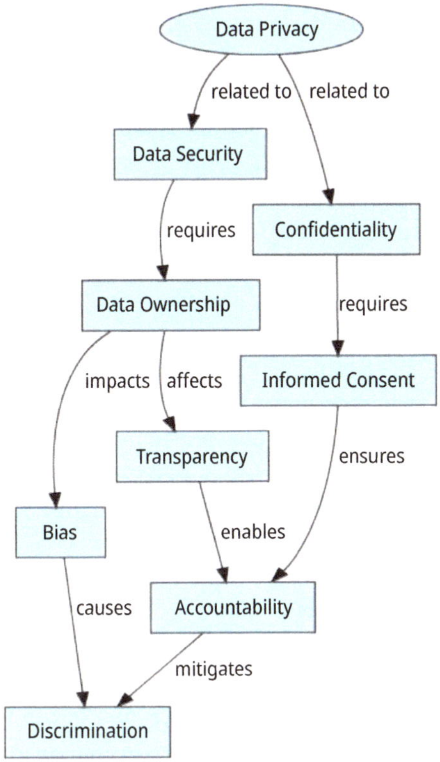

Figure 13: Ethical and privacy concerns in the era of personalized healthcare.

An ethical concern is the probability of genotypic discrimination in phenotypic differentiation. When people have full access to the kind of detailed genetic information, they come across discrimination in various spheres of life including insurance and employment. Concerns of fairness and equity are then posed by the likelihood of individuals with certain genetic makeup for particular diseases getting locked out of the job market or locked out of insurance cover. To tackle this issue there are currently many legislations and laws around the world such as the Genetic Information Nondiscrimination Act (GINA) in United States that seeks to protect individuals from discrimination due to their genetic background.

Informed consent is another important ethical consideration in individualized medicine. Due to the sheer amount and density of genetic information, patients need to be informed of the meaning of genetic testing and the potential uses of this data. In order to achieve true informed consent, it is necessary to safeguard that population that has equal access to comprehensible information on possible advantages and disadvantages of genetic testing. Also there is need to make patients aware of incidental findings, which are genetic findings different from the intended diagnosis that has a large impact on the patient's health.

One of the limitations is the privacy issue relating to the need to mark certain genotypes, which are deemed sensitive, so as to protect them from misuse and unauthorized access. The concern regarding the leak of sensitive information is enhanced by the rising trend of handling and storing health information in digital format and the popularity of cloud storage solutions. Certain essential security standards like grabbing, limiting the access to the genetic information, and the use of secure protocols in transferring it have to be implemented vigorously. In addition to this, to protect the individual affected person's identity and encourage research and innovation, there are requirements to prescribe procedure for transfer of information between researchers and healthcare providers.

Autonomy means freedom in patients' right to make choices concerning their treatment and care. Personalized medicine just contradicts this principle by presenting the patient with numerous factors, which can hardly be interpreted by the patient. The healthcare givers have the duty of keeping their expertise in mind, while recognizing that patients are tested to assist them in gaining the knowledge required in order to make choices that are congruent with their belief systems.

Another problem that has repeatedly emerged in the literature is the issue of fair distribution of personalized primary healthcare services. They also suggest that inequality in access to genetic testing and other personalized treatments actually increase existing disparities in healthcare, as underrepresented groups could be left out of some of the benefits of personalization. Making personalized care available across everyone is important not only in terms of revenue generating but also to take care of any imbalance in terms of revenue level regions predefined by a previously set goal.

6.2 Regulatory challenges in approving personalized therapies

Personalized medicine implementation is associated with a unique set of challenges in meeting regulatory requirements that define personalized therapy approval and commercialization. Unlike conventional medicines, these highly developed individual treatments often involve targeted standards where the information about genetic predispositions, surroundings, and habits of the patient are taken into account firstly. Due to this level of individualization, the existing legislation needs to evolve to ensure that the impacts of these drugs on their targets are positive, safe, and of good quality due to their special features.

In particular, one of the main regulatory concerns, when approving personalized therapies, is the requirement of developing new types of clinical trials. Conventional clinical trials are ISO QS 1901 conformant, meaning developed for assessing treatment patterns expected to be applied to large groups of patients, but the utilization of personalized medicine requires more flexible approaches to clinical trials. For example, basket-and-umbrella trials let check the efficacy of the targeted therapies through multiple biomarkers across different subgroups of patients. In particular, both general and specific trial designs, which include lesser patient numbers and additionally call for different statistical techniques to compare the effectiveness of treatments and safety profiles.

Also, different therapies thus require companion diagnostic tests that can determine which patients will benefit most from the treatment. The use of these diagnostics and therapies as co-developed and receiving simultaneous approval complicates the regulatory process. It means that regulators must ensure the validation of such tools and verify that they provide accurate information for patient treatment. In general, additional regulation combined with overlapping jurisdictions for therapeutics and diagnostics approval increases the level of fragmentation.

Another key challenge lies in defining appropriate regulatory strategies for emerging targeted therapies, including gene and cell-based treatments. As shown in Figure 14, these therapies involve technologies and manufacturing processes that differ significantly from traditional chemical synthesis used in conventional drug development. Consequently, they demand a more research-oriented regulatory approach. Regulatory agencies must establish specific criteria for production standards, quality control, and long-term safety monitoring. Mechanisms like the FDA's Breakthrough Therapy Designation aim to accelerate the approval of such innovative, molecularly targeted therapies without compromising safety and efficacy.

Several ethical and data privacy issues stand as other immense regulatory barriers. Genetic and medical information can be often collected and analyzed for treatment purposes with some regard to the client's individuality. Regulators have the important responsibility that the required data has to be available for the development of personalized medicines while patient's identities have to be protected. Building public confidence in the effective use of the genetic information of each individual person involves accom-

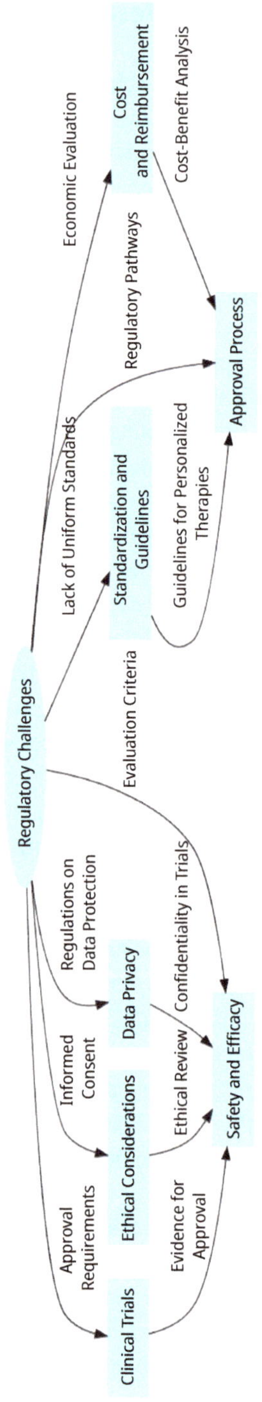

Figure 14: Regulatory challenges in approving personalized therapies.

modating the principle of confidentiality in the use of genetic information in coming up with guidelines for availability of data for the drug development process.

Also, because customized medicine is an international field, regional regulation standards should be brought into consonance. Different regulatory requirements and licensing processes give complex international opportunities for the development and distribution of personalized therapies. Regional bodies such as the ICH are currently in the process of rationalizing requirements and promoting convergence of analysis of individualized therapies.

6.3 Cost and accessibility

The way therapies are going to be broader adopted and incorporated in the theory and practice of medicine is challenged by significant impediments associated with costs and availability of customs-designed medications. Owing to reliance on genetic, environmental, and lifestyle attributes to ascribe care profiles, personalized care medicine requires advanced technology and expertise often. Such aspects put costs of personalized healthcare solutions higher and limit the range of access to them for many clients and providers.

Among all the barriers experienced by advanced-heart-failure patients, the difficulty of kin + cost factors rose to the top, and one of the primary concerns is the high costs of genomic sequencing and genetic testing. Sequencing costs have declined over the last 10 years, but the price of sequencing a human genome is still out of reach for most individuals and healthcare organizations in LMICs. The understanding and analysis of genomic data necessitates integration of bioinformatics as well as personnel, which escalate the expenses for personalized medicine.

Also, the identification of the disease-specific and individualized therapies and treatment regimens require significant amount of investment and testing. Biomedical firms spend a great deal of money in finding and developing individual substances and creating targeted therapies; most of the time, the patented medicines are expensive. Compared to traditional economic model targeted diseases tend to be smaller patient groups with unique genetic makeup, meaning the economic potential of personalized medicine to a particular treatment is likely to be less compared to the cost per patient.

The third significant challenge lies in the issue of availability: where individuals are not offered anything resembling the so-described "personalized care," disparities are likely to increase already existing disparities in health. Perhaps, patients who are from the underprivileged groups receive fewer treatment from different specialists and information pertinent for personalizing of healthcare. Socio-economic factors affect a patient's ability to afford the diagnostic tests and the targeted treatment plans leading to disparities in the benefits of personalized medicine.

Insurance cover and reimbursement policies also form other important aspects of the available and affordable personalized medicine. In some instances, insurance firms do not bear the costs of genetic testing and other personalized therapies because such services are very costly and their use is still under research. This lack of cover stops patients from seeking personalized treatment hence reversing the advances in personalized medicine.

There are plans to resolve issues associated with the cost and availability of the facilities and knowledge of how genomic technologies can be beneficial in the development of the medicine. This requires policy shifts and cooperation between governments, healthcare organizations and industry players to establish cost-effective and technology-enabled care delivery paradigms that support the fair distribution of precision medicine tools.

The focus is on the consumers and the technology and data analysis component advances in terms of cost and future possibilities should eventually make them more accessible. Currently, with increasing improvements in both genomic sequencing and data analysis, personalized medicine becomes more achievable and available to greater numbers of people. Similarly, the use of telemedicine and digital health software also provides a chance to expand the provision of personal healthcare services where the inhabitants are scattered in rugged terrains that are difficult to access.

6.4 Technical challenges

Applying MM entails the sheer number of technical complexities that have to be overcome to deliver on the promise of a new model of healthcare delivery. These issues stem from the case in which huge amount of genetic, environmental, and clinical information have to be put into reasonable and actionable treatment plans. Given the ongoing progress in the field of personalized medicine, it is critical to surmount these technical challenges in order to advance the application and use of PM in clinical settings.

The first major technical issue that has emerged as pivotal for developing personalized medicine is the issue of handling and integrating complex biological databases. The regulatory mechanisms involved in gene expression are complex; hence, the current high-throughput sequencing technologies produce massive amounts of genomic data and the handling, analysis and storage of such data requires advanced computational tools. Given the nature of these datasets, sophisticated tools in bioinformatics and strong computational support are required for analysis and identification of potential trends in this flow of data, which still remains highly challenging and computationally intensive.

Data integration is another great challenge that faces the process of implementing the Big Data strategy. Instead, personalized medicine is based on addressing all types of data related to a patient's genomes, which include proteomics, metabolomics, and clinical history data. The integration of ACM data from various sources presents problems falling under data format, quality, and standard issues. The issue of data convergence also holds some preconditions as it is necessary to establish the set of guidelines and clear formats for the piping of data from various origins to conduct the integration of such data.

Metadata analysis is also challenging, because the majority of genetic variants often described in scientific articles have not been linked to any therapeutic applications. Increased recognition of patients' genomic profiles through the years, the function of discovered disease-related markers, and the way they impacted disease progression and treatment outcomes are not well-understood. Thus, because the understanding of the actual clinical functions of genetic variants is lacking, it is still important to do more study and improve technology for converting genomic information into valuable clinical knowledge.

Further, for successful personalized medicine, effective decision support systems need to be integrated to facilitate informed treatment choices for care professionals using patient-specific data. These systems also must be integrated into current clinical procedures, and this can be technologically complicated, given that such systems are designed to process difficult information in real time and offer practical advice. It is therefore necessary that decision support systems developed are accurate, reliable, and easy to use in clinical environments.

Personalized medicine has other technological challenges, especially concerning data security and confidentiality. This is because alerts on leaks of patients' genetic and health information that leads to identification of patients requires a strong security measure to be put in place to prevent such incidences. Combination of using personalized data for therapeutic and research purposes, which maintains patient privacy and confidence, requires the creation of safe data transfer, storage, and exchange.

The application of personalized medicine requires recent technologies and methods to remain current with the development of this quickly expanding sector. It is particularly important to come up with new diagnostic methods, treatment methods to treat new diseases, and methods for using Big Data to enhance the possibilities of personalized medicine. Further investment in research and development together with active involvement of academia, industry, and healthcare systems is needed to further invest in innovation to overcome existing technical issues.

6.5 Addressing disparities in access to personalized medicine across different populations

The concept of personalized medicine is expected to drive significant advancements in healthcare by offering treatment solutions tailored to a patient's genetic makeup and environmental influences. As illustrated in Figure 15, one of the major challenges lies in the disparity of access to personalized medicine across different populations. These inequities are influenced by various factors, including socio-demographic differences, geographic location, and healthcare system limitations. Recognizing and addressing these gaps is essential to ensuring equitable access to personalized healthcare for all individuals.

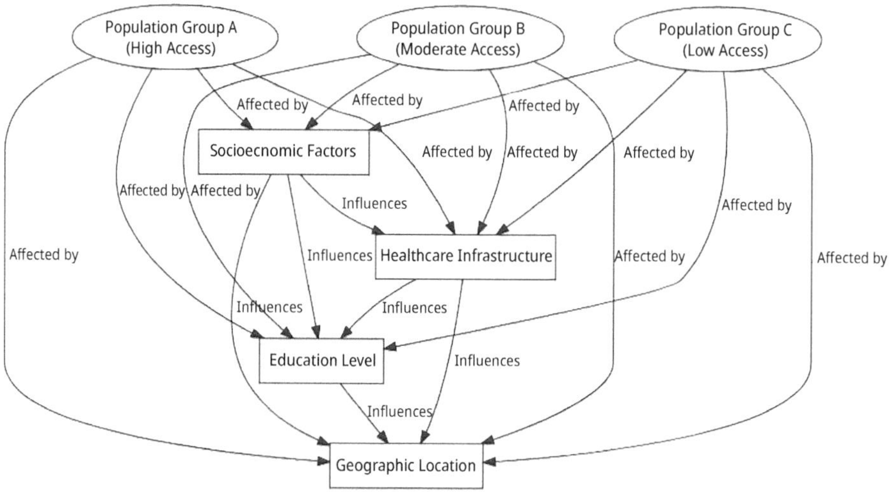

Figure 15: Addressing disparities in access to personalized medicine across different populations.

It is evident that among the main reasons affecting the provision of personalized care is the cost factor at least in relation to genetic testing and other diagnostic procedures. These services can be expensive for those who are not financially well-endowed with health insurance benefits. Thus, the difficulty in directly connecting the client with subjectively customized care is considered to be a major problem for people with low income, as this entails disparities concerning the quality of the care provided to patients and their health in general.

Geographical location is a very important factor in determination of availability of personalized medicine. Remote and unserved communities do not possess adequate physical facilities and expertise in terms of clinicians who can offer targeted treatment. Cohort patients in these areas have to go to extended lengths merely in order to acquire services that are actually capable of conducting genetic tests and providing clients with appropriate medical treatment options. This geographical divide means that those without access to healthcare – which remains a luxury for many in the developing world – cannot benefit from personalized medicine, while the few who can afford it receive further privileges by having their treatments further personalized at expensive urban hospitals.

Also, there is a lack of diversity within genetic studies, which influences how individualized treatment is approached and implemented in different communities. To some extent, most of genetic research that has been done in the past has focused mainly on Caucasian people and less on other ethnic groups. Because of this disparity, more about genetic variations as well as disease markers are unknown between one group and the other; this limits the effectiveness of personalized treatments in non-Europeans. Closing this disparity requires additional research that engages a diverse mix of genetic backgrounds to ensure that customized care benefits everybody, racially.

The tailored approach can also be hampered by, among others, cultural and linguistic differences. Sometimes, patients of different cultural background have difficulty in comprehending medical details to do a genetic test because of such beliefs in their culture. It is proposed that education and understandable communication aimed at improving culture's understanding of the benefits and ramifications of personalized medicine removes these barriers and facilitates personalized healthcare involvement.

Also, there are a number of obstacles to today's widespread use of personalized medicine, primarily data privacy issues. The collection and analysis of genetic information have enough ethical implication and involve security and the privacy of the patients. To alleviate these concerns, strategies that focus on data protection and harmonized policies are called for in order to guarantee that people are willing to participate in custom care services.

Towards this end, the following steps must be pursued in order to level the playing field in access to personalized medicine. The costs associated with genetic testing and the development of personalized medicine should be decreased; this subject should be covered by subsidies, but there are more: attempts to develop medical facilities and increase healthcare accessibility in rural and other unpopulated regions play critical roles in overcoming geographical differences too.

Acceptance of multiculturalism within genomic studies and guaranteed representation of various genetic information for the development of individualized approach to treatment are essential to enhance the general relevance of personalized therapy among ethnically diverse populace. This needs formative funding, implementation, and policies that ensure equal representation of every group in a given study.

7 Conclusion

Personalized medicine is a radical innovation in medical approach that exploits combinations of genomic sciences, informatics, and computing to target medical therapy. This change in paradigm from broad categories of treatment to specific patients' needs promises to revolutionize medicine by minimizing the dangers of reverting the wrong diagnosis and optimizing for the right solutions. The chapter has explained how personalized medicine uses genetic and environmental information to develop treatment plans, setting the drawbacks of conventional treatment approaches including one-size-fits-all.

Another area that benefits greatly from personalized medicine is in preventive and early diagnosis of diseases. Being able to screen patients based on their genetic profile that include biomarkers, the healthcare providers are able to identify patients at high risk of developing the particular condition, therefore diagnosis can be made before their symptoms occur. This approach changes the chances of only providing treatment solution from a treatment solution standpoint by preventing diseases including but not limited to cancer and cardiovascular disorders. Selecting strategies and implementing preventive and corrective action plans based on a risk/need assessment indicates a move up in managing chronic and multifaceted health conditions.

While personalized medicine has reached a high level of penetration there are some difficulties. Socioeconomic, geographic, and systemic differences that affect patients' abilities to receive tailored patient care are also the major challenges to equal care. Eligibility for the cost of sophisticated diagnostic devices and genetic testing limit underserved population that in turn worsens health disparities. Reported matters also draw concern on the ethnic bias and demographics in conducting genomic studies and undermining the generalization of genomic treatments among ethnic and demographic populations.

That is where a wide-scope approach is necessary to solve these problems, alongside measures connected to reduction of prices, advancement of the healthcare system, and attempts to diversify the investigated topics. It is thus only important that the concept of the personalized medicine has to be guaranteed and is to be provided regardless of the patient's financial status. Besides, combining the effective universal strategies studied previously with the targeted methods, the introduction of the concept of personalized medicine with traditional medical seems to bring a ray of hope for enhancing the effectiveness of medical practices as a whole.

References

[1] Suwinski, P., Ong, C., Ling, M. H., Poh, Y. M., Khan, A. M., & Ong, H. S. (2019). Advancing personalized medicine through the application of whole exome sequencing and big data analytics. Frontiers in Genetics, 10, 49.

[2] Ginsburg, G. S., & Willard, H. F. (2009). Genomic and personalized medicine: Foundations and applications. Translational Research, 154(6), 277–287.

[3] Chan, I. S., & Ginsburg, G. S. (2011). Personalized medicine: Progress and promise. Annual Review of Genomics and Human Genetics, 12(1), 217–244.

[4] Dugger, S. A., Platt, A., & Goldstein, D. B. (2018). Drug development in the era of precision medicine. Nature Reviews Drug Discovery, 17(3), 183–196.

[5] Chan, I. S., & Ginsburg, G. S. (2011). Personalized medicine: Progress and promise. Annual Review of Genomics and Human Genetics, 12(1), 217–244.

[6] Ginsburg, G. S., & Willard, H. F. (2009). Genomic and personalized medicine: Foundations and applications. Translational Research, 154(6), 277–287.

[7] Sunil Krishnan, G., Joshi, A., & Kaushik, V. (2021). Bioinformatics in personalized medicine. Advances in Bioinformatics, 303–315.

[8] Nekrutenko, A., & Taylor, J. (2012). Next-generation sequencing data interpretation: Enhancing reproducibility and accessibility. Nature Reviews Genetics, 13(9), 667–672.

[9] Brockmöller, J., & Tzvetkov, M. V. (2008). Pharmacogenetics: Data, concepts and tools to improve drug discovery and drug treatment. European Journal of Clinical Pharmacology, 64, 133–157.

[10] Adam, G., Rampášek, L., Safikhani, Z., Smirnov, P., Haibe-Kains, B., & Goldenberg, A. (2020). Machine learning approaches to drug response prediction: Challenges and recent progress. NPJ Precision Oncology, 4(1), 19.

[11] Ginsburg, G. S., & Willard, H. F. (2009). Genomic and personalized medicine: Foundations and applications. Translational Research, 154(6), 277–287.
[12] Higdon, R., Earl, R. K., Stanberry, L., Hudac, C. M., Montague, E., Stewart, E., . . . & Kolker, E. (2015). The promise of multi-omics and clinical data integration to identify and target personalized healthcare approaches in autism spectrum disorders. Omics: A Journal of Integrative Biology, 19(4), 197–208.
[13] Vasan, R. S. (2006). Biomarkers of cardiovascular disease: Molecular basis and practical considerations. Circulation, 113(19), 2335–2362.
[14] Hood, L., & Flores, M. (2012). A personal view on systems medicine and the emergence of proactive P4 medicine: Predictive, preventive, personalized and participatory. New Biotechnology, 29(6), 613–624.
[15] Di Sanzo, M., Cipolloni, L., Borro, M., La Russa, R., Santurro, A., Scopetti, M., . . . & Frati, P. (2017). Clinical applications of personalized medicine: A new paradigm and challenge. Current Pharmaceutical Biotechnology, 18(3), 194–203.
[16] Beachy, S. H., Alper, J., & Addie, S. (Eds.). (2018). Understanding disparities in access to genomic medicine: Proceedings of a workshop.
[17] Landry, L. G., Ali, N., Williams, D. R., Rehm, H. L., & Bonham, V. L. (2018). Lack of diversity in genomic databases was a barrier to translating precision medicine research into practice. Health Affairs, 37(5), 780–785.
[18] Tawfik, S. M., Elhosseiny, A. A., Galal, A. A., William, M. B., Qansuwa, E., Elbaz, R. M., & Salama, M. (2023). Health inequity in genomic personalized medicine in underrepresented populations: A look at the current evidence. Functional & Integrative Genomics, 23(1), 54.
[19] Tawfik, S. M., Elhosseiny, A. A., Galal, A. A., William, M. B., Qansuwa, E., Elbaz, R. M., & Salama, M. (2023). Health inequity in genomic personalized medicine in underrepresented populations: A look at the current evidence. Functional & Integrative Genomics, 23(1), 54.
[20] Alyass, A., Turcotte, M., & Meyre, D. (2015). From big data analysis to personalized medicine for all: Challenges and opportunities. BMC Medical Genomics, 8, 1–12.
[21] Ng, M. Y., & Weisz, J. R. (2016). Annual research review: Building a science of personalized intervention for youth mental health. Journal of Child Psychology and Psychiatry, 57(3), 216–236.
[22] Ho, D., Quake, S. R., McCabe, E. R., Chng, W. J., Chow, E. K., Ding, X., . . . Zarrinpar, A. (2020). Enabling technologies for personalized and precision medicine. Trends in Biotechnology, 38(5), 497–518.
[23] Kinkorová, J. (2016). Biobanks in the era of personalized medicine: Objectives, challenges, and innovation: Overview. EPMA Journal, 7, 1–12.
[24] Udegbe, F. C., Ebulue, O. R., Ebulue, C. C., & Ekesiobi, C. S. (2024). AI's impact on personalized medicine: Tailoring treatments for improved health outcomes. Engineering Science & Technology Journal, 5(4), 1386–1394.
[25] Hartl, D., De Luca, V., Kostikova, A., Laramie, J., Kennedy, S., Ferrero, E., . . . & Roth, A. (2021). Translational precision medicine: An industry perspective. Journal of Translational Medicine, 19(1), 245.

P. Nagarajan, D. David Neels Ponkumar

Personalized medicine using customized drug delivery systems and patient-specific material solutions, enabled by machine learning algorithms

Abstract: Personalized medicine, as a distinct approach to modern healthcare system, is characterized by the targeted adaptation of medical therapy to a patient's genome, proteome, and metabolome. This chapter outlines the development and the current status of personalized medicine, integrated with focused drug delivery systems and individualized material recipe. The author presents genomic, proteomic, and metabolomic technologies as underlying pillars of diagnostic and therapeutic advancement. Advanced sequencing technologies, high-throughput proteomics, and artificial intelligence technology analyses for enhanced clinical efficacy and resource realization are discussed. In this context, assessments of financial benefits and quality-of-life gains that have been associated with other aspects of targeted methods show their usefulness in reducing the risk of untoward medication interactions, enhancing treatment outcomes and shortening the time needed to deliver care. Recommendations are made for future developments, with special focus on how these enhance the further growth of personalized medicine. These involve precision oncology and molecular analysis, advancement in gene editing, and incorporation of digital health assets. Thus, this chapter provides an extensive overview of their future outlook and outlines the ways that shape the advanced form of individualized medicine, impacting the current picture of healthcare.

Keywords: personalized medicine, genomics, proteomics, metabolomics, artificial intelligence, precision oncology

1 Introduction

The former lumped model in healthcare has been phased out by the modern concept of personalized medicine that is more specific based on an individual's biological makeup [1]. This development stems from the findings in the three related disciplines

P. Nagarajan, Department of Biotechnology, Vel Tech Rangarajan Dr. Sagunthala R&D Institute of Science and Technology, 400 Feet Outer Ring Road, Avadi, Chennai 600062, Tamil Nadu, India, e-mail: nahoobro28@gmail.com
D. David Neels Ponkumar, Department of Electronics and Communication Engineering, Vel Tech Rangarajan Dr. Sagunthala R&D Institute of Science and Technology, 400 Feet Outer Ring Road, Avadi, Chennai 600062, Tamil Nadu, India, e-mail: david26571@gmail.com

https://doi.org/10.1515/9783111503202-006

of metabolomics, proteomics, and genomics, which when combined provide deep insight into the rich network interconnections between metabolites, proteins, and genes [2]. One of its promises is saving the lives of patients, contributing to the efficiency of treatment methods and enhancing the diagnostic accuracy provided by combinations of disparate source of information concerning the patient [3]. Recent advances in high-throughput sequencing techniques have also advanced the scope of this strategy by permitting complex examination of the implications of genetic modifications for health and disease [4]. Personalized medicine is not only an effort to bring improvements to treatment methods but also attempts to avoid diseases through early detection of at-risk individuals due to difference in genetic and molecular characteristics [5].

An important part of large-scale individualized medicine is the development of targeted drug delivery systems to deliver agents with therapeutic activity and minimize the side-effect profiles [6]. These technologies apply proteomics and genetics knowledge to control the method and form of medication administration, according to individual patient characteristics [7]. For enhancing treatment outcomes and reducing the chances of drug-induced side effects, novel carriers can be designed to deliver the drug selectively to the specific target, say tumor tissues [8]. It is easier to design such systems today since nanotechnology and biomaterials have improved to give exact control over the rate of medication release and the targeting strategies [9]. Personalized medicine distribution also improves the general patient experience and patient outcomes since they are less susceptible to negative effects, and receive better therapeutic end-products [10].

Also, it has to be mentioned that the combination of multi-omics data, including genomics, proteomics, and metabolomics, is a major leap forward in the progress of making personal medicine [11]. Here, the omics data are integrated in a way that takes into account the biological context to better understand the systems view of biology and disease [12]. By combining these three approaches, genomic data can be mapped and compared to proteomics and metabolomics, which helps researchers and clinicians define new biomarkers, learn more about the dynamics of the disease process, and create new and precise diagnostic and therapeutic protocols [13]. For example, the association of mutations with the expression levels of proteins and alterations in metabolic profile facilitates the understanding of the development and pathogenesis of diseases [14]. Such an approach allows the definition of the corresponding therapeutic regimens that are applicable singly for each client, depending on the molecular markers, thus providing more effective and individualized approaches to treatment [15].

In this context, it can be said that technology is contributing to the development of personal medicine due to the provision of the instruments necessary for the assessment of other complex bioinformatics and the subsequent implementation of these findings in practice [16]. The opportunity to ascertain whole genomes at a very high speed and a lower cost is now available, which make a significant shift in genomic analysis, enabling advances in sequencing technologies like NGS [17]. Additional comprehension on proteins and its activities are also facilitated by proteomics technologies such as mass spectrometry and protein microarrays in large-scale research [18]. By applying ML and AI algorithms, patterns and predictions to support differentiated treatment decisions are identified [19, 20]. Advances in these technologies are well used in the provision of customized medicine and in improving patients' experiences.

It is expected that research and innovations go on unfolding and strengthening, expanding the application of personalized medicines more in the future [21]. There is a growing prospect for futuristic technologies such as gene editing with technologies like CRISPR-Cas9, use of multiple omics data, and integration of precision oncology [22]. These should enhance the precision of patient care, which includes treatment and diagnosis of illnesses, in the future [23]. It also has its drawbacks, for instance, some issues of the ethical character of data privacy in the process of machine learning; the equitable distribution of precision medicine; and most importantly, the need for proper setting of protocols in this field [24]. Meeting these challenges is important for maintaining the further evolution of personalized medicine as safe and accessible for everyone [25]. It is the accurate, efficient, and customer-oriented treatment that is said to be delivered by the march of what is known as personalized medicine.

2 Historical evolution of personalized medicine

2.1 Early developments in personalized medicine

The genesis of the idea of PM can be dated back to the systematic inclusion of patient characteristics into medical practice though rudimentary means. The notion of individualized medicine can be traced back to the ancient civilizations, where people standards of management of diseases included initial adaptation of management according to patient's distinguishing features and manifestations. The concept of developing dynamically unique medicine for every person has been in practice with the improvement of scientific and up-to-date technology.

The foundation of modern genetics and its relevance to health and disease was established in the early nineteenth century with Gregor Mendel's discovery of inheritance patterns. Mendelian genetics introduced the fundamental principles of heredity, demonstrated through Mendel's experiments on pea plants. These foundational concepts remain essential to the advancement of personalized medicine. As illustrated in Figure 1, the practical implications of genetic information in diagnostics and treatment first emerged in the context of hereditary diseases such as hemophilia and sickle cell anemia, marking the early intersection between genetics and individualized healthcare.

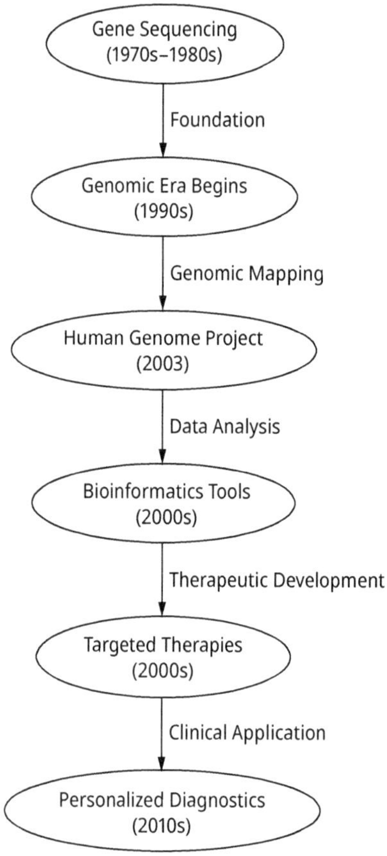

Figure 1: Early developments in personalized medicine.

The sequence of genetic biography began in 1953 when James Watson and Francis Crick found out that DNA was the carrier of genetic information. This made it possible to develop genetic testing that allowed for the identification of genetic variation associated with a number of diseases. Ranking the advancements of molecular diagnostics perspective whilst defining personalized medicine up through early hereditary screening, physicians found a new tool to predict a patient's susceptibility to inherited diseases and adjust preventive care plans accordingly.

Others were developed in 1970s and 1980s, like the polymerase chain reaction (PCR), which paved the way to amplify and analyze specific DNA sequences. PCR technology helped to make the identification of associated genetic markers related to illnesses more direct and also gave a fuller view of the genetic perspective. A growing appreciation of the role of genetic data in individualized treatment also saw the emergence of genetic counselling in this period, which concerned with helping people to comprehend their hereditary risks.

The Human Genome Project thus received a boost, which reached its end by late 1990s with the thinking of personalized medicine. Whole human genome mapping also made an extensive reference for the understanding of genetic differences and its implications for health and disease. It formed a massive foundation for the following rounds of personalized medicine since they provided more understanding of the genetic factors and stimulated the development of new medicines and diagnostic kits, following the genomic data.

2.2 The role of genomics in shaping modern personalized medicine

Genomics have helped the development of the modern so-called individualized medicine, as it has transformed the conventional paradigms of disease management, diagnosis and treatment. The reflection on the novelty of genomics in the medical practice is evident, given that the availability of this sort of technology enables much better understanding of the relevant genetic roots of numerous diseases and the subsequent more refined technique of approaching their treatments.

2.2.1 Newer methods of genomic sequencing

The prospective and innovation of analyzing DNA sequences as well as their capacity has been enhanced by advancements in genomic sequencing, including first-generation Sanger technique to the second-generation NGS. The discipline is said to have experienced a revolution because of NGS technology, enabling the rapid and comprehensive sequencing of whole genomes, whole exomes, and targeted regions. This skill has facilitated the identification of research-based treatment strategies through establishing a variety of disease-associated genetic variants and thereby contributing to the easy development of alternative treatment profiles for each patient.

2.2.2 Discovery of gene signatures and diseases links

Genomic research has explained the complicated relationship between genes and susceptibility to illness by identifying these variations. By organizing variations in the form of insertions, deletions, and single nucleotide polymorphisms, several researchers have been able to associate some genetic anomalies with certain diseases. Based on this information, genetic markers that are important for risk assessment, diagnosis, and development of disease-specific patient treatments have been discovered.

2.2.3 New frontiers in the use of genomic profiling for targeted therapy

Currently, targeted therapy focuses on genomic testing, which analyzes the gene structure of a patient's tumor in the body. It simplifies the selection of the appropriate medication that work, based on the specific genetic abnormalities and variations within cancer cells. Further, the approach of focusing on the specific molecules that cause the disease brings the efficiency of treatments to a higher level, not to mention of the reduction of the likelihood of side effects.

2.2.4 Application of genomics, in conjunction with other omics platforms

The distinction between disease processes and individualized treatments has been considered with the use of genomics, combined with other omics technologies like transcriptomics, proteomics, and metabolomics. To get an improved insight into the molecular mechanisms of diseases, genetic data is integrated with data on protein concentrations, gene transcription rates, and metabolic networks. The holistic approach enables easy development of treatment programs for each patient that involves integration of many biological levels of information.

2.2.5 Ethical and practical implication of genomic medicine

As discussed in the paper, a number of practical and ethical concerns are associated with the extended use of genomics in personalized medicine. As a measure to ensure that free genomic information is protected and that it is treated in a fair way, concerns such as privacy, consent, as well as discrimination have to be met squarely. In addition, data by DNA requires high standards of data management and interpretation to feed into clinical practice for accurate decision-making.

2.3 Key milestones and breakthroughs in personalized treatments

In the following section, we provide a background to the development of personalized medicine from a historical point of view so as to be in a position to highlight the major milestones and achievements that have defined the present-day paradigm of medicine.

The life and work of Rosalind Franklin: breakthrough of DNA structure in 1953

Frederick Sanger developed a sequence system for proteins and James Watson and Francis Crick explained how genetic information is stored by using the double-helix shaped molecule of DNA. This was the precursor to future developments in genetic functional study and individualized treatment.

While recombinant DNA technology can be said to have its evolution as early as 1970s, it is commonly referred to as the Age of Recombinant DNA Technology that began in 1975.

The advent of recombinant DNA technology enabled scientists to manipulate and introduce genetic material into living organisms. As shown in Figure 2, this breakthrough laid the foundation for the genetic development of therapeutic proteins such as insulin and human growth hormone. These innovations marked a significant step toward the production of medicines tailored to a patient's genetic profile, thereby advancing the principles of personalized medicine and genotype-based therapeutic interventions.

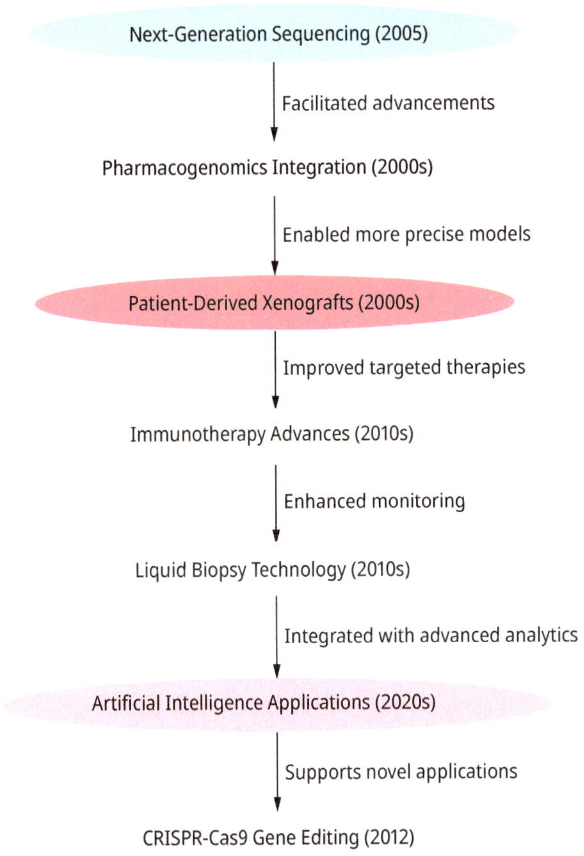

Figure 2: Key milestones and breakthroughs in personalized treatments.

The human genome project: A completion (2003)

The Human Genome Project: In November 2003, after nearly 14 years of work, and with the support of 15 different countries, the international Human Genome Project was complete, with each of these countries successfully mapping out the entire human DNA sequence. This massive achievement paved way for genome-based personal medicine by giving a comprehensive map of human genetic data and identifying genetic markers that are associated with disease.

Introduction of targeted cancer therapy: Targeted therapy – definition and examples

First-line targeted therapy of the type, imatinib (Gleevec), targeted chronic myeloid leukemia treatment, which was one of the most promising trends in the field of personalized oncology. These therapies work selectively on genes and chemical processes because of cancer and help to lessen the side effects of the impacts of chemotherapy as compared to conventional methods.

Pharmacogenomics: The new perspective (2000s).

At the beginning of the new millennium, the phrase pharmacogenomics, the investigation of how a person's genes are affected by the reaction to drugs, emerged into mainstream society. Pharmacogenomics has offered new identification of suitable medication regimens that suit individual genetic make up to enhance therapeutic effectiveness, and lessen the side effects.

Personalized genomic testing: Making strides (Phase 3: Contemporary)

With the appearance of the simple and affordable services of direct-to-consumer genomic testing and internet, people could receive information about themselves, including their genetic risks and ancestry. These tests have given the public useful knowledge regarding genetic information and how it can be used for health purposes.

The current stage of development of AI is the interaction of artificial intelligence and machine learning (from 2010 to the present year).

Biomarkers and diagnostics, big data analytics, and predictive modeling have been revolutionized by the integration of AI and ML into personalized medicine. AI and ML in personalized medicine: Artificial intelligence and machine learning are used to define new pathways in the usage of genetics, and clinical and lifestyle information to identify patterns and potential outcomes of specific reactions to treatments.

It is for these milestones and achievements that the profession of customized medicine has emerged and transformed our way about how we look at illnesses, diagnose them, and treat them. Each of these has made a more complex form of healthcare possible, one that is allowed to cater to the specific genetic and molecular makeup of each patient.

2.4 Influence of technological advancements on personalized approaches

Technological advancement has also been felt majorly in the research and utilization of personalized medicine through the demonstrated progress in the differentiation of medical treatments, based on the patient's profile. It also provides a chance to locate the newer and more precise and efficacious methods in treatment and to comprehend more fully the character of illness.

2.4.1 Technologies in genomic sequencing

The advance globalization of individualized medicine owes it to the high-throughput genomic sequencing technology. Technique such as NGS drastically changed the analysis of large-scale genomic data by providing new knowledge about genetic variants and how they influence treatment responses and disease probabilities. Due to a decrease in sequencing costs, and an increase in speed of data retrieval, genomic profiling is now a regular feature of personalized treatment, providing more precise determination of genes that are linked to particular diseases.

2.4.2 Bioinformatics about computational biology

Computational biology and bioinformatics have come to the aid in explaining such complex biological data. Application of genomic, transcriptomic, and proteomic databases and evaluation of biomarkers are significantly aided by the availability of software tools and algorithm for these purposes. These computational methods enable the integration of multi-omics data, providing a system-level view of disease biology, specific to each patient, and refining the subsequent treatment.

2.4.3 Diagnostic innovations

Novelties in diagnostic tools have also extended the number of opportunities for individualized approach to healthcare. The combination of molecular imaging and high-resolution imaging enables labelling and imaging of disease status and therapeutic outcomes at the molecular level. Advancements in point-of-care testing and wearable health monitors have made it possible to monitor patients' health at close quarters, and quickly alter the management plans, depending on the patient's overall condition.

2.4.4 Artificial intelligent and advanced computing

Personalized medicine now employs ML and AI technologies to fine-tune its analysis of data and work more accurately. Using machine learning, big data is explored, patterns are identified, and the outcomes of various treatment approaches are estimated. AI-boosted technologies, in this conceptualization, facilitate medication innovations, patient identification, and protocol optimization that enhance the effectiveness of individualistic therapeutic approaches.

2.4.5 API integration with digital health platforms

The strategies of chronic illness management and the care-delivery model have shifted significantly as a consequence of the convergence of personal medicine and digital health platforms. SMEs are beneficial for collecting and analyzing patient data from various sources through the use of digital health technologies. Some of these technologies include electronic health records, telemedicine, and mobile health. They enable planning of consultation with the patient and meeting him without personal contact, tracking the condition of a patient's health ongoing, and simple integration of individualized treatment actions into regular treatment activities.

2.5 Case studies illustrating the historical shifts in personalized medicine

According to their mechanism of action, The development of targeted therapies in oncology.

One of the most significant achievements of personalized medicine is the development of targeted therapies for cancer treatment. A landmark example is the introduction of Herceptin (trastuzumab), licensed in the late 1990s, which targets the HER2 receptor a protein overexpressed in a subset of breast cancer patients. This targeted approach marked a turning point in oncology, as HER2-positive patients were able to benefit from a therapy specifically designed for their tumor's molecular profile. The success of trastuzumab offered strong evidence that cancer treatment can be guided by the genetic and molecular characteristics of the tumor, laying the foundation for the ongoing development of other targeted cancer drugs.

2.5.1 New development in the fields of molecular diagnostics in rare genetic disorders

Another history of personalized medicine evolution was the shift in the genetic testing for rare diseases. This was shown by the example of the development of cystic fibrosis (CF). When gene mutations were identified in the early 1990s, genetic testing was developed that enhanced the decision of cystic fibrosis. Due to these advancements, one was able to have proper diagnosing earlier, hence creating room for individual management strategies as well as therapies. CF is a good example on how significant improvements in the detection and treatment of genetic disease have been brought about by genetic research.

2.5.2 The rise of pharmacogenomics

Pharmacogenomics has been critical over time in the historical development of personalized medicine. The case of development and application of warfarin – an anticoagulant medication – is an example of pharmacogenomics and its relevance to individualized treatment. Genomic excerpts from the CYP2C19 gene can also affect the metabolism of Warfarin, and therefore its dosage. Investigations of these variants in the early part of this century resulted in the application of PT testing to determine the appropriate dosage of warfarin, and thus safer anticoagulation therapy. This case illustrates how information relating to genetics can improve outcomes and reduce the side effects of treatment.

Precision medicine, which is a less traditional approach to patient-treatment strategy, has revealed its face in cardiovascular disease.

The trend towards targeted therapy in cardiovascular disease is reflected in the creation of PCSK9 inhibitors. Alirocumab and evolocumab and came into the market in the middle of the past decade for patients with FH and high CV risk, who could be targeted with statins. Through the identification of the PCSK9 gene involved in cholesterol metabolism, targets for personalized medicines based on genetic predisposition to lipid parameters were developed. This case shows how molecular and genomic individualization has bolstered therapeutic management in cardiovascular medicine.

2.5.3 Treatment strategies in specific cancers

One more example of the past modifications to personalized medicine is the treatment of rare disorders such as cancer of Lynch syndrome. Misfires in mismatch repair genes have been associated with Lynch syndrome, a condition associated with increased risk of colon and other forms of cancer. At the beginning of the new millennium, genetic testing for Lynch syndrome emerged. This allowed for targeted monitoring as well as preventive measures with regard to those who are affected genetically and their families. This shift toward targeted screening and prevention shows how genetics info can be applied to other hereditary cancers.

These case studies demonstrate how personalized systems have impacted different fields of practice. Some of these past changes embrace the concept of personalized approach and the following can enlighten of how the medical field has advanced and changed with time:

3 Core concepts and definitions

3.1 Defining personalized medicine

The modern medical practice that goes by the names of precision medicine means that therapy matches the individual patient. This idea refers to the decision-making process, treatment, and therapy, individualized to a patient's genetic predisposition, living environment, and personal traits. Personalized medicine is not identical with the medical model that tends to provide the same treatment for millions of patients, while considering certain peculiarities in an individual in order to enhance the effectiveness of the treatments.

Personalized medicine is basically about integrating a multitude of data – genomic, proteomic, and metabolomic – to drive medical decisions. For instance, genetic data can reveal information about how an individual is likely to develop a specific disease, how most likely the will individual respond to a specific treatment, and the what is the probability of the individual showing adverse effects to medication. In this information, healthcare professionals enhance the protective strategies, organize more directed and effective therapeutic processes, and improve clients' outcomes.

As part of delivering customized treatment, the integration of advanced diagnostic tools and high-throughput technologies is essential for obtaining accurate and comprehensive insights into a patient's condition. As illustrated in Figure 3, this includes the use of high-throughput screening (HTS) to identify diverse biomarkers and next-generation sequencing (NGS) to perform in-depth genome analysis. These technological advancements enhance the understanding of each individual's unique biological characteristics and support the development of personalized treatment strategies tailored to specific patient profiles.

Personalized medicine is larger than the genetic approach to the patient; the concept integrates the patient, the physician, and the healthcare system. It involves blending the objective client-centered health parameters, physician-registered health status, and the context in which the individual exists. To this end, medical therapies are aligned to treat comorbid medical conditions, life styles, environs, and genetic proclivity, among other relevant aspects.

Personalized medicine includes the factors of stratified medicine, which divides patient's group by some criterion, to be more effective. As such, this stratification promotes the effectiveness and safety of medical treatments by targeting patient groups that are likely to benefit from specific therapy techniques.

3.2 Differentiating personalized medicine from traditional medicine

Personalized treatment and conventional treatment are two conventional systems that are different in terms of methods employed and the goals address. It is important

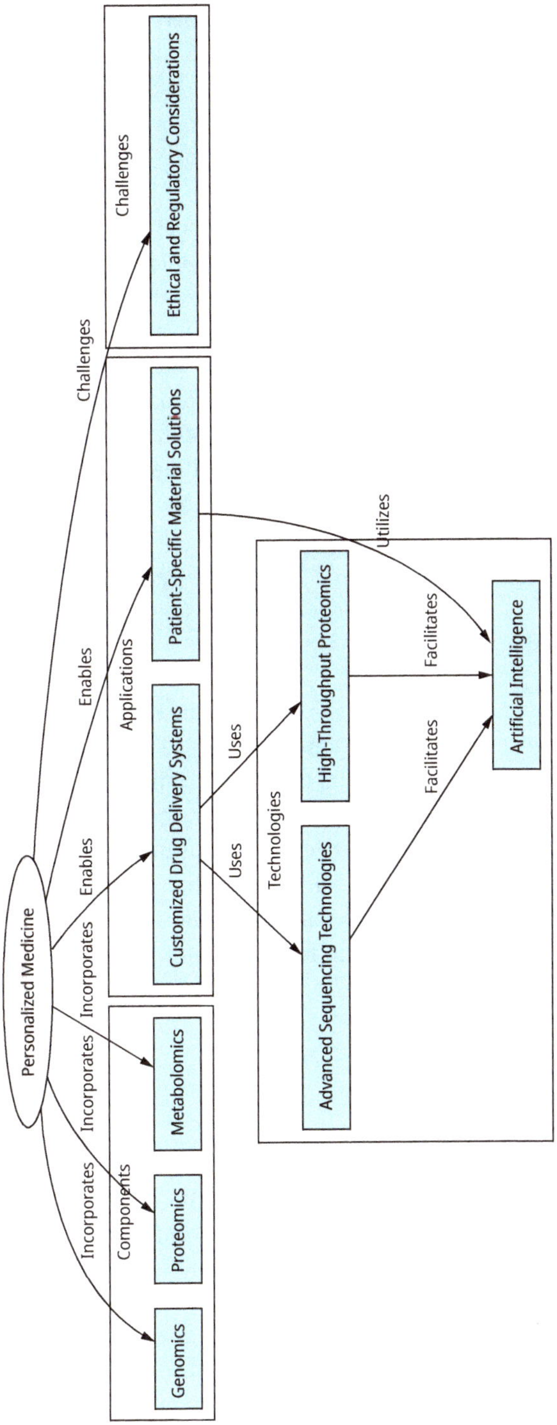

Figure 3: Defining personalized medicine.

to learn the differences between these approaches to realize the progressive character and prospects of the work in the field of personalized medicine.

3.2.1 Treatment and treatment plan/philosophy

Conventional medicine relies on prescriptive principles and most treatment approaches are drawn from controlled scientific studies that involve a large pool of patients. It implements impersonal evidence and technique to solve acute conditions and diseases. Whereas, the latter concept of disease management is governed by the philosophy of 'Personalized Medicine,' where treatment regimes are developed according to the patient's specific attributes and conditions. This approach is based on the knowledge that genetic as well as environmental factors and habits play a certain role in disease distribution and treatment outcomes. Personalized medicine seeks to achieve the best results to be obtained from therapeutic interventions after considering such aspects of an individual.

3.2.2 Role of genetic information

In traditional care system, treatment choices are made in accordance with different signs and symptoms, disease diagnostic categories, approximate age, sex, and color of the patient. Whenever genetic information is to be taken into consideration, it is rarely at the core of a treatment plan. Personalized medicine is closely associated with genetic information. Personalized medicine tries to determine general biomarkers inherent in the human body due to genetic differences that affect diseases and medicines. This, then, is employed to come up with treatments and treatment plans for diseases that are more suitable according to the genetic makeup of the patient.

3.2.3 Diagnosis and risk assessment

The diagnosing and risk evaluation in traditional medicine is not individualized, but generalized; diagnostic criteria and risk factors are the same for most patients. Though often useful for straightforward diseases and illnesses, this approach can easily exclude differences that are present in specific patients. At the human level, tailor-made therapy increases diagnostic precision and evaluation of the individual's risk by using patients' multiple genetic, genomic, and molecular profiles. It ensures that all deviations and inclinations that define disease development and progression are noted, resulting in accurate diagnoses and risk management.

3.2.4 Therapeutic methods and techniques

Medical treatments in traditional medicine are automatically developed in line with treatment plans that result from clinical trials with multiple patient group specimens. These strategies are intended to benefit most patients, but say nothing about the pharmacogenetic variations that define individual metabolism and drug effects. Personalized medicine is aimed at selecting the best possible treatment regimen, based on the patient's individual traits, including genetic profile, biomarkers, and other aspects of lifestyle. It helps them make choices of what treatments to offer and go for the ones that work best for the patient, without many side effects.

3.2.5 Monitoring and adaptation

In traditional applications, change and follow up of management strategies are generally residual and are based on patient's responses to the planned management at successive clinic visits. It is less flexible, but changes occur, as needed; it may be slower and less specific than the case above. Personalized medicine is more of a preventative strategy and involves the use of real-time data and analytics on patients' responses to treatment, so adjustment can be made instantaneously. This means that there is real-time variation in the kind of treatment being applied depending, on the daily performance of the patients and their changing health conditions, thus achieving better patient care as well as health outcomes.

3.3 Essential components of personalized treatment plans

Individualized treatment strategies aim to achieve the least harm and the most benefit from healthcare, dealing with the individual patients' particular characteristics. Several crucial elements are involved in the formulation of these plans:

The optimal patient profiling mechanism is detailed patient profiling.

Many personalized treatment approaches rely on extensive patient profiling, which entails the collection and analysis of comprehensive data related to the patient's condition. As illustrated in Figure 4, this includes demographic information, clinical symptoms, medical history, lifestyle habits, environmental exposures, as well as genetic, proteomic, and metabolomic data. Integrating these diverse data sources enables the creation of highly accurate and individualized treatment plans, as they provide a holistic view of the patient's health profile.

3.3.1 S10Molecular and genetic characterization

Molecular and genetic characterization is essential so as to identify certain molecular and genetic signatures that actually affect the disease process and the response to certain treatment regimes. As the name implies, various methodologies are used, including genomics, gene panels, and RNA profiling to help discover the mutated genes, gene activity, and molecular signaling routes. Such information assists in recommending an individual's specific therapy and developing a treatment model that responds to the molecular characteristics.

3.3.2 Individual therapy models

Integrated individualized therapies are components of the unique treatment. These cover choosing and fine-tuning the therapy according to the patient, his/her phenotype, tumor stage, and previous treatments outcomes. Therapies refer to targeted drugs that are used and individualized chemotherapy. The hypothetical aim of the personalized treatment is to design it in order to achieve the highest therapeutic effect and minimal toxicity based on the patient's biology and clinical characteristics.

3.3.3 Dynamic treatment adjustments

Flexible treatments are also necessary parts of individualized treatment plans since constant patient evaluation and response define the need for modifying the treatments' approaches. This means that after clinical examination, imaging, and biomarkers, the treatment can be altered immediately, since it is a continuous process. This setup also guarantees that the therapeutic management continually adjusts for the patient's changing status and response to treatment, thereby increasing the efficiency and safety of the total treatment process.

3.3.4 Main provisions of the interdisciplinary approach

A prime necessity is enhancing the synthesis of multidisciplinary care structure for the formulation of care strategies. Multidisciplinary teamwork involves working with geneticists, oncologists, pharmacologists, and other specialists to fully address patient's concerns. This structural approach promotes construction of integrated management plans due to the associated systemic relationships between genetic, environmental, and clinical factors.

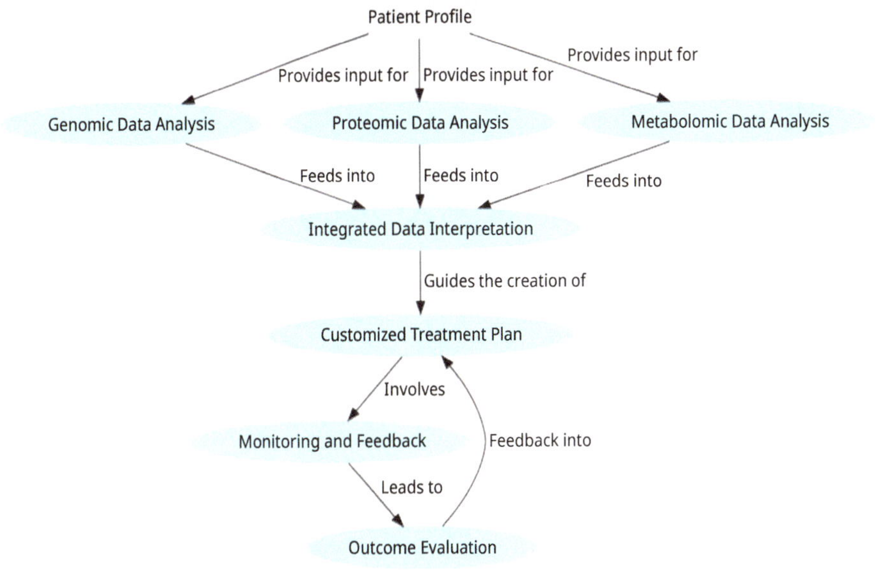

Figure 4: Essential components of personalized treatment plans.

3.3.5 Patient and caregiver involvement

The involvement of the patient and the caregiver in the process of therapy is an essential part of the individualization strategy. An equally important aspect is improving patient's and his family's understanding of the nature of the disease, rationale for individualized management, and the possible course. Optimal coordination and mutual cooperation in decision-making also leads the patient to be more compliant to the set care plan and ensures satisfaction toward the received care.

3.3.6 Ethical and privacy issues

Personalized treatment is focused on the patient's individual needs, and important ethical and privacy concerns can be attributed to the genetic and health information. It concerns their protection and confidentiality through specific consent processes, as necessary for patient information. Patients' rights, together with the rules and laws referring to the practice of personalized medicine, must be respected to ensure that patients' rights are well protected.

Individual patient profiling, which lies at the heart of individualized therapy strategies, involves the collection and analysis of large amounts of patient-specific information. This also includes information about patient history, lifestyle choices, environmental impacts, and genetic, genomic, proteomic, and metabolomic parameters. Therefore, through joining these many data sources, an integrated pathology of the patient is provided, which could warrant more novel and relevant approaches to treatments.

3.4 The role of patient data in personalizing healthcare

The application of patient data in healthcare personalization is an essential element of ASCO's customized medicine implantation process. Usually, patient data is aggregated information that can be subdivided into genetic, phenotypic, clinical, and lifestyle data, all of which are incorporated into treatment plans for the patient, based on his/her unique characteristics.

3.4.1 Genomic data and the influence on personalized medicine

The resources provide necessary information concerning genetic makeup of a specific individual – with a focus on differences that determine how adept a person is to particular diseases and how is his/ her responsive to specific treatments. SNP and gene mutations can also give healthcare professionals an idea of which of the populace is at a higher risk of developing certain disorders. Then, therapeutic treatments can be modified for these patients. For example, genomic profiling helps differentiate specific mutations within cancers and allows the use of special medicines that are more effective, in view of the cancer's molecular characteristics.

3.4.2 Clinical data and its part in decision-making

Digital clinical information, which comprises the patients' medical history, laboratory and radiological findings, and therapeutic response, is relevant for individualized medicine. This information lets clinicians see how patients have reacted to past medication and treatment, how sick they are, and if they are prone to develop any other illnesses. Electronic clinical information can enable clinicians to tailor therapy to a patient's experience and health condition by using clinical information in the decision-making procedures.

3.4.3 Phenotype data and intervention personalization

Phenotypic data describes measurable traits and observable features, the current state of disease and symptoms, as well as the current laboratory values. This data is critical in the aspects of genetic information display as well as outcome of an individual's health. When phenotypic information is introduced, healthcare providers are able to use targeted outcomes to treat displayed symptoms of a disease because the treatment is matched to the phenotype of the affected individual. For example, differences in metabolic enzyme activity can be detected through phenotyping that have implications for certain drugs dosing to enhance the positive effect and reduce toxicity.

3.4.4 Lifestyle and environmental variables in personalization

Factors such as life style, information about diet and physical activity, and exposure to environmental risks make up a considerable part of the concept of personalization of healthcare. Such factors affect the disease occurrence and response to therapies, therefore, their inclusion in individualized management enhances patient care. For instance, lifestyle information can inform the best nutritional changes routines that add onto medication treatments, improving the general wellbeing and helping the treatment to be effective.

3.4.5 Data integration and advance analytics

As many and diverse patient data sources are needed for the purposes of genomic, clinical, phenotypic, and lifestyle integration, enhanced analytics are needed to offer meaningful knowledge of the subject. Structured datasets call for integration and analysis to achieve superior values by using sophisticated analysis like artificial intelligence. By adopting these technologies, it is able to identify relationship patterns and trends that can improve prediction for healthcare management and tailor patient treatment plans.

This new shift of focusing on symptomatic, accurate, and personalized system of healthcare is evident from the proper utilization of patient data in personal health. Clinicians develop specific intervention plans to meet patients' unique needs by integrating an extensive amount of data, leading to improvement in patient status and quality of care.

3.5 Integrative approaches: combining genomics, proteomics, and metabolomics

The combination of genomics, proteomics, and metabolomics provide a holistically and highly detailed idea of biological organizations, as well as the particularities of a specific patient. Each of these "omics" field offers information that can be suggestive from another angle and when integrated, a picture of the full panorama of health and disease emerges.

Genomics: Exploring the concept of gene mapping

Some of the basic understanding of the variation and patterning of genes can be acquired from genomes, which is the means of studying the whole DNA of an organism and all of the genes it contains. Earlier, these differences could not be analyzed comprehensively, but due to advanced high-throughput sequencing technologies, SNP, CNVs, and structural variations can now be studied. This genomic data is helpful in the determination of disease susceptibility, launching of treatment plans, and tailored treatment for individuals. Consequently, decisions on therapy targeting the cancer can be made, once the mutations responsible for tumor development in cancer patients are evaluated through genomic profiling.

Proteomics: Protein function and expression

Proteomics refers to the comprehensive study of proteins, including their structures, interactions, functions, and conformational dynamics. Unlike genomics, which reveals genetic potential, proteomics provides direct insight into cellular processes by analyzing actual protein activity within biological systems. As shown in Figure 5, key areas of focus in proteomics include post-translational modifications, protein–protein interactions, and expression levels, often studied using advanced techniques such as mass spectrometry and two-dimensional gel electrophoresis. By investigating how protein regulation changes during normal and pathological states, researchers can identify early disease markers, discover novel therapeutic targets, and develop clinically relevant biomarkers particularly for diseases such as cancer and cardiovascular conditions.

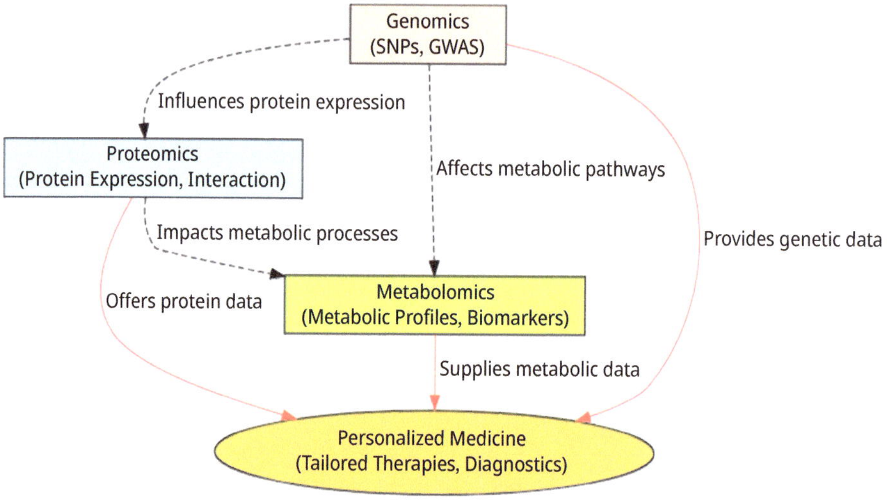

Figure 5: Integrative approaches: Combining genomics, proteomics, and metabolomics.

3.5.1 Metabolomics: Profiling metabolic pathways

Metabolomics is defined primarily by the analytical process of the comprehensive identification and quantification of metabolites, the small chemicals involved in metabolic processes. The specific study area offers information regarding the metabolic processes that are active in health and diseases and are located in cells as well as tissues. To describe alterations in metabolism and to find differences in metabolic profiles related to different conditions, GC-MS and NMR spectroscopy analyze samples. Metabolic data point out alterations in metabolic pathways as also such alterations prior and during disease onset, characterize a disease. The use of metabolomic profiling, for example, has been used to define the biomarkers of the metabolic diseases or to understand the effects of dietary changes.

Integrative approaches: Synergizing omics d

Combining the genomic, proteomic and metabolomic information provide a systematic view of the biological system. The integration of protein expression profiles with genetics variants and changes in metabolic patterns enable the identification of the relationships. For instance, researchers, learn how variation in certain genes affects protein performance and other metabolic processes. They also use a combination of genomics, proteomics, and metabolomics to study disease states. The approach results in increased understanding of the mechanisms of illness. It is equally possible for such an integrated analysis to lead to the discovery of new biomarkers and treatment targets as well as the individualized treatment of a patient, depending on his/her biological make-up.

3.5.2 Systems interoperability and technology solutions: applications and future directions

Integrated omics applications can be expected to cause a big change in the future development of personalized medicine. Using integrative approaches such as genomics, proteomics, and metabolomics, one can enhance the characterization of cancer, predict responses to treatment, and identify novel targets for cancer treatment. In the field of metabolic diseases, integrated approaches can improve illness diagnostics and therapy, if they reveal the necessary amount of knowledge about metabolic dysregulation. Some of these future developments are the development of advanced computational platforms and methods for the compounding and analysis of multi-omics data, and the expansion of the use of omics in treating other diseases and developing customized treatment regimens.

The approach of putting proteomics, metabolomics, and genome into integrative technique is very strong for the development of smart and specific medicine. A lot of fields of research and medicine create a better understanding of the biology and give better and quicker treatment to patients.

4 Technological advancements driving personalized medicine

4.1 Advances in genomic sequencing technologies

Substantial advancement has taken place in genomic sequencing and that alone has revolutionized the whole concept of personalized medicine. Some of these have enabled the study of the human genome in more depth. There is an appreciation of increased protective factors that the analysis of genetic biomarkers has been able to provide, aiding in the understanding of the genes associated with illness and the formulation of personalized treatment.

4.1.1 Next-generation sequencing

High throughput and cost-effective sequencing offered by NGS is another major step forward in genomic analysis. High-throughput sequencing by NGS technologies allows the analysis of millions of DNA fragments simultaneously; therefore, numerous accurate data are obtained. Some of the examples of the said technologies include Illumina sequencing and Ion Torrent sequencing. Because of this technology, sequencing is now much faster and less costly than before; which means that there are more opportunities for genetic analysis in clinical and research facility. The prevailing relationship between genetic profiles and different diseases depends on the identification of the genetic variations in the form of SNP and the structural variations that have been achieved by NGS.

4.1.2 Whole-genome sequencing (WGS)

WGS is used to comprehensively map a complete copy of a person's DNA, including discrete segments contain coding and noncoding regions. Advancements in WGS technology have enhanced both the precision and accuracy of genome mapping and enable the identification of rare and novel genetic variants. Due to WGS providing complete profile of an individual, it is useful in diagnostics of genetic factors to complicated aches and sicknesses. Individual patient approaches, treatments, and genetic disorder identification are all increasingly being powered by this technology, which is already taking place in increasing numbers of clinical contexts.

4.1.3 Whole-exome sequencing (WES)

WES is centered on sequencing only those regions of the DNA that direct protein sequences. Although WES reaches only about 1–2% of genomic DNA, the method allows for extracting information on coding exons, in which most mutation-causing diseases are located. The development of WES technologies has also improved on its sensitivity and specificity making the methodology ideal for tagging genetic sequences that are associated with inherited diseases. WES is used to diagnose and identify disease-related genes when WGS is not practicable because of the cost of information content.

4.1.4 Advanced single-cell sequencing technologies

Single-cell sequencing technologies enable single-cell-level genomic, transcriptomic, and epigenomic data analysis. This method helps to discover gene expression profiles and cell subpopulations masked in a large tissue section, thanks to advances in single-cell sequencing, understanding of the detailed tissues, tumor microenvironments, cellular processes, and the path of the disease. In the field of cancer, single-cell sequencing is fairly useful in defining tumor subtypes and understanding their role in resistance to treatment and tumor heterogeneity.

Next-generation and third-generation sequencing platforms

The limitations associated with the short-read sequencing modality can be surmounted with the assistance of long-read sequencing technologies provided by the Oxford Nanopore Technologies and Pacific Biosciences (PacBio). Previously, using short-read sequencing, only simple regions of variance could be studied, while now complex genomic areas, including repeated sequences and structural changes, can be restored. Long-read sequencing provides the higher-level genomic architecture necessary to understand structural variations and their health and illness implications. It also enhances the quality of genome assembly.

As the improved and broader genetic details became accessible, advancements in genomic sequence technologies provide a rather strong impact to individualize therapy. Using such technologies, people have effectively diagnosed the connection between certain genetic traits and specific illnesses, have enhanced the accuracy of diagnostics, and have been able to draw up personalized treatment. Thus, the concept of personalization of medicine for therapies can be combined with escalating effectiveness as sequencing technologies are implemented in the process of its functioning.

4.2 The impact of bioinformatics and computational biology

Personalized medicine has significantly advanced with the rapid growth of bioinformatics and computational biology, which are essential for managing and interpreting complex biological data. As illustrated in Figure 6, these interdisciplinary fields merge biological science with computational techniques to organize, analyze, and compare large-scale datasets. This capability enables a deeper understanding of molecular mechanisms and facilitates the development of individualized treatment strategies tailored to a patient's unique biological profile.

4.2.1 DMI

Such biological data include genomic, transcriptomic, proteomic, and metabolomic data, of which the success of DM119 is attributed to the contribution of bioinformatics in the management and integration. The high-throughput technologies such as mass spectrometry and NGS require complicated database and data management systems to handle the massive data generated. These technologies enable the data of several sources to be integrated, provide an overall view of the biological systems, and make it easier to identify the relationships and sequences, which are crucial in the synthesis of personal treatment.

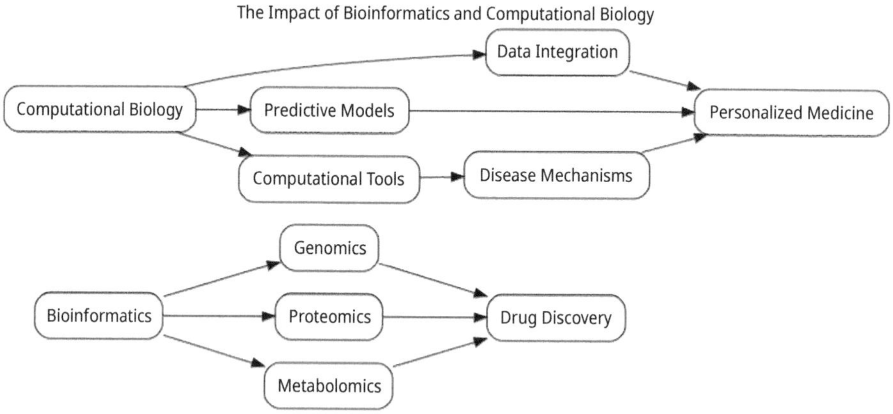

Figure 6: The impact of bioinformatics and computational biology.

More specifically, the attributes are referred to as genomic data analysis and interpretation.

Computational methods play a crucial role in data interpretation at the genetic level of organisms. Raw sequenced data is analyzed using advanced math and computation software to detect genetic variations and relate these variations to biological, functional, and medical effects. Software comprising variant callers, alignments tools, and genome browsers help in the detection of mutations and copy a number of changes and structural rearrangements that determine disease susceptibility and pharmacogenomics. Every human being possesses some form of genetic makeup that responds differently to certain treatments, thus an accurate interpretation of the genomic data can help in building a precise genetic blueprint that outlines the geared therapy and prevention.

4.2.2 Graphics design, logistic regression modeling, and risk analysis

Using specific genomic and molecular markers, computational biology tools are applied to build prognosis models that assess disease probability and anticipate treatment outcomes. Big Data is employed by machine learning and statistical techniques in order to assess various connections concerning the proneness to disease and particular disease types, the disease's evolution, and the treatment outcomes. By estimating such risks and potential impacts of various treatment options, these predictive models, which anticipate the probability of disease occurrence, can provide customized risk appraisal in the clinical practice.

4.2.3 Systems biology and network analysis

Systems approach of computational biology to biology systems can give a key to understanding of topological interconnections between metabolites, proteins, and genes. Gene regulatory networks, protein–protein interaction networks, and coagulation molecular interactions are constructed and evaluated using network analytics. Making changes to the network in disease states and identifying the important regulatory nodes and pathways is made easier by the systems-level approach. Such an all-embracing view of multifaceted biological interconnectedness allows for novel targets of intervention and personalizing the therapy.

The other technical factor that needs to be addressed is the integration of the knowledge representation with clinical decision support systems.

CDSS is increasingly being associated with bioinformatics and computational biology to enhance personalized care. These systems combine patient-related information processes through the use of bioinformatics and provide immediate recommendations on diagnosis, treatment, and patient management. CDSS can assist medical practitioners to make appropriate decisions from the correct clinical/molecular information. Combined in this way, the treatment is improved, the diagnostic accuracy is enhanced, and patient-tailored care is promoted.

4.2.4 Future prospects and issues

Further advancements in data analytics, and data integration and data application are expected in bioinformatics and computational biology in personalized medicine. It is expected that new technologies like huge data analysis using AI methods and advanced calculating algorithms lift the accuracy of analyses and the results obtained. All the same, several challenges remain, like the necessity for homogenization of data formats, integration of numerous forms of data, and management of large data generated by new technologies. To properly implement bioinformatics and computational biology for the improvement of personalized medicine, it is necessary to address these issues.

Computer science and informatics are now geometrically relevant to creating a system for personalizing medication. These disciplines' technologies for data administration, analysis and interpretation make it possible to deliver accurate and customized healthcare. The coalescing of bioinformatics and computational biology further increases the ability to deliver medical treatments that are catered toward the specificities of each patient as technology advances, enhancing the effectiveness and expediting the experience of receiving care.

4.3 Innovations in diagnostic technologies for personalized medicine

Modern diagnostic tools have significantly changed the state of personalization in the medical industry. These technologies enable actual and specific assessment of medical diseases, making treatment and patient care more individualized. The diagnostic technologies that share development of the area of customized medicine are highlighted in the following subtopics.

4.3.1 NGS technologies

For both assessing predispositions to diseases and identifying individuals who are candidates for specific treatments, NGS is a leap forward because it allows for inexpensive and relatively rapid sequencing of whole genomes of genes within the genome. The implementation of NGS technologies provides for complete genome sequencing, essential for generating a comprehensive list of genetic differences related to diseases, and risk assessment of diseases that affect an individual, as well as determine treatment pathways. Such technologies can be coined as whole-genome sequencing (WGS) and targeted sequencing technologies. Relatively rare genetic alterations of specific therapies tailored to one's genetic makeup are some of the uses that have been made possible by NGS. For example, NGS has been useful in characterizing cancer genomes to develop specific therapies built on those specific genetic alterations.

4.3.2 Liquid biopsy techniques

Due to Go sentimental liquid biopsy, it is possible to identify biomarkers related to the presence of cancer in fluids such as blood. Liquid biopsies, different from traditional tissue biopsies that require invasive sampling, detect circulating tumor DNA to describe features of the malignancy for therapy. With the help of this technology, tumors are easily identified at an early stage, and the progression of the disease and the outcomes of the therapy are measured. Due to the developments in Liquid Biopsy technique, tumor kinetics can be observed in real-time and thus based on the available tumor characteristics, treatment strategy can be altered. This has made it possible to improve on sensitivity and specificity levels.

In this present study, the following are the array of research questions: a) What are the current and emerging imaging technologies?

AIRM has significantly enhanced diagnostic potential in individualized therapy by producing detailed and high-quality images of structures and biological processes. In functional imaging, positron emission tomography (PET), Magnetic Resonance Imaging (MRI), and computed tomography (CT) have experienced improvements in spatial resolution. This makes it possible to define the location of the tumor, control the response to treatment, and study the development of the disease. Newer imaging techniques and molecular imaging offer new evidence for disease pathophysiology and increase specificity in therapy management.

4.3.3 Discovery of biomarkers: Steps one and two

Biomarkers are crucial to personalized medicine because they allow identification and verification of diagnoses, prognosis, and predictions about diseases. The proteomics, metabolomics, and genomes high throughput technologies are key that are pushing the discovery of new biomarkers associated with a number of ailments. One of the reasons of higher prevalence is that protein and metabolite biomarkers can be discovered easier by technologies such as mass spectrometry and high-performance liquid chromatography (HPLC). Meanwhile protein biomarkers identification and validation are supported by database and bioinformatics. Applying verified biomarkers in the clinical practice improves prognosis, treatment planning, and diagnosis.

4.3.4 Digital health and wearable technologies

Wearables and digital health enable the continued diagnosis and real-time monitoring of various parameters of the human health, hence broadening the diagnostic frontiers. One of the examples of wearable technology is fitness trackers and smartwatches that can track physiological parameters such as the heart rate, physical activity, and sleep pattern. Information from this data is used in digital health platforms to provide personalized health assessments as well as recognizing possible health issues early. Some of the recent wearable technologies include highly sensitive biosensors that can analyze biomarkers originating from perspiration, saliva, and other body fluids that offer new opportunities for noninvasive health monitoring and personalized therapy.

These advancements in diagnostic modalities are key to moving personalized medicine forward, as they offer the ability to better, faster, and more individually assess the status of people's health. New prospects for the integration of such technologies are expected to be continuously elaborated to provide better substantiation for personalized therapies and optimize the overall outcomes of patients with various diseases.

4.4 The role of artificial intelligence and machine learning

AI and ML are now considered pioneers in the process of roulette of the modern approach to the treatment of diseases – individualized therapy. These technologies have helped in easier assessment of several and complex databases, and predictions and prognosis about the patients, and individual patient management. These technologies involve the use of high-end computers and efficient algorithms to identify features that a normal running procedure overlooks.

In this paper, the two concepts are predictive modeling and risk assessment.

Risk prediction is another critical domain within personalized medicine that leverages artificial intelligence (AI) and machine learning (ML), extending beyond the capabilities of traditional medical decision-making. As depicted in Figure 7, this approach utilizes diverse subtypes of patient data including electronic health records, genomic data, disease-specific characteristics, and lifestyle factors. Supervised learning algorithms are commonly employed to analyze this information and predict health risks. Based on individual characteristics, these tools estimate the likelihood of developing certain conditions, thereby enabling early intervention. For example, ML models can predict the onset of diabetes by analyzing large-scale patient datasets, allowing healthcare providers to initiate timely treatments and implement preventive strategies.

4.4.1 Enhancing diagnostic accuracy

Diagnosis is becoming less of a problem, given that AI is introducing tools that increase both accuracy and efficiency of diagnostics. Among the emerging technologies, ML, with its subfield, deep learning algorithms, produces accurate results in analyzing medical images, including MRI, CT scans, and X-rays. Most of these algorithms can diagnose anomalies with great accuracy, and at times, surpass the skills of the best radiologists in the world. For instance, AI solutions have been designed to detect the first

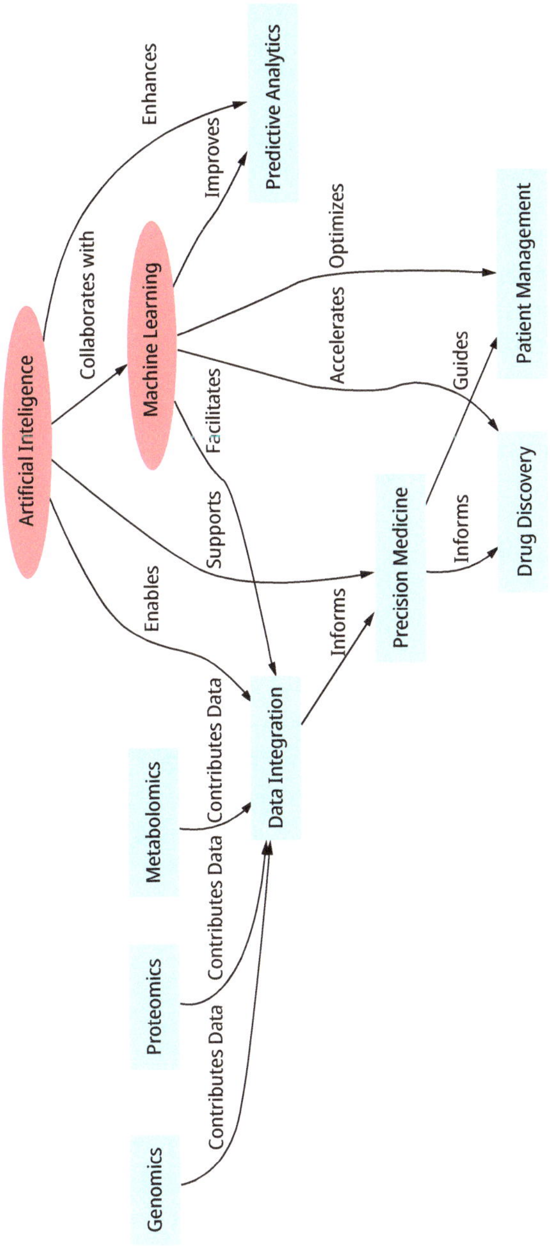

Figure 7: The role of artificial intelligence and machine learning.

signs by analyzing imaging data needed for a diagnosis. AI enables improvement of diagnostic premise and making precise assessments of patients.

4.4.1.1 Treatments based on clients' individual characteristics

Fortunately, to solve such tasks, machine learning models consider patient-unique characteristics, which is an essential element of the individual therapeutic regimens. Thus, the indispensable algorithms would allow individual therapy recommendations based on the data of genetic researches, clinical trials, as well as outcomes of other patients treated until now. In cancer, for instance, machine learning algorithms look at traits of genes present in the tumor to suggest proper medications that are likely to be effective based on the patient's particularity. This method minimizes side effects and optimizes the efficacy of therapy since treatment alternatives are individualized, based on patient's biological characteristics.

4.4.2 Drug development optimization

Through using AI and ML, different stages of the drug discovery and development process are being enhanced. With the help of such algorithms as machine learning algorithms, one is able to explore the large number of records and look for possible drug leads, look for the possible effectiveness of the drug and, at the same time, look for possible side effects. The current approaches of using artificial intelligence in clinical trials help in designing trials with patient response prediction as well as the best design of clinical trials. For instance, AI systems can recognize those biomarkers linked to drug effects, which can help make bespoke drugs and shorten the time and expenses involved in making new drugs market-ready.

4.4.3 Integration with omics data

By integrating AI and ML with genomes, proteomics, and metabolomics, the ability to handle complex biological systems and design of drug treatments is enhanced. These AI algorithms search for relationships between variations in genes, proteins, and metabolism in multiple omics datasets. This integrated approach enhances discovery of new biomarkers, understanding of diseases' origin, and development of personalized treatment strategies. In order to predict patient's response to given medication, for example, proteomic and genetic information can be input into the ML models, thus delivering more personalized and effective treatments.

By enhancing the prospective forecasting capabilities for disease diagnosis and treatment, design of therapeutic plans, development of medicines, and combination of multi-omics, AI and ML play a revolutionary part in personalized medicine. These technological advances are enabling important advancement within the field, allowing for more innovative and individualized advancements within the field. By means of the further advancement and integration of AI and ML into personalized medicine, the field has the possibility to only become more successful as well as efficient.

4.5 Emerging tools and technologies enhancing personalized care

Personalized medicine has thus undergone a paradigm shift since the development of new tools as well as technology that enhance the ability to fashion healthcare to meet individual patient's needs. These changes offer new opportunities for precise diagnosis, effective treatment, and improved patient prognoses.

4.5.1 Next-generation sequencing

The technology of NGS has revolutionized the genomic analysis due to the opportunity of sequencing whole genomes at a relatively low cost and in a short time. Whole-genome sequencing (WGS), whole-exome sequencing (WES), and targeted sequencing – these techniques produce comprehensive genetic data that is instrumental in identifying the corresponding genetic variations associated with different pathologies, large-scale genomic studies, and inclusion of genetic information into clinical practice, being enabled by the principles of high-throughput sequencing of tens of thousands of samples, which contributes to further developments of individualized diagnostic and therapeutic management strategies.

4.5.2 Next-generation proteomics platforms

High-throughput proteomics technologies such as mass spectrometry and protein microarrays are two exemplar ways that have greatly enhanced the analysis of protein expression, modification, and communication. These systems help in the examination of hundreds of proteins at once, revealing new insights into what goes on in cells, as well as what can go wrong. By finding and measure the protein biomarkers, researchers have a possibility to make individualized therapeutics and better understand the development of a disease. These technologies also play a role in identifying many new targets for therapies and designing treatment regimens based on specific protein signatures.

Metabolomics and tools for metabolic profiling

Global analysis of metabolic variations is achieved, thanks to metabolomic approaches, including liquid chromatography-mass spectrometry (LC-MS) and nuclear magnetic resonance (NMR) spectroscopy. These instruments provide information on an organism's metabolic state and establish biomarkers associated with a number of diseases. Metabolic profiling is useful in elucidating disease pathophysiology, following the progression of an illness and assessing the effects of therapy. The complementary approaches adopted from omics data enhance the understanding of individual metabolic differences for personalized therapy methods.

4.5.3 Artificial intelligence and machine learning algorithms

That is why, ML and AI algorithms are applied to an increased degree to assess the extent and complexity of the data, and receive the data of useful importance for individual approach. Large amounts of genetic, proteomic, and clinical data are analyzed using AI-based technologies to define patterns, predict the probability of disease, and prescribe personalized therapy. Some of the machine learning models that are applied in graphic pattern identification, drug designing, and categorization of patients include neural networks and support vector machines. There is enhanced decision-making since these tools perform complicated analytical procedures and provide predictive analytics.

4.5.4 Smart clothing and personalized health solutions

Through information and communication technology, including digital health platforms, physiological parameters and other forms of health information are continually measured through wearable health gadgets like fitness trackers, smartwatches, and others. Information such as the pulse rate, metabolism, activity, and sleeping habits, among others, are captured by these gadgets, with lots of details of an individual's health status. To achieve suitable understanding, there is a connection of the wearables' data to mobile health applications and EHRs, in which timely interferences and personal health management are created. Telemedicine is also made easy by digital health platforms in the same way that remote monitoring is achieved while enhancing patient engagement and tailored treatment.

4.5.5 CRISPR and other gene editing technologies

CRISPR and Cas9 and other molecular scissors are new-generation gene editing tools with a high possibility of altering the genetic code sequence. Researchers study the functionalities of a particular gene, remove abnormalities in the inherited genes, and design new treatment modalities. It covers gene therapy for inherited diseases, targeted cancer therapy, and generation of disease-relevant cellular models for drug screening. While these technologies evolve, they have the possibility to revolutionize the way in which genetic diseases can be treated through personalized medicine.

4.5.6 Advanced imaging techniques

New imaging techniques like MRI, PET scans, and multiple photon microscopy give a high-resolution and often time-lapse snap shots of the biological activity. Motorized imaging systems of this kind allows direct observation of the evolution of the disease, the effectiveness of treatment, and changes in the physiology of the body's tissues in the process of the progress. Incorporation of imaging with molecular information in the clinic improves disease profiling and outcome measures since this combines anatomic and physiologic information into one assessment.

By the steady progression of these novel instruments and equipment, personalized medicines' progress is moving forward as it provides improved diagnostic accuracy and formulating directed treatment and corresponding patient care. These technologies, therefore, hold the potential of expanding the capacity to individualize medical therapies to be of even higher value for strengthening our health and increasing the quality of care.

5 Benefits and impact on healthcare outcomes

5.1 Improved diagnostic accuracy and early detection

Customized medicine deal with early illness identification and diagnosis accuracy, yielding better patient outcomes. Again, the improvements in diagnostic technology and a combination of proteomic, metabolomic, and genomic factors are what are pushing this forward.

5.1.1 Greater accuracy when identifying the disease

In general, the delivery of precise and effective treatment is significantly enhanced through the integration of comprehensive genetic, proteomic, and metabolic data. Advanced tools capable of detecting specific biomarkers and genetic variations associated with diseases enable clinicians to improve diagnostic accuracy. As depicted in Figure 9, genomic sequencing facilitates the identification of rare genetic mutations and variants that are often undetectable using conventional diagnostic methods. This level of precision reduces the likelihood of misdiagnosis and ensures that each patient receives a diagnosis and treatment strategy tailored to their individual biological profile.

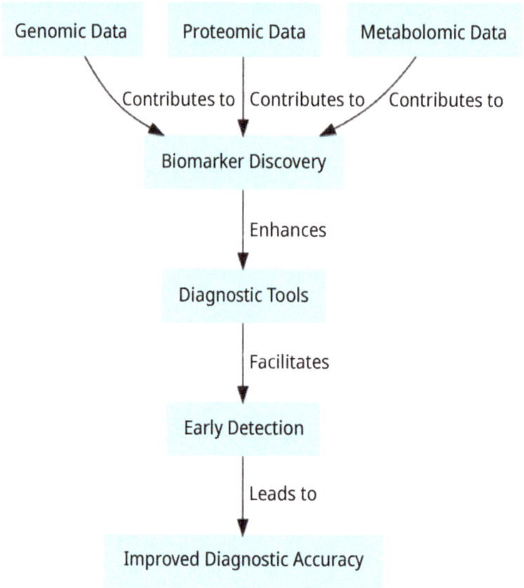

Figure 8: Improved diagnostic accuracy and early detection.

5.1.2 Screening of other silent compensated diseases

Using specific diagnostic procedures, the ability to diagnose the sickness at a preseason stage has been boosted. For such illnesses as cancer, where timely prognosis can radically alter the further evolution of the disease and have a positive effect on the prognosis, the identification of disease is critical. For instance, techniques like liquid biopsy, whereby circulating tumor DNA are used for diagnosis make it possible to diagnose cancer at a clinical-sign stage. Likewise, new advancements in genetic testing of genotypes of inherited diseases allow identification of high-risk individuals for early referrals for treatment.

5.1.3 Factors contributing to high risk for pathogens

Personalized medicine therefore enables the identification of the high-risk groups on the basis of gene link with other individuals. By studying the DNA and ancestors, the doctor can divide the patients into some groups with different risks and set up appropriate screening guidelines. It is thus possible to provide extra time and thorough examination to those people who are in the increased risk group when it comes to definite diseases. For instance, counseling patients with BRCA1, results in better monitoring for breast and ovarian cancer, thereby preventing the occurrence of these diseases.

5.1.4 Enhanced diagnostic precision due to the inclusion of multi-omics data

Combining genomics with proteomics and metabolomics data allows the clinician to pinpoint the disease pathology accurately because it covers different aspects of the abnormalities in the human body. This integrative approach enables researchers to identify biomarkers that perhaps are invisible across any singular omics range. For instance, integration of genomic information with proteomic and metabolomic profiles provides numerous comprehensive interactions and alterations that transpire as precursors of diseases and inflammation. This overall perspective consequently increases the accuracy of diagnostics and promptly identifies diseases.

5.1.5 Supplementary surveillance of disease status

Personalized medicine also enhances the communication on disease status since there is frequent diagnosis and data capture. Wearable health technology devices and remote health monitoring systems facilitate the monitoring of state biosignatures for disease conditions. This enables her to follow up on a given disease's notifications and prognosis, indicating how a disease is changing and how a patient is improving with treatment. For instance, in diabetics, the monitors that measure blood sugar levels offer real-time information to aid in changing the management techniques to enhance the disease state.

5.1.6 Enabling specific client management

The fact is that with the help of increased diagnostic accuracy and early detection of the disease, individual therapeutic guidelines that are tailored to the individual characteristics of each patient's illness are worked out. Clinicians select the most effective interventions after understanding the specific nature of the disease when making a proper diagnosis. Tumor profiling, for instance, can be applied in oncology where through genome analysis of specific mutations can help in targeting by individualized medicines, leaving out deadly and unsuccessful treatment options.

5.2 Tailoring treatment plans to individual genetic profiles

These include enhanced future care arrangements overridden by the patient's genetic plan as a major enhancement of the sector of personalized medicine with vast advantages and better means of healthcare delivery systems. The use of genetic data in the management, treatment and providing care to patients makes it easier for healthcare professionals to achieve and improve on results to improve the tentative outcomes.

5.2.1 Higher accuracy of diseases differentiation

Thus, by assigning treatments based on specific genetic profiles, healthcare providers can achieve more accurate disease diagnoses and tailor treatment regimens accordingly. Genetic testing enables the identification of mutations linked to various diseases, leading to a deeper understanding of illness mechanisms. For instance, in oncology, genomically targeted screening can reveal key genomic alterations that are well-suited for therapeutic interventions, ultimately improving patient outcomes. This precision not only enhances the quality of care based on a patient's genetic makeup but also minimizes diagnostic uncertainty.

An essential factor in optimizing Personalized Biomedical Engineering and Medicine (PBEM) is the accurate selection of drugs and determination of appropriate dosages tailored to individual genetic responses.

This is because the most important strength that is associated with the practice of personalized medicine is the ability to properly select the most effective drug AND its proper dosage. Enzymes are involved in the metabolism and effectiveness of drugs and whether they experience certain side effects in large part depends on genetics. For instance, pharmacogenomic tests determine genetic differences in enzyme like CYP2C19 that influences the metabolism of drugs like clopidogrel. Through such hereditary influences, it becomes easier for the clinicians to administer medication that is right for the patient's genes, and can change the quantity administered with the aim of bringing out the best result with the least side effects.

5.2.2 Improved treatment efficacy

New treatment protocols that are based on an individual's genetic makeup are bound to be effective in the treatment of various ailments. For example, in oncology, different patients have different gene mutations that cause different reactions for some chemotherapies. It allows for the doctors to choose the treatments that are effective against the specific genetic foundations of the tumor and thus raise the chances of successful treatment, leading to higher patient survival rates.

Precautions to ensure adverse drug reactions are at the minimum

Genetic testing evaluates the chances of having particular side effects and, therefore, can reduce the risks of adverse drug reactions. For example, genetic variation in TPMT affects the metabolism of anti-inflammatory and antineoplastic thiopurine drugs used for the treatment of autoimmune and other diseases. Pharmacogenomic testing enables the prescribers to use appropriate drugs and in appropriate dosages, reducing the risk of adverse effects, thereby increasing patient safety without having to rely on trial-based methods.

5.2.3 Individual preventive interventions

Genetic testing is also used in wording treatment goals to genetic test results to not only include therapeutic initiatives but also preventive ones. Known inherited susceptibilities for specific diseases can be established and can pinpoint people who are likely to develop specific diseases, including cardiovascular and cancer. These assessments make it possible to introduce individual preventive interventions such as changes in life styles, routine investigations, and chemoprevention. Thus, that approach is timely in checking disease incidences and managing the same, hence leading to perfect health in the future.

5.2.4 Patient participation and compliance

Treatments that have been developed considering the patients' genetic makeup help to improve patients' compliance to the therapy. If patient's trust the treatment that was developed considering the genetic makeup of their body, they are most likely to understand the worth of the prescribed therapy and follow through on the treatment plan that has been given to them. Having increased compliance improves therapeutic outcomes and generally the satisfaction of the patients.

The incorporation of gene analysis into the treatment process is one of the biggest revolutions in medicine, which has many advantages: better diagnosis, the choice of the most suitable medication, better outcomes, and minimum side effects of treatments. Personalized medicine therefore applies genetic information to target and deliver more efficient and effective treatments, with the patient at the core of the treatment.

5.3 Enhancing drug efficacy and reducing adverse drug reactions

As a result of the improved medicines by disease- and symptom-specific medications, medicine effectiveness has gone higher, while the occurrence of ADRs has reduced, thereby enhancing the success of the therapeutic treatments. Issues related to medicine efficacy and safety of diverse patient groups are solved in personalized medicine that adapts treatments for individual patient characteristics like genotype, proteome, and metabolome.

5.3.1 Pharmacokinetics, drug therapy, and dosage regimen individualization

One of the biggest strengths of personalized medicine is the ability to increase drug dosage according to the features of the individual patient. Pharmacogenomics show

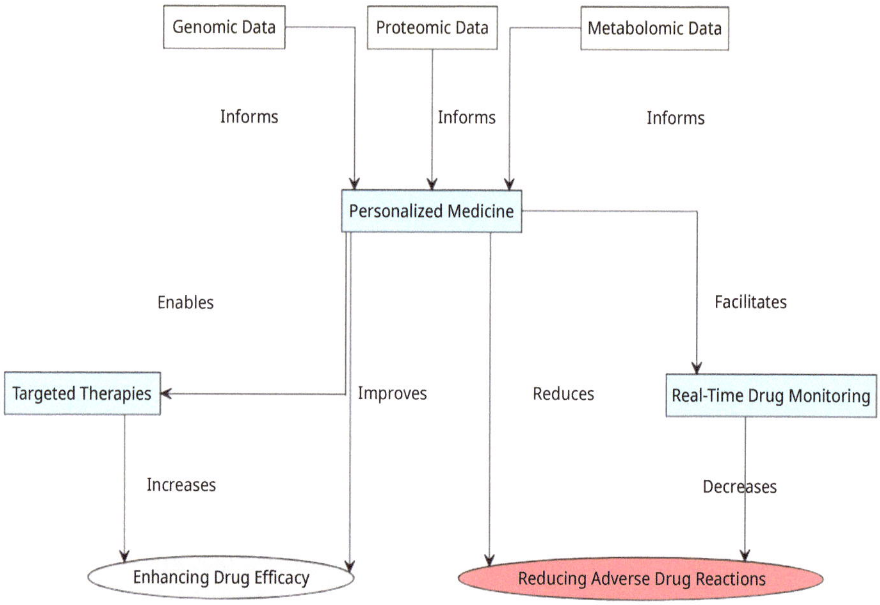

Figure 9: Enhancing drug efficacy and reducing adverse drug reactions.

that genetic differences in patients can significantly affect the metabolic outcomes and drug effectiveness. Due to the various genetic plays as far as metabolism of certain medication is concerned, pharmacogenomic testing helps medical experts to adjust the dosage of the medicine for optimum therapeutic value. To enhance the activity and reduce the risks of side effects that occur, pharmacogenetic genotyping, for example, for polymorphisms in the CYP450 enzyme system can be employed to prescribe dose adjustments of such drugs as antidepressants and warfarin.

5.3.2 Molecular and individualized treatment

Most diseases, and especially cancer, have been upgraded from the traditional way of treating to personal treatment. The ultimate objective of targeted therapy is to selectively interact with actionable molecular targets, which are implicated in futuristic causation of the disease. These treatments often enhance the drug efficacy and decrease adverse effects due to the focus on special genetic abnormalities. There are many examples of small-molecules, such as tyrosine kinase inhibitors imatinib in chronic myeloid leukemia (CML), where it has increased case outcomes and reduced side effects compared to conventional chemotherapy.

5.3.3 Screening and assessment of biomarkers

Detection and monitoring of biomarkers directly play a significant role in improving the impact and minimizing ADRs, which are so largely associated with medication efficacy. Biomarkers can offer details about disease states, predict the effectiveness of treatment, and recognize those who may exhibit adverse reactions. For instance, selection of patients who would likely benefit from trastuzumab (Herceptin) in breast cancer treatment is made possible by the use of HER2 as a biomarker. Evaluating total scores of liver function tests can also help in establishing the chances of developing drug-induced patotoxicity, and providing early treatments to prevent further complications.

Pharmacovigilance and risk management in health and medicine

The rationale being that compared to mass treatment, it would provide individual details regarding drug pharmacodynamics and impact the patient, thus enhancing drug safety monitoring. There is an opportunity to analyze the information obtained from wearable devices through subjective outcome measures and EHRs in tracking the efficiency of treatments as well as in understanding the patient's response to particular medications. This real-time monitoring makes it possible to use risk mitigation techniques and also identify all the problems at an early stage. It would be useful for clinical decision-making and pharmacogenomic data integration into the clinical practice. Possible application scenarios are for example, when considering how pharmacogenomic data integration into clinical practice identifies specific patients who are predisposed to certain diseases, redetermining specific patients who are at risk of certain ADRs. This can lead to better prescription activities as well as early prevention measures.

5.3.4 Enhancing patients compliance and health outcomes

Personalized medicine reduces the deviation probability from the prescribed course of treatment, since it is based on the initial assessment of the patient's characteristics. Hypothesized here is the fact that the treatment tailored toward a patient's individual genetic and metabolic constitution is more effective and better tolerated, hence patients will be more compliant. A better degree of compliance with medications increases the efficacy of treatments and decreases the risk of deterioration of health of a certain ailment. For instance, special focus can be made in the treatment of diseases like diabetes and hypertension with outcome measures that show better control of disease indicators in the long run and improved general health.

Using personalized medicine in clinical practice is of tremendous value since it improves the effectiveness of the drugs and decreases the rate of adverse reaction to medicine. Through the use of genetic, proteomic, and metabonomic information, healthcare providers can accurately dose drugs, choose therapies that are most effective, and manage risks. All these advancements enhance therapeutic intercessions and safety, resulting in enhanced patient success and the general quality of healthcare.

5.4 Personalized medicine's role in chronic disease management

The principle of individual therapy and care is the cornerstone of PM as it approaches treatment and management of chronic diseases based on patient's details. This approach enhances the intervention efficiency, lifts the corresponding patient results, and decreases resource consumption in chronic disease treatment.

5.4.1 Tailored treatment plans

This is the creation of individualized treatment strategies based on patient's genetic factors, environment, and lifestyle. In diseases like diabetes, prescriptions have to be tailored because factors such as the genetic makeup that determine how drugs are metabolized can be addressed. For instance, pharmacogenomic testing would help determine the status of genes that affect the response to antidiabetic drugs, the dosage, and the side effects involved. Structured treatment enhances the way in which patients with disease are cared for because the treatment regimen is based on the patient characteristics of each unique patient.

5.4.2 The development in disease surveillance and control

Personalized techniques improve the ways in which disease is diagnosed and treated by applying new and sophisticated instruments, and continuous tracking of the patient's condition. Computerized wrist-worn health monitors and smart phone health applications give constant data on biomarkers, including glucose, for those with diabetes and hypertension. Such real-time patient data facilitate early modification of the patients' treatment strategies and application of treatment techniques that help control the diseases. For example, the use of the continuous glucose monitors means that hundreds of values are recorded and analyzed and the dosages of insulin is more effectively regulated for patients with diabetes.

5.4.3 Patient engagement and thus medication adherence

Personalized medicine enhances patients' participation and compliance with the recommended treatment procedures through delivery of treatments that are closer to the patients' needs. Through engaging patients in the process of choice making and giving feedback, more are likely to follow through the doctor's recommendations and make necessary changes for disease management. For instance, individualized education and individual management plans for patients focus on explicit behavioral concerns across a health spectrum and increase the motivational commitment to manage chronic conditions.

5.4.4 Minimization of the adverse drug reaction

Due to considerations of patient variability in drug metabolism and effectiveness, customized medicine significantly reduces the probability of side effects in chronic diseases stabilization. Through pharmacogenomic testing, it is possible to identify genetic factors associated with increased risk of side effects' interactions. For instance, patients with heart disease can use genetic testing to select antihypertensive medicines and minimal side effects. This kind of approach makes therapies safer and acceptable and results in better outcomes for patients besides reducing overall the cost of the health facilities.

5.4.5 Inspiring resource management

Custom care enhances efficiency in resource management in chronic disease treatment because it aims at intervention, where it has the most impact. Understanding a patient's biomarkers – including genetic predispositions – can help the medical system locate patients who are at the highest risk for complications to benefit from certain treatments and direct the resources to be used effectively. Such a selective model minimizes the exposure and risks from nonrelated treatment and hospitalization, thus minimizing healthcare expense and maximizing the effectiveness of the disease management program.

This improves the multilayered, advanced, and sophisticated frameworks for predictive and preventive measures.

Using genetic markers, personal history, and lifestyle factors, personalized medicine moves ahead in preemptive measures for sick MESHD chronic disease candidates. These include occasions such as testing for inherited risks for cardiovascular diseases, where early counseling measures, including change in lifestyle or prophylactic drugs could be instituted. Due to the detailed risk assessment data, one can predict cases of chronic diseases and try to prevent them due to the calculation of the unique risk factors.

5.5 Economic and quality-of-life benefits of personalized approaches

Some of the potential economic and quality-of-life impacts associated with decisions in personalized medicine include reduced healthcare costs, improved treatment efficiency, and enhanced patient satisfaction. By tailoring therapies to an individual's genetic and physiological profile, personalized medicine minimizes trial-and-error prescriptions and reduces adverse drug reactions, which in turn decreases hospitalizations and long-term treatment expenses. Moreover, patient engagement in treatment plans specifically designed for their needs leads to better adherence and outcomes, ultimately improving their overall quality of life. These benefits are clearly illustrated in Figure 10, which highlights the dual impact of personalized healthcare on both economic efficiency and patient well-being.

5.5.1 Minimization of adverse drug reactions and healthcare costs as well

Personalized medication aims at avoiding and even predicting unfavorable reactions of the body to medication through tailoring therapies according to the genotype of the individual. Pharmaceutical providers decrease the instance of adverse reactions to medications by eliminating hazardous courses of treatment by heading prescriptions as well as dosage based on the discovered genetic information. As a result, this type of strategy reduces the tendency for costly and risky interventions and hospitalizations, while enhancing patient outcomes. For this reason, people as well as the healthcare systems make significant savings due to personalized therapy.

5.5.2 Improved treatment effectiveness and result

Pharmacogenomics enhances therapy outcomes through making it possible to use treatments that have a strong correlation with the biological aspects of the patient. For instance, molecular and personalized medicines related to cancers have been revealed to improve therapeutic indices and survival in patients diagnosed with tumors. Better treatment outcomes translate into improved health, less disease prevalence, and lower requirement for follow-up and supplementary therapies. This fact helps to solve the issues of economical utility and increase patient satisfaction due to the opportunity of attaining the treatment goal with less amendments.

5.5.3 Screening and early interference

Personalized medicine increases the chances of early diagnosis of diseases, in general, by genetic testing and various predictions. Detection of diseases, including cancer, cardiovascular, and hereditary diseases, means that treatments are initiated before the diseases progress, and reduction of severity is enhanced. Care outcomes deliver conclusions that prevention and early intervention are cost effective as it contrasts with more intensive and expensive interventions in the future, while improving patient's quality of life.

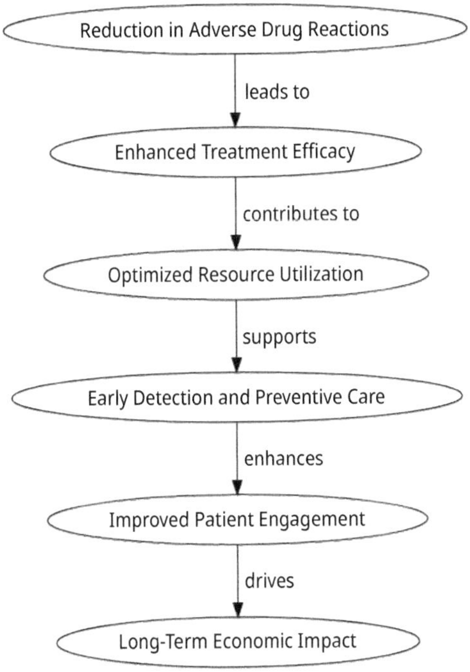

Figure 10: Economic and quality-of-life benefits of personalized approaches.

5.5.4 Better outpatients' involvement and satisfaction

There are always generally individualistic treatments that are given by a physician, depending on the condition of a client and what they need to go through. This notion enhances patient participation since treatment plans consider patients' preference and situation. Incorporating the patient into their care improves their compliance with treatment regimens, decreases the chances for complications, and even results in a more positive experience. Patient satisfaction is added to the quality of life by focusing on health and has a direct impact on the quality of health and on the organization of care.

5.5.5 Economic consequences and healthcare system resilience: current and future

The benefits of personalized medicine operate in the long-term structure of the economic model and includes healthcare sustainability. With timely diagnosis and proper therapies, effective prevention measures, and right treatments that fit each person, costs that are otherwise incurred in the future can be considerably reduced by the concept of personalized medicine. Efficient elimination of chronic diseases cuts emerging healthcare expenses equally as they cut hospitalization and urgent care charges. All these economic impacts enhance the strategies of delivering healthcare.

6 Current trends and future directions

6.1 The rise of precision oncology and targeted therapies

The combination of genomic, proteomic, and clinical data has paved way to enhanced cancer therapy and even development of personalized and/or precision medicine. Due to the fact that they are specifically tailored to target aspects of tumor phenotype, genetic, epigenetic, and pharmacologic approaches to personalized therapy of cancer, they aim to optimize the beneficial effects of treatments and minimize side effects.

Some of the most significant technological developments in genomic profiling

Genetic profiling technology is a key driver in the evolution of precision oncology, as it enables the demographic identification of patients' genetic tendencies. Targeted Exome Capture (TEC), a high-throughput sequencing technique, has revolutionized the analysis of complex cancer genomes by facilitating the identification of specific genetic changes and variations associated with various cancer types. This wealth of genetic information supports the identification of molecular targets suitable for precise therapeutic intervention. For example, the discovery of actionable mutations such as EGFR in non-small cell lung cancer (NSCLC) has led to the development of targeted inhibitors, significantly improving patient outcomes. These advancements are visually summarized in Figure 11, which outlines the role of genetic profiling in tailoring treatments to individual cancer profiles.

6.1.1 Clinical development of specific anticancer agents

Selective agents have now become the mainstay of personalized cancer care. These therapies are intended to target only molecular structures related to cancer, such as mutated proteins functioning in signaling pathways. The targeting used in the case of chronic myeloid leukemia (CML) and HER2 positive breast cancer are perfect examples of success in target therapies. These therapies have the added benefit over conventional chemotherapies in terms of targeting tissues but with a lesser degree of damage to neighboring tissues and thus higher therapeutic index.

6.1.2 Biomarkers into the strategies

Due to the direct involvement in the selection and the subsequent patient monitoring of individually chosen therapy, biomarkers are crucial for precision oncology. Bio-

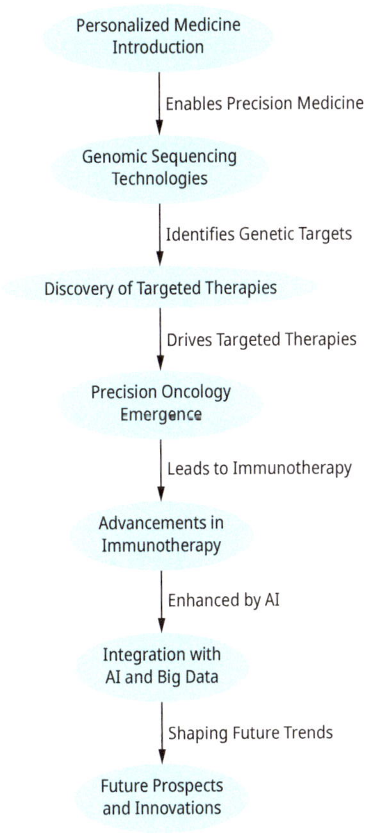

Figure 11: The rise of precision oncology and targeted therapies.

markers such as gene expression, proteins, and lesions specific to tumors and other disease conditions are used for identifying the potential patient candidates who are most likely to benefit from a certain treatment. Some treatment choices are made with a more clear understanding; for instance, when KRAS mutations are found in colon cancer, they can predict against the use of particular molecular drugs. The use of biomarker-based approaches in clinical practice enhances the efficacy and individual approach to cancer treatment since it is based on specific properties of a tumor.

6.1.3 Emergence of immunotherapy

One of the most important advancements of precision oncology is immunotherapy, which uses the body's immune system to eliminate cancer cells selectively. Such a therapy for various sorts of tumors has exhibited some encouraging results because of methodologies such as immunity checkpoint inhibitors, chimeric antigen receptor (CAR) T-cell therapy, and cancer vaccines. Molecules known as immune checkpoint inhibitors make it easier for the immune system to attack cancer by preventing other proteins that can slow down the immune response. Some of these inhibitors are Pembrolizumab (keytruda) & Nivolumab (Opdivo). Within the hematologic malignancies class, identifying CAR T cell treatment, which entails the modification of the patients T cells and enhance their ability to identify cancerous cells, has yielded significant outcomes.

6.1.4 Future paths and trends

As the result of the growth of new developments impacting the subject matter of the discipline, a raise in the future advancements of precision oncology is expected. Of these, one that has received significant attention over the past year is liquid biopsies, which refers to noninvasive approaches toward assessing tumor growth and therapeutic response, for instance, identifying tumor DNA in blood samples. Another progression is the integration of the multi-omics approaches that combines genomes, proteomics data, and metabolomics to develop a holistic view of tumor molecular mechanism and identify new therapeutic targets. In addition, it is expected that advances in machine learning and artificial intelligence enhance the ability to analyze data and enhance health prediction models for better treatment strategies and enhanced patients' health.

6.2 Advances in personalized medicine for rare and genetic diseases

Many progresses have been marked in the diagnosis and specific therapies for rare and hereditary diseases via targeted therapy. These milestones and future directions in this field are outlined in the next section, which focuses on the transformative impact on treatment outcomes and patient wellbeing.

Better genetic diagnostics instruments and whole-genome sequencing

Whole-genome and whole-exome sequencing (WGS and WES) have improved genetic testing by enabling the discovery of genetic roots of rare and inherited diseases. With the help of these technologies, entire coding regions of a person are investigated and new forms of genetic mutations as well as accurate diagnoses of disease that had not existed before, are discovered. In the case of genetics, personalized medicine ensures that one can develop drugs and care approaches that are tailored, based on the exact abnormality in an individual's genetic makeup that lead to an illness.

Targeted therapies and gene editing

This has driven the use of precise therapies and gene modification techniques, whereby diseases that are rare are treated using CRISPR-cas9. Molecular medicines are intended for diseases, making reference specifically to genetic proteins that are conspicuous in rare diseases. For example, ivacaftor for the treatment of cystic fibrosis is aimed at specific genotypes, enhancing patient's prognosis, depending upon certain mutations. Gene editing techniques holds the potential to fix up the mutations at the DNA level, with various prospects of treatment of genetic disorders. Further studies are expected to provide inputs on how to improve these technologies for clinical practice and also for which other genetic diseases they are relevant.

6.2.1 Genetic counseling and risk assessment: new developments

Genetic counselling and genetic risk assessment have advanced to offer better information to patients with rare and inherited disorders. Genomic information and probability models allow genetic counselors to estimate the likelihood of passing on disorders, estimate the risk, and give recommendations for prevention and management strategies. Better genetic expectations lead to improved patient and family education and support for individualized medicine application in practice.

6.2.2 Application of multiplatform analysis

Multidiscipline analysis of multiple genomes, proteins, and metabolites gives a better understanding of inherited and rare diseases. Through integration of multi-omics layers, researchers and doctors can identify new biomarkers, learn the factors behind diseases, and develop stronger therapeutic substances. As you shall soon realize, the characterization of detailed disease characteristics, which is made possible by multi-omics integration, means that treatment regimens are designed bearing in mind the complexity of hereditary diseases.

Personalized treatment approach with reference to clinical trials and drug development

There has also been a change in the design and conduct of clinical trials for rare and genetic diseases by the implementation of personalized medicine. The molecular characterization of patient populations helps the right patient to be identified for trial and improves trial efficiency. Individualized clinical trials involve the use of overdoses on hereditary groups, thus creating a high chance of finding the best treatment and a low chance of side effects. The application of this approach saves time in the development of drugs and gets new treatments to patients faster.

6.2.3 Introduction of digital health resources and telemedicine

Technology-supported health solutions and telemedicine are increasingly prominent in the care of rare and genetic disorders. Mobile equipment, wearable diagnostics, and telemedicine tools are used for constant patient supervision and allow an online meeting with a specialist. These technologies increase the chances of care delivery to patients who have rare disease, especially in areas that are remote. Digital health tools also help individualized treatment management because they give real-time data and also allow for more frequent modifications to treatment plans.

6.2.4 Gene therapy and personalized interventions for the future

The sustained progress in product development of gene therapy and specific treatments offers directions to personal treatment of rare and hereditary disease. Further advancements are being made in the use of this tool, methods of gene delivering, and to address safety and biological questions, concerning gene therapy. The objectives are the expansion of treatable genetic diseases and improvement in individualized therapy safety and efficacy. Using these technologies, genetic disorders that are though to be incurable at the present can be cured.

6.3 Integration of personalized medicine with digital health platforms

Custom medicine, in cooperation with digital health platforms, has introduced significant changes in the provision and administration of healthcare services. Two key areas exemplify how digital technologies are transforming personalized medicine: enhanced analytics and customized therapies, and real-time data collection. Enhanced analytics enable the interpretation of vast amounts of individualized data, leading to more precise therapeutic strategies tailored to each patient's needs. Meanwhile, real-time data collection through wearable devices and remote monitoring tools allows for continuous health tracking and timely clinical decisions. These developments collectively foster proactive and patient-centered care models. The transformative impacts of these innovations are illustrated in Figure 12, which outlines the integration of digital health tools in personalized medicine workflows.

6.3.1 Real-time data collection for monitoring

It means that patient's health status such as heart rate, blood sugar level, and their activity level are continuously collected in digital health platforms with the use of wearable technology and mobile health apps. The real-time data obtained from these devices, when combined with the concept of customized medicine techniques, can provide continuous and contiguous information of the client's health status. By this association, healthcare practitioners obtain relevant information to form new treatment plans and timely manage chronic diseases.

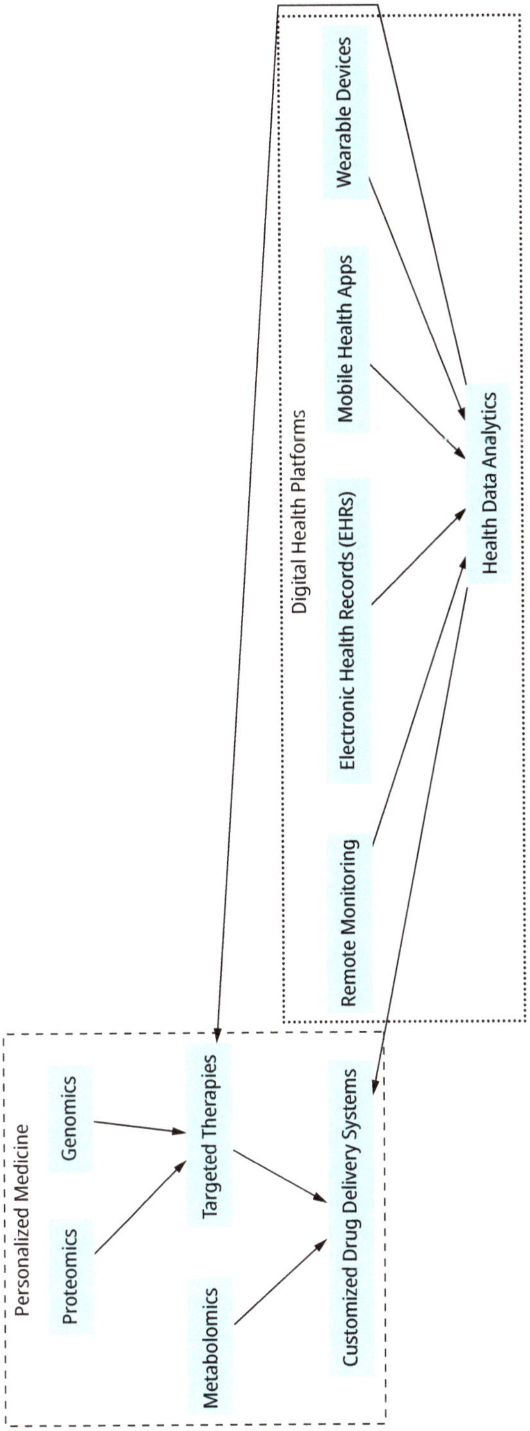

Figure 12: Integration of personalized medicine with digital health platforms.

6.3.2 Personalization with linked data

Integration of information from digital health applications with genomics, proteomics, and metabolomics improves the assessment of the health state of a patient. Electronic health records provide a richer dataset by combining various forms of genetic data and elucidating health preferences and physiological reactions. This broad perspective allows for better matching of traditional and newly developed treatments with patients' needs, making the strategies of personalized medicine more effective.

6.3.3 Remote consultations and telemedicine

Telemedicine is now a widely used solution that helps to provide an individualized approach to patients in those regions that are difficult to reach. Through such solutions, patients can reach a healthcare provider and get advice, without actually visiting this provider. Combining with personalized medicine enables the patient to monitor various parameters and adjust the treatment plan according to the obtained data in a non-face-to-face format. It enhances client satisfaction and healthcare utilization, besides acknowledging effective tracking and management of individualized patient care plans.

6.3.4 Big data analytics and statistical analysis

Several health processes, systems, and interfaces leverage AI and other forms of advanced analytics to develop prediction models based on large-scale health information. Integrated with personal medicine, AI machinery and other algorithm programs help the healthcare professionals to observe patterns that are not detectable when looking at the data of a patient. The use of predictive modeling to anticipate the advancement of ill health, assess prescription solutions, and enhance the delivery of client focus on care processes. It increases the accuracy of the tailored treatment procedure due to concrete recommendations that reflect clinical reasoning.

In this content area, we focus on patient engagement and empowerment.

Mobile health platforms are particularly important when it comes to the question of patient engagement and patient engagement. Patients are allowed to involve themselves in the management process through interfaces and response to feedback. Health IT tools such as health tracking applications, patient places, and teaching tools enable the patient to assess their own condition, comprehend their treatment process, and take appropriate decisions. This enhanced interaction enables patient-centered care and enhances the practice of individualized medicine through compliance by patients with their respective therapies.

6.3.5 The achievement of interoperability with electronic health record systems

Interoperability of EHRs with digital health platforms for sharing health information and enhancing care coordination processes are two advantages of such platforms. Using the data collected by a digital health device with a patient's EHR, it is possible for healthcare practitioners to have a comprehensive view of the entire patient record, their treatment plan, and the most recent data submitted by the device. This integration ensures that original data for the development of patient-specific programs are used and there is continuity of the treatment process, which increases the prospects of making correct decisions.

6.3.6 Prediction and transformation

Some of the studies also assume that while technology of information progresses, so does the integration of health application with personalized medicine. For instance, the development of more elaborate wearable gadgets is one emerging trend; secondly, the use of artificial intelligence and machine learning algorithms; thirdly, the development of digital health assistants for particular diseases. Future advancements of PA consist in the development of improved predictive methods, increased personalization of the interventions delivered through digital technology, and compatibility with genomics and other omics data. All these advances propel the progress of precision medicine, in that it provides increased comprehensiveness and efficiency with regard to medical treatment.

6.4 Trends in personalized medicine research and clinical trials

The interdisciplinary area of the subject, called molecular medicine, is growing very fast due not only, but also owing to the progresses made in the research techniques and strategies, and clinical trials. Such trends are promising with regard to individualized approaches used in both research and clinical care recently, leading to the possibility of enhancing patient's benefits and therapies efficacy.

6.4.1 The integration of multi-omics data

Personalized medicine research has observed a lot of progress in the analysis of multi-omics data including transcriptomics, proteomics, and metabolomics as well as genomes. The synergistic use of such forms of data provides a better understanding of patient heterogeneity and disease etiologies. Multi-omics approaches also hold the promise of discovery of new biomarkers, better understanding of complex biological pathways, and development of more precise treatment goals, for researchers. Clinical trials are increasingly adopting a multi-omics data approach in order to refine patient-subgroup characterization and treatment personalization.

6.4.2 Molecular/precision medicine/genetics: oncology

On top of PM research, there are early frameworks in precision oncology, an area that looks into targeting cancer treatments based on the genealogy and molecular characteristics of a tumor. Recent advances include the emergence of treatments from concept to the development of treatments, depending on the specific genetic alterations and utilization of NGS to identify mutations that are actively managed. It has become increasingly common for genomic profiling for therapy choice and evaluation of the efficiency of combination and target medicines in clinical trials. A more extensive pattern of enhancing cancer treatment and analyzing the outcomes is featured by the growth of precision oncology.

6.4.3 Patient-centric clinical trial designs

K-amplified patient-preference clinical trials are being conducted with increased frequency; patient preference clinical trials are clinical trials for which patients' need and preference are given priority in the clinical trial process. This trend includes distant remote and data collection that are made possible with decentralized clinical trials as well as flexible trial design to allow for modifications based on the interim analysis. Patient-centered methods aim at increasing the connection between trial outcomes and measures of patient-reported outcome and quality of life, enhancing trial accrual/retention and enhancing the application of trials in different types of patients.

6.4.3.1 Fostering of real-world evidence (RWE)

There is also a rising trend in using RWE to complement conventional clinical trial data in personalized medicine studies. In its design, RWE provides details about the effectiveness and safety of treatments in more extensive patient populations and is derived from real-world data, including EHRs, claims data, and patient registries. Trials that incorporate RWE to examine new differential knowledge make valuable contributions, back up systematic decisions with clinical evidence, and offset contextual failures, since personalized medicine is grounded in research.

6.4.4 Advance and recent adoption of digital health technologies

Technology and the accompanying protocols represent an important part of contemporary clinical trials and research in individualized medicine. Some of them are wearables, mobile health applications, and remote monitoring system. These technologies are effective in enabling the seamless and continuous collection of data and in the real-time assessment of physiological signs, signs and symptoms, and outcomes of treatments. It gather lots of data from the patients, enhances patient engagement, and delivers personalized treatments based on distinct patterns in health due to integration of digital health technology. The need and move to incorporate digital health technologies is evidence of how real-time patient data is becoming a valuable tool in deciding how personalized care is delivered.

6.4.4.1 Focus on Underserved Populations Typically Managed by Limited Local Practitioners

Studies on underserved populations are increasingly relevant in personalized care. The identification of specific genetic features associated with rare diseases has driven the development of targeted therapies and advanced diagnostic methods, thanks to progress in genomic technologies. Efforts are also being made to address health disparities and ensure equitable access to appropriate care, particularly for disadvantaged and multicultural groups. Personalized therapies are further strengthened by increasing the representation of minority populations in clinical trials.

6.4.4.2 Emerging trends in therapy

It is remarked that the search for new therapeutic approaches is among the popular topics in the framework of personalized medicine. This includes introduction of various gene therapies, cell therapies, and various immunotherapies that are unique to genetic background and disease profile. New modalities of treatment are already under development and testing, including CRISPR/Cas in oncology, and CAR-T cell therapies. The shift toward new therapeutic endeavors shows the states' concern with the progress of the technique and the methods of individualized treatment.

6.5 Future prospects

New forming and developing is the field of the personalized medicine, which is expected to progress greatly in the near future. With regard to the changes that are awaiting the sphere of personalized medicine, several important future trends can be outlined.

Here, it is important to combine the multiple omics data for successful and effective treatments.

Combining genomics, proteomics, metabolomics, and other omics disciplines into a multi-omics database reveals their growing significance for the future of personalized medicine. Disease etiology and variations among individuals are examined more comprehensively through the integration of information across multiple biological layers. This multidimensional approach enhances patient care delivery, supports the development of individualized treatment plans, and improves therapeutic responses to specific diseases. These advantages are illustrated in Figure 13, which highlights the role of multi-omics integration in precision healthcare strategies.

6.5.1 New trends in artificial intelligence and machine learning

There is an expectation that ML and AI technologies would play an important role in the future of personalized medicine. The assessment of complex information improves due to progressive shifts in AI algorithms and the ability to make better accurate predictions of patient outcomes, the effectiveness of the treatments proposed, and the statistical probability of disease development. The ability to discover more biomarkers, enhance treatments, and enhance clinical determination abilities, all benefit from machine learning models. Personalized healthcare can increase its innovative and effective approach as AI and ML technologies develop.

6.5.2 Potential of precision oncology

Precision medicine is expected to be further defined and practiced extensively as subsequent studies aim at defining further unique cancer-related genetic and molecular markers. New therapies under development offer better treatment for cancer patients such as the ones going for treatment based on tumor genomics and their immune profiles. Technological development in liquid biopsy and tumor sequencing provides additional capacities to track disease development and personalize therapy strategies according to the patient's characteristics.

6.5.2.1 Personalized medicine in rare and genetic diseases

It is forecasted that application of personalized medicine in rare and genetic diseases will increase due to improved genetic testing and gene therapies. Enhancement of gene editing systems like crispr-cas9 paves the way toward the creation of disease-specific drugs. Increased knowledge of rare diseases from genomic studies allow for accurate diagnostic tools with individual therapeutic regimens, resulting in better treatment results of patients with diseases that were hard to manage before.

6.5.3 Links with digital health and wearables

The interaction of personalized medicine with digital health and wearable technologies enhances the surveillance of health in real-time. Smartwatches and integrated

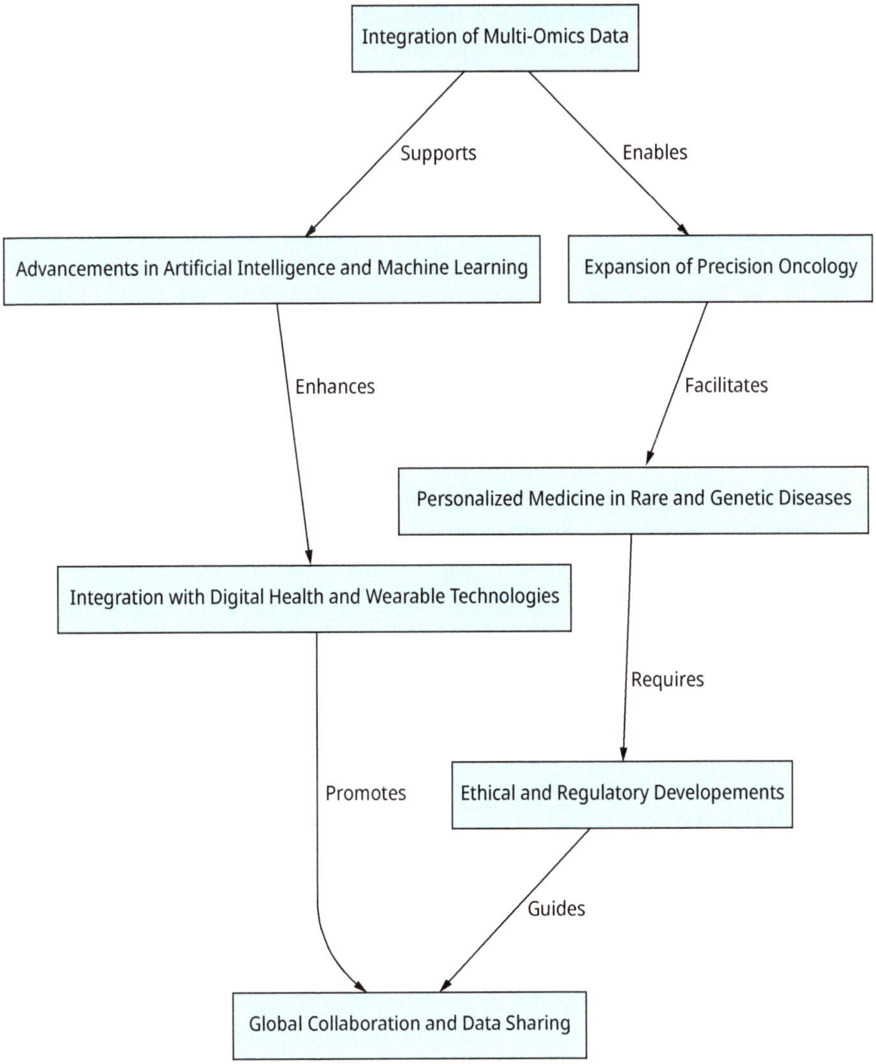

Figure 13: Future prospects.

health applications offer ongoing monitoring of various physiological indices, helping to tailor the management plan in response to the actual experience of the patient. The use of these technologies with EHRs helps with the improvement of effective patient management as well as patient engagement.

6.5.4 Subsidiary of KPMG International Cooperative “KPMG” to pay US$20 million for work related to ethical and regulatory developments

This is why ethical and legislative factors appear to be critical as personal medicine progresses. Subsequent refinements have to do with concerns with regard to data privacy, consent, and fair share equal opportunity for high-quality customized care. It is stated that continuous discussion and the development of strong legal requirements have to remain a focus to ensure the responsible development of the aspects of personalized medicine, where the innovation direction have to be supported with the patient’s protection and equity.

6.5.5 Global partnership and information exchange

6.5.5.1 The future of personalized medicine: global cooperation of researchers, clinicians, and institutions for better data sharing

While efforts for large-scale initiatives enable complex projects to be performed and offer increased varieties and qualities of both genetic and clinical data and transference of findings into clinical practice, such collaborations and data-sharing initiatives help define standards for the global industry of personalized medicine and increase the use of innovative methods across the world.

7 Conclusion

Personalized medicine is widely considered as the new general approach to disease diagnosis and treatment, as well as prevention. This chapter was devoted to the complex perspectives of the concept under discussion as it considers the synergy between personalized medicine and individual drug delivery systems and materials. The development of the concept of personalized medicine is built upon the more recent scientific achievements in genetics, proteins, and metabolites. These improvements involve not only a precise refining of diagnostic and treatment strategies but also an advance in patient outcomes and safety.

This chapter aims to identify the role of pharmacology and the drug delivery system and discuss the advantages of the personalized approach to treatment. Nanotechnology and biomaterials have come up with additional carriers that ensure that drugs are delivered directly to disease sites, hence increasing the efficiency of the procedures carried out and decreasing systemic toxicity. These continue to highlight the need to focus on the specific characteristics of patients when choosing therapeutic approaches for achieving maximum benefits.

I believe that the combining of data from multi-omics is a major step forward in the field of personalized medicine. Integration of genomic, proteomic, and metabolomic knowledge makes disease diagnosis and understanding disease pathways possible as well as the revealing of unknown biomarkers. Such an approach, at the same time, expands the options to create specific therapies and more specific treatment plans, which lead to higher efficiency of treatment, in general. According to the combination of different layers of omics, a better understanding of the biological and pathological levels is provided to design novel diagnosis and treatment approaches.

Molecular diagnostics, improved genomic sequent, high through-put proteomics, and artificial intelligence are among the technologies that have shaped personalized medicine. These technologies allow the possibility of working with extensive data related to biological sciences while providing practical information for use in clinical practice. As technological capabilities become more advanced, they also build on the use of precise and efficient individualized healthcare approaches.

As for the future, the further development of personalized medicine is indicated by present researches and prospective improvements that would enlarge the sphere of application and the effectiveness of interventions. But the field also has to solve issues that pertain to ethics, data privacy, and bias toward providing personalized treatments. Whether it is in response to increased bureaucratic demands or improved pharmacogenomics research of the work, it ensures that these challenges are met with robust solutions, which are vital for the continued success of personalized medicine.

References

[1] Wang, R. C., & Wang, Z. (2023). Precision medicine: Disease subtyping and tailored treatment. Cancers, 15(15), 3837.

[2] Cortese-Krott, M. M., Koning, A., Kuhnle, G. G., Nagy, P., Bianco, C. L., Pasch, A., . . . & Feelisch, M. (2017). The reactive species interactome: Evolutionary emergence, biological significance, and opportunities for redox metabolomics and personalized medicine. Antioxidants & Redox Signaling, 27(10), 684–712.

[3] Castaneda, C., Nalley, K., Mannion, C., Bhattacharyya, P., Blake, P., Pecora, A., . . . & Suh, K. S. (2015). Clinical decision support systems for improving diagnostic accuracy and achieving precision medicine. Journal of Clinical Bioinformatics, 5, 1–16.

[4] Satam, H., Joshi, K., Mangrolia, U., Waghoo, S., Zaidi, G., Rawool, S., . . . & Malonia, S. K. (2023). Next-generation sequencing technology: Current trends and advancements. Biology, 12(7), 997.

[5] Ginsburg, G. S., & Willard, H. F. (2009). Genomic and personalized medicine: Foundations and applications. Translational Research, 154(6), 277–287.

[6] Puccetti, M., Pariano, M., Schoubben, A., Giovagnoli, S., & Ricci, M. (2024). Biologics, theranostics, and personalized medicine in drug delivery systems. Pharmacological Research, 107086.

[7] Sakamoto, J. H., van de Ven, A. L., Godin, B., Blanco, E., Serda, R. E., Grattoni, A., . . . & Ferrari, M. (2010). Enabling individualized therapy through nanotechnology. Pharmacological Research, 62(2), 57–89.

[8] Raj, S., Khurana, S., Choudhari, R., Kesari, K. K., Kamal, M. A., Garg, N., . . . & Kumar, D. (2021, February). Specific targeting cancer cells with nanoparticles and drug delivery in cancer therapy. In Seminars in cancer biology (vol. 69, pp. 166–177). Academic Press.
[9] Caldorera-Moore, M., & Peppas, N. A. (2009). Micro-and nanotechnologies for intelligent and responsive biomaterial-based medical systems. Advanced Drug Delivery Reviews, 61(15), 1391–1401.
[10] Baryakova, T. H., Pogostin, B. H., Langer, R., & McHugh, K. J. (2023). Overcoming barriers to patient adherence: The case for developing innovative drug delivery systems. Nature Reviews Drug Discovery, 22(5), 387–409.
[11] Babu, M., & Snyder, M. (2023). Multi-omics profiling for health. Molecular & Cellular Proteomics, 22(6).
[12] Hasin, Y., Seldin, M., & Lusis, A. (2017). Multi-omics approaches to disease. Genome Biology, 18, 1–15.
[13] Dar, M. A., Arafah, A., Bhat, K. A., Khan, A., Khan, M. S., Ali, A., . . . & Rehman, M. U. (2023). Multiomics technologies: Role in disease biomarker discoveries and therapeutics. Briefings in Functional Genomics, 22(2), 76–96.
[14] Hegele, R. A., & Pollex, R. L. (2005). Genetic and physiological insights into the metabolic syndrome. American Journal of Physiology-Regulatory, Integrative and Comparative Physiology, 289(3), R663–R669.
[15] Hensing, T., Chawla, A., Batra, R., & Salgia, R. (2013). A personalized treatment for lung cancer: Molecular pathways, targeted therapies, and genomic characterization. Systems Analysis of Human Multigene Disorders, 85–117.
[16] Alyass, A., Turcotte, M., & Meyre, D. (2015). From big data analysis to personalized medicine for all: Challenges and opportunities. BMC Medical Genomics, 8, 1–12.
[17] Satam, H., Joshi, K., Mangrolia, U., Waghoo, S., Zaidi, G., Rawool, S., . . . & Malonia, S. K. (2023). Next-generation sequencing technology: Current trends and advancements. Biology, 12(7), 997.
[18] Stoevesandt, O., Taussig, M. J., & He, M. (2009). Protein microarrays: High-throughput tools for proteomics. Expert Review of Proteomics, 6(2), 145–157.
[19] Ahmed, Z., Mohamed, K., Zeeshan, S., & Dong, X. (2020). Artificial intelligence with multi-functional machine learning platform development for better healthcare and precision medicine. Database, 2020, baaa010.
[20] Sahu, M., Gupta, R., Ambasta, R. K., & Kumar, P. (2022). Artificial intelligence and machine learning in precision medicine: A paradigm shift in big data analysis. Progress in Molecular Biology and Translational Science, 190(1), 57–100.
[21] Ginsburg, G. S., & Willard, H. F. (2009). Genomic and personalized medicine: Foundations and applications. Translational Research, 154(6), 277–287.
[22] Srivastava, R. (2023). Applications of artificial intelligence multiomics in precision oncology. Journal of Cancer Research and Clinical Oncology, 149(1), 503–510.
[23] Shiwlani, A., Khan, M., Sherani, A. M. K., Qayyum, M. U., & Hussain, H. K. (2024). Revolutionizing healthcare: the impact of artificial intelligence on patient care, diagnosis, and treatment. JURIHUM: Jurnal Inovasi Dan Humaniora, 1(5), 779–790.
[24] Price, W. N., & Cohen, I. G. (2019). Privacy in the age of medical big data. Nature Medicine, 25(1), 37–43.
[25] Goetz, L. H., & Schork, N. J. (2018). Personalized medicine: Motivation, challenges, and progress. Fertility and Sterility, 109(6), 952–963.

A. Muniyappan, D.M. Kalai Selvi, S. Ramkumar, K. Suresh Kumar,
Ezhil E. Nithila

AI-driven drug design exploring molecular docking and lead optimization using machine learning algorithms

Abstract: AI that has not been integrated with preclinical testing experimental methods is another effective means through which medication development can be hastened. This book chapter examines the several uses of AI in preclinical testing, namely as target and validation identifications, hit identification, molecular docking, lead optimization, toxicity assessment, and biomarker identification. For the purpose of uncovering the disease mechanisms and novel therapeutic measures, the omics information, capabilities of predicting the drug activities and properties, enhancing screening assays, and fusing multiple dimensions of data are all tackled by machine learning. AI tools help scientists identify more effective and safer drug candidates, enhance the implementation of precision medicine, enhance the accuracy, sensitivity, and variability of preclinical testing assays. This work can help understand how AI can change the process of creating drugs and provides a detailed evaluation of the current state of AI implementation during preclinical studies.

Keywords: artificial intelligence, preclinical testing, drug discovery, machine learning, biomarker discovery, precision medicine

1 Introduction

Fluency, precision, and innovative solutions, the formerly unimaginable advancements of the pharma industry are being made possible through the implementation of AI to drug discovery [1, 2]. More efficient strategies are required when conventional drug discovery routes can often be cumbersome and expensive and when identified

A. Muniyappan, Centre for Computational Modeling, Chennai Institute of Technology, Chennai 600069, Tamil Nadu, India, e-mail: muniyappana@citchennai.net
D.M. Kalai Selvi, Department of CSE, RMD Engineering College, Chennai, Tamil Nadu, India, e-mail: dmkalai@gmail.com
S. Ramkumar, Department of Electronics and communication engineering, Sri Eshwar College of Engineering, Coimbatore, Tamil Nadu, India, e-mail: sram0829@gmail.com
K. Suresh Kumar, Department of MBA, Panimalar Engineering College, Nasarathpettai, Varadharajapuram, Poonamallee, Chennai, Tamil Nadu, India, e-mail: pecmba19@gmail.com
Ezhil E. Nithila, Department of ECE, Vel Tech Rangarajan Dr Sagunthala R&D Institute of Science and Technology, Avadi, Chennai, Tamil Nadu, India, e-mail: enithila@gmail.com

https://doi.org/10.1515/9783111503202-007

candidates have high levels of failure rates [3]. AI methods particularly the machine learning (ML)-based algorithms are emerging as strong tools that potentially could change the direction of drug discovery in multiple steps starting from target identification to lead finding and beyond [4].

Molecular docking, which is one of the most crucial tenets of structure-based drug design aims at estimating the preferred conformation of a drug within the context of the target biological entity [5, 6]. This technique is necessary when identifying how potential drug candidates can interact with their targets and subsequent optimizations [7]. It seems that widely used molecular docking techniques suffer from profound limitations, primarily based on a static scoring function; they cannot consider dynamic characteristics of molecular interactions [8]. These limitations are solved by machine learning algorithms since these require large samples to come up with prediction models that are conclusive when it comes to accuracy of binding affinities and efficiency of the docking poses [9].

Lead optimization, the fifth phase in drug synthesis, involves the ability to improve the desirable characteristics of a lead series while reducing the undesirable ones [10, 11]. It generally appears that this process takes at least two cycles of synthesis and testing to achieve, and this process consumes considerable time and resources [12]. With the help of the right navigation in the chemical space made available by the application of ML in the course of leading optimization, better compounds possessing good pharmacokinetics and pharmacodynamics can be discovered [13]. Techniques such as reinforcement learning, generative adversarial networks, and quantitative structure-activity relationship modeling have the potential to predict as well as improve the drug-like properties [14].

The employment of AI in lead optimization and molecular docking has helped reduce the time taken to find new drugs while at the same time increase the probability of finding correct candidates and optimizing them. [15, 16]. AI offers the ability to process and analyze large volumes of biochemical data, to identify patterns and to generate high- "confidence" predictions that can direct experimental strategies [17]. The combination of computational and experimental techniques offers a fresh perspective to drug design; most probably it can significantly reduce the costs and time required to introduce the new therapeutic agents into the market [18, 19]. One of these gaps includes the availability and quality of data. Machine learning models require sizeable, high-quality, and current data to give proper predictions [20]. Most existing datasets in drug discovery are often noisy, sample-biased, and non-standardized; these problems result in degraded model performance and poor generalization [21]. The warning is the desirability of publicly available datasets that are more holistic, better-organized, and larger to help build efficient AI systems [22].

Modern techniques in AI have the capability to revolutionize the pharma sector when machine learning approaches are applied for lead optimization as well as molecular docking [23, 24]. Hence to achieve more efficacious, safe, and specific therapeutic agents, this article explores the approaches, uses, and prospects of this still-emerging field of drug discovery [25].

2 Applications of AI in target identification and validation

2.1 Predictive modeling for target selection

It has been established that identification and confirmation of accurate molecular targets is a critical step that determines outcomes in the subsequent phases of drug development. AI-supported predictive analytics is now an effective tool in achieving this goal. Predictive modeling for target identification also involves synthesis of multiple transcriptomic, proteomic, genomic, and clinical data with a view of post-rating conceivable targets depending on their relevance to a specific disease. These different datasets are heterogeneous in type and the machine learning methods alone are useful for finding some pattern, which could suggest these targets play a role in pathophysiology of a given illness.

Supervised as well as unsupervised learning techniques can be applied for target selection in predictive modeling. In the supervised group, techniques like Support Vector Machines (SVMs), Random Forest, and Neural Networks can be used. These algorithms incorporate gene information from labeled datasets that include data on known disease-related genes and irrelevant genes. In this way, the model can extend knowledge from these examples and determine the propensity of other genes to be associated with the disease of interest, as illustrated in Figure 1.

Another kind of approach is to utilize network-based analysis, that is, biological networks; the data are built from molecular relations; potential data sources are PPI networks, gene regulatory networks, and signaling pathways. Diffusion-based means and network propagation algorithms are further used to rank the candidate targets according to their connection with recognized disease-related genes in these networks.

Genomic, transcriptomic, proteomic, clinical, and medication response data are integrated in various approaches in integrative approaches to achieve a comprehensive understanding of the functions of disease mechanisms and potential therapeutic candidates. Since the formation of integrative models employs sophisticated techniques such as the ensemble learning and deep learning, the different forms of data being integrated can identify forms and relationships that would not have been apparent when the data is looked at singly.

2.2 Network analysis for identifying disease-relevant pathways

Thus, network analysis is a powerful tool when it comes to ratings and defined disease-relevant pathways in the field of development of drugs. Network analysis techniques utilize the biological networks such as gene regulatory networks, protein-protein interaction networks, and signal pathways to provide comprehensive information regarding

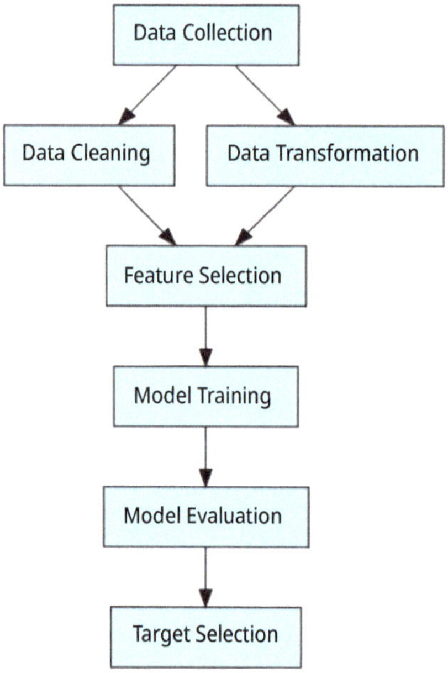

Figure 1: Predictive modeling for target selection.

the interactions of molecular components and to unfold detailed information about the essential mechanisms underlying the development and progress of diseases.

There are many well-known uses of network analysis including one of the most common applications that is to detect pathways that are hypothesized to be abnormal in a given disease state. When combining omics data with network topology, it is possible to define subsets of genes, which should concurrently change their activity levels. These modules can be equivalent to biological pathways modules that have important involvement in various disease processes.

Several algorithms based on the networks have been proposed to rank the disease-centric pathways and modules. For instance, centrality measures typically include betweenness centrality and closeness centrality in which nodes with desirable characteristics of the network are determined based on connectivity. High-centrality node is thought to be essential for the network and represents master controllers of functions related to disease states.

Also, the network propagation algorithms spread messages within the context of the network topology toward the genes associated with other diseased nodes based on functional annotations. These algorithms can simulate numerous focal probes diffusing across a network, thereby predicting additional candidate genes likely to be involved in similar biological pathways as the identified disease genes. These tools highlight significantly overrepresented pathways within a given list of genes, helping to determine whether genetic mutations affect the function and integration of specific biological pathways, as shown in Figure 2.

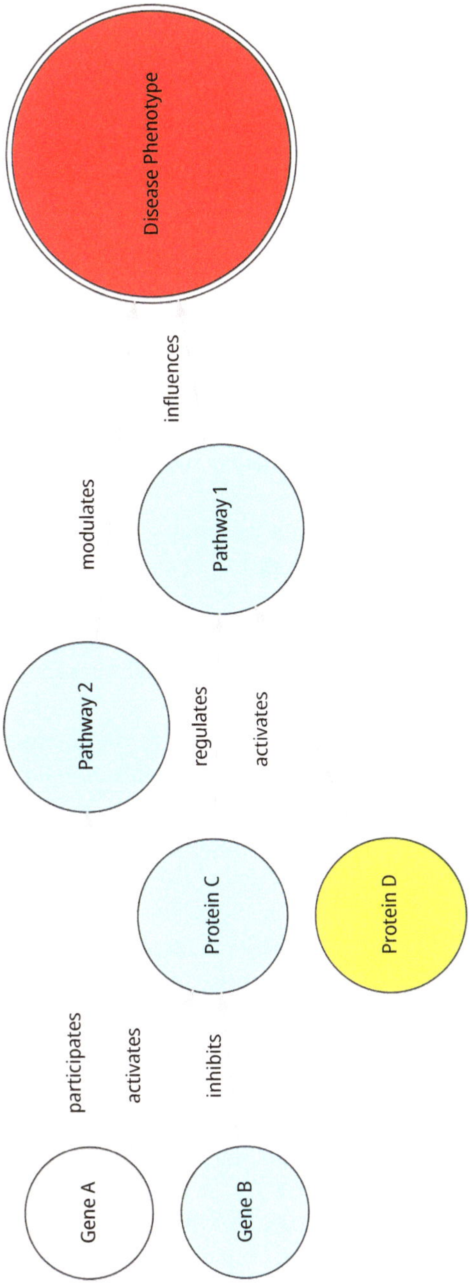

Figure 2: Network analysis for identifying disease-relevant pathways.

2.3 AI-Driven prediction of protein-protein interactions

PPIs are essential elements for quite a number of biological processes and are commonly associated with ailment. Thus, it is crucial to predict PPIs, followed by the comprehension of the cell processes and discovery of potential therapeutic interventions. This paper shows how machine learning techniques and Big biological Data help AI methods become useful for PPI prediction.

Supervised learning methods are also used in machine learning for predicting PPIs and IE aimed at learning from past data with a given set of outputs to determine how this information can be used to predict future output. Used in the training of prediction models, we scale these techniques by means of using datasets with PPI and non-interaction pairs. SVMs, random forests, and deep neural networks use primary protein sequences and structures networks of interactions as features. These characteristics can help the models notice patterns that signal whether two particular proteins are likely to interact.

Another strategy uses network to identify PPIs based on the topological properties of PPI networks by means of network-based algorithms. It has been reported that these algorithms employ proteome-scale graph structures of PPIs, and use network propagation approaches to rank potential interactions. They also predict proteins that, due to their functional association with the known partners, should interact, by mimicking the information flow in the network.

Integrative approaches involve the use of multiple data sets in order to fine-tune PPI prediction, including gene expression data, protein subcellular localization data, and functional annotations. Combining multiple datasets into these models allows them to include additional information about the interactions and reveal new associates of the proteins that are not evident from the analysis of individual types of data.

In silico estimation of PPIs from protein sequences has shown the possibility of using deep learning methods, specifically CNN and RNN. These models possess practical benefits for enhancing the prediction accuracy since they are capable of independently developing a hierarchical representation for the characteristic of proteins and recognizing complicated sequence-structure relationship.

2.4 Genomic data analysis for target prioritization

Recent drug development has come to depend greatly on genomic data, including the case of ranking potential targets. In high-throughput sequencing platforms, large volumes of genomics data are generated, but they hold promising treatment information Recent drug development has come to depend greatly on genomic data, particularly for ranking potential therapeutic targets. High-throughput sequencing platforms generate vast volumes of genomics data, which contain valuable insights into the genetic basis of diseases. Artificial intelligence (AI) plays a crucial role in analyzing and prioritizing these targets by evaluating their relevance to disease pathogenesis, as illustrated in Figure 3.

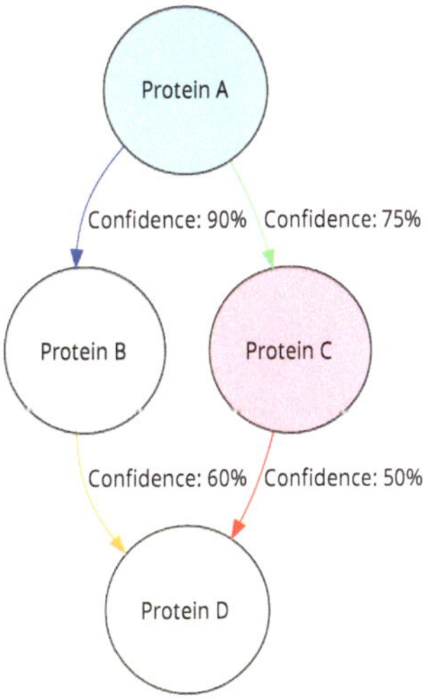

Figure 3: AI-driven prediction of protein-protein interactions.

Combining annotations from GWAS with other genomic functional datasets including gene expression and chromatin accessibility patterns is one approach to using genomic data analysis for target prioritization. Such integrated datasets are used to train algorithms such as logistic regression, random forest, and gradient boosting machines to identify specific genetic variations associated with specific disease symptoms. Those models can point to certain gene regions, which are likely to be implicated in disease causation and then prioritize these according to the patterns and interaction characterized in data.

Other strategies incorporate network analysis of genomic data for ranking potential targets regarding the connection with the known disease-associated genes. Genomic data are combined with protein-protein interactions, gene regulatory interactions, as well as pathways and networks to build models of genetic associations. Then, network propagation algorithms including random walk-based algorithms are further employed to rank genes with known disease genes in terms of functional relevance. Through mimicking the actual flow of signal within the network, these algorithms can predict currently unknown targets that are connected to disease-related signaling pathways.

In data analysis, advanced techniques can integrate data from large-scale functional genomics studies, such as CRISPR-based screens, to identify proteins essential for the survival of disease cells. When these functional genomics datasets are combined with genomic and clinical information, they facilitate the discovery of molecular targets critical to disease sustainability, as shown in Figure 4.

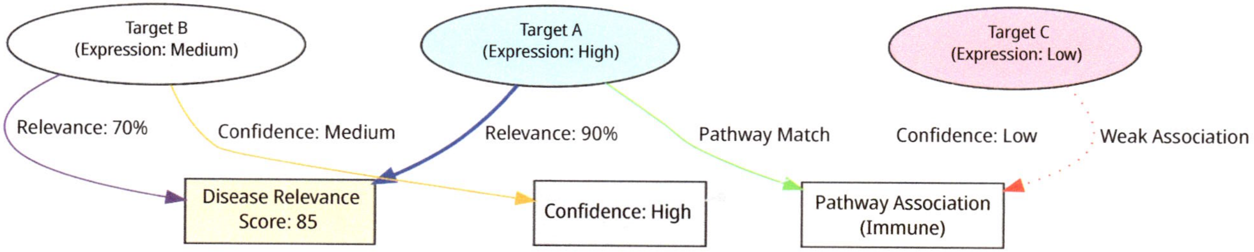

Figure 4: Genomic data analysis for target prioritization.

2.5 AI-based identification of novel drug targets from omics data

Omics technology has become an integrated strategy in analyzing molecular mechanisms underlying numerous diseases in biomedical research. An exciting opportunity in drug design was the discovery of novel targets based on omics information, and AI is a critical component of such efforts.

Another way AI can be used to identify pharmacological targets from the omics data is data integration from proteomics, metabolomics, transcriptomics, and genome data. These integrated datasets utilize machine learning procedures including ensemble in addition to deep learning concerning the mark, molecular markers linked to diverse disease conditions. It is possible to identify the genes and metabolic pathways that are under- or overexpressed in disease conditions, which can be considered potential future drug targets.

Analysis of omics data starting from complex networks offers a promising approach to identify new treatment plans within biological networks. Omics data integrated with functional annotations and pathways are used to construct gene co-regulatory networks, protein-protein interaction networks, and metabolic networks. Network propagation algorithms as well as community detection algorithms on these networks, can be used to prioritize nodes within these networks that are associated with functionally relevant disease processes. These algorithms can help identify potential drug targets relevant to disease-associated pathways and how simulated information flows through what are likely to be highly connected subnetworks.

With the help of AI, new cell type-specific targets involved in the development of illness can be identified on the basis of single-cell omics data. We have single-cell RNA sequencing (scRNA-seq), which provides unprecedented details on the heterogeneity of cells within tissues and tumors. Single-cell omics data is also valuable for computational analysis of clustering algorithms and trajectory inference that can pinpoint cell type-specific markers and disease networks. These markers and networks hit cell type-specific vulnerabilities, potential therapeutic targets of precision medicine.

3 Machine learning techniques in hit identification

3.1 High-throughput virtual screening using ML algorithms

One of the techniques used in the identification of hits for drug discovery aimed at selecting compounds that interact with targets of interest from large chemical libraries is high-throughput virtual screening (HTVS). With the integration of machine learning (ML) algorithms, HTVS has gained tremendous power in improving both the accuracy and speed of estimating the bioactivity of chemical compounds. This advancement is illustrated in Figure 5.

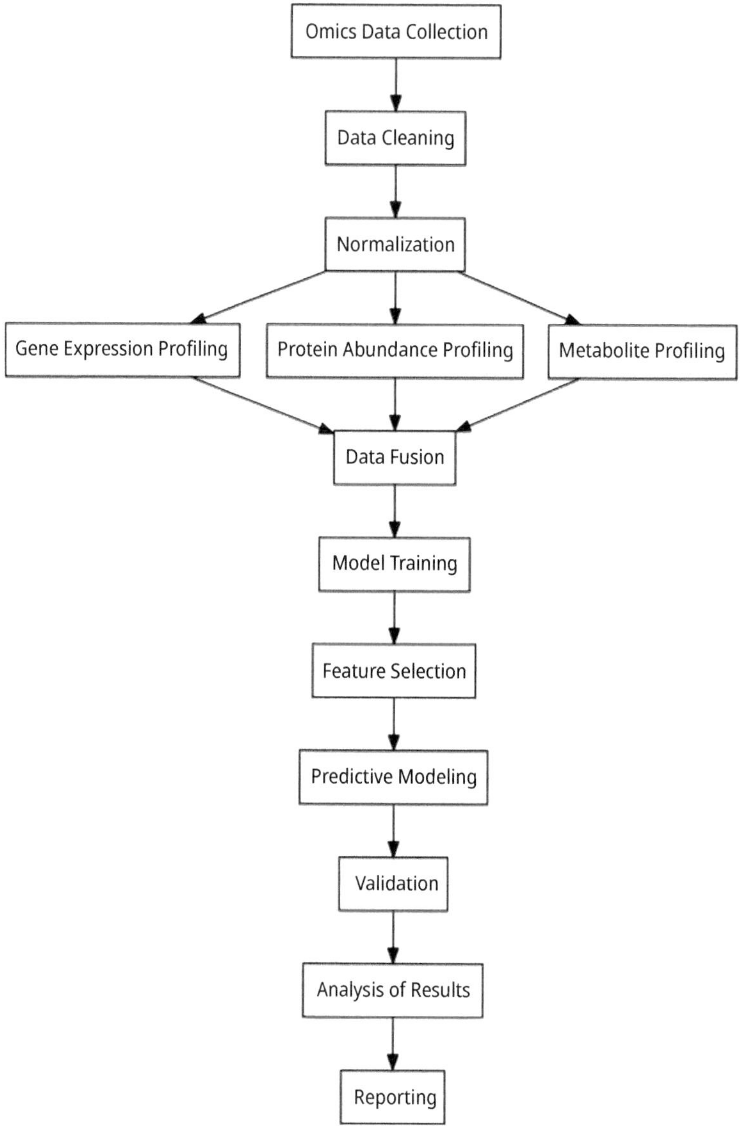

Figure 5: AI-based identification of novel drug targets from omics data.

A typical procedure of HTVS accompanied with ML algorithms is the creation of predictive models, which are constructed on active and inactive compounds databases. Conventional chemical records are supposed to contain the chemical structures along with quantitative data of the corresponding bioactivity, such as ICS values. These datasets are utilized to train the destiny of popular types of machine learning algorithms – SVMs, random forests, and neural networks – in order to eventually identify patterns and associations between chemical characteristics and biological activity. After such training, other possible applications include evaluating the molecular structures of some other chemical compounds and comparing them with molecular structures of active compounds to determine bioactivity possibilities of new chemical compounds.

Another approach to HTVS relies on molecular descriptors and fingerprints translating the chemical compounds into numerical values. Molecular descriptors reflect native properties of molecules, for example size, shape, and charge, while molecular fingerprints express structural details of chemical compounds. Similarity-based methods based on the present invention and kernel methods can then be used to determine the similarity of compounds between the original compounds and those of a chemical library. Compared to other potential style compounds, congeneric structure style compounds that have higher affinity to the receptor and lower binding energy are selected to continue experimental verification.

These are general ensemble learning techniques that have been applied in an effort to improve the performance of the HTVS model, namely, stacking and consensus modeling. Ensemble models combine decisions of several base models that were trained in different techniques of data. Ensemble learning minimizes overfitting and collects data from several sources that are complements of the other to present more accurate and credible predictions that are made from different models.

Molecules and their hierarchical representations, atom interconnections, functional groups, and other complex features make HTVS a suitable area of application for deep learning methods, especially CNN and GNN. Some of these deep learning models can take molecular graphs representation of compounds as input and predict their bioactivity accurately.

3.2 QSAR modeling for predicting compound activity

The primary reason for using QSAR modeling in the discovery of hits in drug development is that their activities can be modeled based on the structures of the compounds. Thus, biological activity measures, for example binding affinities, are related quantitatively to molecular descriptors applying QSAR models, which are based on machine learning.

QSAR (Quantitative Structure–Activity Relationship) modeling begins with the identification of key molecular descriptors that characterize the chemical structure and properties of molecules. These descriptors can include molecular weight, lipophilicity, hydrogen bonding capacity and sites, as well as electronic properties. Furthermore, descriptors derived from molecular structures can be groundbreaking, particularly in understanding spatial configurations and binding patterns with biological macromolecules. This process is illustrated in Figure 6.

Next, training datasets with well-defined chemical activities are employed with chemometric techniques that include neural networks, SVMs, linear regression, and random forests for QSAR model generation. Usually, these files contain chemical

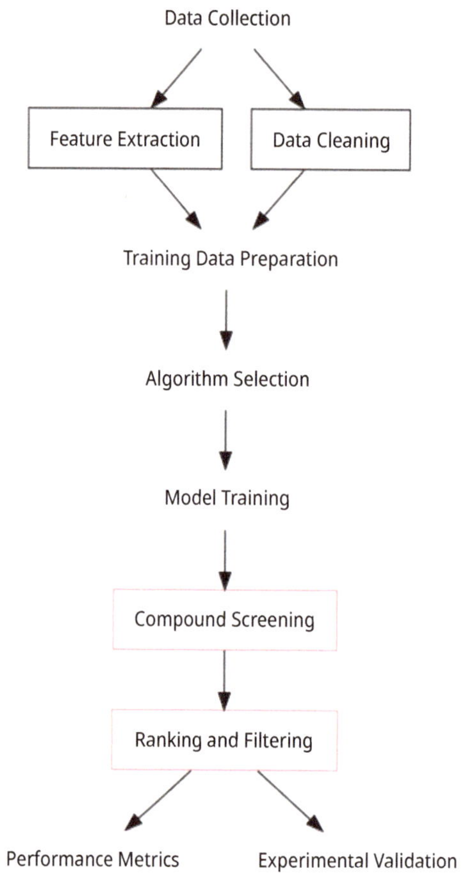

Figure 6: High-throughput virtual screening using ML algorithms.

structures from database tests and the corresponding activity values. The methods optimize expected measures like MSE and minimize prediction errors to find out how the characteristics of input (molecular descriptors) can be mapped to the output variable (compound activity).

These QSAR models can be employed to estimate the activity of new chemical species solely based on their molecular structures as soon as the QSAR models have been built. Quantitative activity predictions of the compounds can be made by the model once the essential molecular descriptors have been entered into the model. This procedure allows for the identification of interesting candidates for further experimental examination.

Other techniques are incorporated in QSAR modeling such as feature selection, which aims to discover the best descriptors that improve the model's interpretative ability and its predictive ability in other areas. In the accuracy of prediction and robustness to noise and outliers, various models of consensus and bagging, in which multiple QSAR models are trained on different subsets of data are used.

3.3 Deep learning approaches for compound similarity search

SMILES-based representations, deep learning approaches have recently garnered substantial interest in the identification of hits for data-driven rational drug design (DRDD), particularly in compound similarity search. These methods utilize multilayered neural networks to capture higher-order representations of chemical structures, enabling more accurate and real-time identification of structurally or functionally similar compounds. This advancement is illustrated in Figure 7.

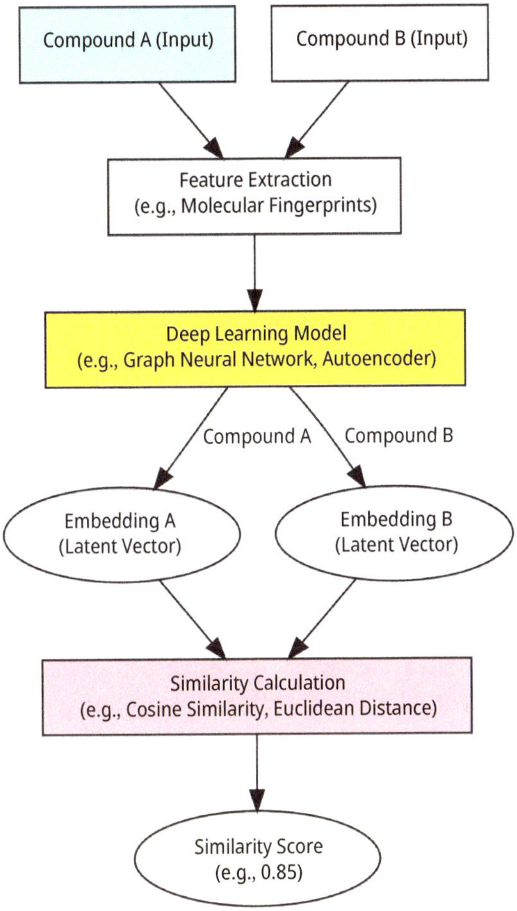

Figure 7: Deep learning approaches for compound similarity search.

The most well-known realization of deep learning for compound similarity search is the Siamese neural network. It involves two neural network streams called the Siamese networks, which operate in parallel with tied parameters for feature learning and map the input compound features into fixed size vectors dubbed as embeddings. These embeddings capture the structural similarities and differences between compounds where the similarity is based on the geometry of the space while the difference is based on the distance in the space. Here, Siamese networks are able to determine structurally related compounds to query compounds simply by calculating the similarity between the respective embeddings of the query compounds and compounds present in a chemical library.

The GNN is another deep learning approach to the compound similarity search. The nature of chemical structure makes use of graphs to explain them; therefore, GNNs suit graph-structured data. To represent intricate interplays of atoms and chemistry of chemical bonds, GNNs operate based on message passing with the viewpoint that node representations are refined based on information about the neighboring data. GNNs transform molecular graphs to low-dimensional vectors allowing the efficient and accurate search for analogs among millions of compounds.

For compound similarity search tasks, CNNs have been prearranged by perceiving chemical structures as a one-dimensional sequence of atom characteristics of molecular fingerprints. CNNs find feature-abstract hierarchies in chemical structures through the convolution of filters and obtain local connections and features at different levels of abstraction. Query compounds are searched structurally similar compounds with high accuracy based on the feature maps of CNNs and SDs with compounds in a chemical library.

Deep learning methods can be integrated with conventional molecular fingerprints, for improving performance of compound similarity search. Scientists can enhance the overall rate of similarity search while using only an equal number of descriptors by proposing deep learning models as features values. This enables them to utilize the complementary aspects of the two approaches.

3.4 Ensemble learning methods in hit identification

Decision fusion methods have emerged as valuable strategies in hit identification for drug discovery, significantly enhancing the accuracy and robustness of predictive models. These methods combine the outputs of multiple individual models in a manner that leverages the unique strengths of each, effectively outperforming any single model used in isolation. This integrative modeling approach is illustrated in Figure 8.

Bagging means bootstrap aggregating, a widely used ensemble learning method where several base models are trained on subsamples of the training datasets where final predictions are produced by voting. Bagging decreases the trend of overfit and increases robustness of the composite model through the addition of randomness into the training process. Bagging is particularly important in hit detection as the technique can be implemented with many machine learning algorithms, such as NN, SVM, and decision trees, in order to improve accuracy of the prediction and find additional patterns in the data source.

Boosting is another method of ensemble learning, which requires using a number of weak learners several times, and putting their solutions together to form one powerful model. Each one effectively increases the overall performance of the ensemble by focusing on the cases where the earlier models misclassified each weak learner gradually. Since gradient boosting methods can support the high-order and nonlinear connections, with various data types, they are applied for hit recognition tasks. Some of these algorithms are XGBoost and LightGBM.

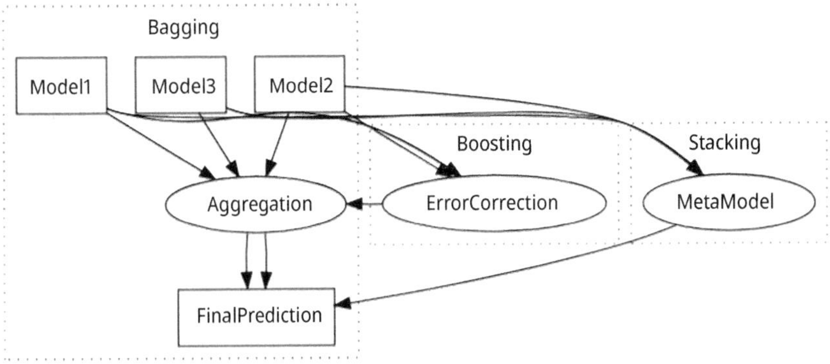

Figure 8: Ensemble learning methods in hit identification.

Stacking, often referred to as meta-learning, is a flexible method of ensemble learning that employs the outcomes of many base models integrated as input features of a meta-learner. The idea is in learning the meta-learner that combines the base models' predictions in a way that takes the advantage and mitigates the disadvantages of the base models. For further improvement of integrated hit detection performance, stacking can be used to combine results of multiple machine learning methods, models specific to the chemical domain, and other heuristics and rules based on expert knowledge.

Also, the identification of hit for the development of drugs incorporates random forest, which is an ensemble learning method generated by the decision tree. More decision trees are created from randomly selected characteristics and samples based on random forests that incorporate to come up with the final result. The use of random forests is suitable to solve detection of hits with high-dimensional data and complicated relations but combined prediction of many decision trees increases prediction accuracy and decreases the probability of overfitting.

3.5 Feature engineering for improving hit identification models

Feature engineering works as a veritable way of enriching hit identification models in drug discovery by incorporating known feature sets from raw data and developing features that can help make a model function optimally. In hit identification, the focus mainly lays on features quality and relevance, which influences the predictive accuracy and reliability of employing machine learning models.

This can be implemented through the extraction of molecular descriptors from chemical compounds in the case of hit identification. Optimized molecular descriptors relate to the size, shape, polar, and electronic character of compounds within the set, which are known to have an impact on their bioactivity. Molecular descriptors are some of the information handles that can be extracted from molecular data; these include molecular weight, topological indices, and pharmacophore fingerprints. When using molecular descriptors that describe the molecular structures of interested compounds in a study, the use of the features in machine learning models help in improving accuracy of predictions.

In the construction of the hit identification models, feature selection approaches are adopted to determine the most relevant and worthy features. It also means that feature selection makes the model less complex, easier to interpret, and improves its predictive abilities as well as rejects unnecessary features that clutter the set. Some of the feature selection techniques include filter methods, wrapper methods, and embedded methods that assess the features based on utilization of statistical measures – the learning algorithm of the intrinsic kind.

Fundamental knowledge and specific experiences are employed to design informative features that include the mechanisms and chemical characteristics linked with the drug activity. For example, a particular chemical molecular substructure, which is observed to bind to a particular biological receptor desired pharmacological effect can be formulated as a feature. In the same way, structure alerts suggesting toxicity can also be introduced for enhancing the model selectivity and safety.

Normalization, scaling, and other operations and transformation functions are used to preprocess data in order to make it fit for use in machine learning. Data transformation increases feature interpretability and helps avoid the impact of skewed distribution of number of outliers in the learning process and the model. For example, taking logarithm of skewed continuous variables measures of the variables within 0 and 1 makes learning more effective.

4 AI Applications in molecular docking and binding affinity prediction

4.1 Deep learning models for protein-ligand binding prediction

In molecular docking and binding affinity prediction, deep learning models have emerged as potent tools in enhancing the accuracy of prediction of interactions of ligands and proteins. Conventional methods of machine learning fail to accurately estimate binding affinities as more complex interactions in protein and ligand structures can be identified by deep learning using neural networks with multiple layers of abstraction.

CNN is one of the widely used deep learning structures in protein-ligand interaction prediction. CNNs are particularly effective in detecting regional features and peculiarities in protein-ligand complexes designed for analyzing spatial information, including molecular structures. CNNs can learn the hierarchical representation of the protein-ligand interactions and extract the proper features for the binding affinity prediction as a result of applying filters for convolution to the molecular input representations. While making an accurate prediction of binding affinities and identifying these interaction residues in the binding site of the protein, the CNN-based models have been shown to be superior.

GNN was another deep learning technique in the prediction of protein-ligand interaction. Since chemical structures are graphs and GNNs are designed to work with graph structured data, GNNs are ideal for this purpose. GNNs are able to predict the binding affinities and also simultaneously capture the chemical and spatial interactions of the protein-ligand interface interaction by compounding information about atoms and bonds in the proximity. The assessment of GNN-based models shows potential in shedding light on the structural aspects of protein-ligand interactions and achieving reliable binding conformations.

The RNNs and long short-term memory (LSTM) network have been used in procedures in which predictions of interaction between protein ligand are involved. Since RNNs and especially LSTMs allow modeling time dependencies and sequence properties it is reasonable to use it to model dynamic processes such as ligand binding and unbinding. Using sequential data representations of protein-ligand complexes, RNN-based models can forecast binding affinities and the poses of ligands with increased precision when targeting flexible complexes of ligands.

The information- and data-fusion approach using a combination of CNNs, GNNs, and RNNs was used for protein-ligand binding predictions. These hybrid models use the advantages of more than one network architecture to create rich representation of protein-ligand interactions and optimize the prediction accuracy. The current work demonstrates the effectiveness of the hybrid deep learning models in achieving better accuracy and reliable binding affinity and binding pose predictions for drug discovery.

4.2 Structure-based virtual screening using ML algorithms

An important component of drug repurposing comprises structure-based virtual screening, which relies on the use of chemical databases to predict how well a novel candidate is expected to bind to a target protein. The machine learning techniques in structure-based virtual screening has enhanced the efficiency and accuracy to a significant level.

Support Vector Machines (SVMs) are among the most widely used machine learning techniques in structure-based virtual screening (VS). SVMs classify compounds as active or inactive based on features derived from chemical structures. In VS applications, SVMs are trained on datasets containing known active and inactive compounds to identify patterns that govern ligand binding. Using these patterns, SVMs can predict the likelihood that new compounds will interact with a target protein, based on the correlation between their molecular descriptors and bioactivity. This approach is depicted in Figure 9.

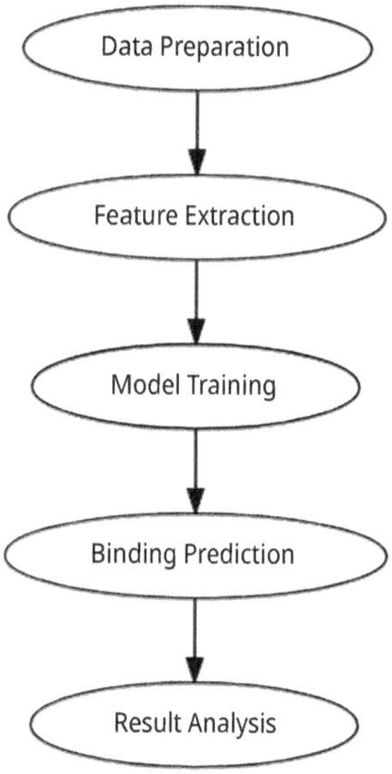

Figure 9: Structure-based virtual screening using ML algorithms.

Structure-based virtual screening also uses another machine learning approach known as the random forests. Random forests, an ensemble learning technique, is composed of multiple decision trees each of which is trained on bootstrap samples of the training data. Random forests are trained for virtual screening that predicts a compound binding affinity to a target protein by categorizing it according to its molecular fingerprints. Decision tree overfitting can be prevented by the use of random forests because the results of several decision trees are combined together in order to enhance prediction rates.

Conventional methods for processing data from structure-based virtual screening include gradient boosting algorithms which include the XGBoost tree and LightGBM tree. Gradient boosting methods eliminate the prediction errors that are associated with a model and increase the value of the model by training a series of weak learner models. Gradient boosting methods can be applied to virtual screening and other tasks in which compounds can be ranked based on their expected binding affinity score to interact with a specific protein. It is possible for gradient boosting algorithms to accurately rank possible medication candidates and generate good predictions of overall performance by aggregating numerous predominantly weak learners.

CNNs and GNNs are two of the deep learning algorithms that have shown promise in structure-based virtual screening. Whereas GNNs can effectively model molecular graphs and encode complex relationships between atomic elements and chemical functional groups at the same time, CNNs can provide 3D analysis of chemical structures and learn multi-scale representations of ligand-protein interactions. By using deep learning, it becomes possible to accurately predict binding affinities and identify potential treatments in large chemical databases.

4.3 Molecular dynamics simulations guided by AI

Molecular dynamics simulations are powerful tools of computational biology for studying the interactions between atoms in biomolecular systems. MD simulations can reveal the kinetic properties of proteins, ligands, and solvent molecules step-by-step by solving Newton's equation of motion of atoms and molecules. The extension of MD investigations has recently revealed much potential, particularly when combined with the use of AI strategies in the simulations.

A sample use of AI in molecular dynamics simulation is the speeding up of what is known as sampling methods. Typically used MD simulations are highly inefficient at sampling the conformational space especially when dealing with large biomolecular systems that have complex energy surfaces. It is also possible to utilize artificial neural network reinforcement and GAN to steer the MD simulations of interest-specific conformational states of molecules. Based upon the previous examples and dynamical modification of sampling techniques in real situation AI-driven MD simulations can sample the states that are difficult to reach efficiently, resulting in enhanced sampling and faster convergence.

Another area of application of AI in MD simulation is the formulation of better sampling techniques. Modified sampling procedures, which include metadynamics, umbrella sampling, and replica exchange are useful in countering energy barriers and sampling more conformations of biomolecules with ease. AI can complement traditional sampling through deep learning metadynamics of the Gaussian mixture model-based enhanced sampling that can introduce knowledge-based biasing potentials sampling protocols. Discussed here is the potential of incorporating principles of efficient sampling schemes with the ability of AI to learn from dynamics of the system and predict appropriate sampling directions to speed up the conformational sampling and improve sampling to subsequently refine the free energy computations.

The force field parameters and the molecular models required for MD simulations could be improved by employing AI techniques. Illustrative force fields commonly applied in interactions of the atoms and molecules rely on factors that are incapable of capturing the structures observed in biomolecular systems. It is found that AI methods can improve the accuracy of the molecular models as well as the force field parameters. Some of these strategies include machine learning=based force field parameterization, QM/MM hybrid simulations aided by neural network. It can be noted that incorporation of AI in force field optimization increases the effectiveness of MD simulation in providing probable molecular interaction and attributes based on quantum mechanical and experimental data.

4.4 Machine learning approaches for predicting binding free energies

Both proteins and ligands are treatments for diseases, and the binding free energies that determine the strength of molecular interactions are critical in selecting potential drugs. ML techniques in particular have garnered considerable interest in this context because the modeling of a relationship between such features and binding affinities can be intricate. A number of approaches have been proposed for computing binding free energies reliably and in reasonable time.

The most commonly used machine learning method for predicting binding free energies is the regression models. These models are built based on a set of protein-ligand complexes to which the corresponding binding affinities are experimentally measured. These basic models are identified as linear regression, random forest regression, and support vector regression (SVR). These models are therefore trained to equate binding free energy with molecular descriptors extracted from protein-ligand complexes. Using data on a variety of molecular characteristics, regression models can predict binding free energies for a number of new and unknown protein-ligand complexes.

Another machine learning method for binding free energy predictions is the deep learning models. RNNs, GNNs, and CNNs reveal the ability to learn abstract representations of molecules together with the ability to incorporate sophisticated interactions within the protein and ligand complexes. In this work, we show that by training models on three-dimensional molecular representation of protein-ligand point-cloud interactions, deep learning models achieve better accuracy and generalization when predicting binding free energies.

For the purpose of the binding free energies predictions, gradient boosting machines (GBMs) and the ensemble regression techniques have been used. Result models stack up outcomes from two models like ordinary regression models for better outcomes and normalcy. The integration of predictions from multiple models constructed from the different data subsets using different algorithms in ensemble learning methods can indeed prevent overfitting, and provide more comprehensive requirements from different sources for accurately estimating binding free energies.

The use of scoring functions based on physics and molecular mechanics, and molecular dynamics, can be combined with the ML techniques and methods to improve binding free energy predictions. Although there are instances of inaccurate measure estimation, certain data, such as interaction fingerprints obtained via machine learning, can be merged with physical energy descriptors in APF architectures. The integration of the benefits from both methods has shown that hybrid models are capable of producing reliable predictions of binding free energies that incorporate the structural and interaction characteristics of protein-ligand complexes at the atomic level.

4.5 Reinforcement learning in molecular docking optimization

There is a method, which is considered to enhance the molecular docking simulations, namely RL, which can enhance and partly automate one of the most critical steps in drug design. In the case of protein-ligand complex, the RL algorithms through the interactions with the environment permit an agent to learn basic dockings.

The application of reinforcement learning (RL) in molecular docking optimization has enabled more efficient conformational space sampling of ligands. RL agents iteratively explore various torsional angles and spatial orientations of ligand molecules, optimizing the client-server communication between the RL system and molecular structures, including quantum mechanical (QM) configurations. These agents receive feedback from scoring functions within the environment, allowing them to select poses with optimal ligand-binding affinities. As illustrated in Figure 10, RL-based docking strategies allow agents to learn from suboptimal docking experiences and continuously refine their approach. This leads to more efficient pose searches and improved docking outcomes with high-affinity ligand geometries.

Specific parameters of RL algorithms can be used to optimize docking protocols and scoring functions in order to enhance the reliability of molecular docking calculations. What is more, RL agents are capable of modifying parameters, such as grid spacing, search algorithms, and scoring terms in view of the docking performance. Through a process of elimination to fine-tune these parameters, the effectiveness of binding affinity predictions of RL agents improves, along with the recognition of the superior ligand poses.

RL could be used to improve and enhance molecular docking where protein-ligand interactions are of interest. RL agents can facilitate optimizing contact and hydrogen bonding between the receptor and ligand; the agents can also learn to modify the positions of protein side chains and ligand atoms. Since RL agents optimally update protein-ligand complexes according to feedback from scoring functions, the bind-

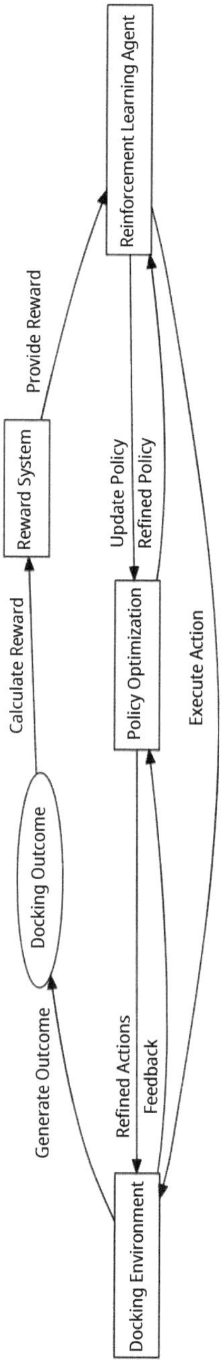

Figure 10: Reinforcement learning in molecular docking optimization.

ing affinity and specificity of protein-ligand interactions can be enhanced, thus providing more accurate molecular binding expectation.

As can be observed, incorporating RL-based algorithms in existing molecular docking programs can help make the docking process more automatable. Docking simulation is one of the important steps within RL agents that can self-adjust the parameter and protocol of docking simulation, including but not limiting to ligand pose generation, scoring function evaluation, and pose refinement. The technology of RL-based data automation of compound libraries can save the extensive time required for large scanning and fast-track the discovery of potential drugs.

5 AI-driven lead optimization strategies

5.1 Generative models for de novo drug design

As potentially capable of generating new chemical substances with required characteristics and activities, generative models seem to have become useful in the de novo drug designing process. There are numerous methods that generative models employ to navigate chemical space and develop new drugs with favorable pharmacological features: GANs and VAEs.

One of the key applications of generative models in de novo drug discovery is the generation of novel molecules with desired properties using probabilistic models such as variational autoencoders (VAEs). VAEs learn compact representations of chemical structure space from input molecular data and are capable of generating new chemical compounds by sampling from a learned latent space, effectively producing structurally novel molecules that still retain essential characteristics of the originals. In contrast, generative adversarial networks (GANs) employ a generator and a discriminator network, where the discriminator distinguishes between real and generated molecules, while the generator aims to produce molecules indistinguishable from real ones. As illustrated in Figure 11, the ability of GANs to learn from large molecular databases allows scientists to synthesize a wide array of new chemical entities with tailored pharmacological profiles.

The generative models allow for optimization of molecular properties and activities not only by adjusting the given parameters but also by redefining them. Although the LMMS is a generative model of molecules, the result generation is done in conjunction with a reinforcement learning function wherewith scientists can direct the outcome of molecular structure generation towards the achievement of some aims such as optimum potency and minimum toxicity pharmacokinetic characteristics. Akin to how a human learns through feedback on the preference of certain generated objects, reinforcement learning agents receive feedback regarding the likeliness of the generated molecules for reinforcement of better quality and diverse structures.

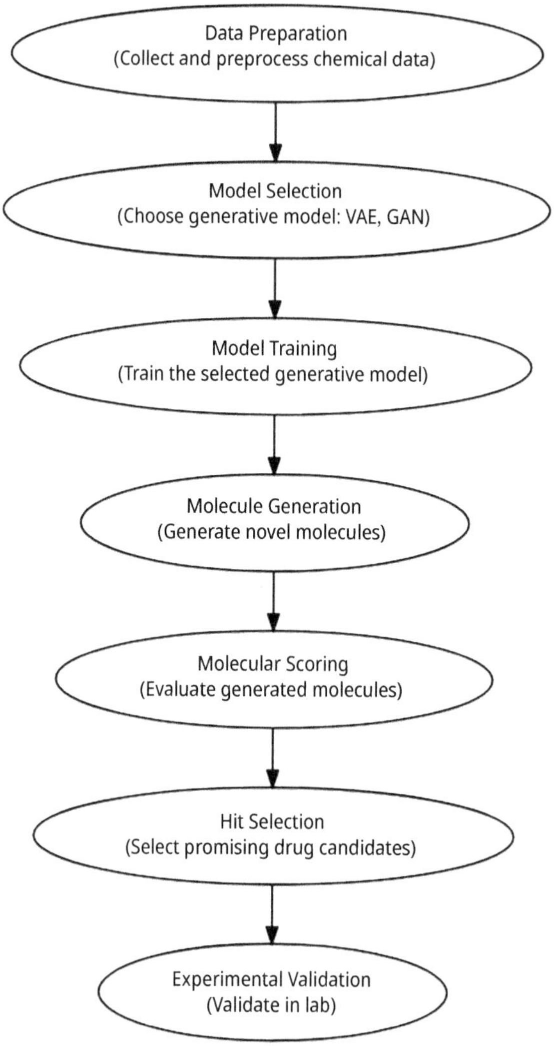

Figure 11: Generative models for de novo drug design.

These generative models can be used to design structure from scratch. Molecular libraries can be designed to hit a certain biological target. This means that by pretraining generative models on target protein structures, the researchers can obtain libraries containing compounds with high affinity and selectivity to the target protein of interest. This focused strategy to de novo drug design can potentially speed up the process of the identification of lead compounds, and also reduce off target interactions, which result in an efficient drug discovery process.

In structure-based design copper-induced gelation and generative models can be employed to create molecules with defined three-dimensional conformation and binding modes. Biased generation by knowledge of protein-ligand interactions and pharmacophore features ensures that generative models generate the compounds likely to interact favorably with the target protein. The structure-guided approach of de novo design increases the probability of developing new leads with improved affinities and selectivity's.

5.2 Multi-objective optimization in lead optimization

Multi-objective optimization (MOO) is also useful in lead optimization to achieve versatile goals for potency at the target, selectivity over related proteins and solubility, as a mixed mission in seeking the best drug candidates. As applied to AI-based lead optimization, MOO approaches provide a structured and time-effective paradigm for addressing the multifaceted relationships implied by the drug design process at hand and for prioritizing the compounds with the most favorable pharma logic processes based on a set of external attributes.

One of the major advantages of multi-objective optimization (MOO) in lead optimization is its ability to provide a diverse set of Pareto-optimal solutions, collectively known as the Pareto front. The Pareto front offers a graphical representation of the trade-offs between competing objectives, enabling researchers to balance conflicting criteria such as potency, selectivity, and ADMET properties. By evaluating candidate solutions based on dominance relationships and Pareto optimality, traditional optimization techniques such as genetic algorithms, particle swarm optimization, and evolutionary algorithms have been adapted to handle multiple objectives. As shown in Figure 12, these methods facilitate the selection of lead compounds that satisfy a project's specific goals and constraints, while also representing different compromise scenarios within the solution space.

It is also notable that MOO techniques include machine learning models for realizing organization of queries in the high-dimensional and nonlinear chemical space. Hence, some algorithms like Bayesian optimization use probabilistic models, including the Gaussian process, with acquisition functions to search for the areas of promising chemical space. As the probabilistic model used in Bayesian optimization is updated upon observations, the method can conclude the areas with promising leads in terms of further optimization within the Pareto frontier. Altogether the same, the reinforcement learning algorithms can bring the adaptive sampling techniques to devise the study of assorted and enlightening areas of the chemical space.

Also, depending on the type of current leads, MOO in lead optimization incorporates domain-specific data and specialists' suggestions in a search for efficient drug prototypes. This enables MOO algorithms to guide the optimization process toward compounds that partially meet medicinal chemistry constraints, pharmacological guidelines, and safety concerns. With MOO frameworks, it is possible to utilize predictive models of molecular properties and activities for the generation and optimization of lead compounds; this allows the effective exploration of the Pareto front and improved speed of identifying promising candidates.

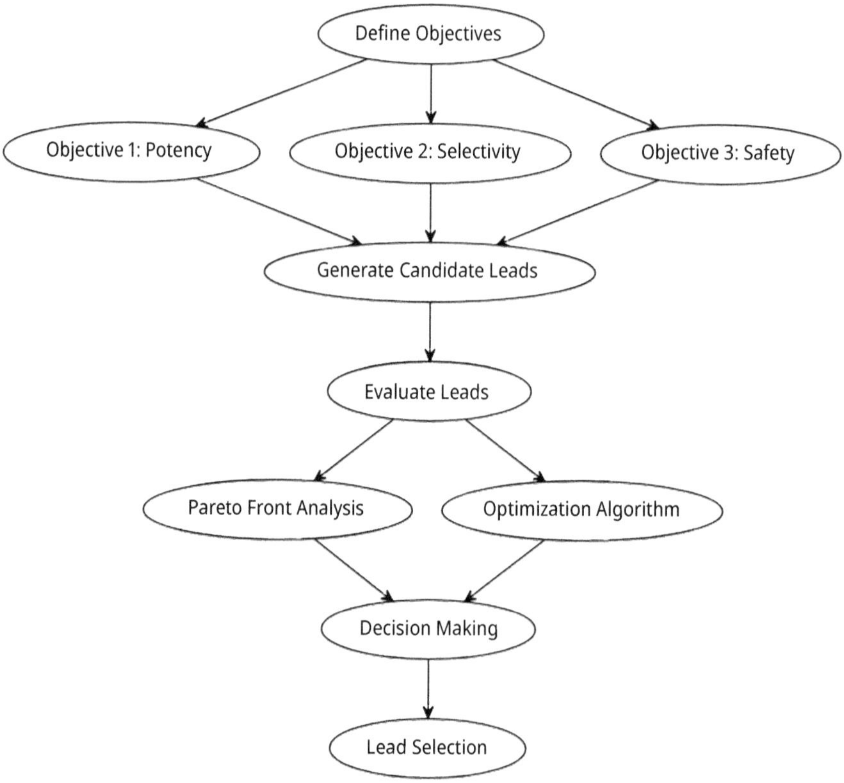

Figure 12: Multi-objective Optimization in lead optimization.

5.3 AI-based structure-activity relationship (SAR) analysis

Another aspect of lead optimization strategies is SAR analysis based on AI, which gives rather effective tools for the explanation of the link between chemical and biological properties. Machine learning-enabled approach to SAR allows for the rational analysis and preplanned investigation of structure-activity relationships with the aim to derive and optimize leads of given pharmaceutical property using the computational SAR data.

Another advantage of using the AI-based SAR analysis is for huge and diverging data sets of chemical compounds and the biological processes associated with them. Such superficial features as fingerprints or structure characteristics, when subjected to machine learning algorithms can reveal complex structure and dependencies in the SAR data. Some of the examples of these algorithms include the random forests, support vector machines, and deep neural networks. These algorithms can pick out as important molecular determinants of potency, selectivity, and efficacy, the least significant differences in chemical structures that are concomitant with changes in biological activity. That has been trained by using a number of datasets containing compound libraries and biological experiences.

The analyses of SAR through artificial neural networks can help the prediction of activities of compounds for new chemical entities and assist in decisions regarding lead compounds essential for the experimental confirmation. Hypothesis-building approaches developed from SAR data ascertain the probability of compounds having a series of specified biological activities due to their structural resemblance to active compounds. Using these predictive models, researchers can select VS libraries selectively, focused CLs, and novel Lead compounds, which have more significant therapeutic application.

It is important to stress that the described methodology of AI-based SAR analysis of compounds can offer important insights on the action of drugs that target specific receptors, as well as define the relationship between the chemical structure of compounds and their interactions. When used to dissect the contribution of each molecular feature of the compound substructures to compound activity, SAR models can reveal pharmacophore elements, binding motifs, and intervals within target proteins. This mechanistic model of SAR makes structure-based rational drug design methods such as structure-based optimization and ligand-protein interaction analysis manipulate compound activity and this augurs well for potency, selectivity, and the drug-like properties.

AI-driven SAR analysis can provide information for successive lead optimization cycles by designing specific compound collections and by providing recommendations for the chosen series of analogs to be synthesized and tested. Combined application of SAR models, medicinal chemistry, computational chemistry, and experimental activities make it possible for the researchers to optimize the lead compounds that elicit a desirable pharmacological effect without causing off-target and toxicity effects. Integral to this process is the use of iterative approaches to lead optimization, so as to efficiently mine chemical space and quickly surface promising leads with desirable drug-like profiling.

5.4 Cheminformatics approaches for lead optimization

Cheminformatics is therefore crucial in lead optimization tactics because it offers a host of computational tools and techniques for analyzing, interpreting, and optimizing chemical information for drug discovery. The use of cheminformatics allows one to predict the properties of substances, draw through the 3D chemical space, and rationally design lead agents with improved pharmacokinetic properties by means of computation.

Chemical data analysis and visualization is another area in which cheminformatics approaches apply within lead optimization. Cheminformatics software and databases offer researchers a big source of chemical structures, biological activities, and physicochemical data. This article shows how similar search, molecular fingerprinting, and clustering can be used to search this chemical space and compare molecules that can be used as lead and what modifications have to be made to get the best result. Further, molecular docking tools as well as the three-dimensional molecular modeling aids in visualization of ligand-protein interactions while leading the design of lead compounds with suitable binding affinities and selectivity.

Cheminformatics enables better prediction of all the activities and characteristics of the compounds through quantitative structure activity relationship (QSAR) modeling. QSAR models use chemical descriptors such as size, shape, and electrical characteristic to model the physicochemical activity of the substances in question. QSAR datasets are employed as sample data in educating machine learning algorithmic such as neural networks, support vector machines, and linear regressions to forecast solubility, toxicity, and chemical activity. QSAR models can also be used in ranking lead compounds for production and testing on the basis of expected pharmacological profile and drug-like properties.

Applying theories of de novo design and virtual screening, cheminformatics allows the rational construction and improvement of lead structures. The large chemical libraries, in some cases, rapidly sample through the methods of virtual screening: ligand-based and structure-based in order to identify the potential lead compounds with the necessary pharmacological effect. Fragment-based design and scaffold hopping are two de novo design algorithms that build entirely new chemical structures, which are optimized to achieve desired properties and activities inherent to design specifications and guidelines. When cheminformatics tools are coupled with molecular modeling followed by machine learning and experimental validation, lead optimization becomes more efficient, and truly novel therapeutic compounds can be developed.

Also, cheminformatics methods enable the examination of HTS data for hit and lead identification with possible therapeutic functions. HTS assays produce large activity data of compounds against numerous biological targets, and, therefore, a corresponding means of data processing and analysis is needed to extract useful information. Clustering analysis, nonlinear regression analysis for dose-response curve fitting, and activity landscape analysis help researchers to successfully predict hit compounds, evaluate lead series, and optimize lead compounds in the chemical space.

5.5 Machine learning models for ADMET prediction in lead optimization

In the strategies of lead optimization, the ML models turn into essential tools for estimating the properties of the lead compounds, which include ADMET. Thus, by assessment of safety and pharmacokinetic profiles of lead candidates, the ADMET prediction facilitates the process of prioritizing and developing molecules that possess qualities deemed propitious for further study and investment.

Estimation of chemical properties, for instance, intestinal permeability and oral bioavailability, is one of the widely applicable uses of machine learning models in lead optimization. When there are large numbers of chemical descriptors and absorption outcomes for which the latter is known, it is possible to use commonly used ML algorithms such as, SVM, random forest, and GBM and learn complex relationship. Through the use of these predictive models, the researchers are able to select for further development, lead compounds with the best absorption profiles.

ML can estimate compound distribution characteristics: BBB penetration and tissue distribution. Employing lipophilicity, molecular size, and hydrogen bond donor and acceptor counts as predictor variables, ML algorithms can estimate the propensity with which the compounds are able to cross the biological barriers to reach their sites of action in vivo. These predictions allow for the determination of how lead compounds distribute in the body and allow chemical modifications to be made in order to improve target interaction and overall activity.

Substances can be predicted based on features like CYP450 interactions and metabolic stability using machine learning models. Through training with compound sets possessing characterized metabolic outcomes, the ML algorithms unveil structural markers related to metabolic activation and metabolic deactivation and estimate the probability of metabolic clearance in vivo. Currently, ML models have become an integral part of lead optimization approaches to predict the properties of the lead compounds, specifically ADMET characteristics. By analyzing the lead candidates' safety and pharmacokinetic characteristics, ADMET prediction helps discover and optimize active compounds suitable for further development and enhancement.

Machine learning models assist in lead optimization by predicting some characteristics of the absorption of chemicals such as oral bioavailability and intestinal permeability. Other approaches, for instance, SVMs, random forests, and GBMs, can be learnt from a data set of some compounds for which the absorption characteristics are known and look for more intricate associations between the causal chemical descriptors and the absorption results.

The applicability of the framework and the ML models are demonstrated in the ability of the models to predict toxicophore properties like mutagenicity, hepatotoxicity, and cardiotoxicity. Because of this, the training of an ML algorithm on datasets of compounds, which have a known toxicological profile allows for the determination of structural alerts and pharmacophore patterns, which correlate with compound toxicity and odds of reaching toxicity. These predictions allow researchers to categorize lead compounds as to which of them are safe to continue development before clinical trials in order to avoid potential dangerous reactions and safety issues.

6 Integrating AI with experimental methods in preclinical testing

6.1 AI-driven analysis of high-throughput screening data

HTS assays furnish a massive volume of information on compound activities against many targets, for offering unique prospects into the relationships of chemical compounds to biological systems. HTS data are big and often complex and pose problems for analysis and interpretation. The availability of low-cost, high-throughput screening data is a significant advantage of AI-oriented approaches, which seize valuable information from HTS data, select potential lead compounds, and prioritize compounds for further in vitro studies.

The application of developing classification models to identify hit chemical structures with desired biological activity is one of the foremost uses of AI organized for HTS data analysis. For instance, some of the machine learning algorithm models that can be trained on the HTS datasets to classify substances as active, based on their structural and activity profile include but not limited to: support vector machines (SVMs), random forests, and deep neural networks. These algorithms can work faster than lead identification because while the activity of new compounds is unknown, algorithms can make this prediction and prioritize hits based on the further validation and the labeled data.

The identities of SARs and pharmacophore characteristics that initiate chemical actions are unveiled with the help of AI analysis of the outcomes of HTS operations. AI approaches can also define the molecular nature of a hit compound's potency, selectivity, as well as efficacy based on chemical structures of hit compounds and the activity map corresponding to them. The "rational design and development of lead compounds with better pharmacological profile" and therapeutic utilization of compounds are based on understanding SARs at molecular levels.

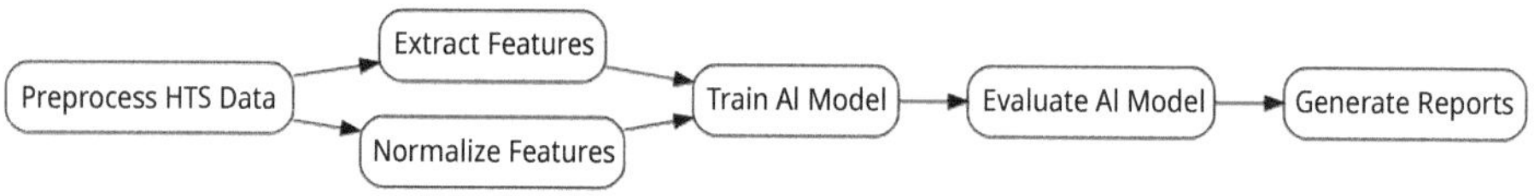

Figure 13: AI-driven analysis of high-throughput screening data.

Clustering analysis and dimensionality reduction techniques are essential for uncovering and visualizing complex patterns and relationships within high-throughput screening (HTS) datasets. These methods allow researchers to classify compounds based on structural similarities, facilitating the identification of chemically related compounds that exhibit comparable biological activity profiles. Dimensionality reduction algorithms such as Principal Component Analysis (PCA) and t-distributed Stochastic Neighbor Embedding (t-SNE) enable the visualization of HTS data in low-dimensional spaces, making it easier to interpret complex multivariate data. As demonstrated in Figure 13, these techniques provide valuable insights into compound clustering and activity mapping.

This means that by employing AI in analysis of the HTS data, it is possible to develop multiple sources of information, including chemical structure database, biological pathway database, protein interactions, etc. From the information collected independently from various sources, an AI system can deduce latent connections between the compounds and targets and possibilities for drug action profiles. These two complementary analyses of HTS data increase the understanding of the compounds' activities and biological effects to make better selections of lead compounds with favorable pharmacological profiles.

6.2 Predictive modeling for toxicity assessment

Toxicogenomics is a new and valuable tool in preclinical testing; predictive modeling of toxicity assessment helps to determine possible adverse effects of drug candidates at the initial stages of the drug development process. These models employ AI methods to parse structural arrangements, biological information, and chemical characteristics to predict and rank potential compounds to possess benign toxicology to facilitate their subsequent examination.

Chemical structure- and physicochemical property-based toxicity prediction represents a vital application of predictive modeling in modern toxicity assessment. To decode the complex relationships between chemical descriptors and toxicity endpoints, various machine learning algorithms including Random Forests, Support Vector Machines (SVMs), and Neural Networks are applied to datasets comprising compounds with known toxicological profiles. These models enable the prediction of toxic effects, such as hepatotoxicity and cardiotoxicity, based on features like molecular structure, key molecular fingerprints, and other physicochemical properties. As illustrated in Figure 14, such predictive models are instrumental in identifying potentially hazardous compounds early in the drug development pipeline.

The accuracy and robustness of toxicity predictions and the toxicity assessment by predictive modeling enable the incorporation of many types of data such as animal, clinical, and in vitro toxicological data. The algorithms can accumulate information from different sources and, depending on them, reveal certain risks that can be left unnoticed if an individual dataset was analyzed. This integrative approach to toxicity assessment improves the assurance of the compound safety profile; it advances the decision-making process for drug development.

Toxicity models used to predict toxicity could address the mechanistic toxicity knowledge and biological modes of action in toxicity prediction for more accuracy models and quantitative interpretation of results. AI algorithms can integrate path-

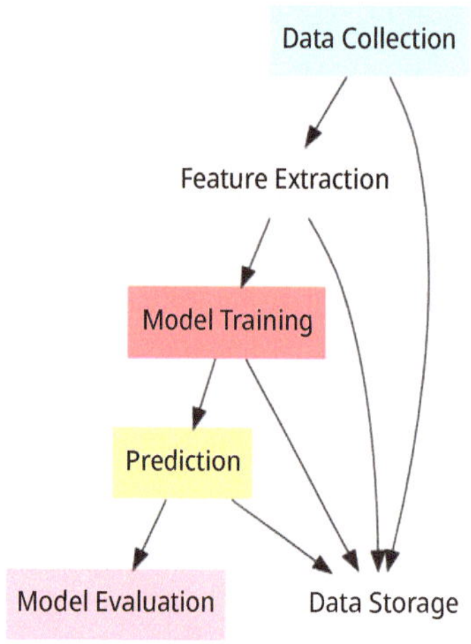

Figure 14: Predictive modeling for toxicity assessment.

ways, protein targets, and cellular responses to determine toxic contributors affecting compound and expound on the origin of toxicity effects. This systems biology manner of viewing toxicity pathways allows scientists to set up experiments, design treatment protocols, and fine-tune lead candidates to guard against potential safety issues while boosting therapeutic impact.

Also, use of taxonomic database for predictive modeling has the advantage of sorting through compound libraries to prioritize lead candidates with suitable toxicity profile to be advanced into development process. Thus, by using predictive models to assess the toxicity potential of compound libraries, investigators can more efficiently screen compounds with low scores for experimental testing, eliminating the requirement for a vast array of animal testing and enhancing the chances of finding safe drugs for patients. This approach to toxicity assessment brings the advantage of advanced preclinical assessment and the subsequent pinpointing of compounds with improved risk-profile characteristics for clinical advancement.

6.3 AI-based biomarker discovery in preclinical studies

AI-driven biomarkers in biomarker discovery is a revolutionary platform for preclinical studies as it provides new instruments for biomarker selection and confirmation, which predict disease prognosis and therapeutic outcomes in patients as well as reveal toxic effects of potential drugs. With the help of AI approaches, researchers can effectively process a large number of biological data, estimate the significant molecular patterns, and discover new biomarkers that help explain the disease progression and reveal new treatment options.

Omics data such as genomes, transcriptomics, and proteomics data are important applications of AI-based biomarker discovery in the preclinical study phase. In order to identify patterns, correlations between molecular markers and phenotype, omics data is used to train machine learning algorithms. Some of these techniques are random forest, support vector machine, and deep neural network. AI systems can identify previously unrecognized markers of disease status, treatment outcome learning from gene clusters and protein levels, as well as metabolic pathways.

The analysis of multi-scale biomolecular organization, combined with comprehensive and validated biomarker signatures obtained through AI-supported biomarker discovery and the integration of multiomics data, provides deeper insights into disease mechanisms.

AI systems are able to identify imaging biomarkers from imaging data such as CT, PET, and MRI scans on disease pathophysiology and treatment efficacy. As we describe below, image features and interesting patterns contained in medical pictures and imaging biomarkers related to disease progression and tumor characteristics, are mined by using CNNs to train deep learning architecture on imaging databases. Machine learning techniques make it possible to noninvasively diagnose and monitor disease biomarkers in preclinical studies utilizing structural, functional, and molecular imaging data.

Biomarker identification and validation workflow using AI occur in the same way as in traditional sciences with a few differences where it provides noninvasive biomarker discovery, which can include circulating biomarkers in blood, urine. Perhaps, biofluid samples subjected to advanced machine learning analysis on molecular qualities could take researchers to biomarkers that are evidence of disease status and therapeutic response. The primary value of using noninvasive biomarkers is that it offers advantages for preclinical and clinical studies, which are patient comfort, cost, and availability.

6.4 Integration of AI with phenotypic screening platforms

The combining of AI and phenotypic screening platforms is a major step forward in preclinical analysis, providing scientists with the chance to discover complex associations and potential drug targets. Phenotypic screening is the process of testing of small molecules on cell readouts that can reveal information about a biological process, a disease model, activity, and toxicity. The application of AI techniques in phenotypic screening assays improves the speed, selectivity, and reliability of the process enabling faster discovery of therapeutic leads.

Another significant application of AI in phenotypic screening lies in the processing and interpretation of high-content imaging data. High-content screening (HCS) assays generate massive volumes of cellular images that reflect morphological, spatial, and temporal changes in response to chemical treatments. Convolutional Neural Networks (CNNs) and other deep learning architectures can be trained on these HCS datasets to identify phenotypic patterns, quantify cellular responses, and predict small molecule activities. The use of AI in analyzing cell morphology, subcellular localization, and temporal dynamics enables the identification of compounds that induce desirable phenotypes, thereby accelerating the discovery of novel therapeutics. As shown in Figure 15, these capabilities mark a transformative step forward in AI-driven phenotypic drug discovery.

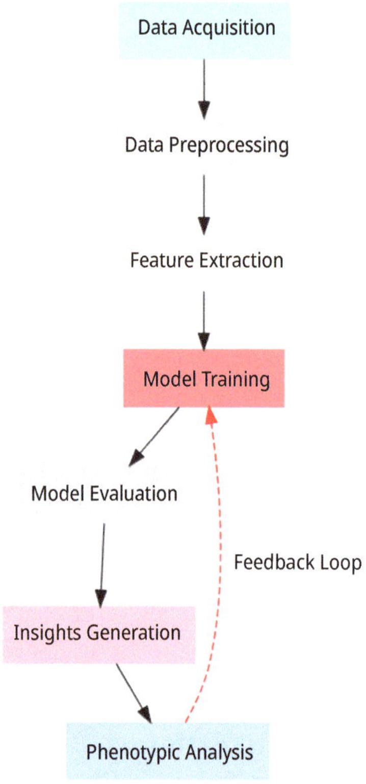

Figure 15: Integration of AI with phenotypic screening platforms.

Using phenotypic screening data along with the support of AI tools and techniques, the intricate organization as well as the relationship between many different heterogeneous cellular populations can be decoded. Decoding cells into clusters according to phenotype behaviors allow AI algorithms to discover subsets that represent different biology and cancer responses to treatment. These characteristics make an unsupervised analysis approach suitable for identifying rare phenotypes, heterogeneity in drug response, and the cellular processes that drive disease progression. Using phenotypic analysis with AI combined with molecular profiling methods the researchers allow extension the knowledge about the diseases to develop a hypothesis for the further investigation of other targets.

To enhance the sensitivity, repeatability, and scalability of phenotypic screening assays, there is an opportunity to incorporate AI algorithms into the course of designing and developing such assays. Academics and professionals can improve the chances of noticing related phenotypic variations through fine-tuning the chemical libraries, screen variables, and assay conditions suggested by complicated reinforcement learning and Bayesian optimization. Potential new lead compounds are determined more quickly, and experimental costs and time likewise are reduced through the use of AI to modify the assay process.

6.5 Machine learning approaches for predicting in vivo efficacy

ML techniques have become very useful in semiconductor testing for evaluating the in vivo effectiveness of drug candidates and providing meaningful information about their possible therapeutic effects in the drug discovery process. This way, utilizing ML techniques, the researchers can process large biological datasets, define biomarkers for the prediction of compounds' activity in vivo and build accurate models of compounds' effectiveness.

Multifaceted integration of multiomics data for predicting in vivo efficacy is one of the most common ML approaches toward the discovery of molecular pharmacosignatures. When preclinical data such as genomics, transcriptomics, proteomics, and metabolomics are integrated into ML algorithms, researchers can decide on biomarkers that predict the efficacy of treatment as well as disease progression. These biomarkers also serve as a source of practical information about the molecular processes involved in the effectiveness of drugs and help analyze who would be most responsive to the treatment.

Machine learning can be employed to see the effect of structure-based drug-target interactions and determine the potential of compounds by their mode of action. Through the use of descriptions about drug-target binding affinities, pathway activation, and network perturbation, trail ML algorithms are capable of estimating the probability of compounds regulating biological pathways that are associated with diseases. This mechanistic viewpoint of drug targeting allows selecting lead scaffolds with appropriate target interaction potential and favorable therapeutic effect among the most promising candidates for additional preclinical trials.

Complementarily, ML techniques can be applied to extracting efficacy data from preclinical animal models to make clinical predictions in humans. When generating and training efficacy-based ML models, such as dose-response curves, survival rates, tumor growth inhibition data, etc., researchers can build predictive models that can transpire the efficacy of compounds from preclinical to clinical phase. These models include pharmacokinetics, pharmacodynamics, and disease diversity to model patient response to drug treatment to gain reasonable predictions of how human patients would respond in relation to efficacy and toxicity during clinical trials.

To predict the therapeutic efficacy of a compound in specific patient populations, machine learning (ML) tools can analyze data from clinical trials, electronic health records, and patient registries. Certain ML classification models are capable of forecast-

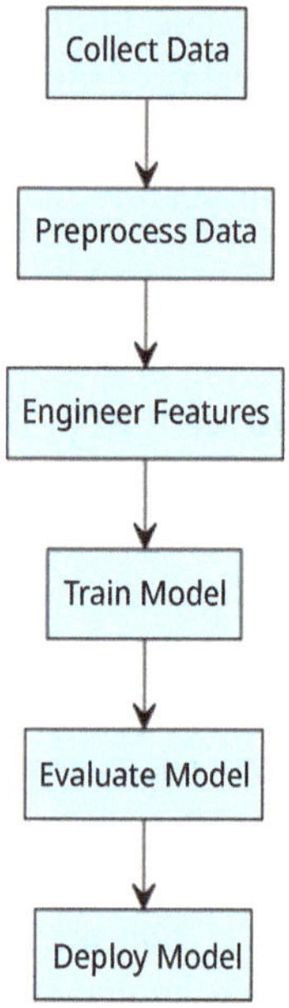

Figure 16: Machine learning approaches for predicting in vivo efficacy.

ing therapy response and disease progression by examining variables such as patient demographics, medical histories, and previous treatment outcomes. These predictive capabilities support personalized medicine by enabling the stratification of patients and tailoring of therapies, thereby enhancing treatment effectiveness and reducing potential adverse effects. As illustrated in Figure 16, this approach plays a crucial role in aligning therapeutic strategies with patient-specific characteristics.

7 Conclusion

These possibilities of AI application in drugs design, as well as in lead optimization, and molecular docking by using ML algorithms, are great steps forward for the pharmaceuticals industry. The transformative nature of AI has been illustrated in this chapter at several prior-to-clinical trial phases. Machine learning tools provide unparalleled speed and accuracy, thus becoming vectors of innovative approaches to identify and validate targets using only predictive models and networks. The integration of various omics information with the most innovative and modern approaches to machine learning has advanced the identification of potential therapeutic targets to greater limits. Lead compound optimization and compound activity prediction are therefore amenable to HTVS and QSAR modeling. There are areas that need further consideration, for example, the question of having large and stable data sets and the question of how to interpret the applicable prediction models. To maximize the impact of AI and machine learning in drug discovery, there is need to build on the mutual cooperation between computational and experimental chemists and biologists. Computational drug design using AI has a bright future in the further development of pharmaceuticals.

References

[1] Gupta, R., Srivastava, D., Sahu, M., Tiwari, S., Ambasta, R. K., & Kumar, P. (2021). Artificial intelligence to deep learning: Machine intelligence approach for drug discovery. Molecular Diversity, 25, 1315–1360.

[2] Pasrija, P., Jha, P., Upadhyaya, P., Khan, M., & Chopra, M. (2022). Machine learning and artificial intelligence: A paradigm shift in big data-driven drug design and discovery. Current Topics in Medicinal Chemistry, 22(20), 1692–1727.

[3] Shiammala, P. N., Duraimutharasan, N. K. B., Vaseeharan, B., Alothaim, A. S., Al-Malki, E. S., Snekaa, B., . . . & Selvaraj, C. (2023). Exploring the artificial intelligence and machine learning models in the context of drug design difficulties and future potential for the pharmaceutical sectors. Methods.

[4] Ivanenkov, Y. A., Polykovskiy, D., Bezrukov, D., Zagribelnyy, B., Aladinskiy, V., Kamya, P., . . . & Zhavoronkov, A. (2023). Chemistry42: An AI-driven platform for molecular design and optimization. Journal of Chemical Information and Modeling, 63(3), 695–701.

[5] Rayhan, R., & Kinzler, R. AI-driven advancements in molecular design: Revolutionizing drug discovery and material optimization.

[6] Tripathi, M. K., Nath, A., Singh, T. P., Ethayathulla, A. S., & Kaur, P. (2021). Evolving scenario of big data and AI in drug discovery. Molecular Diversity, 25, 1439–1460.

[7] Niazi, S. K., & Mariam, Z. (2023). Computer-aided drug design and drug discovery: A prospective analysis. Pharmaceuticals, 17(1), 22.

[8] Qureshi, R., Irfan, M., Gondal, T. M., Khan, S., Wu, J., Hadi, M. U., . . . Alam, T. (2023). AI in drug discovery and its clinical relevance. Heliyon.

[9] Nandi, S., Bhaduri, S., Das, D., Ghosh, P., Mandal, M., & Mitra, P. (2024). Deciphering the lexicon of protein targets: A review on multifaceted drug discovery in the era of artificial intelligence. Molecular Pharmaceutics, 21(4), 1563–1590.

[10] Niazi, S. K., Mariam, Z., & Magoola, M. (2024). Optimizing lead compounds: The role of artificial intelligence in drug discovery.

[11] Peña-Guerrero, J., Nguewa, P. A., & García-Sosa, A. T. (2021). Machine learning, artificial intelligence, and data science breaking into drug design and neglected diseases. Wiley Interdisciplinary Reviews: Computational Molecular Science, 11(5), e1513.

[12] Chopra, H., Baig, A. A., Gautam, R. K., & Kamal, M. A. (2022). Application of artificial intelligence in drug discovery. Current Pharmaceutical Design, 28(33), 2690–2703.

[13] Sahoo, A., & Dar, G. M. (2021). A comprehensive review on the application of artificial intelligence in drug discovery. The Applied Biology & Chemistry Journal, 2(2), 34–48.

[14] Singh, S., Kumar, R., Payra, S., & Singh, S. K. (2023). Artificial intelligence and machine learning in pharmacological research: Bridging the gap between data and drug discovery. Cureus, 15(8).

[15] Singh, S., Gupta, H., Sharma, P., & Sahi, S. (2024). Advances in AI-assisted approaches in drug screening. Artificial Intelligence Chemistry, 2(1), 100039.

[16] Tripathi, A., Misra, K., Dhanuka, R., & Singh, J. P. (2023). Artificial intelligence in accelerating drug discovery and development. Recent Patents on Biotechnology, 17(1), 9–23.

[17] Andrianov, A. M., Shuldau, M. A., Furs, K. V., Yushkevich, A. M., & Tuzikov, A. V. (2023). AI-driven de novo design and molecular modeling for discovery of small-molecule compounds as potential drug

candidates targeting SARS-CoV-2 main protease. International Journal of Molecular Sciences, 24(9), 8083.
[18] Parvathaneni, M., Awol, A. K., Kumari, M., Lan, K., & Lingam, M. (2023). Application of artificial intelligence and machine learning in drug discovery and development. Journal of Drug Delivery and Therapeutics, 13(1), 151–158.
[19] Chen, W., Liu, X., Zhang, S., & Chen, S. (2023). Artificial intelligence for drug discovery: Resources, methods, and applications. Molecular Therapy-Nucleic Acids, 31, 691–702.
[20] Tiwari, P. C., Pal, R., Chaudhary, M. J., & Nath, R. (2023). Artificial intelligence revolutionizing drug development: Exploring opportunities and challenges. Drug Development Research, 84(8), 1652–1663.
[21] Serghini, A., Portelli, S., & Ascher, D. B. (2023). AI-driven enhancements in drug screening and optimization. In Computational drug discovery and design, (pp. 269–294). New York, NY: Springer US.
[22] Jiménez-Luna, J., Grisoni, F., Weskamp, N., & Schneider, G. (2021). Artificial intelligence in drug discovery: Recent advances and future perspectives. Expert Opinion on Drug Discovery, 16(9), 949–959.
[23] Batool, M., Ahmad, B., & Choi, S. (2019). A structure-based drug discovery paradigm. International Journal of Molecular Sciences, 20(11), 2783.
[24] Lee, J. W., Maria-Solano, M. A., Vu, T. N. L., Yoon, S., & Choi, S. (2022). Big data and AI methodologies for computer-aided drug design (CADD). Biochemical Society Transactions, 50(1), 241–252.
[25] Kumar, R., Sharma, A., Alexiou, A., & Ashraf, G. M. (2022). Artificial intelligence in de novo drug design: Are still there? Current Topics in Medicinal Chemistry, 22(30), 2483–2492.

Amit Sharma, G.L. Karthik, Rakhi Kamra, S. Praveena, Venkatesan Hariram

Machine learning models for predicting drug toxicity and side effects

Abstract: An extensive examination of machine learning models created specifically to meet these challenges was provided in this chapter. The topic begins with a thorough analysis of the established techniques for predicting drug toxicity, emphasizing their drawbacks and the need for more advanced strategies. It then transitions to a detailed examination of various machine learning techniques, with a focus on supervised and unsupervised learning methods, including neural networks, autoencoders, and hybrid models. Emphasis was placed on the advantages these methods offer in improving prediction accuracy and robustness. Additionally, the chapter addresses crucial evaluation and validation techniques, with a particular focus on external validation, to assess model generalizability and reliability. Finally, ethical and regulatory considerations are discussed, underscoring the importance of adhering to established standards to ensure responsible model deployment. The purpose of this chapter was to provide researchers and practitioners in the field with a comprehensive understanding of state-of-the-art machine learning techniques for drug toxicity prediction.

Keywords: drug toxicity, machine learning, neural networks, autoencoders, external validation, ethical considerations

1 Introduction

The conventional approaches used in early screening of these side effects have been conclusions made based on past records and experimental research, which mostly involve animals and test tubes [3]. While these approaches have offered certain bene-

Amit Sharma, School of Computer Application, Lovely Professional University, Phagwara, Punjab, India, e-mail: profamitsharma@gmail.com; amit.25076@lpu.co.in
G.L. Karthik, School of Biosciences, Engineering and Technology, VIT Bhopal University, Kotrikalan Village, Ashta Tehsil, Sehore District 466114, Madhya Pradesh, India
Rakhi Kamra, Department of Electrical and Electronics Engineering, Maharaja Surajmal Institute of Technology, Janakpuri, New Delhi 110058, India, e-mail: rakhikamra@msit.in
S. Praveena, Department of Electronic and Communication Engineering, Mahatma Gandhi Institute of Technology, Gandipet, Hyderabad, Telangana, India, e-mail: spraveena_ece@mgit.ac.in
Venkatesan Hariram, Department of Practice of Medicine, Vinayaka Mission's Homoeopathic Medical College and Hospital, A Constituent College of Vinayaka Mission's Research Foundation (Deemed to be University), Salem, 636308, Tamil Nadu, India, e-mail: venkathompath@gmail.com, orchid I'd: https://orcid.org/0000-0003-2237-7071

https://doi.org/10.1515/9783111503202-008

fits, they are often limited in terms of scalability, sensitivity, and the capacity to address multiple interactions, which are only uncovered during human testing [2]. Increasingly, people acknowledge the need for complex tools to manage the colossal amounts of information generated in modern drug design processes, as the requirements for enhanced and refined strategies for drug safety assessment continue to grow [3]. Due to their ability to model complex relationships and discover subtle interdependencies within large datasets, machine learning appears to be a plausible solution to these problems [8].

Traditionally, predicting drug toxicity has been a complex process that involved the use of rules of thumb, statistical methods, and experimental validation [5]. Drug assessment at the initial stages, especially during the lead identification and optimization stages, has been aided by SAR models, QSAR models, and other in vitro tests [6]. These analytic tools are often not effective for prediction, particularly when considering new chemical entities that challenge bioassay models [7]. Concerns including intraspecies differences, the inability to address delicate interactions occurring at the cellular level, limitations of preclinical models, and learning disadvantages are some of the obstacles encountered [8]. The analysis of tagged signals using traditional methods was inherently restricted, thereby providing compelling reasons for the development of enhanced methods that combine heightened predictive capabilities with a closer correlation to human physiological states [9].

With the entry of machine learning, the subject of drug toxicity prediction has undergone a total overhaul with powerful tools capable of learning from massive data sets and recognizing patterns that other methods might not detect [10]. Supervised and unsupervised learning approaches were employed through machine learning models, offering a framework for evaluating large and intricate data with numerous measurement features [11]. Such analysis was beneficial and could contribute to increased accuracy in prediction and the discovery of novel knowledge [12]. When learning was conducted using supervised data to categorize negative impacts and make predictions on the same, methods such as neural networks and ensemble techniques proved useful [13]. Categorization, together with methods of dimensionality reduction, was clustered for feature extraction and utilized to detect anomalies and improve the predictions made within the unsupervised learning process [26]. The use of these machine learning approaches was expected to overcome the limitations of conventional techniques and enhance the overall reliability of drug safety evaluations [15].

It has been agreed that stringency in the assessment and validation processes was central to the success of the machine learning algorithms for estimating medication toxicity [16]. In assessing the performance of these models regarding robustness and generality, external validation, especially, was crucial [17]. To evaluate the performance of models in various conditions and to ensure that they are not memorized for certain subsets of data, testing of newly developed models on unseen sets of data, which were not previously used for the building of the models, is conducted in different forms of validation [18]. In addition, several validation techniques, such as benchmarking and cross-validation, help determine the accuracy, reliability, and consistency of the model [19]. As a result, suitable validation criteria must be used to properly validate the validity of models and ensure they are reliable in real-life drug safety cases [20].

These issues become even more pressing when machine learning models are gradually used as foundational technologies for predicting drug toxicity [21]. These technological concerns include data privacy, algorithm correctness, bias, data, and transparency, all of which must be properly addressed [22]. In the context of the industry, regulatory agencies demand the documentation and validation of the safety of a machine learning model [23]. The latter also includes the appropriate treatment of patient information and compliance with the rules regarding informed consent [24]. Addressing these concerns helps researchers rebuild the lost public trust in ML and ensures that the technology used in drug safety and public health is properly validated [25]. The merging of high-level statistics with ethical and regulatory principles underlines the need for suitable development and operational values of technologies and practices.

2 Supervised learning techniques

2.1 Linear models

Linear models represent a fundamental concept used in supervised learning, often applied to the determination of risks and side effects of drugs. These models are developed based on the notion of achieving a direct relationship between the predictor variables and the target outcome. Linear models are relatively easy to interpret in the interest of drug safety and are thus well suited for the described setting.

Linear regression strives to find the relationship between a continuous dependent variable and one independent variable by estimating through the use of a fitted line formula derived from the observed data. In the case of quantifying the toxicity levels of drugs based on their chemical and biological characteristics, linear regression can be utilized. Linear regression models are described, and their connections and uses in different situations are shown in Figure 1. The impact of each variable is easy to explain since the model has a linear relationship between the input variables and the output. Despite being quite simple and following a linear path, linear regression can be used as a reference to compare the results of more complex algorithms.

If the outcome variable was categorical, I used logistic regression, for instance, when predicting the probability of a certain medicine producing undesirable effects. In this model, the logistic function, which takes the linear sum of input data and transforms the resulting figure into a probability score within the range of 0–1, was

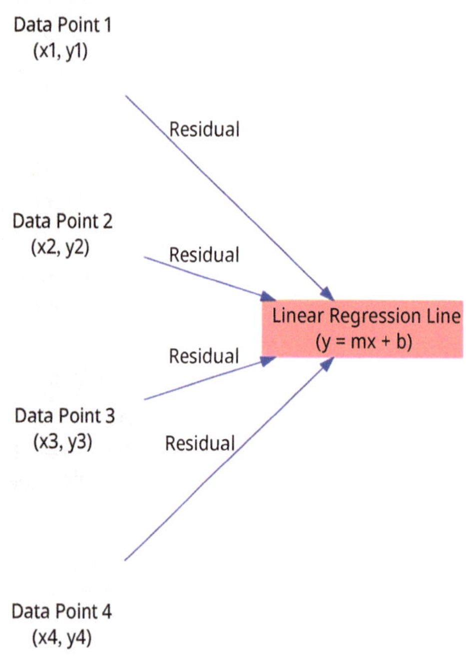

Figure 1: Linear models.

used for the prediction of binary outcomes. When there was a curvilinear relationship between an independent variable and the probability of toxicity, logistic regression proved viable owing to its measure of interpretability and accuracy. The coefficients generated in the context of logistic regression shed light on the comparative importance of the characteristics in predicting negative effects.

It was used in ridge and lasso regression, both of which are linear regression algorithms that work to improve model accuracy and handle issues such as multicollinearity and overfitting. To complement this, ridge regression reduces the coefficients of less important characteristics by adding a penalty proportional to the square of the coefficients' size. Conversely, lasso regression contributes to the construction of sparse models where some coefficients are equal to absolute zero because it involves a penalty factor equal to the absolute value of the coefficients. This lasso regression feature selection property has proven more useful in estimating the most relevant characteristics when it comes to drug toxicity prediction.

Consequently, integrating a degree of robustness into the linear model was critical; the ability to generalize well beyond the training set made regularization techniques very important and included ridge and lasso regression. Regularization was useful in reducing the complexity of the model by making large coefficients expensive and hence minimizing overfitting. These techniques find application where high dimensions are likely to result in overfitting, as was the case with drug toxicity prediction. The variance was therefore controlled by regularization to ensure that it generalizes well to unseen data, a key factor for toxicity prediction, and improves the model's performance and robustness.

While products of first-degree polynomials are adequate for predicting intensions regarding drug toxicity, they do not describe nonlinear dependencies quite as effectively. Figures 3d and 3e show that linear models are less optimal than decision trees and neural networks in terms of flexibility in representing feature interactions. They are good instruments for initial analysis and for building more complex models because of their clarity and simplicity. Concerning medication safety, evaluating linear models together with other advanced methodologies provides a comprehensive understanding of their strengths and weaknesses.

2.2 Decision trees

Decision trees, which constitute an elementary machine learning technique used for categorizing and predicting the toxicity and side effects of medication, were employed. The ensuing graphical representation of the decision-making process affords a clear and easier understanding of various correlations between the input variables and the toxicity outcomes. Consecutively partitioning the data into subsets based on the most relevant features creates the decision tree; a decision tree is a structure akin to a tree with nodes and branches symbolizing decisions and their likely consequences.

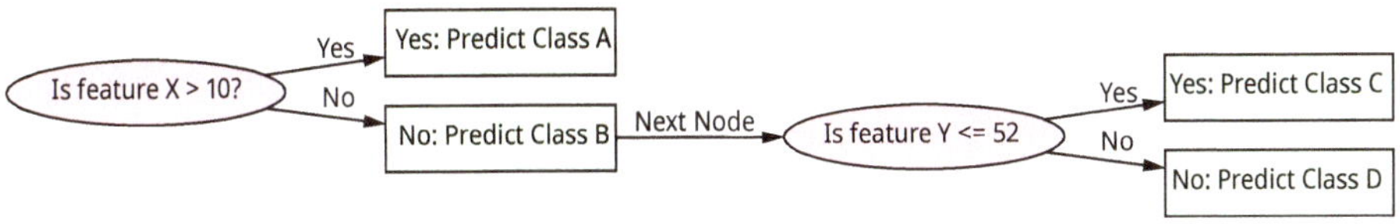

Figure 2: Decision trees.

A decision tree is made up of three basic components: nodes, branches, and leaves. Nodes represent decision points where feature values are employed to divide the data, and they are also known as splitting nodes. Figure 2 depicts the structure of decision trees, especially their nodes and branches, for reaching decisions depending on the input features. The final categorization is depicted as a leaf, while the branches represent the implications of the decisions made. The best attribute through which the data at each node of the decision tree has to be split must be identified to achieve the goal of either maximizing information gain or minimizing impurity. The criteria used to determine these splits include variance reduction, entropy, and the Gini index.

Several algorithms are commonly used to build decision trees, including:

CART (Classification and Regression Trees): With this technique, basic measures are used for the cost function, including mean squared error for regression problems and the Gini index for classification problems, for the construction of binary trees. To address this, pruning is applied, and an exhaustive search is performed to find the best split in the node.

ID3 (Iterative Dichotomiser 3): The ID3 program incorporates information gain and entropy to create decision trees. To reduce data uncertainty, it categorizes data by the feature that provides the biggest information gain.

C4.5: Compared with ID3, C4.5, for example, selects splits according to the gain ratio, which is the difference between it and information gain. In addition to tools for managing continuous and categorical data, it also includes tools for pruning the tree to improve generalization.

Decision trees are useful in predicting medication toxicity because of their ability to process both numerical and categorical data. They can model complex relationships between biological information, chemical characteristics, and patient attributes. Decision trees, for example, can be utilized to sort materials based on how prone they are to stimulate adverse reactions and to identify which features of medications are most predictive of probable deleterious effects.

The use of a decision tree also comes with advantages, such as interpretability and the possibility of visualization, so both practitioners and academics are able to understand the outcomes well. Besides, it is capable of handling missing values and can be used in problems of regression as well as classification. Instances of overfitting, where applicable, were largely observed, and this was most prevalent when dealing with noisy as well as deep trees. It can also be unstable; for instance, if one introduces a small change in the data, the tree diagram changes significantly.

To address some of these limitations, several enhancements and extensions to basic decision trees are commonly used:

Pruning Techniques: Thus, prior to and after construction, the simplification of the tree and an increase in its generality were carried out. Among them, reduced-error pruning and cost-complexity pruning are effective and maintain the same level of prediction accuracy for models.

Group Techniques: Individual tree decisions can be made collectively with many decision trees, where predictive performance and stability are enhanced through random forests and gradient boosting. To decrease variation and improve the system's ability to remain stable against perturbations, these techniques recombine the outputs from multiple trees.

2.3 Random forests

Random forest is a robust type of ensemble learning, where several decision trees are integrated to enhance the resilience and accuracy of predictions. It achieves this by producing the final output through an aggregation of results from a number of decision trees, which reduces the possibility of overfitting often associated with single decision trees.

In other words, random forests create a large number of decision trees during the training phase and, for classification and regression problems, provide the average of all trees' frequent class labels. Random forests are illustrated in Figure 3, which demonstrates how several decision trees are utilized to improve prediction accuracy and the reliability of predictive models. Each decision tree in the forest is constructed using boot strap sampling to train the samples and features with replacement. There is no doubt that the randomization procedure ensures that the trees have different characteristics and sample various aspects of the data.

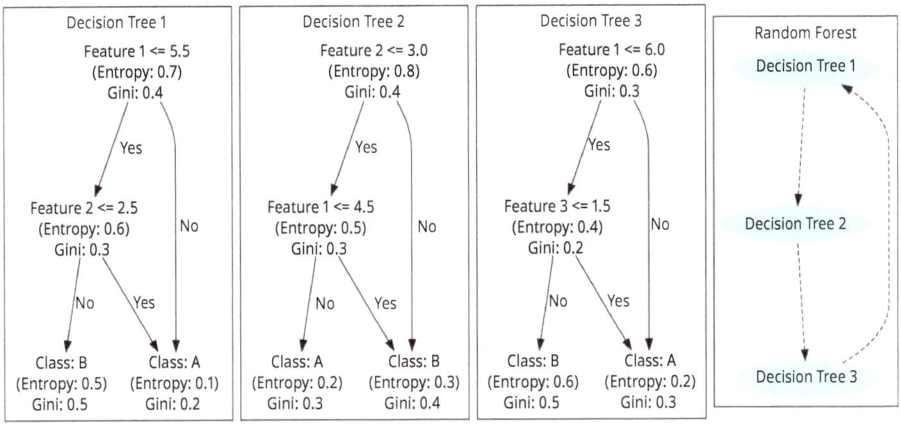

Figure 3: Random forests.

Bootstrapping: To construct several distinct subsets of the training data, random forests make use of bootstrapping. To minimize variance and increase model generality, separate decision trees are fitted on the subgroups.

Randomness of Features: During the tree construction, whenever a split occurs, some random features are considered for that split. Due to this ambiguous characteristic of the features, there is less correlation between the trees, which helps make the ensemble better.

Aggregation: This gives the final forecast as all the decision trees' forecasts are summed up. In regression-based problems, simply averaging all the productive trees was adopted, while in classification-based problems, the final class label of the decision was determined by tallying the count of trees.

Improved Accuracy: Random forests are usually created from multiple decision trees; thus, the algorithm typically provides higher accuracy than a single decision tree because it reduces the problem of overfitting and can accommodate multiple complex structures in the data.

Robustness: Random forests efficiently handle noise and outliers compared to single decision trees due to the use of an ensemble approach, thereby improving generalization, scope, reliability, and robustness.

Feature Importance: Random forests provide valuable information on the significance of features, making it possible to identify which ones have the greatest influence on the outcome. This can be particularly useful in determining the fundamental relationships between the toxicity of medication and other factors.

With respect to the prediction of drug toxicity, the model random forests indeed have the capability of dealing with bio spectral and chemospectral information. Random forests, therefore, can find correlations between drug properties and adverse effects because of its capability to handle many features and unbalanced datasets. For example, it can be utilized for the estimation of toxicity using chemical structures, protein biological assays, and/or patient demographic information.

Model Complexity: Random forests can become very slow and reach a point where the number of trees causes the training and prediction times to increase significantly.

Interpretability: Albeit the random forests package offers the ability to calculate feature importance, it remains less interpretable in terms of feature selection and presents a much more complex model compared to simpler models. This can be a problem when trying to derive additional information from the ensemble of learned decision boundaries.

2.4 Support vector machines (SVM)

SVMs can be described as a powerful class of SLAs that are used for solving regression and classification issues. Here, SVMs provide a reasonable approach for discriminating between dangerous and nonhazardous chemicals in the scope of medication toxicity estimation, taking into account their chemical and biological properties.

In the case of Support Vector Machines, it was to identify the best hyperplane to place data points into different classes. A hyperplane that maximized the gap between the nearest points in each class, referred to as support vectors, was created. The main aim was to find a hyperplane in a feature space that separates positive and negative drug toxicity cases optimally. SVMs give better prediction results due to the use of such support vectors to ensure that the classifying border was as clean as possible.

Because of kernel functions, SVMs can handle nonlinear relationships, which is one of the main strengths of the model. SVMs map input characteristics into higher-dimensional space, where the data can be made separable through linear techniques

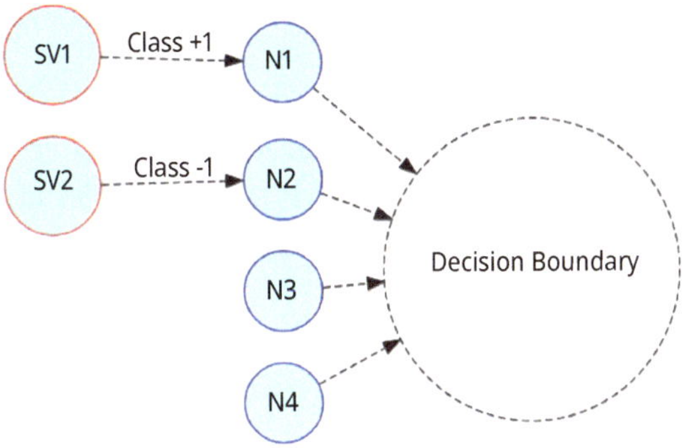

Figure 4: SVM.

using the kernel trick. Polynomial, sigmoid, and radial basis function (RBF are the three common types of kernels. The classification model is depicted in Figure 4, where the decision boundary and the support vectors present in the dataset are also shown. For example, while predicting drug toxicity, the RBF kernel is highly valuable since toxicity consequences are often nonlinearly related to chemical features. It allows SVMs to capture complicated structures in toxicity data that could be overlooked by a system based on linear models.

In drug toxicity datasets, the number of toxic compounds is considerably smaller than the number of nontoxic ones; thus, datasets are usually imbalanced. This can be addressed by using techniques such as tuning class weights, where the impact of the minority class during training is increased. This approach helps improve the model's ability to identify rare toxic events because it does not have a tendency to predict the majority class.

Such additional factors, known as the regularization parameter (C) and other kernel-specific parameters, are tunable parameters with significant influence on SVM. The cutoff value has shown that the trade-off between removing classification mistakes and maximizing the margin was achieved using the regularization parameter. Another measure that affects model performance at scale is the balance of this setting with other kernel parameters. The best hyperparameters are generally obtained through techniques such as grid search and cross-validation, ensuring that the SVM model generalizes effectively on new data.

SVMs have proved useful in medication toxicity prediction; by this, one was able to classify substances according to their chemical and biological qualities. Accurate prediction and identification of possible toxicity were achieved at the highest level. However, the use of SVMs was not entirely problem-free. Working with large datasets poses a problem due to the high computational cost, and the performance significantly depends on the kernel as well as the choice of hyperparameters. Noisy data, and especially a very large feature space, are significant issues with SVMs, which actually lower their performance.

2.5 Neural networks

Supervised learning applications saw a major boost through the development of new artificial neural networks that offer dependable means by which complex patterns and relationships in data are modeled. These models, in which layers of nodes are interconnected, are based on the neural networks that exist in the human brain. Each node processes its inputs through mathematical operations before presenting them via an activation function to produce an output.

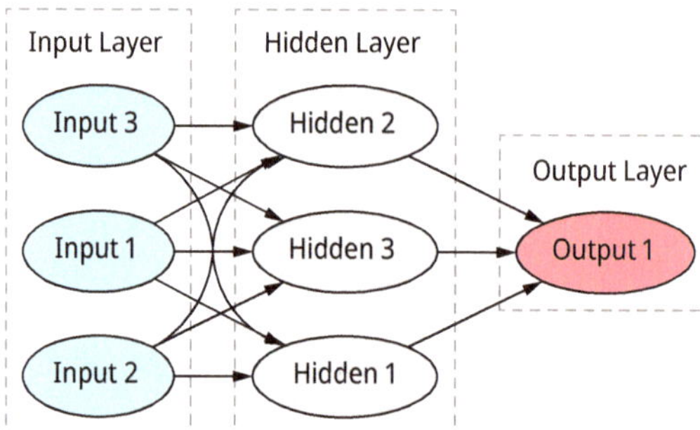

Figure 5: Neural networks.

Neural networks are primarily composed of three basic layers: input, hidden, and output layers. In Figure 5, there is a depiction of a neural network framework where the different layers of neurons and their data processing and learning features are shown. Intermediary calculations are produced by the data features in the raw data by the hidden layers after being forwarded to the input layer. According to the data analyzed, the output layer of the model provides an estimation of the next values. Activation functions that we saw before, including ReLU (Rectified Linear Unit), sigmoid, and tanh functions, add nonlinearity to a network and allow the network to capture complex relationships that cannot be captured by linear models.

Among the simplest ones are feedforward neural networks (FNNs), which do not contain any cycles in their linkage. Contrary to this, data streams flow one way only within FNNs, ranging from the input layer to the output layer, inclusive of the hidden layers, with no feedback systems. These networks perform well with basic data feeds but struggle with more complex data feeds involving dependencies.

The filters here are applied to specific regions of the input data to enhance the functions of feedforward neural networks; these are called convolutional layers, and the networks are called convolutional neural networks. This allows the network to learn features such as edges, textures, and forms, which are very helpful when working with pictures. The CNNs' pooling layers thus help to down sample and render more abstract representations to act as simpler computational models. In several spheres where spatial relations are crucial, such as medical imaging and medication toxicity prediction, CNNs deliver the best results.

To seize sequential input, Recurrent Neural Networks (RNNs) contain feedback, which allows information to remain active across time. Some of the problems that can be fit into this architecture include tasks that require time series data, such as predicting medication interactions given previous data. LSTMs and GRUs, which are additional developments of RNNs that address issues of vanishing and exploding gradients, are utilized. This advancement makes it possible to learn more easily over sequences that span a long time.

Neural networks are trained by adjusting the weights of connections between nodes to minimize the error function, which reflects the difference between the values estimated and those in the real world. The main approach for this optimization is backpropagation, in which gradients are calculated, and optimization methods such as Adam are used. Regularization is a method used to enhance the ability to generalize to unseen data and diminish over learning; two methods are dropout and weight decay.

That is by no means to say that employing the powers of neural networks does not present certain difficulties. Especially for comparatively small datasets, it turns out to be critical to have sufficient processing power as well as a large amount of manually labeled data. In addition, the issue of model interpretability remains a concern, as the complex architectures of neural networks often result in a limited capability to explain certain predictions. In the future, research efforts aimed at overcoming these problems will likely be driven by further advancements in algorithms and explainable artificial intelligence (AI).

Artificial neural networks are under constant enhancement, and new methodologies can revolutionize expectations for drug toxicity and side effects. As these models advance in complexity, the broad versatility of this data will also expand the usability of these models in pharmaceutical research and safety assessment.

3 Unsupervised learning techniques

3.1 Clustering algorithms

Clustering algorithms, one of the most basic unsupervised learning methods, were then employed to categorize data points that resemble one another in terms of their features. Unlike supervised learning, where data samples are classified, clustering seeks patterns within the data samples without necessarily requiring them to be classified. These algorithms can reveal latent structures that are essential for understanding medication toxicity and side effects and are highly effective in conducting exploratory hypothesis generation.

Clustering: One of the most used clustering methods was clustering: k-Means. Although the approach minimizes the variance within each cluster, it ultimately categorizes the data into different clusters. The following stages are taken by the algorithm: it starts by placing centroids randomly, then it assigns every data point to the nearest centroid, computes the centroid value again in terms of the average of the points belonging to a certain centroid, and continues this process until reaching proper optimization. The choice for which k-Means was made quite successful involved employing criteria such as the Elbow Method. However, with clusters of different densities and shapes, k-Means was not so efficient despite being one of the simplest methods.

A hierarchy of clusters was created via hierarchical clustering, which can be applied in two ways: either bottom-up. In the agglomerative technique, each data point forms its individual cluster that merges according to its distance to nearby clusters until this criterion is met: there exists no pair of clusters that can be merged. In contrast, in the divisive technique, every data point initially starts as a single cluster and is recursively divided into further subclusters. A dendrogram, a tree-like diagram that depicts the process of merging or splitting, was the most frequently used outcome. Nevertheless, hierarchical clustering is useful because it can demonstrate the hierarchical structure of the data, but it is computationally expensive for large datasets.

The density of data points was accordingly applied by the density-based clustering method, DBSCAN, to distinguish between clusters. The technique was based on observing that clusters are areas characterized by high density and separated by areas of low density. Figure 6 outlines the flow of a variety of clustering algorithms, detailing the approach of several of them to cluster the components of the datasets. Two parameters are needed for DBSCAN: These two important parameters are MinPts, which defines the minimum number of points to form a dense zone, and ρ, the maximum distance between the points belonging to a single cluster. The system divides the data into three categories: noise points, which are those outside the neighborhood; boundary points, which are those within the neighborhood range but not within the dense range; and finally, core points, which are within the dense neighborhood zone. As seen, its performance was dependent on the set parameters, but overall, it was excellent in finding clusters of any shape and handling noise points.

Average Shift: Given a set of data points, clustering is an exploratory method that seeks the peaks of a probability density function. The algorithm also controls how a data point is constantly shifted toward the mean of other data points that fall within a specific range, and this is done repeatedly until the algorithm converges. Clusters are essentially sets of points that coalesce into a single point. While using the Mean Shift algorithm, there is no need to specify the number of clusters to be identified, and it is flexible in identifying clusters of different sizes and shapes. However, it can be computationally heavy at times, especially when dealing with large datasets.

Unlike k-Means clustering, which requires dimensionality reduction before implementation, Spectral Clustering makes use of eigenvalues of a similarity matrix for that purpose. This involves the calculation of a similarity matrix based on the dispar-

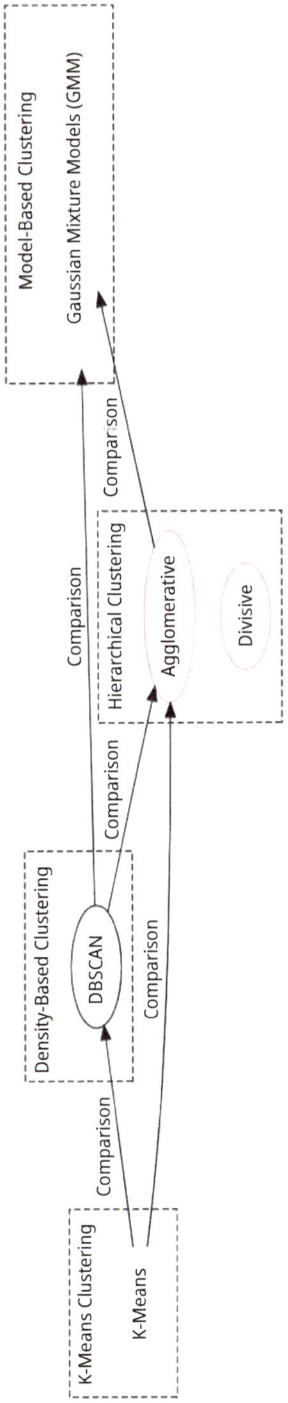

Figure 6: Clustering algorithms.

ity between two instances, the calculation of the Laplacian, and a resultant eigendecomposition. The resulting eigenvectors represent the data in a lower-dimensional space, for which traditional clustering algorithms can be applied. Finally, Spectral Clustering is particularly useful when working with data that includes clusters with different structures and is capable of detecting nonconvex clusters. The method is based on the use of matrix operations and, as a result, tends to be slow for very large datasets.

There are several uses of clustering algorithms in drug toxicity prediction, for example, in finding new patterns in chemical data, categorization of patient groups based on their response, and discovering latent relations between physicochemical properties of drugs and their toxicities. Just like all algorithms, there are benefits as well as disadvantages to each; hence, the approach selected should align with the specific details of the data as well as the goals of the analysis. Because clustering results often apply to specific domains, analyzing clustering findings usually requires domain knowledge and procedures often referred to as validation measures.

3.2 Dimensionality reduction

The process of reducing a dataset's features while preserving its core form and material is referred to as "dimensionality reduction." This is particularly crucial when dealing with demanding datasets with many facets, which is typical across the domains WADILAC is designed to accommodate, such as drug toxicity prediction, which often entails datasets with numerous features. While reducing the impacts of dimensionality, dimensionality reduction also plays a role in improving the visualization of complex data, as well as increasing the computational efficiency and performance of a model.

Of the techniques most widespread for applying dimensionality reduction, principal component analysis. The principal components, which are a set of new orthogonal components with the maximum variance, are derived using PCA from the selected features. Figure 7 depicts the process of dimensionality reduction, which reduces the amount of data but retains its characteristics in a certain dimension. The maximum variance is contributed by the first principal component, the second largest variance is contributed by the second principal component, and this pattern continues. Most of the variation, therefore, stays with the clustered features while reducing dimensionality by mapping data onto these components. PCA is particularly useful in cases of finding patterns and displaying high-dimensional data.

Thus, the t-Distributed Stochastic Neighbor Embedding (t-SNE) method reduces the number of dimensions in the data sample and brings it down to 2, so it can be visualized. Following the intention of other methods to preserve nearby instances close to each other in space when the space is reduced to fewer dimensions, t-SNE tries to preserve the local structure of the data. This method has proven quite useful

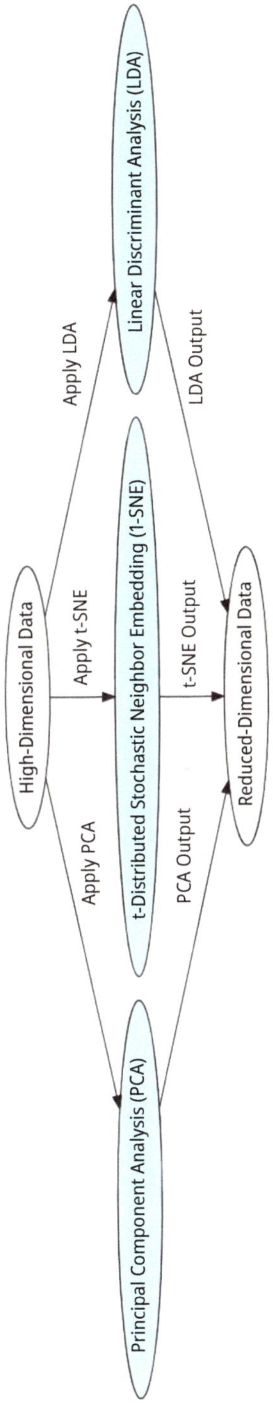

Figure 7: Dimensionality reduction.

for identifying patterns and clusters in complex and large data sets. However, while this technique is very efficient for exploratory data analysis, it is computationally demanding and performs poorly on very large data sets.

There exists a dimensionality reduction technique known as linear discriminant analysis (LDA), which has the objective of maximizing the distance between classes in the dataset. To achieve this end, the supervised LDA aims at finding another space of a lower dimension where the class separation is higher – an aim that differs from the unsupervised PCA, which focuses on variance. That is, LDA aims at the maximization of the ratio of between-class variation to within-class variance to reduce the dimensionality. This method performs well when used in classification tasks where the intention is to reduce dimensionality with the aim of improving the class structure of data to assist upcoming models.

The term "autoencoders" is used when neural networks designed for unlabeled data or dimensionality reduction tasks are being referred to. An encoder transforms the input and possibly maps it to a lower-dimensional space, while a decoder uses this result to reconstruct the original input. This process is known as an autoencoder.

In this study, ultra-short-term EGCG destryptophanogenic profiles are aimed to be derived from EEG by using an autoencoder, followed by subsequent analysis of the resulting profiles. The autoencoder becomes capable of performing effective data encoding since it is trained to minimize the reconstruction error. Extending this concept and making it more flexible and stable, probabilistic components – variational autoencoders – are introduced within this framework. Autoencoders can be very specific in dealing with complex, nonlinear data structures.

Feature extraction and feature selection are the two main approaches to dimensionality reduction. This can be done without changing the general feature space of the data set and thus is called feature selection, as it only selects the most important features from the entire set. On the other hand, feature extraction involves representing the data in a new, smaller feature space. The latter includes methods such as PCA and autoencoders that build a new feature space rather than selecting from existing feature spaces. Therefore, considering these issues, the strategies used in the study had to be distinguished to select the most appropriate methodology for the specific conditions of the investigation.

Dimensionality reduction techniques have been widely employed in drug toxicity prediction to enhance the model's accuracy, reduce the complexity of large datasets, and improve visualization. These techniques assist in reducing complexity and overfitting through the predefined reduction of dimensions in the methods mentioned above. The following are the limitations of dimensionality reduction: Despite their powerful descriptiveness, PCA and LDA, for instance, presuppose linear relationships between variables in an effort to fit complex and nonlinear relations. Since t-SNE is a nonparametric method, it may not be suitable for subsequent machine learning applications, even though it excels at visualization. In the original dataset, feature selection reduces dimensionality while maintaining the original feature space. On the other hand, feature extraction entails reshaping the data into a smaller feature space. The latter group includes methods such as PCA and autoencoders, which create new features instead of selecting from preexisting ones. It is essential to comprehend the differences between these techniques to choose the best methodology for the particular requirements of the investigation.

Drug toxicity prediction uses dimensionality reduction techniques extensively to improve model performance, simplify large, complicated datasets, and enhance visualization. These methods help to reduce computational complexity and overfitting by lowering the number of dimensions. However, there are restrictions associated with dimensionality reduction. For instance, to capture complex, nonlinear patterns, techniques like PCA and LDA assume linear correlations among features. Furthermore, because t-SNE is nonparametric, it may not be appropriate for downstream modeling tasks, even though it works well for visualization.

3.3 Anomaly detection

Anomaly detection is one of the methods of unsupervised learning that serves to determine values and trends outside the normal data distributions. These are called outliers, which are usually connected to important events and/or anomalies in the data, such as drug toxicity, fraud cases, and system failures. In the case of drug toxicity prediction, anomaly detection can be very useful when it comes to identifying rare side effects that may not be captured by the chosen predictive model.

In this context, anomaly detection techniques focus on the difference between normal and anomalous data, the nature of which was unknown in advance. Various approaches are utilized in this process:

Statistical Methods: Most of the data were arbitrarily assumed to follow a certain distribution by using statistical methods. Z-score and the Modified Z-score are examples of measures that determine by how much an observation deviates from the mean of a given distribution. In this work, anomalous observations are considered to be those observations that are above a particular value. These techniques are not suitable for multidimensional data types.

Distance-Based Approaches: Anomalies are identified through techniques such as k-Nearest Neighbors (k-NN), a distance measure that indicates the distance of a given anomalous point. The measures that define the anomalies involve the isolation of an object from its nearest neighbors at the periphery of the data space. This idea was subsequently developed into techniques including the Local Outlier Factor (LOF), which measures the local density of individual data points, as well as the Isolation Forest, which measures the isolation factor of points identified as outliers.

Density-Based Methods: Distance-based methods use distances between points to determine the level of their density in a given area to detect anomalies. Let us examine points in areas with low density. The DBSCAN algorithm (Density-Based Spatial

Clustering of Applications with Noise) marks such sites as outliers. These approaches are very useful when working with datasets of different densities and shapes for the search of outliers.

Methods Based on Models: In model-based techniques, data are fitted into a statistical model, and outliers are determined from the model-fitted graphs. BCM was used in methods like Gaussian Mixture Models (GMMs) for estimating the probability of the data. Low-probability points are called failures. Another method is One-Class SVM, in which outliers are identified with the help of a single judgment boundary drawn through most of the data samples.

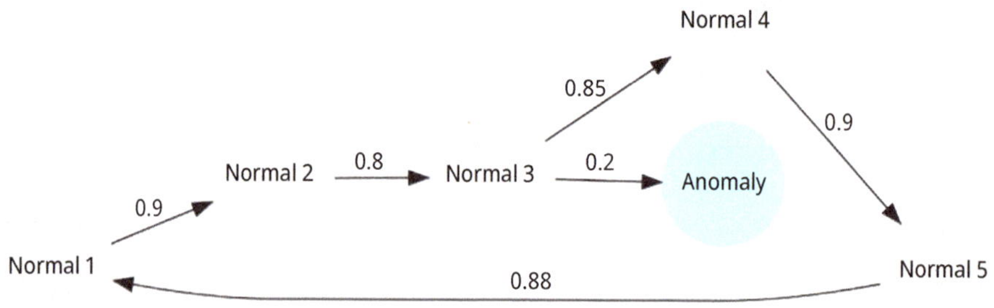

Figure 8: Anomaly detection.

Techniques Based on Reconstruction: Using models such as autoencoders, reconstruction-based methods apply a lower-dimensional representation to reconstruct the data. In other words, anomalies are detected based on reconstruction errors, which are the differences between the original data and the data that is recreated. A large value of reconstruction errors suggests that the data point is an outlier. This method works well for detecting unstructured patterns from large storage volumes.

Structural models in drug toxicity prediction can also use anomaly detection approaches, where an important method identifies effects that are not typical for drug toxicity models. For instance:

Detecting Rare Adverse Reactions: Anomaly detection can also be applied to clinical trial information to identify patterns of rare adverse reactions that have not been identified during the pre-clinical and clinical development phases.

Identifying Novel Toxicity Patterns: When anomaly detection was applied to molecular data, entirely different patterns of toxicity were discovered, which were not detected by other methods, and led to the identification of new toxicophore groups and toxic mechanisms.

Enhancing Safety Monitoring: Data analysis, in conjunction with anomaly detection, can assist in the early detection of unfavorable side effects, thereby providing better chances of treatment and increasing patient safety.

Despite that, anomaly detection has its own set of flaws and shortcomings. Unclean datasets are magnified in Figure 8, where points outside the expected pattern are identified as anomalies. Issues related to the high-dimensionality of the data were one of the major factors that made it difficult to discover noticeable irregularities. In addition, the data distribution and parameters of the data set used in applying the anomaly detection techniques have a significant impact. The objective of future work is to develop new anomaly detection algorithms that possess increased scalability, robustness, and versatility, and work better on large, complex data. Another research topic is the enhancement of the accuracy and reliability of anomaly detection systems through integration with other machine learning techniques, including ensemble methods and deep learning.

3.4 Association rule learning

An effectively used type of unsupervised machine learning technique for discovering both more and less obvious relationships in rather large data sets is association rule learning. Exploratory data analysis is when this technique is very useful, as the goal of such analysis is to uncover relationships that are not likely known between variables.

Association Rule Learning was initially applied mainly to the task of finding characteristic relationships between the occurrence of one item and that of another. These are usually stated in the logical form of an "If . . . Then . . ." statement. For example, if a drug has a high binding affinity, then it MUST have caused an adverse reaction. The process of association rule mining was explained with the help of Figure 9, where frequent item sets are employed to develop rules that determine relationships between variables in a set. The core concepts in association rule learning include support, confidence, and lift:

Support: It measured how often an itemset appeared in the dataset as one of the validity measures. The relative frequency of the itemset was used in its calculation.

Confidence quantifies the probability that the consequent (or "then") part of a rule is true given that the antecedent (or "if") part of the rule is true. Relevance measures the rule's reliability and is obtained by using the relative support of the itemset and that of the antecedent.

Lift: Lift first subtracts the predicted support of a rule from the existing support and then-divides it by the predicted support, which is contingent on the presumption that all the components of a rule are independent of each other. It demonstrates the proportion of the outcome with regard to its likelihood of happening, provided the antecedent is given.

Among all the mining association rules approaches, the Apriori algorithm is recognized as one of the most widely used algorithms. It operates in two phases:

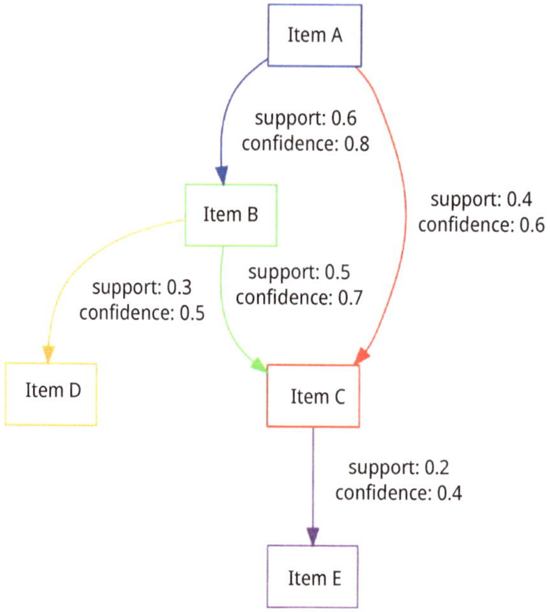

Figure 9: Association rule learning.

Generation of Frequent Itemsets: It involves periodically producing frequent itemsets, which are itemsets that satisfy a minimum support value. The process starts with individual items and builds upward toward forming larger itemsets, eliminating those that do not meet the support threshold.

Rule Generation: Once the frequent itemsets have been identified, the algorithm formulates association rules based on the confidence value of the possible rules that can be generated from the frequent itemsets derived. The rules that have crossed the specified confidence level are the set of rules to be considered valid.

Apriori was quite effective when a large number of items are involved, but it has limitations because many itemsets must be created and tested.

As a result of improved computing performance, the FP-Growth (Frequent Pattern Growth) method provides an Apriori solution. FP-Growth constructs a compressed data structure, frequently called the FP-tree, instead of candidate itemsets. This tree compresses the significant frequency information of items and makes the frequent itemsets immediately extractable from it without the generation of candidates. Compared to Apriori, this method significantly reduces memory and computational overhead costs, which are major issues in that algorithm.

A promising approach for predicting drug toxicity using Association Rule Learning has been shown to identify patterns and relationships among drug attributes and effects. For instance, formulas can be established that lead to effective correlations between chemical characteristics, reactions, and clear toxicity indices. Such findings assist in the downstream ranking of compounds and guide efforts to better understand adverse effects during the initial stages of drug development.

Association Rule Learning was beneficial, but it has limitations. Therefore, the choice of support and confidence values plays an important role in the quality of discovered rules and can impact the further applicability of the results. Thirdly, such rules always increase in number with the size of the dataset, which makes their management and interpretation seriously problematic. Scientists often apply practices of rule pruning, filtering, and post-processing as means to enhance the effectiveness of the rules in actual applications to address these issues.

Data mining is one of the most crucial and promising tools used in current developments in association rule learning. The primary objectives of future developments are to extend the scalability of the method, integrate it with other machine learning techniques, and make the results more comprehensible. As with all fields, the constant effort is to advance the methods and strategies with regard to processing vast, intricate data.

3.5 Autoencoders

Autoencoders are one of the powerful classes of unsupervised learning mechanisms used to encode and reconstruct data. The general usage of these neural network designs is for the purposes of detecting anomalies, feature learning, and dimensionality reduction. Figure 10: Autoencoders are unsupervised learning neural network architectures that attempt to take input data, compress it, map it into a new representation, and then decode it back to the same input data format. This method efficiently learns data representations. Autoencoders have many applications in working with complex data structures since they can extract key features of the given data by applying efficient compression.

The autoencoder comprises an encoder and a decoder, though other layers can be added as required. The encoder consequently codifies the input data into a more diminished latent space, while the compressed input is used by the decoder to reconstruct the input data. To train the latent space so that it can capture the most important features of the data, the main paradigm is to minimize the distance between the input image and the output image. In most implementations, the encoder and the decoder have the same structure; the encoder maps the inputs through multiple transformations to a lower-dimensional space, while the decoder transforms the reduced vector back to the original dimensionality of the inputs.

Several variations of autoencoders cater to different data and application needs:

Vanilla Autoencoders: The least complex type, with each layer connected to every neuron in the subsequent layer, utilizing a bottleneck architecture to identify the compressed form of the data. These auto encoders are general-purpose feature extractors and can also be used for data denoising.

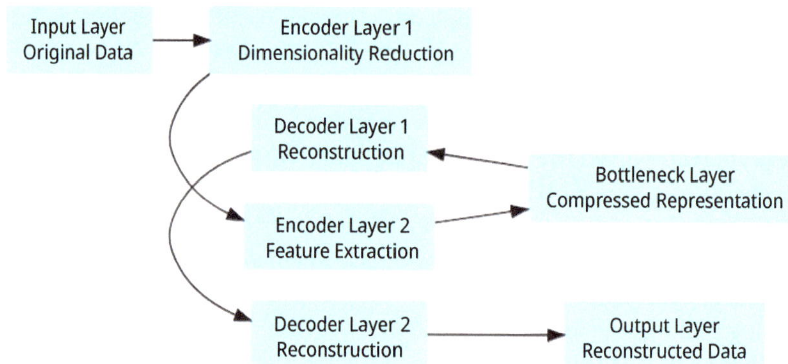

Figure 10: Autoencoders.

Convolutional Autoencoders: Try utilizing convolutional layers in the encoder and the decoder to process spatial data, such as images. Convolutional Autoencoders should be implemented because of their capability to learn hierarchical representations as well as reconstruct high-dimensional image data with minimal computational resources.

VAEs: They extend conventional autoencoders by providing a new probabilistic encoding method. In particular, using VAEs to train a distribution across the latent space could mean that, in lieu of being able to obtain fresh samples from the observed data distribution, new samples that belong to the learned distribution could be generated. When it came to emulating generative modeling and creating new data instances, this method was of great value.

Denoising Autoencoders: Designed to reconstruct clean data based on inputs that have been corrupted. The original signal can then be reconstructed by denoising autoencoders, which are trained on noisy data; they can be applied in areas that require anomaly detection and image denoising.

When using sparse autoencoders, the model is forced to seek a more complete and concise representation of the data by limiting the variance in the latent space. This approach is beneficial when computing sparse representations and selecting features.

Autoencoders have significant applications in the context of drug toxicity prediction:

Dimensionality Reduction: Autoencoders have valuable aspects in the representation and visualization of large densities of chemicals and bioinformatics through dimensionality reduction. Autoencoders enable the detection of important features that cause toxicity by reducing the depth of data to a latent space.

Feature Extraction: Autoencoders can be useful for increasing efficiency in the predictive models employed in drug development because they help extract valuable features from raw data – such as gene expression patterns and molecular descriptors.

Anomaly Detection: Peculiarities and outliers are well captured by denoising and sparse autoencoders in drug safety data. These models can recognize patterns of toxicity while being able to understand normal patterns in the data.

Despite their advantages, autoencoders face several challenges:

Model Complexity: Autoencoders suffer from the problem of hyperparameter and network structure selection in many aspects. Overcomplicated models distort the data and do not work as well, whereas when a model is too simple, it omits some data characteristics.

Interpretability: The embeddings of the autoencoder models are opaque and thus make it complex to explain the features that are being used to predict toxicity. The absence of interpretability tools makes it important to develop methods for visualizing and analyzing latent representations.

Computational Resources: Deep and convolutional forms of autoencoder training indeed require a great deal of memory and computing power. These difficulties can be mitigated by the use of proper hardware acceleration as well as efficient training algorithms.

Autoencoders remain a prominent direction in the development of new methods within the framework of unsupervised learning, providing useful instruments for encoding and feature extraction, as well as anomaly detection at the present stage. They are also promising for finding latent relationships in drug toxicity prediction and enhancing the safety assessment of newly developed chemicals.

4 Ensemble learning methods

4.1 Bagging

Bagging is a type of ensemble learning algorithm designed to combine different base learners to make overall prediction models more robust. Bagging, a method created by Leo Breiman, aims to reduce variance and increase the actual precision of models prone to overfitting.

The basic concept of bagging is to create multiple bootstrap sample subsets of the initial training data, derived from repeated random sampling of the original data, with replacement being made for each draw. Figure 11 demonstrates the Bagging ensemble method, whereby multiple models are created, and the final prediction is a weighted mean of all the models built in a process designed to minimize variance and thereby prevent overfitting. We trained a different instance of the basic model with each subgroup. For these separate models, to develop a final output, the average of the results was typically taken for regression problems, while voting was used for classification problems. To achieve a more stable and generalized outcome, this method utilizes the diversity of the base models, which serves as the foundation for this aggregation operation.

Data Resampling: For each bootstrap sample, the original data set was resampled with replacement to create a new data set. Even though it contains twin samples, each bootstrap sample also has the same number of instances as the initial data set.

Model Training: To build a base model, a decision tree, each of the bootstrap samples was trained independently. This leads to a situation where there are many models, all of them trained on somewhat different data sets.

Aggregation: The majority voting works in the following way: the base models operate independently to make a prediction of a class, and whichever class is most frequently predicted by the individual base models is considered the final prediction for classification problems. For regression tasks, the final prediction is obtained by averaging the predictions made by all the models that have been designed.

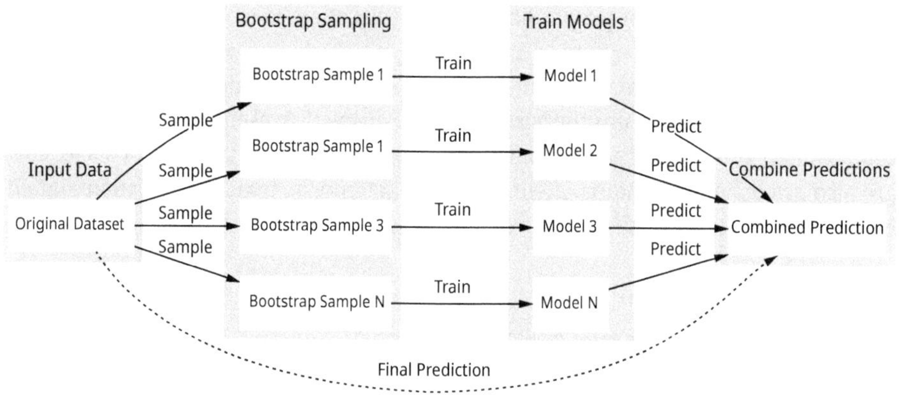

Figure 11: Bagging.

Variance Reduction: Thus, by taking the average of the mistakes made by each base model, bagging reduces the variance of the model. High-variance models, which include deep decision trees and other models with high variability, can benefit from this more than any other.

Improved Accuracy: Normally, the bagging technique achieves a better estimated error rate compared to any of the base models. This is because the aggregated ensemble's special feature of learning ability across different partitions enhances the general comparative performance.

Robustness: When bagged, the model seems to be more robust to noise in the training dataset. The influence of the noisy cases was also reduced in the final aggregated predictions because every base model predicted the targets based on different subsets of the data.

Bagging has demonstrated its utility in the field of drug toxicity prediction through several applications:

Model Stability: In general, where noisy data is available in drug safety assessment, bagging can be used to produce more stable data by combining a number of models, thereby providing greater accuracy in the results.

Feature Importance: When using decision trees as base models, bagging was also capable of offering feature importance to provide a degree of decision insight, leading to variable importance for toxicity predictions.

Handling Imbalanced Data: If the sample size of an observational set had more toxic cases than nontoxic ones, bagging was extremely useful. It was used to minimize class imbalance problems since bagging involves training on multiple subsets of a given data set.

Computational Complexity: The development of more than one base training model was computationally intensive and time-consuming, especially if the data was large and the base learners were complex. These challenges can be overcome by implementing the process more efficiently and by having some steps processed in parallel.

Diminishing Returns: Bagging is beneficial and does improve model performance, but the change that occurs as we increase the number of base models diminishes. There is always a compromise between the number of models to simulate and the time it takes to do so.

Interpretability: In case each model is placed in a group, an increase in the ensemble's opacity is observed; that is, the understanding of particular forecasts becomes less clear. For this reason, methods for interpreting ensemble models and explaining the importance of features can be helpful.

Bagging is an important type of ensemble learning that has consistently proven effective in improving predictive accuracy and model robustness by integrating different types of base learners. In drug toxicity prediction, its utilization leads to providing stable quantitative predictions and facilitates the creation of safer pharmaceutical interventions.

4.2 Boosting

Boosting is an old method known as ensemble learning that uses many weak learners to create a strong learner, which, in other words, enhances the performance of prediction models. A typical example of the development of powerful predictive models is the boosting process shown in Figure 12: A number of weak learners are trained, and new strong learners are built, where each new learner achieves a better outcome than the previous one by correcting the errors of the earlier learner. The main premise of boosting is to learn from the errors made by earlier models by using weak learning algorithms on reweighted copies of the data in succession. The efficiency of this process is realized through the iterative procedure of integrally combining the results of multiple learners and, as a result, achieving overall enhanced efficiency and robustness in the constructed model.

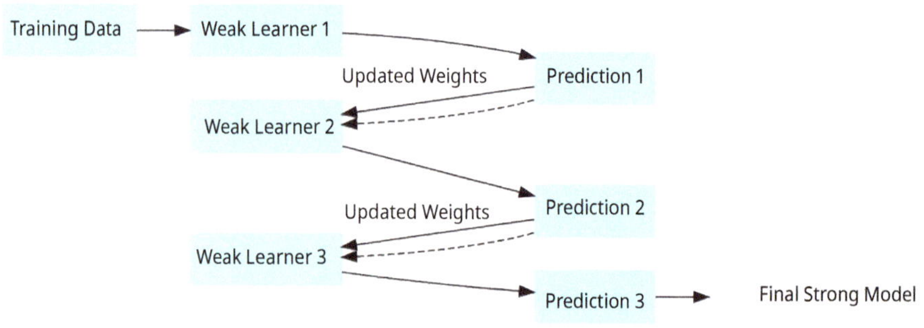

Figure 12: Boosting.

The boosting algorithm operates in several stages:

Initialization: A base model, often a weak learner, was developed from the original data by building a decision tree with a few levels. The first model was used as the basis for developing the subsequent models.

Error Calculation: After the training of the base model, the errors (or residuals) are computed. For the primary model above, the errors are computed after the base model is trained. These residuals show where the model has underperformed; that is, they reveal the data that needs more focus to be effectively predicted.

Reweighting: As with the previous round, the errors obtained are used to alter the training dataset. Misclassified cases are given higher weight than correctly categorized cases. This change ensures that the successive model gives more focus to the hard cases.

Model Training: As for the reweighted dataset, a new weak learner was built. This learner focuses solely on the cases that the earlier model classified incorrectly to correct its errors.

Aggregation: Of all the learners, it was the final weak learner that was obtained by combining the predictions of all the weak learners. For the aggregate of learners, weighted voting (for classification) and weighted averaging (for regression) can be used, with each learner's rating being proportional to the accuracy it displayed.

Iteration: When a certain number of rounds have been completed and no further improvement is made, the procedure is repeated. The above model becomes more sophisticated in the way it describes the complexity of the patterns in the data during each iteration.

Several boosting algorithms have been developed, each with unique characteristics:

The AdaBoost (Adaptive Boosting) algorithm was one of the first and most widely used boosting algorithms. To focus future models on these harder scenarios, AdaBoost adjusts the weights of misclassified examples. All the individual learners are combined into a final model, and the models are weighted, with higher-performing models being given more weight.

Gradient Boosting Machines (GBMs): GBMs train each new model to predict the residuals of the prior models in a step-by-step process. It is generalizable for all kinds of data and loss functions because this method performs gradient descent to minimize a given loss function.

XGBoost (Extreme Gradient Boosting): an improved gradient boosting flavor that includes techniques designed to prevent overfitting as well as improve efficiency. Due to its flexibility and rapid calculation, XGBoost can easily solve vast, complex problems and handle large datasets.

LightGBM (Light Gradient Boosting Machine): LightGBM has been developed to be incredibly fast and scalable by employing histogram-based methods for boosting. This is especially useful for handling big data as well as feature spaces with a high number of dimensions.

CatBoost (Categorical Boosting): Focuses on the efficient management of categorical features through the application of approaches used for coding and processing categories in variables. The original CatBoost aims to minimize the amount of preprocessing and enhance the raw evaluation of a model based on datasets that contain categorical features.

Boosting methods have demonstrated significant utility in predicting drug toxicity and side effects:

Enhanced Predictive Accuracy: When a number of weak learners are combined, boosting algorithms enable them to capture the complex patterns and interactions in toxicity data, thereby enhancing the prediction accuracy.

Handling Imbalanced Data: Some of the improvement methods, especially those with complex initialization strategies, are efficient in managing instance imbalance problems, as are common with toxicity datasets where dangerous effects are infrequent compared to nontoxic ones.

Feature Importance: Using boosting algorithms, information on the importance of individual features can be provided, which helps in understanding the factors responsible for drug toxicity and subsequently aids further research.

Despite their strengths, boosting methods face several challenges:

Overfitting: Boosting algorithms have limitations, including the over-fitting problem, which arises frequently when using complex base models in boosting. To counter this danger, it is necessary to make use of regularization techniques and carefully tune the hyperparameters.

Computational Complexity: The training of multiple models in an MTL manner is a computationally expensive and time-consuming procedure. Some of these concerns include inefficient implementations and the inability to handle multiple operations simultaneously.

Interpretability: The combined nature of boosting models complicates interpretation. Methods for displaying models and their assessment are crucial for identifying the individual impact of learners and general model trends.

Boosting is a powerful and flexible technique in the ensemble learning model that significantly improves predictive performance. Thus, when applied to predict drug toxicity, the proposed method enhances the detection of potential side effects and supports the development of safer and more efficient drugs.

4.3 Stacking

Stacking generalization is one of the most complex ensemble learning methods, where multiple models are used to enhance the predictive accuracy of a particular model. This method produces the final prediction by utilizing the advantages of many base learners, often surpassing the performance of individual learners. Figure 13 presents the stacking process, where a number of models provide their prediction results, which are then used to formulate the final prediction after aligning with each other. The main concept behind stacking is to achieve better accuracy and robustness of the ensemble by learning the final decision-making process from many base learners.

The stacking process involves two main stages: the basic level and the meta level.

Base Level: At this point, several base models are trained on the same data set. These models consist of different methods, such as decision trees and support vector machines. All base models apply the patterns they have learned to generate outputs.

Subsequently, a meta-learner utilizes predictions as input features of basic models found in the literature. This meta-learner learns how the predictions of such underlying models should be combined to give the final result. It is often a more complex model, such as the logistic regression model. To ensure the meta-learner is well generalized, it is trained on a different validation set.

Stacking employs different training strategies to optimize the ensemble:

Layered Stacking: Includes placing one model on top of another. The first layer is built from the base models learned directly on the original data, while the other layers are built from the predictions of the previous layers. This can reflect the effect of one model on another but creates a high risk of over learning.

Stacked Generalization: This work involves employing base models that were trained with the original dataset to train the meta-learner using the predictions of the base models. To ensure that the meta-learner has an unbiased prediction, cross-validation is often used to partition the data into training and validation sets for both the base and meta-level learners.

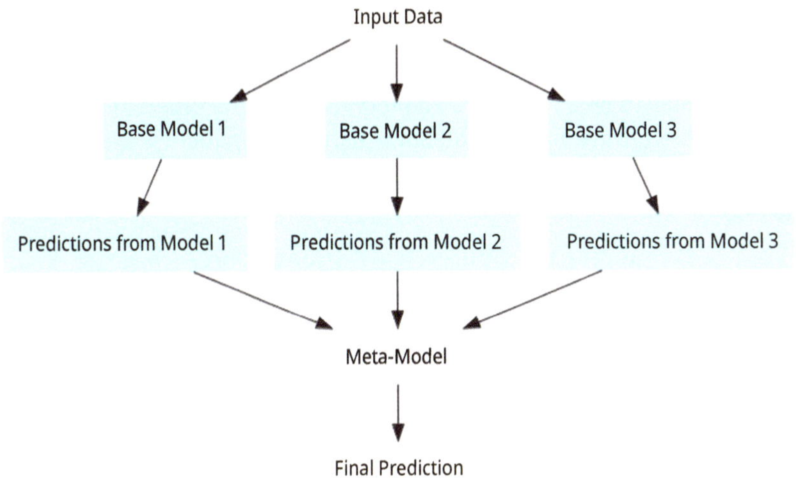

Figure 13: Stacking.

Stacking provides several benefits in the context of drug toxicity prediction:

Improved Accuracy: Stacking is when the output of several base models is combined in such a way that the accuracy of the predictions is higher compared to the accuracy of predictions from each model separately. This is particularly helpful in drug safety, considering the fact that precise prediction of side effects is important.

Robustness to Overfitting: External combination techniques decrease the likelihood of developing an overly complicated model, as several models with different characteristics are created and incorporated. This is a diverse approach that assists in generalizing better on unseen data.

Flexibility in Model Choice: It makes it possible to use base models from different classes so that different kinds of algorithms, which account for diverse empirical aspects, can be used. For instance, integrating models that perform especially well when applied to linear interactions and models that handle nonlinear interactions improves performance.

Despite its advantages, stacking presents several challenges:

Computational Complexity: Training a meta-learner and several base learners can require a significant amount of computational resources, especially when working with large datasets and complex algorithms. Good organization and utilization of resources, as well as putting them into practice, are essential.

Risk of Overfitting: Though stacking can control for overfitting, the meta-learner model is prone to overfitting if the validation test is not well executed. The meta-learner should, therefore, be able to generalize well to new data, and cross-validation techniques must be used to achieve this.

Model Selection: L1 base models and G meta-learner were key to the success of stacking with L2 models. The efficiency of the ensemble highly depends on the choice and placement of these models.

Stacking continues to be an effective method of combining different models in machine learning to improve basic predictions. Its use in predicting drug toxicity makes it evident that it has the capacity to enhance safety evaluation and prevent adverse drug reactions.

4.4 Voting classifiers

Voting classifiers offer a simple form of ensemble learning to improve classification accuracy: Instead of relying on a single model's prediction, voting classifiers use multiple models. In Figure 14, depicting the voting classifiers, a certain number of models contribute to the final prediction in an ensemble method, either by voting or majority rule. Voting classifiers take advantage of the variety among the different models by combining the outputs of many base classifiers to arrive at a final decision.

Majority Voting (Hard Voting): In majority voting, the class with the highest vote count is chosen as the final prediction. Every classifier in the ensemble provides its opinion on which class should be assigned. This method uses a simple majority rule and was earlier most efficient when the base classifiers are diverse and effective. Majority voting is particularly important when models have distinct advantages and disadvantages, as their combined results offer a reasonable outcome.

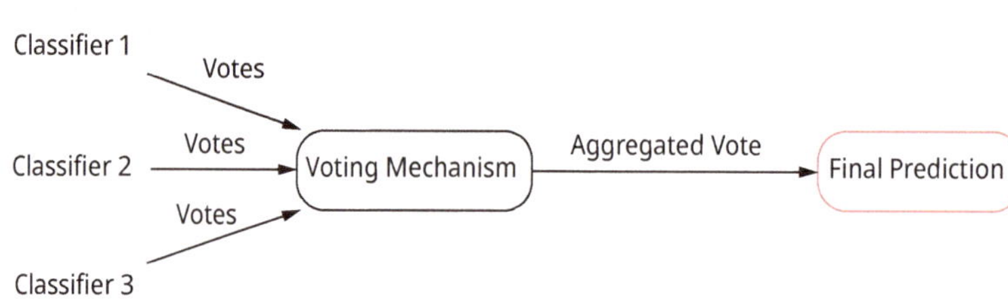

Figure 14: Voting classifiers.

Averaging (Soft Voting): Finally, the last prediction was made by considering the class with the highest average probability value. Soft voting consists of averaging the predicted probabilities of all the classes for all base classifiers. This method can enhance the model's capability to handle uncertainty and was useful while working with probabilistic data. When confidence in the predictions of the base classifiers differs, "soft" voting turns out to be better than "hard" voting.

Weighted Voting: In this method, every base classifier is assigned a weight corresponding to its accuracy as well as its consistency. The sum of the weighted votes of the maximum response from each classifier gives the final result. This technique allows for every single model's performance to be incorporated into the voting process, which in turn enhances the functionality of the entire ensemble. There is also the possibility of weighting based on measures such as the F1 score or accuracy.

Stacked Generalization (Stacking): Even though stacking is not a voting approach as such, it involves training a meta-classifier on the predictions of several base classifiers. In stacking, a higher-level model derives characteristics from base models and uses them to make the final decision. This strategy can work better than elementary voting methods by leveraging the advantages of several models.

Voting classifiers have notable applications in predicting drug toxicity and side effects:

Combining Diverse Models: When it comes to enhancing prediction capability, voting classifiers are capable of incorporating a variety of base model forms, such as decision trees, support vector machines, neural networks, among others. Voting classifiers can handle more data patterns in the voting process and improve overall toxicity prediction by utilizing multiple diverse models on the data fed to the algorithm.

Handling Class Imbalances: In drug toxicity prediction, where adverse effects are significantly less frequent than nontoxic outcomes, voting classifiers can aid in addressing class imbalance, as different voting classifiers handle them differently. This aggregation is useful to ensure that the minority classes, such as rare toxic effects, are given adequate consideration.

Enhancing Predictive Performance: Ensembling strategies can reduce bias and volatility, which stem from individual models, through the amalgamation of multiple base classifiers. The results show that using the ensemble approach in toxicity evaluation offers higher reliability and makes the overall prediction of toxicity assessment more accurate.

Despite their advantages, voting classifiers face several challenges:

Computational Complexity: With the integration of multiple classifications, the computational overhead depends on the number of classifiers, where large data sets and complex models increase the time required to execute. To address this challenge, implementation and optimization are inevitable.

Diversity of Models: This means that the number of base models available determines how well voting classifiers perform. Potential voting gains are reduced if the base classifiers are rather similar and make related errors. For the best performance of the ensemble, the model pool must cover a large variety of models.

Interpretability: Although voting classifiers enhance predictive capability, deciphering the outcomes can be difficult, especially when there is a need to understand how and to what extent the different models impact the decision. Methods for identifying and characterizing the work of individual models and the ensemble as a whole are key to increasing transparency.

Election classifiers are a powerful ensemble learning method for enhancing predictions. Their use in drug toxicity prediction offers a way of utilizing various models and handling multifaceted data to improve the results of toxicity tests.

4.5 Hybrid models

Ensemble learning techniques have been formulated into various categories, one of which is known as hybrid models, considered an advanced approach in ensemble learning because it integrates and takes advantage of various categories of machine learning. As a combination of various disjointed algorithms, using a hybrid model has the main objective of increasing predictive capacity and stability while working to overcome some of the weaknesses associated with other models. Hybrid models are incorporated, as shown in Figure 15, where more than one machine learning model is used to make a prediction, and adding more than one model to the work can improve the accuracy and flexibility of the model. This approach is especially valuable for domains that involve various, and often nonlinear, data types and patterns, such as drug toxicity prediction.

Most hybrid models are made of two completely separate modules that work in parallel to interpret inputs and make predictions. These elements comprise many types of first-level models, each with special advantages – for example, decision trees, neural networks, and support vector machines. Through stacking and mixing processes, which are all parts of the hybrid methods, the model integrates the elements, making it helpful to study the data more profoundly.

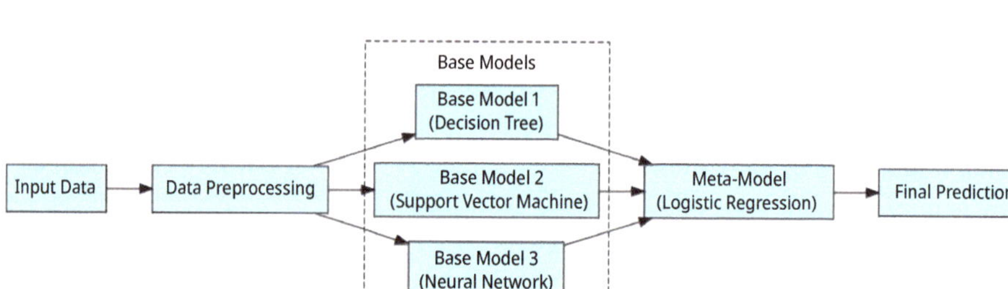

Figure 15: Hybrid models.

Stacking: Stacking means training a number of base models on the same training sample and using their outputs to train a meta-model. The base models can vary significantly with respect to each other and, in the following, could include neural networks, decision trees, and ensemble methods. The meta-model, often a less complex model, such as logistic regression, tries to find the right proportion by which the base models can generate an output. This approach improves prediction capability because the best features of several models are harnessed.

Blending: Similar to stacking, but base models are used to integrate predictions via a weighted average for aggregation. Blending usually involves the use of a holdout data set in the training of the meta-model. Most of the time, blending can be faster compared to stacking, but it may result in slightly lower accuracy. During blending, the required models are successfully combined and are typically effective at various aspects of the prediction task.

Voting: Voting, in which the forecast outcomes of several models are integrated through a simple average for regression and by employing voting based on the majority rule for classifications, was used for voting. In voting ensembles, an equal weight was provided as all models were considered equal to the last output, and the total output was the consensus of the models. Voting can be implemented using simple models such as decision trees, support vector machines, and neural networks, among others.

Boosting and Bagging with Hybrid Elements: Boosting and bagging can also be combined with other categories of machine learning techniques to form hybrid models. For instance, integrating boosting algorithms, such as GBMs, with neural network selection methods can improve the performance of the models. Additionally, using multiple versions of a single model, which are combined through bagging techniques, provides high stability and helps minimize overfitting.

Multi-View Learning: It is called multi-view learning when integrating models trained with different views of data. Each view represents a unique aspect of the data and the feature sets present in the current situation. Multi-view learning can help improve predictions, as in addition to the usual model's predictions, additional models trained on these views are taken into consideration, providing a broader understanding of the data.

Hybrid models offer several advantages in the context of drug toxicity prediction:

Enhanced Predictive Accuracy: Ensemble learning, where many models are used, has the ability to detect the complex relationships and forces in toxicity data that other models cannot identify. Consequently, undesirable medication responses as well as potential side effects are determined much more accurately.

Improved Robustness: There are always advantages to using hybrid models since they neutralize the shortcomings inherent in a particular base model. For example, if one model is very sensitive to noise and another one overfits the data, then the integrated models result in a more stable and reliable predictive model.

Handling Diverse Data Types: As seen, drug toxicity prediction uses heterogeneous formats of data such as chemical characteristics, biological tests, and records. Combined models can incorporate different types of information through various base models, ensuring that the analysis of toxicity is comprehensive.

Complexity in Implementation: Using and developing hybrid systems present additional challenges compared to single-model systems, particularly when selecting base models and determining ways to aggregate their outcomes. Such complexity also results in longer training times for models and a general demand for more computational resources.

Model Interpretation: The performance evaluation of individual models under the hybrid framework was not very easy to comprehend. As the final model for practical use, it was essential to ensure that the result of combining all the mentioned models was understandable and had some kind of transparency to track the cause-and-effect connection between the model's output and particular elements of the system.

Data and Resource Requirements: Hybrid models are, by-default, more complex and result in increased data demands as well as heightened requirements for computational platforms due to the presence of multiple models and training. These combined factors indicate that efficient management of resources and accurate handling of data are mandatory for successful implementation.

The kind of study that holds potential for enhancing the performance and robustness of existing machine learning systems belongs to a class of models known as hybrid models. These models create an effective system for overcoming challenges related to the forecasting of adverse medication effects and improving drug safety assessments based on the integration of various approaches.

5 Evaluation and validation techniques

5.1 Cross-validation

Since it allowed us to estimate and ensure the performance of the machine learning models and their robustness, cross-validation was a mandatory procedure. This approach reduces the chances of overfitting by splitting the dataset into several parts, providing a better measure of performance, as well as the robustness and effectiveness of the model when determining efficiencies.

Cross-validation helps in breaking the dataset in to training and validation sets systematically, which enables testing the model on a new set and training it on the first set. Figure 16 pictorially explains cross-validation as a method of measuring the performance and accuracy of the generalization of a machine learning model by dividing the available data in to sub-sections for training and validation areas. This is done multiple times through employing different divisions of data for training as well as validation. This generalization is important when making actual predictions in the real-world by generating new data, as the main aim of the exercise is to assess the efficiency of the model on unseen data.

k-Fold Cross-Validation: In this case, the dataset was partitioned into k folds that are of equal proportions. For k-1 folds, an 8–3 train/verify split was performed on the data, and for the last fold, the roles were reversed to form an 8–3 verify/train split.

Once in the k iterations of this process, every fold becomes the validation set. This means that the average of the outcomes of each of the k rounds gives the final measure of performance. The k-Fold Cross-validation involves the use of sets of different data subsets for training and validation; therefore, it provides a good test.

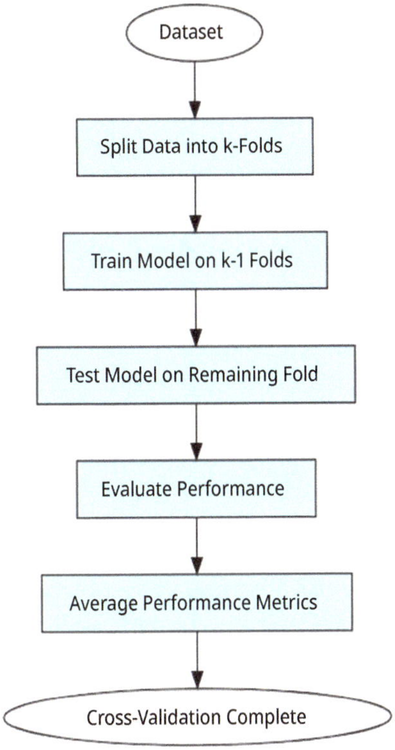

Figure 16: Cross-validation.

A cross-validation method that was a modification of k-fold cross-validation, in which k represented the number of observations in the dataset, was known as leave-one-out cross-validation (LOOCV). The last observations are used for training, and each point is used exactly once as part of the validation set. Small datasets find tremendous value in LOOCV's highly accurate model performance estimation with minimal bias. However, due to the large number of iterations, it requires significant processing power to solve the problem.

The modification of k-Fold cross-validation that ensures the percentage of class label proportions in each fold is equal to the original set is termed stratified k-Fold cross-validation. This technique is especially important for datasets used in the study where it is possible to encounter issues with class imbalance, which means some classes contain no samples. Since it is essential to maintain proportionate distributions within the folds and obtain more reliable performance estimates, cross-validation by strata is applied.

k-Fold Group: When individual data samples are grouped into distinct sets, such as health records of experiments, cross-validation is utilized. To avoid the problem that arises when the same groups of data are exploited during training and validation, the model is trained on one group and tested on the rest of the groups. This approach can be useful when analyzing data points from a group, especially when many of those data points are interconnected.

Time-Series Cross-Validation: This means that the data was divided in accordance with time, and no observation that falls in the future of any element of the data set was used for training. Thus, for better evaluation, there is more than one way to train and validate, which are known as rolling window techniques. The Time-Series Cross-Validation preserves the temporal characteristics of a sequence, which is particularly effective for forecasting and sequence prediction.

As has been observed, especially when evaluating ML models for predicting medication toxicity, cross-validation proves to be crucial. Sophisticated assessments provide a clear evaluation of a model's performance on different subsets of toxicity data sets, thus ensuring the model's stability and flexibility to adapt to various situations. Cross-validation was employed to assess the potential of the model in terms of achieving accurate predictions for adverse drug responses when tested on a variety of datasets.

Demand for Computation: This can become computationally intensive, particularly when operating on large datasets. Some limitations are encountered in terms of the high demand for computing resources required for repetitive training, validation and language translation, which limit implementation in resource-constrained settings.

Choice of k: The selection of k folds in the number of cross-validation influenced the separation of bias and variance. Increased k values entail a higher computing cost, while, at the same time, fewer k values result in large variances in the performance estimates. The variable of k, as used in the analysis, was very essential to arrive at the right balance in the assessment of any model.

Information Disclosure: Performance evaluation involves a validation process that should not be included in any way during the training phase. This is why it is crucial to handle and split data appropriately to exclude contamination between the training and validation sets.

As far as methods for validating models are concerned, cross-validation was reliable and provided useful information about how the machine learning models worked and how well they could generalize. It enables researchers as well as practitioners to be confident in using different types of cross-validation techniques to ensure that models for the prediction of drug toxicity have good performance, are reliable, and are valid to deliver accurate results in different situations.

5.2 Performance metrics

Evaluation measures are very important in assessing the performance of developed machine learning models, especially in drug toxicity prediction. They provide information about model quality, including accuracy, precision, and how well the given model handles imbalanced sets of data. These metrics help in selecting and improving the models and their parameters to meet the specific demands of toxicity prediction.

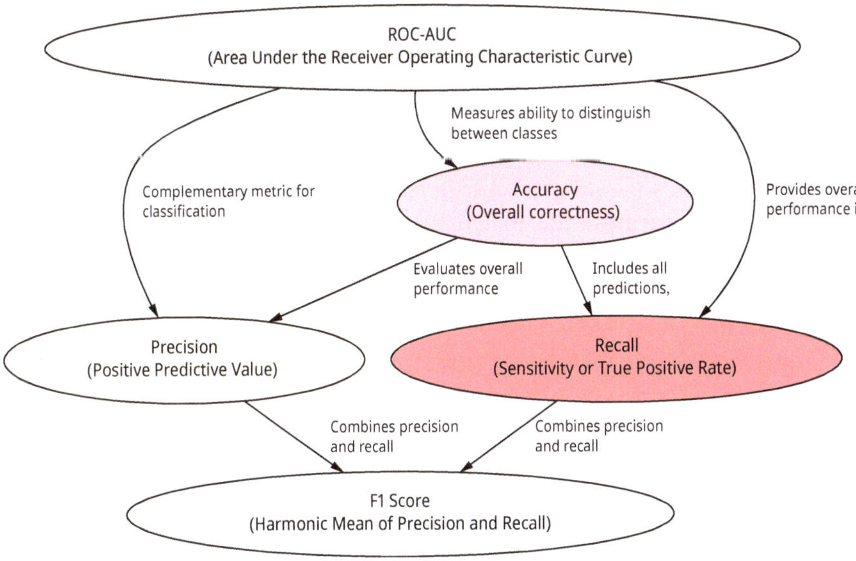

Figure 17: Performance metrics.

Accuracy pertains to the proportion of correctly predicted cases to all cases. It provides a measure of the overall correctness of predictions. However, accuracy by itself can be misleading, especially when there is an unequal distribution of data points across some classifications.

The accuracy of a given model was described by the number of valid positive predictions out of all positive predictions. When the cost of false positives was high, as in the case of medication toxicity where averagely nontoxic chemicals are erroneously classified as toxic, thereby necessitating further investigations, this was important.

Recall shows how effective the model is in identifying all the related cases. This was significant for achieving the primary goal of identifying, to the largest extent possible, true positives, even at the cost of having more false positives. In drug toxicity prediction, high recall is desirable to guarantee that all toxic effects are predicted without the suppression of any possible toxic compounds.

Specifically, both recall and accuracy are evaluated to develop the F1 measure. This means that it presents a balance between the ability to achieve high accuracy of the model with respect to true positives and the ability of the model to identify all the relevant occurrences, as it indexes this entire measurement by a single score. In Figure 17, multiple characteristics are demonstrated to assess the efficiency and effectiveness of the machine learning algorithms. When evaluating models where it was impossible to achieve high accuracy and recall at the same time, the F1 score proved rather valuable.

The true positive rate and the false positive rate at the varying thresholds were displayed by the ROC curve. The averaged accuracy of the positive class versus the negative class was used to assess the model's overall ability to classify between the two classes using AUC. Finally, AUC stands for Area Under the Curve, which means that a model that does a better job of distinguishing between harmful and nonharmful substances has a high AUC value.

Another detailed measure of a model's performance is the so-called confusion matrix, which contains the number of true positives, true negatives, false positives, and false negatives. It is from this matrix that other performance measures are derived, and these can help in determining where the model went wrong.

True positives, true negatives, false positives, and false negatives are the four outcomes of predictions that the Matthews Correlation Coefficient considers to provide a balanced accuracy of a model. It is very useful in evaluating models when assessed on imbalanced data and provides a richer measure of how well models perform compared to accuracy alone.

In particular, logarithmic loss was used to evaluate the accuracy of probabilistic forecasts by comparing the difference between the expected probability and the realized outcomes. Lower log loss values indicate better performance of the model on various probabilistic predictions, as it brings the probability estimations closer to the actual outcomes of the model.

The Brier score assesses the accuracy of probabilistic forecasts by finding the mean square of the difference between the forecast probability and the observed outcome. Lower values indicate higher calibration of the probability estimates. It evaluates the extent to which the predicted probabilities align with the realities of the findings.

These performance evaluation metrics provide a comprehensive assessment of the machine learning models, offering a measure of the models' utility in predicting drug toxicity. The selection of these metrics, depending on the particular problem, makes it possible to assess and develop models in the most effective way required for further usage.

5.3 Hyperparameter tuning

The most important point in the tuning phase of an ML model is called hyperparameter tuning, which is the procedure to select the most effective hyperparameters.

Often known as hyperparameters, these values are assigned prior to model training and do not undergo modification based on learned data. For predictive tasks like medication toxicity prediction, the unavoidable settings of such parameters are an obligatory step to achieve good results, as they affect the training process and the accuracy of models.

Parameters like the strength of the regularization, learning rate, and model complexity of the machine learning algorithms are all hyperparameters. In neural networks, the number of layers and units that exist, the type and parameters of the kernel in support vector machines, and the number of trees and depth in random forests are some of the best-known hyperparameters. It can therefore be seen that even the choice of hyperparameters affects the ability of the model to fit the data 'too well,' and to generalize to new datasets.

Grid Search: Grid search is an exhaustive approach where each possibility of hyperparameters is examined in combination with each of the fixed hyperparameter values. Generally, by applying this strategy, a grid of hyperparameter values is defined, the model is trained on each of them, and the combination that provides maximal performance according to some measure is selected. It is depicted in Figure 18 how hyperparameters affect the performance of a model during the training process; in other words, hyperparameters help improve the performance of a model through optimal training tuning. Even though grid search is performed, there is high complexity involved with a lot of hyperparameters, and the process can be computationally costly and time-consuming.

The first approach to the setup of the hyperparameters was called random search, in which the values of hyperparameters are chosen randomly according to a pre-determined distribution. Random search was computationally less intensive compared to grid search, where not all possible combinations are compared. Sometimes it produces better results, in addition to being more effective than a grid search at navigating through the hyperparameter space, especially when there are many hyperparameters.

Bayesian optimization: This technique utilized probabilistic models to guide the search for the best hyperparameters. Finally, Bayesian optimization was very efficient in searching for the best hyperparameters by progressively constructing a probabilistic model that estimates the utility of different regions of the hyperparameter space. Unlike grid and random search, this method optimizes both exploration and exploitation simultaneously in a bid to find the best hyperparameters with the least amount of evaluation.

Gradient-Based Optimization: A selection of hyperparameters that can be optimized with gradient-based methods includes learning rates in neural networks. These include what is known as gradient computations of hyper-parameter values in relation to the performance of the model in question. Performance-based ranking is especially helpful for hyperparameters, whereby they are optimized to an optimal value selected from a continuous array.

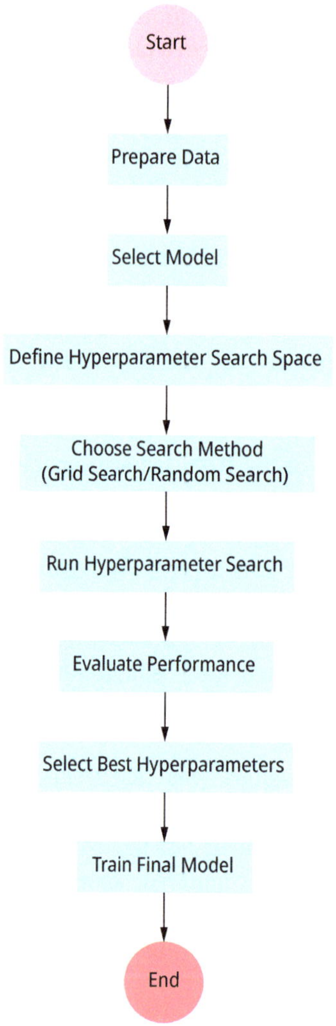

Figure 18: Hyperparameter tuning.

Automated Machine Learning (AutoML): AutoML systems naturally integrate hyperparameter tuning algorithms into their loops. Such frameworks combine grid search and random search, as well as nonconvex functions such as Bayesian optimization, to create an automated process that performs hyperparameter tuning alongside model selection and preprocessing. AutoML has been helpful in tasks such as hyperparameter optimization with reduced intervention and the enhancement of model performance.

Usually, to determine the effectiveness of hyperparameter tuning, cross-validation techniques are used. In cross-validation, the population is split into training and validation data. Further, the model is trained on the training set with different parameter settings and evaluated on the validation set. Thus, precision and recall values, F-measure, points 0.75–0.25 of the ROC curve, and AUC are examples of such assessment measures. There is a need to ensure that efforts made to optimize the parameters for the model do not cause overfitting of the validation set. This issue can be corrected by using several test sets to evaluate the final model.

Computational Cost: Optimizing hyperparameters can be very computationally expensive and can take a lot of time, especially if one is carrying out a grid search. To achieve the best tuning, both the extent of the search of the hyperparameter space and the time complexity must be balanced.

Dimensionality of Hyperparameter Space: This makes it almost impossible to determine the best settings, especially with so many hyperparameters to choose from. Some of the approaches include dimensionality reduction techniques and efficient search algorithms to control the complexity.

Overfitting and Model Selection: It was also important to ensure that the model was not overfitting when tuning the hyperparameters for the validation set. We needed two different test sets to determine the final generalization performance of the model and to check that the chosen hyperparameters provide good performance on unseen data.

Hyperparameter optimization was a key driver of the accuracy of an individual machine learning model for drug toxicity predictions. It was, therefore, possible to achieve improved model accuracy and predictability, which, in turn, allowed for improved reliability of insights influencing drug safety assessment.

5.4 Model robustness and stability

Other things being equal, model reliability and stability are major areas to consider when evaluating machine learning models, especially in computations for drug toxicity. Robustness refers to performance when exposed to inputs different from the source of training or to changes in the environment, while stability refers to consistent and reliable performance across different periods of time and various datasets.

Sensitivity analysis was the procedure used to examine changes in the mechanisms of defining predictions of the model with regard to changes in input parameters. This method helps in identifying which characteristics have a significant impact and how changes in the above-mentioned aspects affect the overall functioning of the model. Figure 19 shows the model's stability, which can also be characterized as good resistance to initial conditions and input variations. On that note, one learns about sensitivities – that is, how much the model's outputs change when inputs are slightly adjusted – so that their perception of the model's robustness enables them to test how well the model holds up to such treatment and whether minor shifts cause forecasts to vary significantly.

Adversarial Testing: Adversarial testing involves incorporating adversarial examples in an attempt to determine how the model performs under adverse conditions.

This method is especially significant in drug toxicity prediction, where the model encounters different data points and may generate incorrect predictions. Adversarial inputs, for instance, should not significantly affect the performance of robust models.

Stress Testing: Stress testing is an evaluation of a model in worst-case scenario situations. This approach checks whether the model performs optimally in high-stress scenarios and ensures that the model remains reliable when run under a range of conditions by "stress-testing" the model, for example, by passing it data that is noisy or in some way out of distribution to the model.

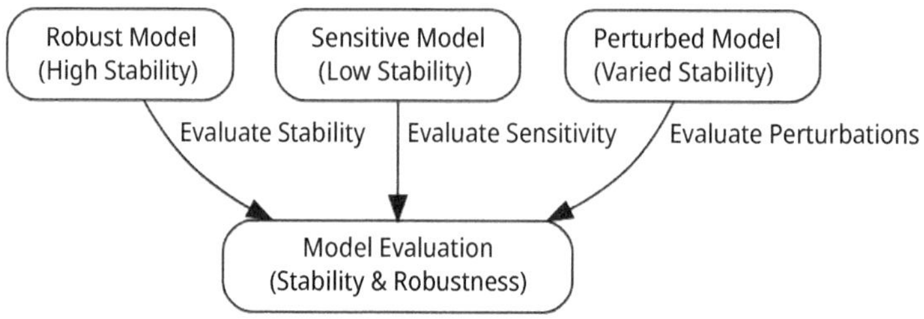

Figure 19: Model robustness and stability.

Data Perturbation: In data perturbation, certain forms of random noise are added to its input to observe the results experienced by the model. This approach assists in establishing the reliability and variability of the model output, that is, whether the model is overfitting on some trivial variations of the data.

Cross-Validation Across Datasets: As for improved cross-validation techniques, it is possible to work with several datasets and obtain additional information concerning a model's performance, including but not limited to working with diverse samples. The results obtained for different datasets also confirm the model's ability to generalize and its consistency across various situations.

Consistency Checks: The second type of analysis was called consistency checks, which include checking whether the model was accurate in its predictions across different kinds of runs and datasets. One also gets a sense of whether the model has behaved similarly across the different experiments or if the model's performance was erratic.

Error Analysis: The error analysis procedure was broken down to a detailed level to reveal the reasons for instabilities in model predictions. When misclassifications occur, there can be systematic reviews to determine issues that affect model stability and correct them.

Robustness Metrics: This section briefly discusses robustness and stability indices as quantitative measures of a model's stability. These metrics quantify model performance given specific types of perturbations and variations, thereby offering a foundation for comparing model robustness.

Regularization Techniques: By controlling overfitting problems and promoting the emergence of general models, techniques such as dropout and weight decay are beneficial to the stability of the model. They ensure the model works properly and remains dependable under any situation and for any dataset when these methods are applied.

Complexity in Stability Assessment: Measuring model robustness and stability is not easy and thus requires intricate methods of testing and a detailed working through. The development of procedures for uniform evaluation of these aspects has been beneficial and helps make the assessment process more efficient.

Handling Evolving Data: With data, the models should be adjusted to fit new formations and trends as they emerge. Possible future studies should, therefore, concentrate on achieving the adaptability of models to ensure they remain stable despite changes in dynamic data and unpredictable environments.

Balancing Performance and Robustness: At times, it's difficult to find the right balance between predictive capability and ensuring how robustly built our model is. A key challenge when seeking to advance the model was the problem of regularity: guaranteeing that improvements in the former jeopardize the latter as little as possible.

This makes the questions of model robustness and model stability critically important for achieving the desirable characteristics of machine learning applications. The quality of models used in drug toxicity and similar forecasts significantly increases when stricter evaluation methods are applied, and corresponding issues are resolved.

5.5 External validation

The external validation proved to be a critical step in machine learning models, especially to confirm high accuracy not only on the data set against which the models were trained but also on other data sets that had not been used in constructing those models. Figure 20 shows the external validation of the model in terms of the model's ability and generalization to an independent dataset. This procedure helps in understanding the stability and transferability of the intended models, which is useful in areas of interest such as medication toxicity prediction, where accurate and consistent performance across a variety of datasets is desirable.

The threat of overfitting, a situation such as attaining extremely high results on training data but being incapable of generalizing to new unseen data, was checked by external confirmation. External validation, on the other hand, involves testing the model on data that the model has not seen before, giving a true picture of how the model performs in terms of accuracy with new and diverse data. This was especially important in drug safety, where models used for predicting adverse effects have to be robust to the variations of populations and experimental conditions of the test.

Independent Test Datasets: Possibly the most direct way of external validation is to implement a new test data set that has not been incorporated into the training part of the model. Ideally, this dataset should be collected from another source under different conditions to make the evaluation relevant and satisfactory.

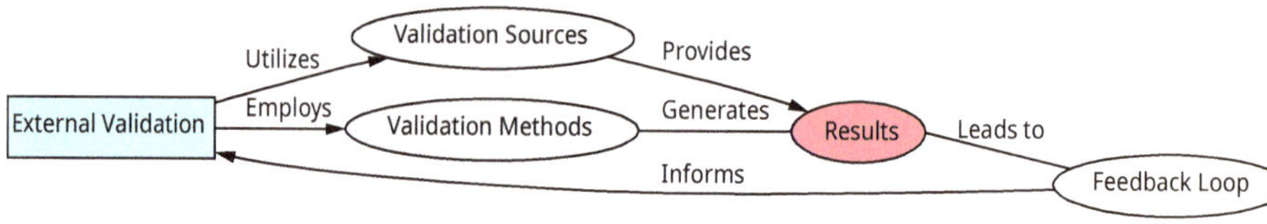

Figure 20: External validation.

Cross-Site Validation: In this way, the model was tested at one site employing this technique. Each site provided a dataset that was used to assess the model's efficiency. The cross-site validation process thus complements the achievement of these objectives and helps in the assessment of the model's dependability and interchangeability across different sites and across different sub-groups in terms of the demographics identified above.

The process of using one learned model in another dataset that was related to it is known as transfer learning. At a time when there is limited publicly available data for the target domain, this method is particularly useful. Some knowledge of the model's flexibility and its applicability must begin by evaluating it on the new data set.

External Benchmarking: According to external references, it is useful to compare model performance with well-known standards established by other sources. This method requires comparing the model with existing, widely integrated tools within the current field.

Real-World Data Application: Subsequently, it was possible to apply the model to actual data, specifically clinical trial data, and assess the validity of the predictions made by the model. This helped evaluate the model when it was applied in practice, thus offering an overall check on the usefulness of the model.

Dataset Differences: Check differences between training and external validation datasets, including data gathering, sampling techniques, subjects, externality, environmental conditions, and diversity. It was, therefore, important to ensure the external dataset resembled the training set as closely as possible to warrant accurate validation.

Sample Size and Quality: Depending on the size and quality of the external validation dataset, it was possible to obtain more validation results. The generality of the model would not have been assessed properly had the sample been limited. The size of samples and good data, in this case, are critical for sound external validation information.

Reproducibility and Consistency: To this end, it was important to ensure that the outcomes of the EXTERN process are reproducible and comparable across different datasets. Outliers indicate issues with the external validation process of the model being used by the program.

Regulatory and Ethical Considerations: External validation in drug toxicity prediction candidates was required to conform to regulatory and ethical standards. It was crucial for the regulation and purpose of the external validation to meet these standards to pass regulatory approval.

External validation of the dependability and generalizability of machine learning models is important to establish for certain critical applications, such as medication toxicity. Model evaluation by researchers helps ensure that their models are effective, precise, and can perform satisfactorily regardless of the conditions that prevail in real-life situations.

6 Advanced topics and future directions

6.1 Deep learning architectures

Neural network architectures embedded through multiple layers to analyze complex data patterns are known as deep learning architectures in the field of machine learning. As these structures are simplified specifically for processing large amounts of data to find relationships between them, they have been applied in different fields such as medication toxicity prediction. Deep learning architectures are depicted in Figure 21 to display diverse structures and the various applications of deep learning approaches in solving complicated problems, as illustrated in the four domains. New deep learning models are now advancing the frontier of data interpretation and enabling the anticipation of analytical results.

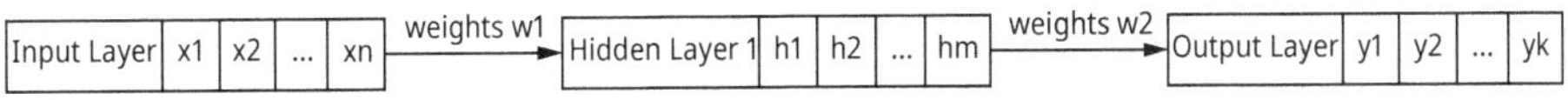

Figure 21: Deep learning architectures.

CNNs are particularly designed for data that has a mesh-like property, such as photographic data. In this case, it is necessary to apply filters that capture local patterns, such as edges, and this is achieved in the convolutional layers. Then, pooling layers are utilized as a way of reducing dimensionality and extracting features. CNNs' performance has been noted in tasks such as object identification, picture classification, and, more recently, drug interaction prediction and molecular structure analysis. CNNs are very effective at learning hierarchical features in datasets related to drugs, and the use of these networks allows for the detection of detailed spatial relationships.

Since RNNs contain feed-forward loops that preserve information across time steps, they are designed for sequential data. Chronic care points that involve time series, such as monitoring drug interactions over time or a patient's treatment history, are well suited to this architecture. Concerning the vanishing and exploding gradient issues in standard RNNs, LSTM networks and GRUs help identify long-range dependencies and improve sequential prediction results.

Autoregressive models introduce problems related to the inverse and self-attention processes to address this issue, which was quite an achievement in solving sequential data. Contrast this with the fact that, whereas transformers are able to process the whole input at once with ease, RNNs are much more sequential, and sequential data is not very parallelizable. Some existing models, such as BERT (Bidirectional Encoder Representations from Transformers) and GPT (Generative Pre-trained Transformer), are true examples of how design captures detailed relationships and contextual nuances. Hence, by using Transformers, it became possible to obtain a better contextual understanding of sequences, including patient records or chemical reactions; this can enhance medication toxicity prediction.

Applying Discriminant and Generator neural networks in a zero-sum matrix, Generative Adversarial Networks (GANs) are produced. While the discriminator checks the authenticity of the developed artificial data, the generator creates it. This adversarial training enables GANs to offer very realistic data samples in the scientific applications of data augmentation, drug interaction modeling, and probable side effect investigation. The capability of predicting pharmacological characteristics and designing molecular structures has been observed using GANs that learn from prior information.

Probabilistic autoencoders are applicable to data compression and feature learning, while VAEs stand for probabilistic ones. VAEs take the autoencoder concept further by incorporating modeling of data distribution and resemble generative models that can generate new samples, which have not been experienced before, from the learned distribution. In the DTI prediction context, we can use VAEs to generate the manifold of chemical properties, assist with data generation, and reveal additional, unobservable factors leading to toxicity. These models are very useful in situations where new drug candidates have to be developed and their possible side effects studied.

Scalability and Computational Efficiency: With the increase in artificial intelligence, and especially deep learning models, training and deployment necessitate significant resources. Future research for this model is concerned with the refinement of the models' computational efficiency and extendibility by means of superior optimization strategies and hardware offloading, including specialized accelerators and distributed computing paradigms.

Interpretability and Explainability: Nevertheless, the deep learning models' advantages in making predictions are accompanied by their inability to explain the outcomes. New trends in Explainable AI are focused on providing an appropriate explanation of how deep learning systems make decisions, as well as how they can meet the requirements of regulators.

Integration with Domain Knowledge: Extending deep learning techniques with knowledge about a certain domain can enhance the quality and readability of the models. Future directions consist of fusing chemical informatics, biological data, and clinical knowledge with deep learning models to improve their relevance for drug toxicity prediction.

Generalization to Diverse Data: One pertinent problem was getting deep learning models to extend over various datasets and real-world situations. In domain adaptation and transfer learning studies, the primary goal was to develop methods capable of dealing with differences in data distribution and enhancing the model's ability.

The enhancements of deep learning architectures are still ongoing, providing enhanced tools to address the challenging issues related to "drug toxicity prediction." In expanding these types of models and dealing with these issues, investigators achieve superior predictions, publish new findings, and refine risk assessments in pharmacologic testing.

6.2 Transfer learning and domain adaptation

Transfer learning also addresses problems related to low sample sizes and other related issues that stem from the irregular distribution of data. Thanks to these methods, models can transfer information from one task to another and therefore facilitate performance on similar but different tasks. Figure 22 shows the process of transfer learning and the adaptation of the gained knowledge to another but related domain. Crucially, this feature is most valuable in areas such as predicting medication toxicity, where large, labeled datasets are scarce, and models need to be robust across domains.

Transfer learning is the act of enhancing a model that works on a new but related task by using a model that was developed on a large and, in many cases, different data set. This method uses features or similarities between the source task and the new task that one wants to perform. In practice, transfer learning often refers to changes in a model's architecture for a specific assigned task. Using prior knowledge reduces the need for extensive training on new data each time, which is achieved in this way.

Feature Extraction: The method identified in this paper uses a pre-trained model for feature extraction. As for the new job, the higher levels of the model are changed, whereas the lower ones, more basic in their functionality, are retained. Some of the characteristics were then extracted from the pre-trained model and input into a new model developed to achieve the intended job.

Fine-tuning: It involves fine-tuning a model with certain weights adjusted to serve another purpose apart from the one it was trained for. Typically, the last layers of the model are trained on the target data, while the earlier layers of the model are locked to reutilize the domain invariances learned from the source domain data. The likewise referred-to fine-tuning allows the model to adapt the learned representations to the specific details of the new job.

It used to be called "domain-specific pre-training" when a model was pre-trained on a domain similar to the target one. A model learned on chemical compounds, for instance, can be further trained using data on the toxicity of drugs. This intermediate pre-training was important because it introduces features that, depending on the target dataset, need to be fine-tuned.

Domain adaptation is concerned with the process of transferring learned information from one source domain to another related but distinct domain. To address the shift between the source and target domains, it involves modifying either the model or its parameters. Thus, the objectives in the present work are focused on enhancing the model's ability in the target domain and reducing the domain shift.

Domain-Invariant Feature Learning: This approach was employed with the view of learning features that are invariant to the given source and target domains. The so-called domain adversarial training promotes the model to seek representations that reduce the differences between domains to make the learned features more transferable.

Instance Weighting: Example-based schema adaptation includes the modification of weights of the instances from the source domain during the training process according to their similarity with the target domain. This technique is beneficial in terms of correcting the deceptive focus of the main model toward the parameters of the target domain.

Data Augmentation: To address the domain shift of the target domain, splintered data augmentation procedures produce additional samples. Due to the ability to change the values with completely synthetic data, these methods can fill the gap between the source and the target domain and lead to performance enhancement.

Model Regularization: Auxiliary functions applied to strengthen the recovery regularizations are utilized to impose constraints that ensure the model performs well across domains. It is easy to lose performance when transitioning between domains, but methods like domain-specific regularization assist with this.

Transfer learning and domain adaptation offer significant advantages in drug toxicity prediction:

Leveraging Pre-Trained Models: As we explained before, transfer learning can help utilize models trained on chemical datasets and then fine-tune them to perform specific toxicity prediction tasks with limited data. This approach shortens the time required to develop models and increases the accuracy of predicting outcomes.

Handling Data Scarcity: a similar approach, with models trained on related but more abundant data sources, can be utilized in situations where annotated toxicity

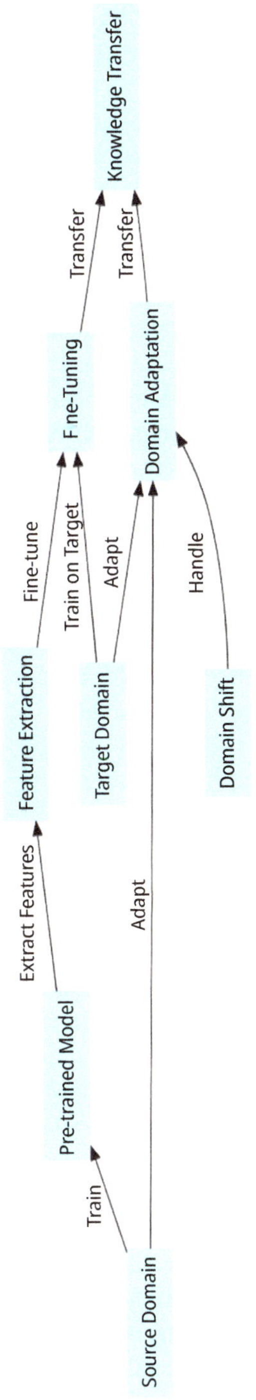

Figure 22: Transfer learning and domain adaptation.

data is scarce by means of domain adaptation. This method enriches the generalization potential and the ability to provide highly accurate predictions regarding new therapeutic opportunities for the model.

Cross-Population and Cross-Study Adaptation: Domain adaptation and transfer learning allow the use of models in different samples and under different study conditions. This is important to ensure that toxicity forecasts are accurate and generalizable to various patients and experiments.

Domain Discrepancies: A major concern remains how to handle the wide disparities that exist across the domain differences between the source and target. Promising approaches for mitigating domain shift and improving model transferability are still evolving.

Model Interpretability: It is important to ensure that when using transfer learning and/or domain adaptation, the interpretability of the models does not suffer but is even enhanced, as is desirable in practical scenarios such as drug safety.

Scalability and Efficiency: The key was to build frameworks for transfer learning that were effective and adaptable for mass use in learning from large amounts of data.

Transfer learning and domain adaptation are really useful and important in attempting to improve machine learning models for the prediction of drug toxicity, as well as in other areas of study. These methods are, therefore, useful tools for handling the limitations of data by transferring knowledge and prior insights from related contexts and tasks.

6.3 Explainable AI and model interpretability

When developing and utilizing machine learning models, XAI and model interpretability were vital, especially in critical domains such as medication toxicity prediction. Given both the practical complexities of machine learning models and the ethical concerns, it was increasingly important to understand how the models make decisions and to ensure the forecasts are not only accurate but also understandable.

Despite the higher generalization ability obtained through deep neural networks and other complex models, the relationships the models learn are complex themselves and operationalize the machine learning model into a "black box" that contains very intricate internal mechanisms. It is crucial to build trust in these models by providing the reasons behind certain decisions made by the model, especially in areas such as adverse drug response prediction, where decisions have a significant impact. In Figure 23, we evaluate the correlation between XAI and interpretability, showing how interpretation methods help the recipients of the AI to trust the model's decisions by providing routine insights into the model's procedures. It is essential to focus on making the decision-making process more understandable to stakeholders so that, by using models, one complies with the rules and ensures that ethical AI is being employed.

Feature importance was also calculated; for example, the score shape suggests which features affect the performance of that model the most. These techniques assist in explaining the nature of the impact made by the different input constituents on the output by emphasizing emphasis on the importance of specific characteristics of the model, which would otherwise be complex.

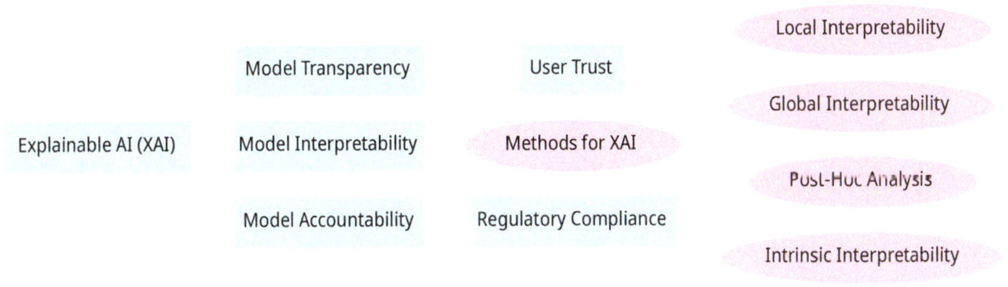

Figure 23: Explainable AI and model interpretability.

Model Visualization: Visualization techniques were used to provide a more intuitive understanding of how the model functions. For example, visualizing learned filters in Convolutional Neural Networks within classification models was beneficial for understanding how the model processes data and makes decisions.

Saliency Maps: Saliency maps help determine which areas of an image play a key role in making a specific decision in the context of image data. Saliency maps add clarity to models applied to the analysis of medical image data by emphasizing areas of interest.

Some models, where the basic structure is the neural network, contain the possibility of focusing on specific elements of the input data at the time of making preliminary estimates. In the course of investigating attention weights, researchers gain deeper insights into which data constituents are crucial for the further functioning of the model.

Rule-Based Explanations: As a result of that, one can derive human-understandable rules from complicated models and hence provide a reasonable specification of how the models behave. These methods, if effective, could make the decision-making process of complex models easier and, hence, more useful for people who are not experts.

Trade-off between Complexity and Interpretability: Often, interpretability and complexity of a model are interrelated. This is because as the capacity of a model to interpret high levels of complexity increases, its transparency often decreases. This was the reason that, with Extension 4, deep neural networks and other extremely complicated models could provide predictions that are more accurate but have less transparency. One of the most critical bottlenecks in the process of AI systems' creation was the question of explainability versus performance.

Context-Specific Explanations: It is important to establish the context and target audience for aims and objectives to determine both the relevance and utility of the explanation. For example, some explanations that are suitable for a data scientist may not be appropriate for a regulatory authority. Adjusting the explanation to the level of knowledge and understanding of different individuals requires intervention during consideration of the mode of how it would be explained.

Evaluation of Explanations: Assessing the quality and usefulness of the explanations was not always easy. Deciding whether the explanations provided were adequate and understandable by the targeted audience was conditional and had to undergo verification.

Regulatory and Ethical Considerations: It can now be seen that there are enhanced pressures for transparency and interpretability in AI systems, especially in high-risk fields such as drug efficacy. Meeting both of these requirements with accuracy and performance is a challenge when ensuring that models meet the following specifications.

Development of New Explainability Methods: Further work is required to develop techniques that can advance explainability, especially with the increasing sophistication of models. Each of these advancements in this subject area can result in more interpretable models that still maintain accuracy.

Integration with Domain Expertise: For this, it was crucial to have explanations from domain experts that are both accurate and semantically relevant. Domain expertise, along with AI explanations, can provide further context and also make the model more trustworthy.

Advances in Interactive Visualization: Additional features that are useful and could improve usability and understanding of model behavior include animated tools that allow users to observe changes in the model and control them independently. Future evolutions in this area offer less mediated means of engaging with and understanding complex models.

Ethical Guidelines and Standards: It is very important to clearly define the ethical principles and rules concerning explainable AI, as these systems are aimed at being integrated into decision-making models. Formulating integrated paradigms that engage both the technical and ethical levels enhances responsible artificial intelligence.

The foundations of Explainable AI and model interpretability are extremely important in ensuring that artificial intelligence is transparent, trusted to be employed ethically, and understood. With the development of the field, current research and development inquiries are still working on these issues and improving the understanding of how complex AI systems work and their applications, especially in life-relevant fields such as drug toxicity prediction.

6.4 Interaction with computational chemistry

The combination of machine learning with computational chemistry seems to be one of the most groundbreaking achievements in the science of finding new drugs. As indicated in this study, incorporating machine learning algorithms with the scientific and statistical computations of computational chemistry allows for more accurate predictions of drug toxicity, while also improving the efficiency and accuracy of predictions of chemical properties. This cross-functional approach enables each field to capitalize on the other's greatest attributes in solving issues related to pharmaceutical research.

Machine learning models improve computational chemistry as the prediction of molecular properties and interactions becomes more efficient than with existing approaches. Figure 24 shows an integration of computational chemistry with machine learning algorithms to improve predictive capability in molecular dynamic simulations. Molecular behavior and the structure of compounds are clarified through computational chemistry, and large data sets can also be processed by machine learning methods to identify patterns and predict trends. Higher resolution on ensuing drug interactions, binding profiles, and possible toxicogenomic outcomes is made available since the techniques are merged.

Prediction of Molecular Characteristics: Using computational chemistry data, it was possible to predict a number of properties of molecular compounds, such as pharmacokinetics, lipophilicity, and solubility. This, in turn, makes it possible for researchers, for instance, to easily sample and select medication candidates that possess the required qualities.

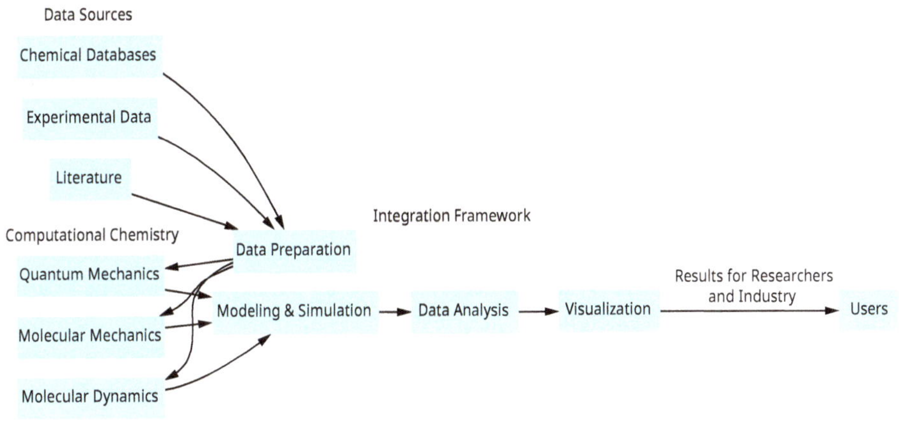

Figure 24: Integration with computational chemistry.

QSAR Modeling: QSAR models utilize data in computational chemistry to determine links between the chemical structure of compounds and their biological activity. The current techniques used in QSAR modeling do not account for advanced forms of relationships, such as nonlinearity and high dimensions; hence, the application of machine learning techniques would enhance and provide a better means of achieving the required goals, such as determining the efficacy and safety of drugs.

Molecular Docking and Binding Affinity Prediction: Molecular docking, a technique within computational chemistry, is able to predict how drugs are likely to interact with targets in biological systems. Other methods can build upon these predictions through machine learning analysis of docking results to determine relevant features that define binding strength and improve the discovery of promising drug candidates.

Toxicity Prediction and Risk Assessment: The combination of machine learning with computational toxicology can enhance the understanding of how chemical structures impact biological systems, providing better predictions of adverse drug reactions. This approach assists in the identification of possible toxicity threats, thus minimizing drug development failures at later stages.

Generative Models for Drug Design: GANs and VAEs are employed to design new chemical compounds with specific properties. These models can also scan chemical space more systematically and propose new drug entities in light of computational chemistry concepts.

Deep Learning in Molecular Dynamics: Molecular dynamics simulations and protein-ligand interactions are predicted using deep learning-based methods for drug stability. These methods can process molecular dynamics data for greater comprehension of the interactions of potential drugs under diverse situations.

Integration with High-Throughput Screening (HTS): Large chemical libraries are assessed by computational chemistry in high-throughput screening processes, and the resulting information is analyzed by machine learning algorithms. This integration accelerates the drug discovery process and helps to identify numerous possibilities for therapeutic approaches.

Data Integration and Quality: Integrating data from computational chemistry and machine learning involves working with different data sets in various formats and with differences in data quality. Having high-quality and well-annotated datasets has become a necessity to train machine learning models and make reliable predictions.

Model Interpretability: Even after some time, it was relatively challenging to visualize the results that the machine learning algorithms produced, especially when dealing with chemometric data. For real-world NDE applications, methods for interpreting and graphically representing ML models, which provide a coherent account of formulations from computational chemistry, must be devised.

Computational Resources: As one would expect, it can be rather resource-intensive to integrate computing techniques such as machine learning with those involving computational chemistry. To manage massive interactions and computational results, complex computational facilities and efficient algorithms are required.

Regulatory Considerations: Since machine learning models are now incorporated into drug development, it is crucial to ensure that these models comply with regulatory requirements. Due to the above factors, certain rules for using machine learning with computational chemistry in drug discovery have been elaborated to make the approaches valid and acceptable.

Machine learning combined with computational chemistry is believed to have a bright future for drug discovery and development. It has been shown that the integration of predictive modeling with detailed chemical analysis –such an interdisciplinary approach in drug discovery, development, and optimization – can be useful in identifying the desirable properties of drug candidates, making accurate decisions about drug properties, and conducting better safety assessments. With innovative and progressive technologies and methodologies, the combination of these fields is expected to foster more innovations and enhancements in drug discovery processes in the future.

6.5 Ethical and regulatory considerations

There are important and growing considerations surrounding the development and deployment of machine learning algorithms for predicting medication toxicity. As these models become increasingly integrated into clinical practice and pharmaceutical research decision-making, it is essential to address ethical concerns and comply with legal and regulatory requirements. Ensuring that these technologies are developed and implemented responsibly and effectively is critical to maximizing their benefit while minimizing potential harm. This is underscored in Figure 25, which highlights the core challenges and frameworks guiding ethical AI use in toxicology prediction.

Security and Privacy of Data: Some of the problems associated with the collection and processing of the described data are treated as critical ones as far as privacy is concerned, such as chemical characteristics, for instance. Lack of transparency and accountability: This is because deep learning and computer vision applications require privacy and should not be biased. For the purpose of personal data protection and compliance with the GDPR and HIPAA guidelines, participant data anonymity and the secure storage of participant data are essential.

Bias and Fairness: In cases where the training data is not inclusive of various groups, the training of a machine learning algorithm may either amplify existing biases or create new ones. The elimination of bias involves the selection of datasets with the right diversity in demographics and clinical factors, as well as the use of nondiscriminatory algorithms.

Transparency and Accountability: Some models can be hard to decipher due to the intricacies related to how they arrive at the conclusions reached. AI methods must be transparent and explainable to appease various stakeholders involved and ensure

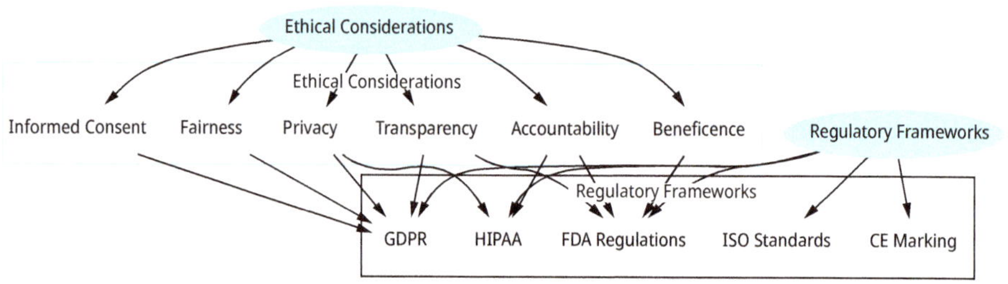

Figure 25: Ethical and regulatory considerations.

that accountability is achieved. Patients and regulatory institutions ought to be capable of comprehending and comparing model decisions.

Informed Consent: The use of data for training machine learning models ethically requires obtaining the consent of the data subjects. This includes explaining how the data can be utilized, the dangers involved, and the advantages of the research. The consent process must be explained in detail so that individuals have an opportunity to refuse.

Compliance with Regulations: Like other drug discovery and development models, the machine learning models for drug toxicity have to meet specific regulatory requirements and guidelines from the FDA and EMA. These regulations govern the assessment and approval of prediction models for safety and other performance criteria before they are used in clinical practice.

Validation and Verification: To ensure that machines are learning correct and reliable information and that the models are valid, regulatory bodies require validation and verification. This includes presenting the outcomes of applying the model to diverse datasets, documenting the validation process, and providing evidence of how the model performs across different conditions.

Documentation and Reporting: Auditable evidence of the development process, model parameters, and validation must be available as part of applications to the regulatory authorities. This enhances the general understanding of the intended use of the model as well as its compliance with the set regulatory requirements.

Post-Market Surveillance: Another element that can be learned is that after the model has been put to use, AI must be constantly monitored for its performance, efficiency, and safety. Monitoring and updating occur post-market, where the user provides feedback, monitors specific model outcomes, and revises the model when problems arise.

Ethical Review Boards: Consulting committees for ethics ensure that all matters concerning ethics are considered and also provide advice when designing machine learning models. These boards are there to ensure that the use of the model and studies involving it conform to societally and ethically acceptable standards.

Developing Standards for AI Ethics: Industry pillars also took responsibility for promulgating guidelines and norms for ethical AI as a giant step toward proper machine learning models. The mentioned standards are useful in enabling researchers and practitioners to address ethical issues and maintain public confidence.

Advancing Explainability Techniques: Further advancements in research in the area of explainable AI help enhance the interpretability of the results of machine learning. Improved explainability means better compliance with regulations as well as ensuring that proper ethical decisions are made.

Integrating Ethical Considerations into Model Design: Including ethical factors in the model design process assists in preventing potential problems before they occur. This proactive approach requires assessing the model's consequences, where model selection also takes into account unethical issues, conformity to societal standards, and adherence to the law.

Due to the significance of medication toxicity, it was becoming urgent and important to consider ethical and regulatory aspects when developing and deploying machine learning models. To implement these models, researchers and practitioners aim to enhance legal concepts of ethics and/or privacy, in addition to fairness, openness, and compliance, when improving medication safety and the public's health. This approach helps regulatory bodies assess the model's suitability for its intended use and ensures compliance with established standards.

7 Conclusion

The application of machine learning models in the field of drug toxicity represents a new direction for drug discovery and drug safety evaluation. In this chapter, the author has presented the differing modes of thinking, transitioning from traditional methods to advanced machine learning approaches, and how these modern techniques can overcome the flaws inherent in conventional methods. Historically used approaches, although essential, provide insufficient means for capturing multivariable, multiscale interactions typical of biological systems. Hence, there is a need to turn to machine learning, as it offers more advanced algorithms and employs a data-driven approach that allows for better predictions and the ability to work with more complex data.

In the discussion, it has been explained what a specific type of machine learning is, as well as providing an overview of the classification algorithms and the neural networks that are used in supervised machine learning, and the autoencoders more appropriate for unsupervised machine learning. Each of these methods has certain benefits: from increasing forecast precision by using neural network topologies to identifying hidden features by using autoencoders. The strengths of these approaches are then aggregated in hybrid models to achieve even better drug toxicity prediction and to make the analytical tool more versatile. Machine learning brings remarkable versatility to the different methods mentioned above due to the fact that these technologies are widely used to enhance the efficiency of drug safety evaluations.

The chapter has underlined how the basic assessment and validation procedures are as important as methodological innovations. Various strategies of model selection were required to guarantee the validation of the stability and accuracy of the models, and therefore, the need for evaluation on external datasets was critical in confirming that the models perform well when tested on new data that have not been seen before. To turn the theoretical models into effective practical intervention approaches that inspire confidence in practical settings, proper validation was necessary. The use of machine learning technology was appropriate only when there were no questions regarding its ethical concerns. The validity and accuracy of a predictive model are directly associated with data protection, minimized bias, and compliance with regulatory laws.

In the future, newer advancements are expected in the field of drug toxicity prediction. Ongoing improvements in the use of Machine Learning algorithms, changes in regulations, and ethical standards by which AI models are designed and created contribute to the development of even more efficient and reliable models. The continuous integration of these technologies has several benefits for drug safety, minimizes the effects of adverse reactions, and overall helps patients. Such enhancements can be achieved when both researchers and practitioners engaging within this domain embrace them while adhering to advisable ethical practices and valid methods that lead to a safer pharmaceutical system from which the public can benefit.

References

[1] Horii, I. (2016). The principle of safety evaluation in medicinal drug-how can toxicology contribute to drug discovery and development as a multidisciplinary science? The Journal of Toxicological Sciences, 41(Special), SP49–SP67.

[2] Wu, F., Zhou, Y., Li, L., Shen, X., Chen, G., Wang, X., . . . & Huang, Z. (2020). Computational approaches in preclinical studies on drug discovery and development. Frontiers in Chemistry, 8, 726.

[3] Vamathevan, J., Clark, D., Czodrowski, P., Dunham, I., Ferran, E., Lee, G., . . . & Zhao, S. (2019). Applications of machine learning in drug discovery and development. Nature Reviews Drug Discovery, 18(6), 463–477.

[4] L'heureux, A., Grolinger, K., Elyamany, H. F., & Capretz, M. A. (2017). Machine learning with big data: Challenges and approaches. IEEE Access, 5, 7776–7797.

[5] Gombar, V. K., Mattioni, B. E., Zwickl, C., & Deahl, J. T. (2007). Computational approaches for assessment of toxicity: A historical perspective and current status. Computational Toxicology: Risk Assessment for Pharmaceutical and Environmental Chemicals, 183–215.

[6] Berhanu, W. M., Pillai, G. G., Oliferenko, A. A., & Katritzky, A. R. (2012). Quantitative structure–activity/property relationships: The ubiquitous links between cause and effect. ChemPlusChem, 77(7), 507–517.

[7] Thomford, N. E., Senthebane, D. A., Rowe, A., Munro, D., Seele, P., Maroyi, A., & Dzobo, K. (2018). Natural products for drug discovery in the 21st century: Innovations for novel drug discovery. International Journal of Molecular Sciences, 19(6), 1578.
[8] Valic, M. S., & Zheng, G. (2019). Research tools for extrapolating the disposition and pharmacokinetics of nanomaterials from preclinical animals to humans. Theranostics, 9(11), 3365.
[9] Oberkampf, W. L., Trucano, T. G., & Hirsch, C. (2004). Verification, validation, and predictive capability in computational engineering and physics. Applied Mechanics Reviews, 57(5), 345–384.
[10] Ceriotti, M. (2019). Unsupervised machine learning in atomistic simulations, between predictions and understanding. The Journal of Chemical Physics, 150(15).
[11] Kotsiantis, S. B., Zaharakis, I., & Pintelas, P. (2007). Supervised machine learning: A review of classification techniques. Emerging Artificial Intelligence Applications in Computer Engineering, 160(1), 3–24.
[12] Usama, M., Qadir, J., Raza, A., Arif, H., Yau, K. L. A., Elkhatib, Y., . . . & Al-Fuqaha, A. (2019). Unsupervised machine learning for networking: Techniques, applications and research challenges. IEEE Access, 7, 65579–65615.
[13] Dara, S., Dhamercherla, S., Jadav, S. S., Babu, C. M., & Ahsan, M. J. (2022). Machine learning in drug discovery: A review. Artificial Intelligence Review, 55(3), 1947–1999.
[14] Gupta, R., Srivastava, D., Sahu, M., Tiwari, S., Ambasta, R. K., & Kumar, P. (2021). Artificial intelligence to deep learning: Machine intelligence approach for drug discovery. Molecular Diversity, 25, 1315–1360.
[15] Zhang, L., Zhang, H., Ai, H., Hu, H., Li, S., Zhao, J., & Liu, H. (2018). Applications of machine learning methods in drug toxicity prediction. Current Topics in Medicinal Chemistry, 18(12), 987–997.
[16] Green, L. W., & Glasgow, R. E. (2006). Evaluating the relevance, generalization, and applicability of research: Issues in external validation and translation methodology. Evaluation & the Health Professions, 29(1), 126–153.
[17] Cawley, G. C., & Talbot, N. L. (2010). On over-fitting in model selection and subsequent selection bias in performance evaluation. The Journal of Machine Learning Research, 11, 2079–2107.
[18] Tantithamthavorn, C., McIntosh, S., Hassan, A. E., & Matsumoto, K. (2016). An empirical comparison of model validation techniques for defect prediction models. IEEE Transactions on Software Engineering, 43(1), 1–18.
[19] Keevers, T. L. (2019). Cross-validation was insufficient for model validation. Joint and Operations Analysis Division. Victoria, Australia: Defence Science and Technology Group.
[20] Cocoros, N. M., Arlett, P., Dreyer, N. A., Ishiguro, C., Iyasu, S., Sturkenboom, M., . . . & Toh, S. (2021). The certainty framework for assessing real-world data in studies of medical product safety and effectiveness. Clinical Pharmacology & Therapeutics, 109(5), 1189–1196.
[21] Jia, Y., McDermid, J., Lawton, T., & Habli, I. (2022). The role of explainability in assuring safety of machine learning in healthcare. IEEE Transactions on Emerging Topics in Computing, 10(4), 1746–1760.
[22] Vo, A. H., Van Vleet, T. R., Gupta, R. R., Liguori, M. J., & Rao, M. S. (2019). An overview of machine learning and big data for drug toxicity evaluation. Chemical Research in Toxicology, 33(1), 20–37.
[23] Idakwo, G., Luttrell, J., Chen, M., Hong, H., Zhou, Z., Gong, P., & Zhang, C. (2018). A review on machine learning methods for in silico toxicity prediction. Journal of Environmental Science and Health, Part C, 36(4), 169–191.
[24] Mensah, G. B. (2023). Artificial intelligence and ethics: A comprehensive review of bias mitigation, transparency, and accountability in AI Systems. Preprint, November, 10.
[25] Tatineni, S. (2019). Ethical considerations in AI and Data Science: Bias, fairness, and accountability. International Journal of Information Technology and Management Information Systems (IJITMIS), 10(1), 11–21.

P. Nagarajan, D. David Neels Ponkumar

Machine learning innovations in biomedical materials from drug discovery to personalized medicine

Abstract: Machine learning (ML) has also introduced new paradigms to biomedical materials science for new material discovery, material optimization, and customized treatment. Pertinent to this, this chapter presents a comprehensive review of the progress and applications of ML in fabricating biomedical materials, including drug discovery and patient-specific treatment. Some of the unambiguously critical areas of ML, which are transfer learning, predictive modeling, and high-throughput screening, are discussed from the point of view of how they can be useful in improving material performance and efficiency. Despite this, the focus is on presenting use cases, especially the application of ML toward improving the systems used for drug delivery, the scaffolds employed for tissue engineering, and the "tailoring" of medical treatments. Besides offering solutions to such crucial questions as data integration, model interpretability, and computational requirements, the chapter provides readers with an idea of the future directions for research and advancement. This chapter shows how ML is a revolutionary method in biological material science and in improving patients' lives by bridging the gap between computational and clinical utilization.

Keywords: machine learning, biomedical materials, drug discovery, personalized medicine, material optimization, data integration

1 Introduction

Having emerged as a disruptive technology in the biomedical materials research domain, machine learning (ML) was already revolutionizing how materials are selected, designed, and fine-tuned for specific healthcare applications [1]. The manufacturing of biomedical materials has for a long time mainly relied on basic methods, experimental methods, and random processes, which often required a lot of time and resources [2]. As a result, the incorporation of ML into the analysis of complex data to identify relationships that are not easily distinguished using traditional approaches

P. Nagarajan, Department of Biotechnology, Vel Tech Rangarajan Dr. Sagunthala R&D Institute of Science and Technology, 400 Feet Outer Ring Road, Avadi, Chennai 600062, Tamil Nadu, India, e-mail: nahoobro28@gmail.com
D. David Neels Ponkumar, Department of Electronics and Communication Engineering, Vel Tech Rangarajan Dr. Sagunthala R&D Institute of Science and Technology, 400 Feet Outer Ring Road, Avadi, Chennai 600062, Tamil Nadu, India, e-mail: david26571@gmail.com

https://doi.org/10.1515/9783111503202-009

was bound to bring a revolution [3]. Using ML algorithms, it was possible for researchers to reduce the time taken in searching for new materials and enhance the ones available to meet medical needs [4]. This chapter identifies that ML was changing the course in the biomaterials field and analyzes its contributions to the state of the art and its possibilities to advance the subject in the future [5].

Something like big data and the previously unimaginable speed with which it was capable of analyzing information was one of the perks of using ML in the field of biological materials research [6]. Biomedical research generates vast amounts of data from multiple sources, including but not limited to Omics technologies, clinical trial data, and high-throughput experiments [7]. These datasets can be provided to ML algorithms for analysis to determine correlations that inform the creation of medicinal materials [8]. There are strategies, like predictive modeling, which can predict future material performance while avoiding time-consuming and costly empirical tests [9, 10]. This capacity results in enhanced medical solutions by shortening the development process and enhancing accuracy in material expectations [11, 12].

One major advancement in the employment of ML to biological materials was transfer learning. Called transfer of concept, this method assisted in mapping current models to new problems, where the model was derived from one job but applicable to another, although different [13, 14]. The technique of transfer learning in biomedical materials research used models with superior datasets from general material science and applied them to biomedical material applications [15]. To determine the behavior of the biomedical materials, including drug delivery systems and tissue scaffolds, the model intended to predict the chemical aspects can be adjusted [16]. In this strategy, new knowledge was derived from previous knowledge and experience so that new models do not have to be designed as much, and time and effort are saved [17].

To fully achieve these objectives, the integration of a machine learning approach into biological materials research has some issues that must be addressed [18]. This is especially a problem for data, as biomedical research often employs data in multiple formats and of varying quality from various sources [19]. It is obvious that, in order to construct stable ML models capable of providing exact and valuable information, various datasets must be combined properly [20]. In addition, the validation and acceptance of results by the scientific community require, in turn, that ML models are understandable [21]. To tackle this problem, researchers have to find ways to increase the interpretability of ML models to enable proper validation of the outcomes and to incorporate the findings into practice effectively [22].

Possible future trends of applied ML for biomedical materials science point to discovering new algorithms, enhancing data integration strategies, and addressing the problem of ethics [23]. Research into and application of ML technology can only progress; thus, there is more potential for refining the models that can be constructed to explore increasingly complex data sources and gain a deeper understanding of how various materials behave [25]. Additionally, the integration of other advanced technologies, such as real-time data collection and high-resolution imaging with ML, enhances material identification and optimization even further [25]. It is crucial to redefine other ethical issues, such as addressing false positives and algorithms, to ensure that all applications of ML on biological materials are ethical both at inception and in practice. All of these topics are reviewed in this chapter as well as possible future investigations highlighted, and the importance of ML for the biological materials science field is stressed.

2 Introduction to machine learning in biomedical materials

2.1 Historical overview of biomedical materials development

The developmental process of biomedical materials can be divided into distinct milestones, which make a substantial impact on medical science and the treatment of patients. The start was in the early part of the twentieth century with the introduction of biocompatibility, the starting point toward designing implants that could be compatible with the tissues of the body. The first biomaterials had certain requirements involving mechanical properties and the biocompatibility of stainless steel and titanium, so they were chosen. These materials lacked design flexibility for integration with body tissue, which resulted in problems such as inflammation, infection, and implant rejection.

The evolution of polymer-based materials took a significant turn in the mid-twentieth century, driven by the demand for flexible, durable, and biocompatible solutions in medical applications. Polymers such as silicone and polyethylene emerged as ideal candidates due to their low cost, mechanical strength, and biocompatibility, making them suitable for a wide range of medical devices, including valves and joint replacements. This era also marked the beginning of tissue engineering, as researchers started exploring materials that could support and stimulate tissue regeneration. The introduction of biodegradable polymers, such as polylactic acid (PLA) and polyglycolic acid (PGA), revolutionized the field by enabling the development of scaffolding materials that naturally degrade in the body, reducing the need for follow-up surgeries. These advancements are depicted in Figure 1, which illustrates the timeline and material progression in biomedical polymer development.

In the latter twentieth and early twenty-first century, the development of biomedical materials applications shifted from simply bio ceramics and biopolymers to bioactive and smart materials. Bioactive materials consist of bioactive glass and hydroxyapatite that are intended to be bioreactive with the surrounding biological systems on a molecular level, enhance bone formation, and improve the apposition of implants with the neighboring tissues. New types of materials, such as shape memory alloys and hydrogels, came into existence, offering functions that respond to stimuli, including temperature and pH impulses. These developments led to the formation of more advanced equipment and treatment plans for various diseases and illnesses, in re-

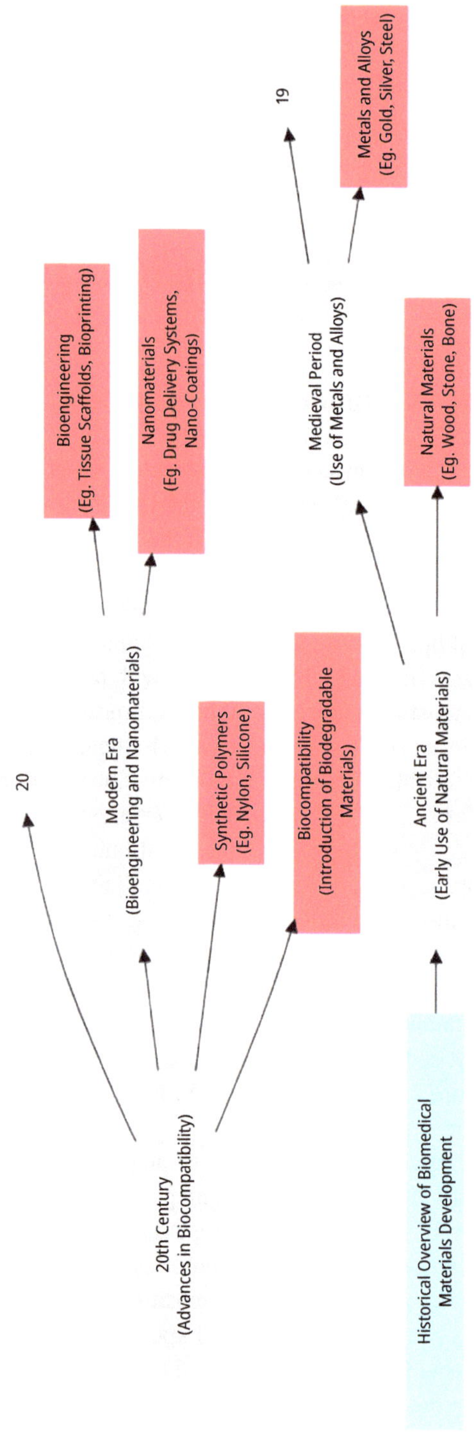

Figure 1: Historical overview of biomedical materials development.

sponse to the continually increasing requirement for materials that would suit the distinct identity of people.

Further advancement was brought by nanotechnology, implying the ability to design and build nanomaterials with better properties and features. The application of numerous nanostructured materials, like carbon nanotubes and nanoparticles, has been researched mainly for biological purposes and includes tissue engineering, medication delivery, and imaging techniques. Therapies can therefore be made more accurate and efficient in a clinical approach since cells can be altered by tackling materials at the nano level.

As the complexity of biological issues grew, the limitations of traditional material development approaches became clearer. The needs of present-day medicine, which imply the synthesis of different branches and the analysis of large amounts of information, are too complex for trial and-error approaches that were the driving force behind most of the progress. This realization led to the incorporation of computational and ML approaches into the modern biomedical materials development age. This development can help researchers predict and enhance the properties of materials and accelerate the discovery of new ones; in addition, it is helpful for providing patients with targeted and faster care.

The aforementioned historical development shows that biomedical material science is a fast-growing field that has required frequent innovation to meet changing healthcare demands. As the field unfolds, the application of ML is the luminous future that presents itself as the most promising area to dramatically advance new biomedical materials in medicine.

2.2 The emergence of machine learning in material science

The incorporation of ML into material science signifies a shift in the long-standing approach to materials and science, especially in biomedical material science. In the past, the discovery of materials used in biomedical applications was based on the accumulation of knowledge derived from experience and from identifying materials which, upon testing, meet the desired performance criteria in the development process. Although these are the traditional approaches to new product development, they have been known to bring about long development cycles and high costs; success has depended on mere luck rather than organized ingenuity.

ML has become an exciting new tool in the world of material science, allowing the large-scale analysis of data and more accurate predictions of materials' characteristics and properties. Algorithms of the supervised as well as unsupervised learning type have been useful when it comes to the task of finding a pattern within large data that would go unnoticed. This capability is significantly valuable in biomedical materials, as the relationships involving materials and other living structures are always rich in diversity and quite complex.

ML has started picking up as an important tool in material science because of the opportunity it offers in speeding up the search for, as well as the fine-tuning of, materials. In biomedical applications, it means the identification of materials that are safe for the human body, the modeling of those materials and their behavior in biological milieu, and the fine-tuning of materials to perform for specific medical therapies. Implicit in material design and discovery, ML has decreased the dependence on the conventional costly and time-consuming experimental processes in anticipation of the outcomes of material reactions with biological systems.

This has boosted the growth of ML in material science due to the availability of large-scale, high-dimensional data sets involving data from proteomics, genomics, and other biological fields. These data sets provide a vast collection of data for the training of ML models, enabling accurate prediction-making, and nurturing creativity in the creation of these materials. Advancements in the capability of using ML to analyze and assess large data sets have opened up new ways through which novel materials with tailored properties are synthesized, enhancing precision and efficiency in the biomedical field.

2.3 Key differences between traditional and ML-driven approaches

The more conventional strategies for designing biomedical materials rely on the use of experience and experiments. These conventional methods depend on testing and adjusting various properties of the material through successive trials and errors. It was quite tiresome, and many resources were consumed as well as time taken in trying to establish appropriate materials. The process generally follows a linear progression: although it was possible to make predictions in design, innovation occurs in an iterative process of conceptualization, synthesis, testing, and evaluation, with comparatively low levels of foresight.

As opposed to this, approaches under ML-IfA utilize superior computational procedures to optimize and expedite material creation. In contrast to conventional techniques, the ML-aided approaches employ mathematical models to forecast characteristics of the fabric and its response without actual experimentation. There is a wide range of data used to train these models, which encompasses data regarding molecular structure, biological activities, and noted experimental results. As data and correlations between these patterns are understood by the ML algorithms used, it is possible to accurately and efficiently estimate the performance of the new material.

Handling and analysis of data are one of the key differentiators between ML-driven techniques and traditional techniques. While traditional approaches are likely to be severely limited by small datasets and manual analysis, the ability to recognize refined interconnections and enhance material properties also suffers. Large quantities of data can be processed and analyzed using various methods based on ML, wherein such processed information reveals complex patterns different from those seen through other analytical methods such as regression analysis. Decisions in design should be made with an increased level of understanding of material performance and their relationships due to this ability.

Once again, the authors pointed to the fact that ML-based approaches were more cyclical than fixed, as were most of the traditional approaches. In conventional work processes, every round of material testing involves physical experimentation, which was tiresome and expensive. Another advantage of ML-equipped processes was that, due to simulations and predictive modeling, the number of experiments required becomes vastly fewer than in other cases. It seemed that this iterative approach was often supported by algorithmic optimization strategies, which recalculate the design of visited materials on an ongoing basis according to predicted results.

The procedure of material creation potentially encompasses several types of data, such as proteomics, genomics, and chemical informatics, more conveniently enriched by employing the help of the technique known as ML. This integration allows for a generally coherent understanding of material performance and interactions that was impossible to attain through more conventional approaches, which could focus predominantly on limited data sets and experimental variables.

2.4 Importance of data in biomedical material science

Data was crucial to the advancement of biological material science as the foundation for creating and testing ML models. As it was observed, the data quality and quantity that are accessible are directly associated with the effectiveness of ML algorithms regarding the material characteristics, design, and searching for new materials. In biomedical material science, information can be variable and comprise biological activity, molecular organization, patient individual data, and consequences of experiments.

In order to accurately train and validate machine learning (ML) models, the availability of high-quality, well-annotated data is essential. This is particularly true in the domain of biomedical materials, where the biological and chemical complexity of operational environments demands comprehensive datasets. These datasets must capture intricate details about a material's structure, formation, and interaction with living systems, thereby enabling deeper insights into its performance under various physiological conditions. The diversity and scale of such data facilitate the detection of subtle patterns and interdependencies relationships that would likely remain hidden using traditional, non-ML approaches. Consequently, this leads to improved prediction accuracy and greater efficiency in the design and discovery of advanced biomedical materials. These relationships and the need for data richness are illustrated in Figure 2, which highlights the role of data depth and breadth in enhancing ML-driven materials innovation.

Systems biology enabled by integrated multiomics analysis with proteomics, metabolomics, and genomics sheds light on the detailed interactions of materials in biological systems. By incorporating these distinctive data inputs, the operational ML models can achieve a deeper understanding of the processes behind material performance and biocompatibility. This multistage approach enhances the objectives of individualized treatment and pharmaceuticals by reducing the difficulty in creating materials that are useful and effective in biological systems.

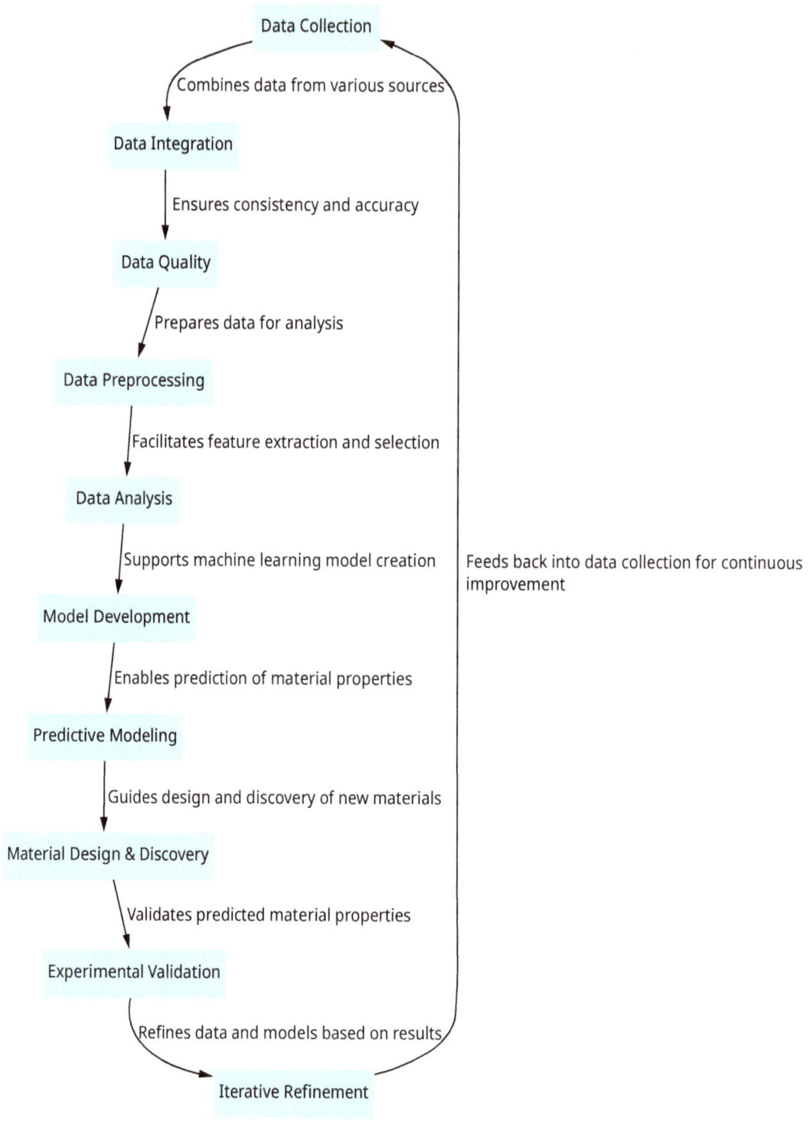

Figure 2: Importance of data in biomedical materials science.

It was also crucial in ensuring the credibility and reliability of data that feed into ML models and data preprocessing. To address issues of missing values, uneven data, and blended biased information, data has to go through a process of cleaning, normalization, and validation. This indicates that by undergoing good preprocessing of the data, the efficiency of the model is enhanced and there are fewer chances of making wrong decisions, which is an important factor in biomedical uses due to safety and accuracy concerns.

2.5 Challenges in adopting machine learning for material development

Before the true potential of ML as a foundational tool for the creation of biomedical materials can be achieved, several more challenges must be faced within the present field. Suddenly, the volume and variety of data required to achieve efficient ML applications became a problem. To create powerful ML techniques, it's necessary to use datasets that are appropriate, diverse, and of high quality. Nevertheless, owing to considerations related to the nature of biological systems, sources of variability in the methodology of data collection, as well as the general limitations on research materials in the course of experiments, attaining such data can be problematic in the sphere of biomedical materials. This means that unreliable model predictions are brought about by these limitations, and this is something that impacts the results of material development.

Another significant challenge was the traceability of the models to interpretation. Higher levels of ML algorithms, like deep learning models, work in the "black box" format, where predictions are made without necessarily providing an explanation of the system's workings. It becomes even harder to assert and believe in the outcomes in cases of opacity regarding the mechanism of choosing the ML models. The mathematical model had to be accurate to ensure that the predictions made by the model could be used for the development of biomedical materials. In the Holobloc process, model interpretability was important as the final results of the model must reliably match the experimental outcomes and the norms and standards set for the approval of different biomedical materials.

As expected, one of the challenges of using ML was how to incorporate it into the current operations effectively. The conventional material development process often relies on registered methods and processes, which limit the application of utilizing ML radically. To include ML in these procedures, this part implies key changes in workflows such as the implementation of data management systems, changes to experimental approaches, and personnel training. These changes can be costly and may meet a lot of resistance from stakeholders who favor the traditional mode of planning.

The requirements for computing power rendered by ML models can be vast. The considerable computing demands of fast-evolving advanced ML algorithms such as high-volume and complex data set training, and high-complexity neural networks, are significant. Additional costs involved in the operation and maintenance of these high-performance computing facilities, along with the rare demand for skilled personnel to run these systems, pose a big challenge for research laboratories with limited funding.

There are two more relevant categories of issues, which include ethical and regulatory factors. Biomedical materials utilizing ML have to be in line with strict governing regulatory protocols and ethical best practices to protect users, particularly their patients' sensitive information. Anticipating and managing the regulatory factors concerning ML-driven innovation requires understanding some of the regulatory concerns such as data protection, validation of the models, and the question of demographic bias in the model prediction. It is important for these applications of ML to fit into this regulatory compliance while at the same time remaining ethical in their implementation.

Solving such issues entails the joint cooperation of researchers, practitioners, and regulatory authorities in finding ways of improving quality, model explainability, and integration of ML in work environments as well as interacting with the ethical and legal challenges. Overcoming these challenges is crucial to pushing progress in ML applied to biomedical materials creation and bringing this potential to life.

3 Machine learning algorithms for biomedical materials

3.1 Overview of supervised learning techniques in material science

Supervised learning is one of the principal branches of machine learning (ML) that has found significant application in materials science, particularly for the prediction and optimization of biomedical materials. In this approach, input features such as structural, chemical, or physical attributes are mapped to known output labels, enabling the model to learn specific patterns during training. This learning paradigm proves especially valuable for tasks like material classification, property prediction, and performance optimization, where labeled data is available. The strength of supervised learning lies in its ability to build predictive models (P-models) that can generalize effectively to new, unseen datasets, which is a key requirement in designing next-generation biomedical materials. The framework and workflow of supervised learning in materials research are depicted in Figure 3, demonstrating its importance in accelerating material discovery and innovation.

3.1.1 Regression techniques

Multiple linear regression was applied to predict a continuous outcome originating from input features. In material science, regression models are employed to forecast the mechanical strength, thermal conductivity, and drug release characteristics of ma-

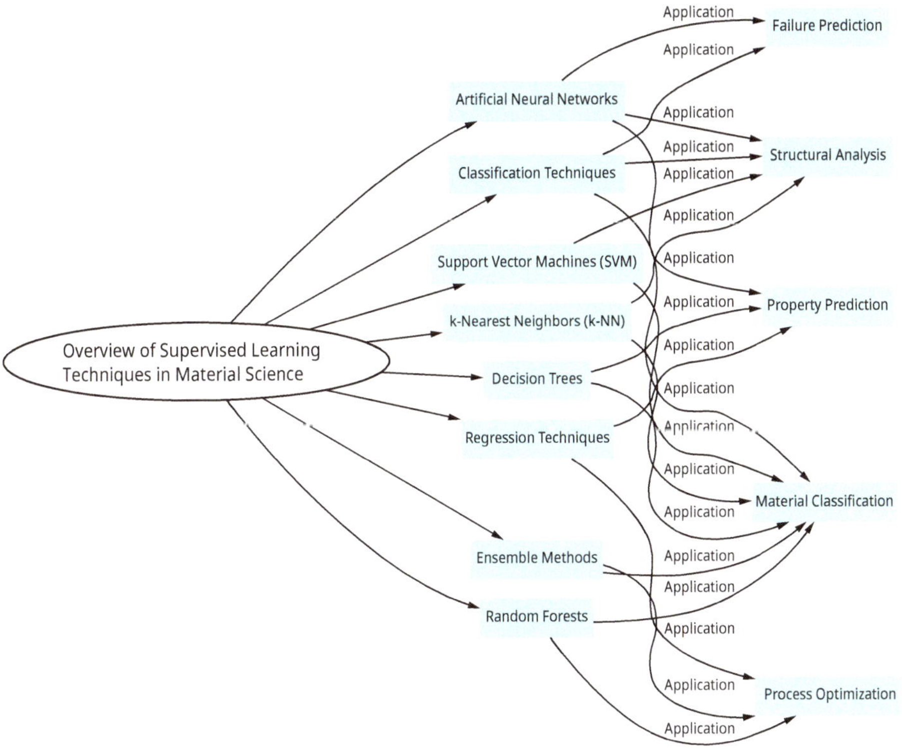

Figure 3: Overview of supervised learning techniques in materials science.

terials. A simple type of regression, known as linear regression, involves finding a linear relationship between the input variables and a continuous target variable. Other, more complex forms are support vector regression and polynomial in nature and can work in predicting material qualities and identifying nonlinear relationships.

3.1.2 Classification techniques

The collected data was automatically classified using classification algorithms into a priori classification types. Classification models were applied for organizing materials in the field of biomedical materials and devices concerning safety, effectiveness, and compatibility. Tools like random forest, decision trees, and logistic regression are often employed. Decision trees and random forests provide more sophisticated methods through developing numerous decision rules and managing data aggregations to enhance the efficiency of the classification. Nevertheless, logistic regression was employed for binary classification issues.

3.1.3 Neural networks

Since neural networks include complex and nonlinear associations and grow more popular over time, especially deep learning models, they are increasingly used in various fields. In material science, neural networks are employed to predict the behavior and other characteristics of complex materials based on large datasets. While RNNs can be used for processing and analysis of successive data, such as material testing time series data, convolutional neutral networks (CNNs) are particularly useful when the data that needs to be processed and evaluated are spatial like material imaging data.

3.1.4 Ensemble methods

There are many models employed in ensemble techniques in order to improve prediction reliability and accuracy. The base model predictions are then aggregated in ways such as boosting, bagging, and stacking with the aim of coming up with much more reliable and accurate results. Indeed, compared to single models with their weaknesses, ensemble techniques have the strengths of multiple algorithms and increase the efficiency of the predictive model in the context of biological material science. For instance, some widely applied group methods such as random forests and gradient boosting machines have been employed to predict material performance and improve material characteristics.

3.1.5 Support vector machines (SVMs)

Support vector machines (SVMs) are used in cases of both regression and classification problems. SVMs work by finding the best hyper plane to split classes in the feature space, with the least possibility of error. SVMs are applied in material science to predict how a particular material reacts in certain conditions based on different characteristics. SVMs are better able to deal with both linear and nonlinear connections well because of the versatility of the kernel functions used.

Overall, supervised learning methods provide a suitable starting point for material property classification and prediction, design improvement as well as further development of the biology-based material science field. In this way, scientists build prediction models that help in the discovery of new materials, enhance the performance of existing materials, and participate in the development of innovative medicinal applications.

3.2 Unsupervised learning for clustering and classifying materials

As one of the most important methods of micro/macro structure analysis, repairing, and classification, unsupervised learning provides extremely important information about biomedical materials without previously defining their characteristics. This particular type of ML deals with transforming data into machine-processable structures where the internal structure of groups and relations between them is implicitly defined. There are several key methods that fall under the umbrella of unsupervised learning in biomedical material science applications, all of which support material characterization and prospective uses.

3.2.1 Clustering techniques

As one of the categories in unsupervised learning, clustering algorithms take similar materials and group them based on existing features. Despite a few differences, one technique that was used most to classify data into *k* subsets, whereby the variance within the cluster was minimized, was known as *k*-means clustering. This was a versatile approach within which the materials can be categorized, depending on characteristics including medication release patterns, mechanical features, biocompatibility, among others. Another method, called hierarchical clustering, constructs a tree-like structure with the clusters stacked one on top of another, and this allows the analysis of material correlation at different levels of detail. The significance of some biological applications requires the identification of groups of materials with similar characteristics, which has become possible owing to these clustering methods.

3.2.2 Data visualization

Many applications are provided features by dimensionality reduction for decoding and representation of relational datasets. Another well-known method widely used to tackle the high-dimensionality problem of the data, while retaining most of the variation, is called principal component analysis (PCA). In material datasets, PCA can reveal such essential aspects as trends between different attributes. Along the same line, the nonlinear dimensionality reduction tool called t-distributed stochastic neighbor embedding is used and provides a two-dimensional representation of high-dimensional data. These techniques assist in how to categorize and choose the material since it is not easy to identify clustered data.

3.2.3 Applying the concept of material discovery and optimization

Some of the applications of techniques for unsupervised learning are illustrated concerning the optimization and discovery of new materials. For instance, they show new materials with desired characteristics for specific biological uses by categorizing materials not seen before. From high-throughput studies, complex information was analyzed to reveal the behavior and interactions of materials through dimensionality reduction and feature extraction. Most of these methods also assist in enhancing material characteristics since they identify trends and relations that guide the development of materials with required characteristics.

3.3 Applications of deep learning in biomedical materials

In the field of biomedical materials, deep learning, a branch of ML characterized by the use of multilayered neural networks, has shown revolutionary potential. In particular, it is perfectly suitable for a range of tasks within this domain because it allows for the fine representation of relationships and identification of nonlinear dependencies based on large datasets.

Therefore, using predictive modeling of material properties means that the information provided was extracted from the data accumulated in the course of global communication.

CNN-based deep learning algorithms have been successfully applied to predict the properties of biomedical materials. By training on large datasets that encompass a wide range of material characteristics and performance outcomes, these models are capable of accurately forecasting mechanical strength, thermal conductivity, biocompatibility, and other critical attributes. This predictive power significantly reduces experimental costs and time, allowing researchers to focus their efforts and resources on the most promising candidates. Moreover, it supports more informed decision-making early in the materials development pipeline, improving the efficiency of R&D processes. The architecture and application of CNN models in biomedical material property prediction are illustrated in Figure 4.

3.3.1 Design of drug delivery systems

Drug delivery systems have evolved and improved using deep learning. Specific product systems and different time sequences are processed using RNNs and transformers so that materials can be developed that are capable of releasing therapeutic agents in a highly targeted manner. They can predict the kinetics of drug liberation through different materials to enhance formulations for particular therapeutic objectives as well as enhance the quality of patient care.

It was also important to identify new biomaterials

There are also known approaches to search for new biomaterials with the required set of properties with the help of deep learning methods. New material forms are designed when generating VAEs and GANs to train new forms of well-known biomaterials from existing data. The existing algorithms possess the capability to propose

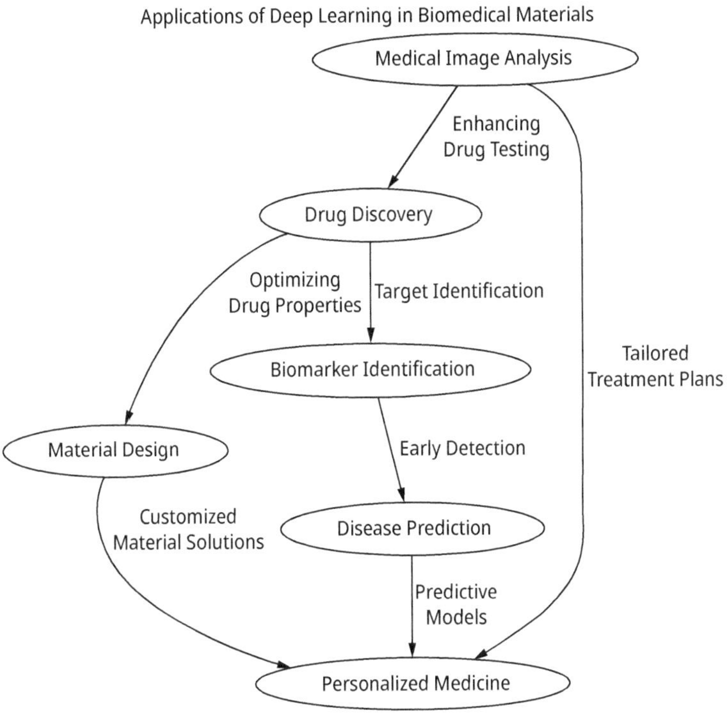

Figure 4: Applications of deep learning in biomedical materials.

hitherto unknown novel material options, extending the range of potential materials used in implants, regenerative medicine, and tissue engineering applications.

3.3.2 The use of images in the determination of material characteristics

In particular, image analysis tasks are well solved with the help of deep learning models, CNNs in particular, which are very beneficial when characterizing materials. In biomedical materials science, these models are employed to manipulate microscopy images and other image-enhancing data in terms of material morphology, defects, and features. Deep learning minimizes the extraction of information from complicated images and improves the time required to characterize the materials.

3.3.3 Improvement of methods used in preparing materials

The experimentation of material synthesis processes has also been fueled by deep learning algorithms in order to derive from the existing data, the best conditions for producing a particular material. Neural networks can capture the functional dependencies of material properties with synthesis parameters, thereby identifying optimal dependent values that are associated with given independent values of synthesis parameters. It simplifies synthesis processes and also enhances the quality of the resulting products so that they are not characterized by a wide variation range.

3.3.4 Bordering on personalized biomedical materials

In the area concerning personalized medicine, deep learning is used to customize biological products to suit certain consumer requirements. Deep learning models can develop individual content for prostheses, implants, and medication delivery by using patient-relevant records including their genetic and disease records. Customization is safe and effective because material qualities align with each patient's concerns during medical procedures.

3.3.5 Improving material degradability

Artificial neural networks of deep learning have been used in optimizing the biocompatibility of materials used in medical applications. These models enable the analysis of quantitative data obtained from in vitro and in vivo experiments and can determine factors affecting the material/biological system interface. This application helps in the development of materials that prevent negative reactions and enhance patient outcomes.

The potential of deep learning in biomedical materials is enormous and quite transformative, and it is a vital tool for the discovery, design, and optimization of biomedical materials. When such deep learning approaches are employed, researchers are well-positioned to solve multifaceted problems in biomedical material science to arrive at a myriad of inventions that help augment the effectiveness of therapy and, in effect, the delivery of care to patients.

3.4 Reinforcement learning for material property optimization

The important method that helped to optimize the characteristics of the material was called reinforcement learning (RL). They use cyclical learning in dealing with complex decision-making problems. RL algorithms are employed in the context of biomedical materials, where by designing in higher dimensions, the performance of the material was enhanced in terms of mechanical strength, impartibility, and drug distribution rates.

At its core, RL was an agent with learning capability, which enabled the agent to acquire decision-making ability with reference to an environment, and the feedback received was in the form of a reward. In this way, the agent can fine-tune the strategies based on the semantics of the way its actions appear. The material design space was the environment in material property optimization, in which every design decision affected a material with certain constant properties as well as performance and other characteristics. The RL agent self-consistently explored this space in order to learn the parameter setting that gives the desired properties of the material.

Among the benefits derived from RL in material property optimization was that the technique fit well for complicated multiobjective optimization challenges. It can often mean that biomedical materials have to have several properties which can counteract each other. For instance, a material that is intended for use in a drug delivery system should meet characteristics of controlled release as well as biocompatibility. It is important to note that RL algorithms can easily extend the capability to achieve multiple goals in parallel by outlining proper reward functions that incorporate those trade-offs and would prove more useful for material design.

In the case of applying RL in this domain, a reward function is usually chosen to optimize exactly these material properties. The RL agent engages with simulation to judge the utility of different material designs. In the course of the interaction, the agent finds out which design features contribute to improved performance based on the reward signals accumulated. There are also more optimistic approaches to RL that include deep RL, which helps take advantage of a neural network in approximating reward functions and decision policies.

There are a number of significant case studies in biomedical material optimization, and one amongst them was in the designing of drug delivery systems. RL algorithms have been implemented to find the rate of drug release for drug delivery, the stability of the delivery system, and the compatibility of the system with biological tissues. Indeed, this decision-making process can be formulated as iteratively trying out different formulations and configurations through RL models and assessing their outcomes to determine the best configurations that would help achieve particular therapeutic objectives.

Nonetheless, RL faces certain challenges in optimizing material properties based on its application. The specificity of the RL models and the demand for a high amount of computational backbone can also be a disadvantage. The simulation of an environment sufficiently faithful to the real world and the assembling of suitable data to train RL models are crucial for advancing optimization work. To deal with these challenges, there is constant improvement of computation techniques, more data acquisition, and the creation of new models.

3.5 Transfer learning in the context of biomedical materials

Transfer learning has emerged as a very useful tool when it comes to biomedical materials, especially because it helps in training new material and increasing the accuracy of predictions. This approach implies the reuse of previously learned patterns and features and the use of prior knowledge in solving new problems but in a slightly different task domain. In the area of biomedical materials, transfer learning solves several critical issues; these include limited data, the model's ability to generalize, and the complexity of biological mechanisms.

One of the key advantages of transfer learning in the field of biological materials lies in its ability to address the persistent challenge of sparse data availability. High-quality, labeled datasets required for training reliable machine learning models are often costly and time-consuming to obtain, especially in the biomedical materials domain. Transfer learning overcomes this by allowing models trained on large, related datasets such as chemical science databases to be adapted for biomedical applications, enabling accurate predictions even with limited new data. This strategy not only accelerates the development process but also minimizes the need for extensive new experimental datasets. The concept and application of transfer learning in this context are demonstrated in Figure 5.

Apart from solving the problem of data deficiency, transfer learning enhances model generalization since the information is obtained from related activities. For instance, one can use the data on thousands of chemical compounds that have been fed into a large number of models in order to predict the characteristics of biological materials, which are also composed of those chemical compounds. As with transfer learning, large patterns can be learned first and then refined when the model is fine-tuned on a smaller set of data. This leads to enhanced precision and specificity of the predictor models that are intended for the biomedical field.

The difficulty in accurately modeling biological systems was the next challenge where transfer learning can be of great help. Living phenomena and material properties are good candidates for relational learning since they represent complex and context-sensitive behaviors that cannot be easily designed. This allows the use of the learned models in another biological context and incorporates sophisticated features and relationships that would otherwise have taken considerable time to be learned from scratch. This capability for transfer and transformation of such advanced models improves the comprehension and prognosis of material performance in different biological contexts.

Some practical examples from the biomedical materials field are described below, which prove the efficiency of transfer learning. For instance, models generated from big data of molecular structures and biological activities have been used in a modified way to predict the efficacy of a newly designed drug delivery system scaffold. Likewise, transfer learning has been used to repurpose image recognition models for identifying material properties from micro images of biomedical materials, enhancing the exactness of material characterization.

Nevertheless, like other applications of this technique, the application of transfer learning in biomedical materials had some issues. The criteria for model selection and their suitability for the fields of the biomedical domain are crucial factors, mainly because the pretrained models are already trained. Fine-tuning models require some extra tuning in order to avoid overfitting the models and to ensure that the knowledge transferred to the new environment is in the right form of tuning.

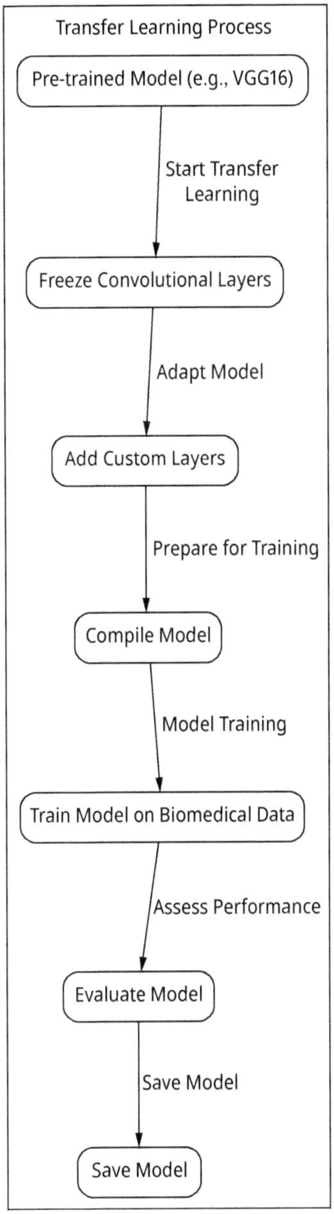

Figure 5: Transfer learning in the context of biomedical materials.

4 Data acquisition and preprocessing in biomedical material science

4.1 Types of data in biomedical materials

In the biomedical field of material science, different types of data are part and parcel of the development, optimization, and use of the materials. All types of data are valuable and help to achieve a good understanding of the structural characteristics of materials and their behavior in a biological environment. The following sections outline the primary types of data utilized in biomedical materials science:

4.1.1 Genomic data

Genomic data refers to details regarding the content of genes possessed by an individual. This data type involves DNA and gene sequencing, gene expressions, and genetic single nucleotide polymorphisms. In the context of biomedical materials, published works regarding genomics were used to ascertain genetic factors concerning material interaction, diseases, possible treatments, and approaches to personalized medicine. Thus, the knowledge of genetic reactions to the materials helps designers create superior materials tailored to specific genetic profiles and improve compatibility and therapeutic outcomes.

4.1.2 Proteomic data

Proteomic data incorporates the measurement of proteomes, which involve structure, function, and the processes by which the proteins interact. This data type offers information on the degrees of protein expression of cells that have interacted with biomedical materials. Data from proteomics are used in determining the effects of materials in biological systems, in the identification of biomarkers, and in analyzing the action of materials. Protein microarrays, as well as mass spectrometry, are standard methods for proteomic data production that are mandatory for material performance and safety evaluation.

4.1.3 Metabolomic data

Plasma metabolomic data deals with the large-scale characterization of metabolites in a biological system. This includes lipids, sugars, and amino acids that participate in the metabolic activities of cells within the body. Metabolomic data were employed to evaluate the alteration of metabolism by the materials in biomedical materials science, biomarkers related to metabolism were detected, and the influence of the metal–organic framework on metabolism was investigated. Such data were useful for assessing the physiological impact of materials and for designing the materials with reference to a particular metabolism.

4.1.4 Material property data

Material characterization data contain mechanical, chemical, and physical property data of biological materials. This information consists of attributes such as surface characteristics, mechanical characteristics, thermal characteristics, and the composition of the material. Understanding these characteristics is important in the creation of specific biomedical applications like those mentioned above, including tissue engineering scaffolds and medical implants. To characterize these qualities, mechanical testing, microscopy, and spectroscopy techniques are applied.

4.1.5 Biological interaction data

Biological interaction data is data about the connection between biomedical materials and biological units such as cells, tissues, and organs. This data type encompasses cytotoxicity status assays, internalization experiment outcomes, and in vivo bioluminescence data. Biological interaction data helps in understanding how the material is accepted, effective, and safe in biological systems. This is how material sciences come in handy for optimizing the designs of materials so that they have a positive impact on therapeutic processes and, conversely, do not contribute to adverse effects.

4.1.6 Clinical data

Clinical data can therefore be described as information derived from patient populations, especially in terms of demography, clinical results, as well as the consequent reactions to the treatment. This data is essential in bridging the knowledge between material developments within the laboratory and their practical use in the practice of medicine. Clinical data is also useful for assessing the effectiveness and safety of biomedical materials when used in practice, for developing strategies for the individual treatment of patients, and for making decisions about the authorization of biomedical products.

All kinds of data are important for the growth of biomedical materials research. New and effective biomedical solutions are generated through the integration and analysis of the many data sources mentioned above, which provide comprehensive knowledge of material characteristics, responses, and functions.

4.2 Data cleaning and preprocessing techniques

Before using data in the concrete field of biological material science as a foundation for applying machine learning (ML) algorithms, data cleaning and preprocessing become critical steps. These strategies have a direct impact on the accuracy and reliability of the resulting models, as they ensure the data is accurate, consistent, and analysis-ready. In the context of biological materials, where data is often complex and heterogeneous, proper sample cleanup and treatment are essential. Without these measures, predictive insights and meaningful interpretations cannot be achieved. The importance of these preprocessing techniques is illustrated in Figure 6.

4.2.1 Handling missing data

This data was incomplete because sometimes experiments are only partially completed erroneously. Several techniques are employed to address this problem:

Imputation techniques: Where the data is quantitative, mean or median should be used for replacing missing values. In qualitative data, most often value imputation should be used. The final technique is called regression imputation, which seeks to assign a value to the missing field based on other fields.

Many imputations: Another limitation of this approach was that for the imputation process uncertainty, many datasets are created using different imputed values, and the results are averaged.

Deletion methods: Specifically, if the proportion of missing data was low and the number of cases that should be dropped due to the presence of missing values could not be considered negligible, the rows that contained the missing values were deleted.

4.2.2 Data normalization and scaling

Features in biomedical datasets are mostly quantitative and can be scaled differently, making up the structures that can influence the results from a ML system. Normalization and scaling techniques are used to standardize the data:

Min-max normalization: In this method, features are rescaled with the help of the minimum and maximum values to bring the data to a desired scale, commonly in the range of 0 to 1.

Z-score standardization: Standardization of features reduces the range of choice regarding how the mean should be subtracted from the chosen set of data and defines a new measurement of skewness and dispersion that is scaled based on the standard deviation.

Robust scaling: This technique scales the data using statistical measures such as the median and interquartile range (IQR) that are not considerably affected by outliers.

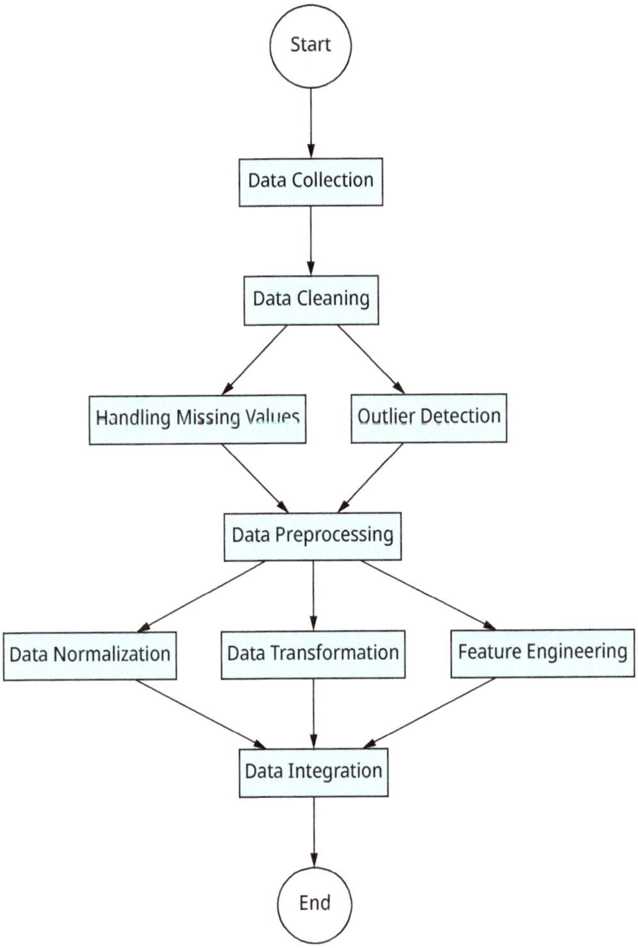

Figure 6: Data cleaning and preprocessing techniques.

4.2.3 Feature engineering and selection

Effective feature engineering and selection are critical for improving model performance and interpretability.

Feature extraction: New features can be created from previous features using mathematical transformations like dimensionality reduction (PCA).

Feature selection: Techniques such as recursive feature elimination and mutual information are used to address the c Spirits problem and the b Number of Spirits problem, enabling the reduction of dimensionality and increasing the efficiency of the model.

Handling categorical data: To support the analyses, categorical variables are first encoded, often through a process such as one-hot encoding, to ready them for application to an AI model.

4.2.4 Coordination and harmonization of data

Biomedical material science often involves integrating data from multiple sources such as experimental results, literature, and clinical databases:

Data integration: Integrating data from different data sources entails the matching of schemas to allow proper alignment of the data format as well as the proper definition of the variables.

Data harmonization: Techniques like data mapping and data standardization were applied to eliminate the complexities in the different approaches used in data representations so as to transform it to a common form in order to make it compatible with other datasets.

4.2.5 Outlier detection and how to handle it

Outliers can significantly impact the quality of data and the performance of ML models.

Statistical methods: Outliers can be identified using statistical methods, for instance, employing *Z*-scores and graphical illustrations.

Robust methods: To address the issues raised by such outliers, strategies such as undertaking a more reliable regression method like robust regression or implementing models that are not as influenced by outliers can be employed. Sometimes, using the information, it is possible to perform a simple analysis and check if these observations are actual deviations from the norm in the measuring process.

Feature extraction: New features can be derived from existing data through techniques such as dimensionality reduction (e.g., PCA).

Feature selection: Identifying the most relevant features through methods such as recursive feature elimination and mutual information helps reduce dimensionality and improve model efficiency.

Handling categorical data: Encoding categorical variables using techniques such as one-hot encoding ensures that categorical features can be effectively utilized by ML algorithms.

4.2.6 Data integration and harmonization

Biomedical material science often involves integrating data from multiple sources such as experimental results, literature, and clinical databases.

Data integration: Combining data from disparate sources requires matching schemas, aligning data formats, and ensuring consistency in variable definitions.

Data harmonization: Techniques such as data mapping and standardization are used to reconcile differences in data representation and ensure compatibility across datasets.

4.2.7 Outlier detection and handling

Outliers can significantly impact the quality of data and the performance of ML models.

Statistical methods: Outliers can be detected using statistical techniques such as *Z*-scores and IQR to identify data points that deviate significantly from the norm.

Robust methods: Techniques like robust regression for outliers can help mitigate their impact on the analysis.

Manual review: In some cases, a manual review of outlier cases was necessary to determine if they represent valid variations in data collection.

4.2.8 Data transformation and encoding

Data transformation techniques prepare data for analysis by converting it into formats suitable for ML algorithms:

Log transformation: Used on data sets exhibiting exponential growth to reduce variability and increase normality.

To account for nonlinear relationships between the variables, interaction terms are formed.

Encoding techniques: Converting textual data belonging to certain categories into forms intelligible to ML systems.

4.3 Handling missing data and inconsistencies

Mainly, missing value treatment and dealing with inconsistency were the major issues in biological material science data gathering and preprocessing stages. As biological and material data contain large amounts of information and, as such contain nuances, it means that the quality and reliability of ML can be significantly influenced by gaps and differences. These challenges were crucial to fulfill in order to ensure an accurate and relevant study.

4.3.1 Causes of missing data and inconsistencies

There are many possible causes for missing data in biomedical material science data such as experimental errors, data entry errors, and the limitations of measurement instruments. This happens due to differences in procedures involved in data collection, and differences in measurements, among other factors, in datasets from different authors. Knowledge of the causes of missing data and other inconsistencies is important in choosing proper approaches to address them.

4.3.2 Strategies for handling missing data

Several methods are available for addressing missing data, each with its advantages and limitations:

Imputation techniques: Imputation is a process in which missing values in a dataset are estimated using statistical means. Mean imputation and median imputation are common techniques that involve replacing the missing values with average values, respectively. Some of the more sophisticated techniques include the use of methods such as multiple imputation and *k*-nearest neighbors (*k*NN) imputation, where the algorithm makes educated guesses using pattern recognition techniques to try and make a better-educated guess at the missing data.

Model-based approaches: Model-based imputation techniques, like regression imputation, create missing value expectations based on relationships established in data analysis. These techniques make use of available data to develop models for predicting gaps, and as such, imputations made using these methods are often more semantically suitable than average ones.

Data augmentation: Sometimes, bootstrapping activities can be applied in an attempt to mitigate missing data problems. This approach mainly involves the formulation of additional input feature data points whenever distribution patterns are noticed and can be utilized to improve the resiliency of the ML model.

4.3.3 Managing inconsistencies:

Addressing inconsistencies in biomedical material data requires a combination of techniques to standardize and harmonize the dataset:

Data cleaning: Data cleaning is the process of screening the data and removing or correcting any errors. These steps involve formatting and finalizing data in such a way as to make units of measurement consistent, making decisions that clarify differences between datasets, and fixing data entry mistakes. It is possible to use algorithms to solve some problems and organize the work of an analyst in such a way that focuses on critical cases.

Normalization and standardization: Data normalization and standardization help to bring sets and variables in the same category into a common frame of reference. Operations that include standardizing features to a scale or a standard format assist in lowering variability as well as enhancing comparability.

Data integration: Since the data was compiled from various sources, some of the common issues, such as consistency of data formats and variables, became critical. Record linkage entails matching datasets according to unique common elements and processing any disparities to generate an integrated form of data ideal for analysis.

4.3.4 Impact on machine learning models

The way that queries containing uncertain and omitted profiles are dealt with can affect the effectiveness of machine learning models. The forecasts produced by models learned on datasets comprising unsolved issues are skewed. Hence, in an effort to ensure that the data used for training and validation was correct, up-to-date, and uniform, proper preprocessing measures had to be exercised.

4.3.5 Best practices and tools

Utilizing appropriate methods in data preprocessing, such as exploratory data analysis, missing value treatment, and data merging, improved the quality of biomedical material datasets. There was essentially preprocessing of data to clean it in a way that was easy to use and useful for a specific model; hence in data cleaning and preprocessing, specialized software and libraries can be very effective.

4.4 Feature engineering and selection for material properties

Feature engineering and feature selection are two critical steps in building effective models in biomedical material science, especially when model interpretability is a priority. These steps encompass procedures such as feature creation, feature transformation, and feature selection, all aimed at refining feature attributes and aligning them with relevant material properties. The goal is to enhance the accuracy and relevance of model prediction. These processes are essential for improving model performance and understanding how specific features contribute to predictive outcomes, as shown in Figure 7.

Feature engineering can be defined as converting raw data into features. In simpler terms, feature engineering is about utilizing decreased dimensional data in order to make features. In this case, feature engineering, particularly with biological materials, requires one to pull knowledge from the material science field as well as have an understanding of the ML algorithm being used. There are several crucial phases in this process:

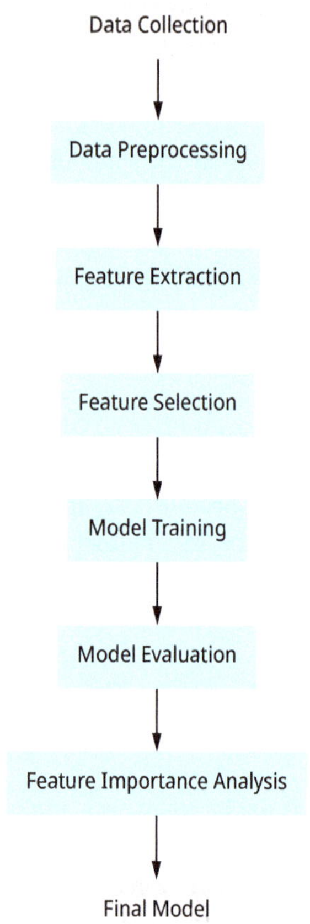

Figure 7: Feature engineering and selection for material properties.

Extraction of domain-specific features: This was because the biomedical material data frequently involves inherent measurements and attributes such as chemical components. While it is beneficial in measuring a large set of characteristics, features such as molecular fingerprints were required to be extracted from these measurements to capture the underlying patterns critical to material behavior. Enhanced approaches, which include cheminformatics, are applied to convert raw data into the feature set.

Creation of composite features: Often, the automation of individual features does not provide a complete picture of the material characteristics. Other approaches include creating composites by either using two features, which makes these features beneficial. In other words, an improved feature representation of different chemical descriptors, such as integrated measures of the reactivity of the material, leads to improved material performance prediction.

Normalization and scaling: Normalization and scaling were also done during feature engineering to ensure that the scale was comparable on some level. This was especially important for models that utilize the relative magnitude of the input parameters and include SVMs. Scale issues are very important in determining convergence and accuracy in model making.

Feature selection's goal was to decrease the number of computations required to build a model and increase its accuracy by choosing the best features out of a large number of potential features. There are a few effective ways to choose features:

Statistical methods: Those factors which were highly related to the material quality attributes can preliminarily be identified with some statistical tools like the correlation matrix tool and hypothesis testing tool. In the construction of the model, the features which bear a high correlation coefficient with the attribute, such as tensile strength, have to be prioritized.

Wrapper methods: These methods estimate the quality of a model for each of many different feature subsets and select the best subset based on the achieved model quality. If one wants to build a model based on the continual shifting of the feature subsets, the forward/backward selection and the recursive feature reduction techniques are commonly used.

Embedded techniques: These techniques involve the selection of features during the training of the model. Inherently, feature selection is performed using the process known as least absolute shrinkage and selection operator. These algorithms downplay minor attributes, evaluating the value of the attributes based on their worth to the overall predictive significance of the framework.

Dimensionality reduction: For that, one can apply techniques like PCA. It also helps in reducing the dimensionality of the management of high dimensionality to enhance the model's interpretability.

Integration of domain knowledge: Using knowledge of the domain to enhance feature selection procedures goes a long way in improving the relevance of the features chosen. Choosing the group of attributes most fitting to the biological environment should be done based on specialists' knowledge of the behavior and properties of the material.

4.5 Challenges in data integration from multiple sources

A major issue that resonates in biomedical material science is the problem of data acquisition from many sources, which affects the success and precision of ML models for materials' construction. These challenges are attributed to the type of data, the quality of data, and, more importantly, the possibility of integrating the details.

The first major complication was that the data sources themselves greatly differed from one another. Biomedical material science uses data of different types and origins, for example, genomics, proteomics, databases with chemical data, and measurements. Moreover, the format, granularity, size, and level of detail between various data sources are unique, making it hard to reconcile these related datasets. Data can be numerical values, categorical labels, and text annotations, and the differences between these data types have to be bridged using elaborate preprocessing.

Another challenge was to ensure that the data collected was accurate and homogeneous. Quality, completeness, and sources of information differ, and this causes variations in the credibility of the information collected. Measurement precision, lack of data, and variability in the environment of the experiments are some of the imperfections that harm the ML model and jeopardize the integration. Because of these challenges, ensuring that data quality differences do not dominate the linked data and offering robust and reliable linked datasets require much data cleaning and normalization processes.

Privacy and security of data are two more topics that come up with the use of integration from various sources. Many biomedical datasets contain sensitive patient data, and combining datasets from various sources means that there is likely to be leakage of patient identifiers' information. It would also be pertinent at this time to note that ethical standards and the privacy of the patients must be upheld to the letter, and data protection statutes such as GDPR. To address these challenges, there is a need to adopt safe ways of integrating data and also anonymizing data that pose a threat to individual privacy.

There are considerable issues with data alignment and synchronization when combining data from several sources. Data obtained at one time may differ in approach and experimental conditions. In essence, the three involved processes pertain to the synchronization of time series data, normalization of the referenced units of measurement, and checking for consistency when combining the different data sets. If misalignment difficulties occur, the ML models may make wrong assumptions and predictions.

Finally, the integration process must address the challenges of scalability of the large numbers of datasets and the computation cost of data integration. Applying the proposed framework to multiple and very large datasets raises high computational requirements such as increasing time and memory needs. Proper methods of data management consist of the applicability of suitable frameworks for processing and appropriate data storing strategies due to these challenges.

5 Model development and validation in biomedical materials

5.1 Building predictive models for material properties

Creating and designing decision-making predictive models for material properties in biomedical materials science require several fundamental stages, which are vital for enhancing the efficiency and reliability of the constructed models. These stages include choosing the right algorithms to be used, determining which features should be included in the models, training the models, and ensuring that they have been tested and verified.

5.1.1 Algorithm selection

Some of the problems associated with data from many sources include data security and privacy. Patient-specific data is often incorporated into biomedical data, and combining datasets from multiple sources can lead to compromising a patient's right to privacy. Data protection laws such as the GDPR and HIPAA have to be upheld to ensure that ethical standards and patient privacy are maintained. To address these privacy issues, it is necessary to apply suitable methods of data integration and anonymization of such data.

5.1.2 Feature definition and engineering

It was therefore important to understand that feature selection and engineering are critical when designing good predictive models. Facts are obtained by processing the memorized raw data that form the attributes concerning an object with regard to which a forecast was made. Structural characteristics in the case of biomedical materials include molecular descriptors, material formulations, structure factor details, and conditions of the experiment. Feature engineering is a process of selecting the best features to use, then extracting features in a way that makes them usable for the model to make the best prediction more accurate and easily understandable, as demonstrated in Figure 8.

5.1.3 Model training

The fine-tuning of the parameters describing the predictive model was necessary for the materials' property prediction; thus, the training process requires the use of a dataset. To ensure that the model was trained on many different instances to be able to generalize when faced with new data, this method requires the splitting of the dataset into the training set and the validation set. During training, structural hyperparameters such as learning rate, layer density, and regularizing factors are set in order to achieve optimality. To ensure that the data model can be valid for other data subsets and not overfit the model, there are techniques like cross-validation used.

5.1.4 Model validation and evaluation

For this reason, validation and evaluation enable the prediction model's accuracy and dependability to be categorized. The metrics used to evaluate the performance of models under consideration were given by accuracy, precision, recall, and $F1$-score for classification problems and MSE, RMSE, and R^2 for regression problems. In addition, the quantitative charts of learning curves and residual diagnostics explain how the model works and what potential enhancements are required. As it was done before, to confirm model reli-

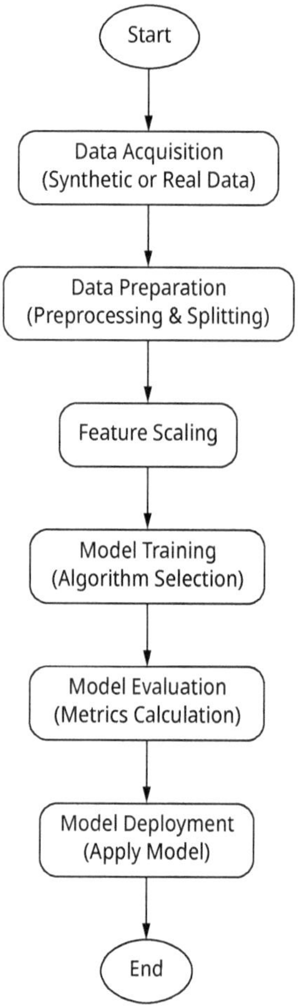

Figure 8: Building predictive models for material properties.

ability and have the possibility to check its ability to work in other conditions, it was crucial to observe its efficiency when using both training and validation sets.

5.1.5 Model interpretation and refinement

Interpretation of the results returned by the predictive model involves the identification of different feature tendencies and overall patterns as well as the presence of bias. Some techniques include SHapley Additive exPlanations (SHAP) values, partial dependency plots, and feature significance analysis to understand the reasons and ways different traits affect the kind of predictions made by the model. To increase the accuracy and reliability of the model, it should be improved based on these interpretations by altering the features and employing other techniques from new data.

5.2 Cross-validation techniques for ensuring model robustness

When developing and validating predictive models using ML on biomedical materials, cross-validation was essential in creating, testing, and verifying its broad applicability and reliability. This process of dividing the data into subsections was used with the aim of testing the performance of the model on some data to reduce overfitting and hence increase the accuracy of the results given to unseen data.

This cross-validation was one of the most frequently used cross-validation techniques. Would Witold Bebernitz from the Psychology Department at view model positively affect the forecast horizons if Witold Bebernitz from the Psychology Department at this method creates *k*-folds of similar size of the dataset and pulls out onefold in the process? When training occurred on $k-1$ folds, the evaluation was on the last fold. Each fold was used as the test set at one point in the experiments, and this process was administered *k* times. For the comprehensive performance measurement, each iteration was concluded with the averaging of its results. For data sets that are relatively small, as can be found in biomedical studies, the *k*-fold cross-sectional test was very useful in the process since it tries to estimate the ability of the model to perform well even when it was given subsets within the data.

There was another variation also known as leave-one-out cross-validation (LOOCV): even though all the sample points constituted the training sample, one point from the sample was used for testing. What this method did was essentially an extreme form of *k*-fold cross-validation, where indeed *k* equals the number of data points. Because each data point was analyzed, LOOCV provides nearly an unbiased assessment of the model' performance; it is computationally intensive and time-consuming, especially for large datasets.

Another modification of *k*-fold cross-validation that ensures each fold of data contains the same percentage of classes as that of the original data set is known as stratified *k*-fold cross-validation. This method is valuable primarily for those datasets where some classes are significantly less represented than others. To minimize the influence on the evaluation metrics and to obtain more appropriate estimates of the model's performance, stratified *k*-fold cross-validation should be used while preserving the ratio of class proportions.

On the other hand, cross-validation and hyperparameter tuning are brought together in nested cross-validation. This method employs an outer cross-validation loop favorable to the assessment of the model's accuracy and an inner cross-validation loop optimal for tuning hyperparameters. This method helps in selecting the best hyperparameters and also evaluates the performance of the model. There are some cases where the patterns are very complicated in biological material applications, where overfitting was a cause, and in backgrounds where model parameters impact the showcases tremendously; then nested cross-validation has been more helpful.

Another form is called leave-p-out cross-validation, where all the rest of the data points are used for training, while p data points are used for testing. This method provides a middle ground between the less accurate method of *k*-fold cross-validation and the computationally expensive method of LOOCV, which makes it viable in circumstances where the dataset is gigantic. It prevents the interaction shares of the calculation time from becoming completely dominated by one factor, thus allowing one to achieve a better relative flexibility of the model assessment comprehensiveness.

By using these forms of cross-validation, it was guaranteed that the models being used to predict results in biomedical material science are accurate, reliable, and generalizable to new data. These methodologies give an all-round performance assessment of models, prevent overfitting, and make the models more accurate. The proposal emphasized that integrating good cross-validation practices into the creation process was critical to the development of dependable ML solutions required to navigate the formidable biomedical materials science terrain.

5.3 Model interpretability and explainability in material science

In material science, transparency and understanding of models are key attributes of ML, particularly when constructing as well as validating the ML models of prediction for application in biomedicine. Ascertaining and deciphering what is being opted for by ML models is identified to be essential for the reliability, openness, and operationalization of these systems.

The term interpretability means the ability of a human at any level to explain the rationale behind a model's predictions. In material science, where the stakes involve designing materials with specific properties, interpretability is crucial in reviewing the model results and verifying that they coincide with experimentation. Interpretability of the developed models, written in plain language, ensures that researchers are able to assess whether the model is arriving at reasonable and scientifically plausible decisions based on the features of the samples, as illustrated in Figure 9.

The logical continuation of interpretability within the framework is based on the ways of giving fine-grained explanations for model predictions. It refers to the manner in which causal relationships between input factors and outcomes have been explained or made. In biomedical material science, explainability is used to isolate and determine exactly what factors lead to material characteristics so that further experimental substantiation can be done on the basis of an improved understanding of mechanisms. This is particularly important where predictions about the model result in materials design and optimization because the feedback received helps researchers improve the model.

Several approaches can enhance interpretability and explainability in ML models used for material science:

Feature importance analysis: Techniques, such as SHAP values, sensitivity analysis, and feature importance scores, are used to estimate the impact of each feature in the decision made by the model. Unlike the previous approach, researchers can pay

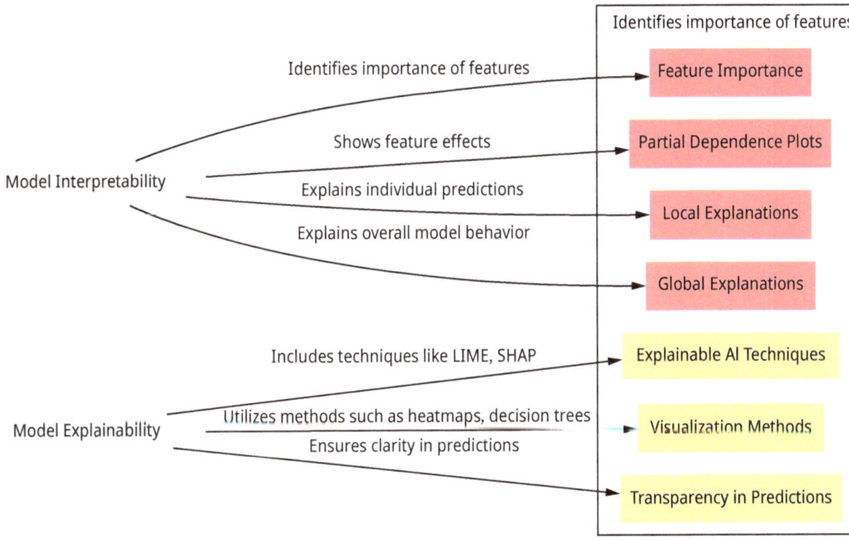

Figure 9: Model interpretability and explainability in materials science.

attention to the most significant aspects of material design and gain some understanding of the model's step-by-step decision process if the measures that contribute to the outcome are identified.

Model simplification: The structural nature of some models includes inherent interpretability, implying that models such as linear regression can easily be explained. Although these models do not always simulate the precise nature of material systems, they afford a clear understanding of the accountability of inputs and their relation to outputs. Other forms of model simplification can also be extended to practically any model in order to improve the model's interpretability.

Tools for visualization: As an example of visualization, we can mention heatmaps and partial dependency plots, which represent the changes of input variables in the model. These tools enable the behavior of the model to be made more understandable and further facilitate the ability of the researcher to visually verify any exceptional characteristics and outcomes.

Model debugging and error analysis: Looking at failures and slips, it is possible to find essential data about the specific model's weaknesses and strengths. By understanding why model failures occur, the field can identify further needs for model development and executable approaches to enhance interpretation.

Integrating domain knowledge: Expert input into the modeling, as well as the inclusion of domain-specific expertise, can enhance the interpretability of the built models. Models that are associated with confirmed scientific and material science theories ensure that the results obtained by the model are comprehended by the existing real knowledge system.

Nevertheless, the discussed approaches do not eliminate the challenges of interpretability and explainability. Real-world models, particularly the deep neural networks, pose a challenge in terms of interpretability. Probably, the most enduring problem is to simultaneously maintain model complexity and the interpretability of the results achieved in an interrelated process; thus, further work on the necessary techniques and tools is needed.

5.4 Techniques for reducing overfitting in ML models

If an ML model achieved impressive results on training data but could not on new data, this used to be termed as overfitting. Reducing overfitting was important in the development of reliable and accurate ML models necessary for contexts where small prediction margins often make a big difference such as in the field of biomedical materials. Overfitting was a problem that can be resolved by a number of methods to improve the ability of ML models to generate general results.

5.4.1 Regularization methods

There are three general categories of methods for reducing model complexity and hence preventing overfitting: (i) reduce data dimensionality; (ii) work with probability models; (iii) use regularization techniques where certain restrictions are put on model parameters. There are two types of regularization: the first is known as L1, and the second is known as L2. To ensure sparsity and feature selection, L1 regularization imposes a penalty proportional to the absolute value of parameter estimators. L2 regularization reduces the magnitude of weights and makes models simpler as it includes a penalty proportional to the square of coefficients in the loss function. Both approaches correct large models that fit the training data too closely, which prevents model complexity and improves generalization.

5.4.2 Cross-validation

Cross-checking was another technique we used to divide the data into various folds in order to measure the competency of a model and to minimize overtraining. The data was randomly divided into k different groups for the cross-validation such that the training was done from $k - 1$ groups and the testing was done from the remaining group. Each of them was a test set once during k iterations of this operation. To maintain separation between the training and the validation dataset, cross-validation guarantees that the model is tested on new data for the first time and provides a better estimate of the model's performance.

5.4.3 Early stopping

One of the training techniques discussed here was about stopping early – with the aim of avoiding overfitting, one monitors the model's performance on a validation set. When the model gives bad results on the validation set, it is likely that the model has started overfitting, and hence the training is stopped. In this way, the effects of using this method are that the model is prevented from learning paradigms specific to the training data, which results in poor generalization on new data. This helps in achieving a balance between the criterion for accurate models and evaluation on generalization by training the model for the optimal duration.

5.4.4 Data augmentation

The general idea of data augmentation is to create new instances out of the current data by somehow transforming it so that, at its essence, the size and variety of the training data set are artificially boosted. Thus, there are steps for the augmentation of the data in terms of biological materials, where noise is added to the records, and they are scaled and rotated. In other words, augmentation of the data contributes to improving model robustness due to the reduction of the probability of overfitting to some of the training data samples.

5.4.5 Ensemble methods

Ensemble methods do not fit the models to an extreme, which leads to the development of an overfitted model and also improves the general performance by coming up with a general conclusion of each model developed. Bagging, boosting, and the stacking methodology involve creating another layer of models and making the predictions be in one. To have higher stability and reduce the variability, many times a model is generated with different subsamples of data through bagging (bootstrap aggregating), and then their prediction average is taken. Whereas, in stacking, the results of different models for classification are combined using another model known as a meta-learner. In boosting, different models are learned progressively, and each tries to rectify the errors made by the previous model. Ensemble allows for leveraging the preferred and stronger aspects of several models and, at the same time, addresses the problem of over-empowerment for improved generalization.

5.4.6 Pruning techniques

Recurrently, some form of model simplification was performed using pruning to reduce model complexity by eliminating the least useful features. In neural networks, the pruning process means removing the neurons that have limited impact on the particular model. This avoided model complexity, reduced the need for computation, and was likely to perform well on unseen data. Pruning assists in eliminating high variance and highlights the existence of features that are most important in the model, enhancing interpretability.

5.5 Benchmarking and comparing ML models in biomedical materials

Comparing and benchmarking ML models is very essential in biomedical material science since it allows the evaluation of the effectiveness of models by testing them against a benchmark model to determine their suitability for specific applications. This process involves a comprehensive evaluation of different models against a set of predetermined parameters in an effort to assess their ability to accurately predict the properties of materials, their behavior, and their interactions.

One of the important aspects of benchmarking ML models was the development of performance metrics. Many measures used in the field of biological materials research include accuracy, precision, recall, F1 score, and AUC-ROC. These measurements provide insight into how efficient models are when it comes to addressing activities such as material-type classification and property estimation of appropriate materials. Choosing the right metrics is what enables the depiction of the model's ability to meet the specific needs of the biomedical application in question, as demonstrated in Figure 10.

Another aspect that was considered during benchmarking was the extent to which the employed data sets were standardized. To benchmark, what was required were datasets that are recognized in the research literature and pertain to the issue area. The application of datasets standardizes the methods utilized in the assessment of models and facilitates extensive cross-study comparison. For the same purpose, to provide a realistic assessment of the generalization capabilities of models, the datasets should also be separated into training, validation, and test sets.

In most cases, techniques for cross-validation are applied to make benchmarking more reliable. For example, in k-fold cross-validation, the patterns are split into k sets, and the model is run and verified k times by using the sets of patterns. This actually helps to reduce the risk of overfitting as well as ensuring that the performance of the model is steady for any chosen data division.

The other dimension in model comparison was the scalability and computational cost of the developed ML models. This was especially true for biological materials,

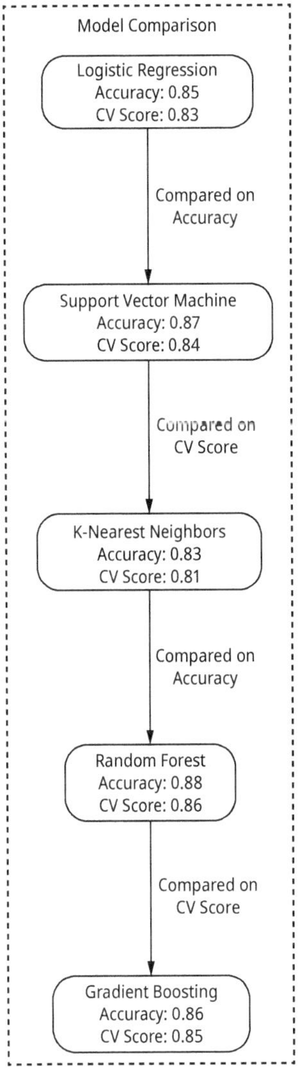

Figure 10: Benchmarking and comparing ML models in biomedical materials.

where often the data were large scale and the models introduced were sophisticated for various techniques' computational needs. In the practical applications of a model, essential measures like the time taken to train the model, the time taken to test the model, as well as the resources used, are essential. In real-world scenarios, models giving a good trade-off between time complexity/computational requirements and prediction error rate are desirable.

The model interpretability was the key consideration when evaluating the ML algorithms. In the context of biomedical materials, it was important to understand why those models gave those predictions to make sure that they are consistent with the experiments. Raw output models, which make interpretations easy for one to understand, are more easily accepted in the regulatory world and in clinical practice compared to methods that are entirely black-boxed.

Last but not least, the benchmarking and comparing of the ML models should be done with the repetitive nature of the task in mind. It was fundamental to guarantee that models supply dependable outcomes, whether working within a unique experiment by different research groups toward the objective of asseverating findings and developing new knowledge. These included building robust methods to ensure reproducibility, sharing of code and data, and following the best practices in model validations.

6 Applications of machine learning in biomedical material discovery

6.1 Predictive modeling for material biocompatibility

Using ML and a data-driven approach, we were able to create predictive modeling for material biocompatibility, which marks a major leap in the adoption of AI-based approaches in the biomedical material discovery field. Another important feature experienced in tissue material was biocompatibility, which referred to compatibility with the biological environment where it was used without adverse effects. There has been a gradual application of ML approaches to define and forecast the biocompatibility of materials, which has proved to be an exquisite technique to enhance patient benefits and the rate of material innovation.

The very essence of predictive modeling for material biocompatibility lies in the creation of suitable methods that analyze how new materials behave in biological environments and the creation of methods that can predict the given behavior. These models use information derived from material properties, results from biological assays, and clinical findings to make inferences about biocompatibility. The combination of various kinds of data allows for the use of features that are not easily detected by conventional approaches, improving the capacity to model material interactions with biological tissues.

The most widely used technique in biocompatibility prediction modeling is the use of supervised learning algorithms. These models are trained with labeled datasets containing details of the biocompatibility of different materials. The model is trained in paradigms that correlate chemical composition, mechanical properties, and surface features to biocompatibility results. These features are then input into the model. Once trained, the model can predict if other unknown and unspecified materials are biocompatible, which saves time in the evaluation and minimizes the necessity of sophisticated in vitro and in vivo testing.

The other technique was the use of methods in which the learning performed was without prior knowledge of the results, and the method aims at recognizing clusters of related materials when it comes to biocompatibility. This method can show hidden patterns and relationships that can be beneficially used in the selection and optimization of materials. For example, in our research, clustering algorithms can classify composite materials based on their chemical structure and the resulting biocompatibility and help focus on materials of interest while further tweaking their characteristics.

Other techniques of deep learning, such as neural networks, have also been applied in biocompatibility prediction modeling. These models can replicate the complicated and indirect relationships that exist between material properties and biological responses. These sophisticated models are able to detect layers of intricate structures and relationships that a basic model would fail to see, given their ability to handle large amounts of data at multilayer abstraction. This method is particularly useful for datasets with many features and where it is hard for standard models to achieve high precision.

Coanalysis of multiple omics data, including genomic data, proteomics data, and metabolomics data, adds a higher level of complexity to the predictive modeling. When the cellular response and molecular interactions are included in the ML models, a better evaluation of the material compatibility can be determined more effectively. This multifaceted approach also results in the ability to predict material behavior in different biological environments and enhances the accuracy of biocompatibility queries.

In this case, we should outline that the use of predictive modeling for material biocompatibility was not without advantages and disadvantages. Only high-quality and representative data can be used in training the given model to reach the same level of quality. Biased data also leads to wrong forecasts and the general non-consideration of the downside risk. Because feedback was an important component of verifying results and ensuring clinical applicability of the models, when reasoning about predictions was essential, some other factors made the interpretability of ML models a cause for concern.

6.2 Machine learning in the design of drug delivery systems

As demonstrated by the various applications presented in this manuscript, ML has evolved as an essential instrument in the design and enhancement of drug delivery systems and is shaping novel solutions to various issues faced by pharmaceutical sciences. In this field, the use of ML sheds light on the development of sophisticated delivery systems, increasing drug performance, safety, and targeting.

A significant implementation area in drug delivery system design is in the area of drug formulation and release profiles. A drug designed by previous procedures, which are not always logical, is rather based on experience and failure in some experiments. Regression analysis and optimization methods in ML, depending on the results of the analyzed data from earlier experiments and simulations, can result in the best formulation parameters. This approach can help to determine the ideal release profile characteristics of the drug delivery systems to ensure improved drug efficacy and reduced toxicity effects.

Another major area of use was observed in the synthesis of nanocarriers and intelligent drug delivery systems. The adverse effects of nanocarriers and liposomes and/or their interaction with biological systems can also be predicted using the advanced features of ML models. Using big data on the performance of nanocarriers, ML can pinpoint the factors affecting the efficiencies of drug encapsulation as well as its release and targeting mechanisms. This information enhances the formulation of nanocarriers to enable efficient drug delivery to particular tissues to enhance the results of targeted treatment.

It was crucial to reduce gaps in the ways that medications are administered and routes for their distribution using the concept of ML. For instance, deep learning can be used to analyze patient data alongside previous performance data to recommend delivery routes for the patient that would enable the right concentration of the medicine to be delivered efficiently, with fewer side effects. Delivery systems should be made to face the necessary demand and context of every patient group by identifying the patient groups with similar characteristics through clustering and classification algorithms.

In the application of controlled and sustained drug release, there are ML algorithms that help develop systems for the controlled release rates of drugs, considering the physiological signals. For instance, smart delivery systems with sensors and stimuli-responsive materials attached to them can change dosage according to the surroundings such as pH. From these systems, it is also possible to feed data to the ML models for the improvement of their output to match the expected response to drug delivery and the ability to manage these systems with enormous precision, as illustrated in Figure 11.

Further, the use of ML was in the process of generating replicas of the drug delivery systems for understanding their operation and their functioning under specific conditions. Performing experiments on simulations coupled with using ML algorithms allows people conducting the research to foresee the manner in which certain delivery systems perform in various bioenvironments, their stability, as well as their ability to interact with other drugs. This predictive capability can be utilized in designing new systems that would exhibit safer operation and more reliably reliable outcomes.

Last but not least, the integration of ML with high-throughput screening expands the discovery of new drug delivery systems. Screening data can then be effectively analyzed by the use of ML algorithms to pin point potentials for further development. By analyzing large amounts of data and finding patterns that are hard for people's eyes to notice, ML can speed up the process of discovery and enhance the development of drug delivery systems.

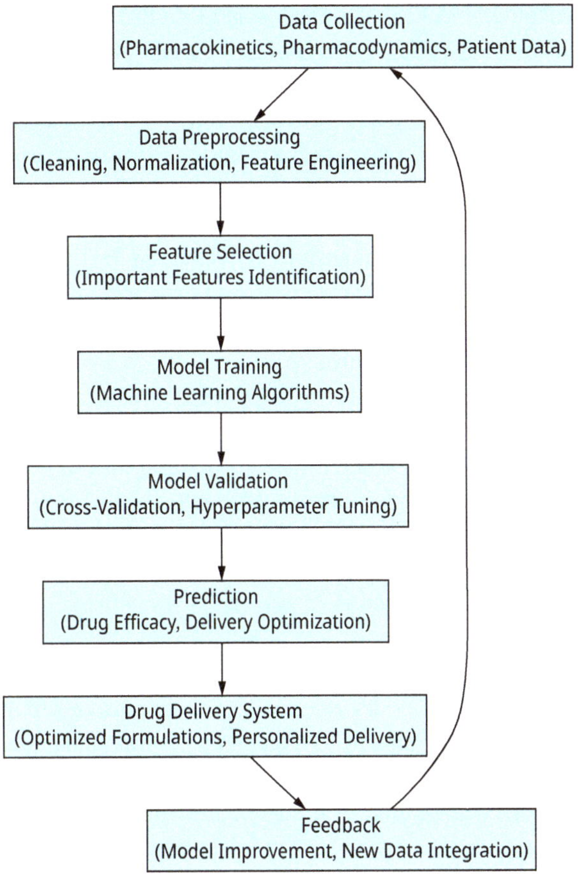

Figure 11: Machine learning in the design of drug delivery systems.

6.3 Role of ML in tissue engineering and regenerative medicine

In the areas of tissue engineering and regenerative medicine, ML has turned out to be a revolutionary approach, thereby enhancing the rate at which the development of biomedical materials and treatments is made. Using ML algorithms, one can analyze complex biological information, predict the behavior of materials, and develop new-generation regeneration solutions. This section looks at all the ways that ML is being used in many fields, especially how it improves the outcomes and increases the speed of research and development.

In tissue engineering, ML plays a critical role in constructing the biomaterials used to create a tissue scaffold. ML algorithms select materials for scaffold properties such as porosity, mechanical strength, and biocompatibility by analyzing large data-sets originating from both material science and medicine domains. By using supervised learning, unsupervised learning, and RL, scientists develop prediction models that are used to guide the fabrication of scaffolds pertaining to specific tissue forms and uses. This method tends to enhance the efficiency and reliability of the scaffold design, and hence better tissue regeneration outcomes are achieved.

Here, we also see that ML was beneficial in regenerative medicine in terms of processing and analysis of omics data. Data from transcriptomics, proteomics, and genomics shed understanding on the mechanisms of tissue growth and repair. The availability of various ML algorithms, such as clustering and dimensionality reduction, provides a chance to identify significant biomarkers of tissue regeneration and signaling pathways. In this way, personalized regenerative therapies, as well as targeted therapies, are developed, adding to the effectiveness of the interventions.

One of the most impressive applications of ML in regenerative medicine was shown in cell-based therapy optimization. Through processing data from preclinical studies and cell culture, ML predicts the behavior and efficiency of different cell types in regenerative medicine. Indirectly, these models assist in the selection of optimal cell sources, culture conditions, and delivery strategies that subsequently enhance the outcomes of cell-based therapies for tissue repair and regeneration.

Computational components enable the realization of personalized medicine paradigms within regenerative medicine, also through the use of ML algorithms. Using genetic information, patient history, and the response to previous treatments incorporated into the ML models, regenerative therapies are personalized. Such customization also makes it possible to deploy exact therapy approaches based on specific individual features that are ideal for treating the client in question.

It scales the original methods of imaging data in tissue engineering through the application of ML. Due to the increased use of microscopy in image diagnosis, there is a massive output of data, which cannot be analyzed using conventional means. Imaging datasets can then be analyzed using ML algorithms to obtain information on scaffold integration, tissue regeneration, and scaffold structural changes overtime. This capability provides the specifics for the monitoring and assessment of tissue engineering results to enhance improvements to materials and methods.

Consequently, progress in ML for tissue engineering and regenerative medicine has occurred, although there are still challenges to face. These include the complexity of biological systems, the need to deal with large-scale, high-quality information, and the use of decision-making artificial intelligence models as part of experiments as well as clinical procedures. As a result, to ensure that ML continues to enhance regenerative medicine for the development of this field, these problems require further research and collaboration between different disciplines.

6.4 Optimizing material synthesis processes with machine learning

The usage of ML has cemented itself to revolutionize the processes of material synthesis positively by providing enhanced solutions to one of the biggest obstacles: the precision and discovery of biomedical materials. By using data derived from experimental data and other available information, together with predictive modeling, ML helps determine the best synthesis conditions, outline the material development process, and minimize the time and resources required for experimentation.

Among the most important tasks of ML in the improvement of material synthesis, one of the most required aspects was the determination of synthesis parameters. Conventional approaches to synthesizing the reaction conditions are highly time-consuming and laborious. This method uses regression models and optimization algorithms to forecast the change in mechanical and kinetic properties of materials and the yield as influenced by different factors like temperature, pressure, concentration, and reaction duration. Reflection on experimental history that involves the use of ML models enables the prediction of synthesis conditions that would yield the necessary material characteristics.

Another key application that was made was using ML in the design of experiments in general. Methods like factorial design and response surface methodology of experimental design can be used with improved ML to select the most useful experimental situations. Along the same lines, as the experiments are conducted, the ML algorithms could provide direction on which other subsequent experiments could be conducted in an attempt to gain as much information as possible. This approach limits the number of experiments to be conducted and targets high-potential synthesis conditions, therefore fast-tracking ideation and optimization.

Another important aspect was the automation of the synthesis itself through the use of ML. Through integration with high-throughput experimentation platforms, material synthesis and characterization were automated using ML. The synthesis procedures can be run by means of ML models to control the behavior of robotic systems, to observe the reaction conditions continuously, and to modify the parameters with respect to the received feedback. This automation increases the capability of creating material and also enables one to search for a large number of parameters at great speed, resulting in improved discovery of such material, as illustrated in Figure 12.

By using ML models, there was an improvement in the efficient and accurate evaluation of the characterization of material and its quality. Notable ML applications in the context of analytical techniques comprise spectroscopy, microscopy, chromatography, deep learning, computer vision, and result analysis. These models are capable of detecting patterns and abnormalities, which signal potential changes in material quality. The analysis carried out by ML enables early identification of problem areas, thus ensuring consistent and quality production of material in the process.

Besides, ML also helps to process multiple data modes for joint optimization. Most material synthesis processes also entail crucial data from data sources such as experi-

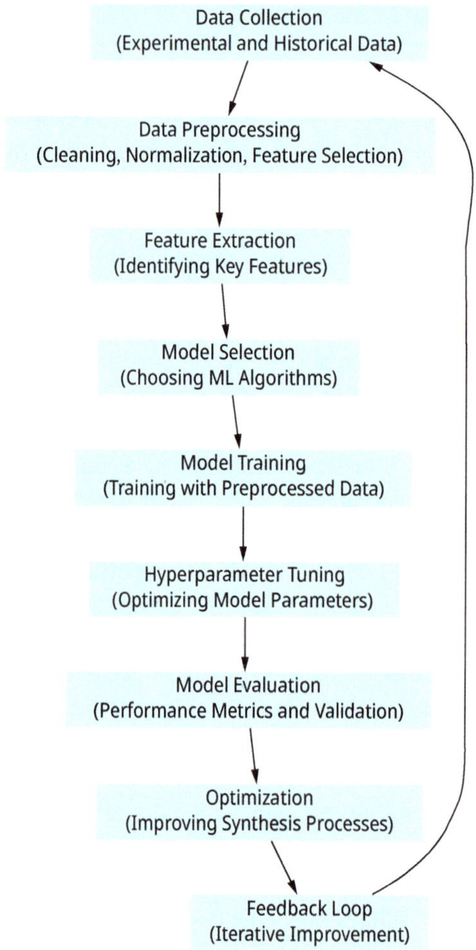

Figure 12: Optimizing material synthesis processes with machine learning.

mental outcomes, simulation information, and theoretical models, among others. Different data types can be easily incorporated to update the ML algorithms to build a comprehensive picture of synthesis and its effect on material properties. This approach allows one to predict more nuances and to create better optimization strategies.

There are challenges to optimizing material synthesis with the use of ML as follows: The accuracy of any form of "prediction" using ML largely relies on the quality and volume of training data. The use of partial data creates wrong conclusions and leads to incorrect conditions during synthesis. Some material synthesis processes are complex and need efficient computational power and highly developed ML models, which are rather difficult to perform in certain research environments.

6.5 Case studies

Examples of using ML techniques in biomedical material discovery consider the ability of such approaches in the identification, selection, and validation of new tissues and materials. These cases demonstrate a conceptual application of ML for facilitating material development and improving therapeutic benefits.

6.5.1 Important considerations in the identification of new drug delivery techniques

A rather good example relates to the use of ML in developing and improving pharmaceutical systems. For these purposes, deep learning algorithms were used to explore huge data sets of molecular structures and drug release curvatures. By applying these models, new polymeric materials for drug encapsulation with improved efficiency and controlled release were also studied. The time to find and confirm new materials decreased noticeably due to the ML-based approach that helped to create new delivery systems with better therapeutic outcomes.

6.5.2 The prediction of biocompatibility for biomaterials

Another example of applying computational methods in expression profiling was related to the prediction of biomaterials' biocompatibility for implantation. ML models were trained on large data sets, which contain data about the properties of materials, biological reactions, and clinical results. Such models also contributed to the biocompatibility estimates of prospective new material candidates to reduce undesirable biological responses. This case of ML assistance helped in the faster shortlisting of safe and effective biomaterials, corrections for which would earlier call for sundry in vivo tests – these corrections proved beneficial for the quicker advancement of healthcare implants.

6.5.3 Tissue engineering scaffolds optimization

In tissue engineering, ML techniques have been used to develop models with better designs of the scaffolds used for tissue regeneration. Scientists used RL algorithms to continuously shift scaffold parameters, including CEM, porosity, and mechanical properties, in response to simulated biological activity. This approach helped to design scaffolds with better performance in supporting cell and tissue regeneration. In general, it was possible to benefit from ML to define the most efficient and specific scaffolds for tissue engineering applications by optimizing the search for scaffold designs.

6.5.4 Work to produce materials useful for the implementation of personalized medicine

A case study in the biomedical application of personalized medicine demonstrates how ML can help match specific biomedical materials to specific individuals. Using data on patients and their individual genomic profiles and disease states, the ML models determined the materials whose characteristics matched the respective patients' needs in therapy. For instance, ML-based solutions applied to drug development created individual drug carriers based on the patient's pharmacogenomic data. It enhanced the medical treatment's efficiency and safety and showed how the concept of ML can elevate patient experiences in terms of materials.

6.5.5 A high-throughput screening of antibody-drug conjugate

In seeking potent ADCs, the application of the ML technique has helped reduce the high-throughput screening process. Using ML on big data sets containing interactions of antibodies and drugs, researchers have been able to identify candidate ADCs with desirable characteristics such as high binding affinity and cytotoxicity. Namely, the ML approach allowed for an almost instantaneous hit list generation, thus greatly enhancing the prospects for new ADC therapies in cancer targeting.

6.5.6 Application of regenerative medicine

A case study in regenerative medicine shows how one can apply this field of ML to predict the properties of materials used in regenerative therapies. Scientists use ML models with the intention of predicting the mechanical strength, degradation characteristics, and biological responses of new biomaterials. These models, based on historical data and simulations, offer key information about the behavior of various materials and open a way toward the creation of new materials more suited to specific regenerative applications.

7 Conclusion

Biological materials research has been revolutionized by incorporating ML into the discovery, development, and use of new materials. This chapter has made the following urgent advancements, which have occurred within the sector, evident: firstly, the use of ML has customized medical treatments; secondly, there has been optimization of characteristics; and thirdly, the use of ML has made the discovery of materials faster.

Looking at the contribution of ML to biomedical material discovery, the following discoveries can be noted. In recent years, the employment of predictive modeling and techniques of high-throughput screening has led to a drastic reduction in the time necessary to accomplish the screening of potentially favorable materials and a minimum dependency on the conventional, time and money-consuming empirical screening methods. Specifically, the use of transfer learning has led to many new and specific biomedical applications, which have been adapted based on existing models to create new solutions for the delivery of drugs and tissue engineering as well as for personalized medicine. All the above demonstrations show the effectiveness of applying ML in fulfilling specific medical requirements to transform material, enhance patients' quality of life, and improve health-care technologies.

There are a few issues that need consideration to unleash the full potential of ML in biomedical materials science. Data acquisition from different sources, model explain ability, and demand for computational power are considered the challenging aspects that need further development. The main focus must be on data quality as well as on creating the proper means to integrate the models into the respective systems. Improving the explanation of these models within both the academic and applied communities is necessary for the sciences as well as for the application of ML findings in the field.

At the same time, the application of ML in biological materials research looks rather promising. Further advancements in the identification of material properties and their optimization are possible due to contributions from new algorithms and technologies, including deep learning and online data analysis. The integration of material property monitoring with advanced technologies, including modern modeling techniques and detailed imaging, can enrich the characteristics of materials and their performance. Also, questions of an ethical character – such as algorithmic prejudice and data protection – are crucial to providing for the proper advancement and deployment of forward-thinking ML systems.

References

[1] Xie, S. (2023). Perspectives on development of biomedical polymer materials in artificial intelligence age. Journal of Biomaterials Applications, 37(8), 1355–1375.

[2] Gokcekuyu, Y., Ekinci, F., Guzel, M. S., Acici, K., Aydin, S., & Asuroglu, T. (2024). Artificial intelligence in biomaterials: A comprehensive review. Applied Sciences, 14(15), 6590.

[3] Pasrija, P., Jha, P., Upadhyaya, P., Khan, M., & Chopra, M. (2022). Machine learning and artificial intelligence: A paradigm shift in big data-driven drug design and discovery. Current Topics in Medicinal Chemistry, 22(20), 1692–1727.

[4] Batra, R., Song, L., & Ramprasad, R. (2021). Emerging materials intelligence ecosystems propelled by machine learning. Nature Reviews Materials, 6(8), 655–678.

[5] Singh, A. V., Rosenkranz, D., Ansari, M. H. D., Singh, R., Kanase, A., Singh, S. P., . . . & Luch, A. (2020). Artificial intelligence and machine learning empower advanced biomedical material design to toxicity prediction. Advanced Intelligent Systems, 2(12), 2000084.
[6] Kerner, J., Dogan, A., & Von Recum, H. (2021). Machine learning and big data provide crucial insight for future biomaterials discovery and research. Acta Biomaterialia, 130, 54–65.
[7] Mathé, E., Hays, J. L., Stover, D. G., & Chen, J. L. (2018). The omics revolution continues: The maturation of high-throughput biological data sources. Yearbook of Medical Informatics, 27(01), 211–222.
[8] Suwardi, A., Wang, F., Xue, K., Han, M. Y., Teo, P., Wang, P., . . . & Loh, X. J. (2022). Machine learning-driven biomaterials evolution. Advanced Materials, 34(1), 2102703.
[9] Feng, C. X., & Wang, X. F. (2003). Surface roughness predictive modeling: Neural networks versus regression. IIE Transactions, 35(1), 11–27.
[10] Weiss, S. M., & Indurkhya, N. (1998). Predictive data mining: A practical guide. Morgan Kaufmann.
[11] Xu, Y., Liu, X., Cao, X., Huang, C., Liu, E., Qian, S., . . . & Zhang, J. (2021). Artificial intelligence: A powerful paradigm for scientific research. The Innovation, 2(4).
[12] Holzapfel, B. M., Reichert, J. C., Schantz, J. T., Gbureck, U., Rackwitz, L., Nöth, U., . . . & Hutmacher, D. W. (2013). How smart do biomaterials need to be? A translational science and clinical point of view. Advanced Drug Delivery Reviews, 65(4), 581–603.
[13] Iman, M., Arabnia, H. R., & Rasheed, K. (2023). A review of deep transfer learning and recent advancements. Technologies, 11(2), 40.
[14] Kora, P., Ooi, C. P., Faust, O., Raghavendra, U., Gudigar, A., Chan, W. Y., . . . & Acharya, U. R. (2022). Transfer learning techniques for medical image analysis: A review. Biocybernetics and Biomedical Engineering, 42(1), 79–107.
[15] Goetz, A., Durmaz, A. R., Müller, M., Thomas, A., Britz, D., Kerfriden, P., & Eberl, C. (2022). Addressing materials' microstructure diversity using transfer learning. Npj Computational Materials, 8(1), 27.
[16] Ferracini, R., Martínez Herreros, I., Russo, A., Casalini, T., Rossi, F., & Perale, G. (2018). Scaffolds as structural tools for bone-targeted drug delivery. Pharmaceutics, 10(3), 122.
[17] Collobert, R., Weston, J., Bottou, L., Karlen, M., Kavukcuoglu, K., & Kuksa, P. (2011). Natural language processing (almost) from scratch. Journal of Machine Learning Research, 12, 2493–2537.
[18] Alber, M., Buganza Tepole, A., Cannon, W. R., De, S., Dura-Bernal, S., Garikipati, K., . . . & Kuhl, E. (2019). Integrating machine learning and multiscale modeling – Perspectives, challenges, and opportunities in the biological, biomedical, and behavioral sciences. NPJ Digital Medicine, 2(1), 115.
[19] Graw, S., Chappell, K., Washam, C. L., Gies, A., Bird, J., Robeson, M. S., & Byrum, S. D. (2021). Multi-omics data integration considerations and study design for biological systems and disease. Molecular Omics, 17(2), 170–185.
[20] Khan, R., Usman, M., & Moinuddin, M. (2024). From raw data to actionable insights: Navigating the world of data analytics. International Journal of Advanced Engineering Technologies and Innovations, 1(4), 142–166.
[21] Carvalho, D. V., Pereira, E. M., & Cardoso, J. S. (2019). Machine learning interpretability: A survey on methods and metrics. Electronics, 8(8), 832.
[22] Carvalho, D. V., Pereira, E. M., & Cardoso, J. S. (2019). Machine learning interpretability: A survey on methods and metrics. Electronics, 8(8), 832.
[23] Basu, B., Gowtham, N. H., Xiao, Y., Kalidindi, S. R., & Leong, K. W. (2022). Biomaterialomics: Data science-driven pathways to develop fourth-generation biomaterials. Acta Biomaterialia, 143, 1–25.
[24] Guo, K., Yang, Z., Yu, C. H., & Buehler, M. J. (2021). Artificial intelligence and machine learning in design of mechanical materials. Materials Horizons, 8(4), 1153–1172.
[25] Hong, S., Liow, C. H., Yuk, J. M., Byon, H. R., Yang, Y., Cho, E., . . . & Lee, H. M. (2021). Reducing time to discovery: Materials and molecular modeling, imaging, informatics, and integration. ACS Nano, 15(3), 3971–3995.

Spoorthi P. Shetty, N. Pragadish, Ashish Verma, K.N.V. Satyanarayana

High-throughput screening for novel medical materials: machine learning-enabled approaches

Abstract: The adoption of machine learning in biomedical research in the context of drug delivery system characteristics, drug release profiles, and the optimization of nanoparticle systems is quickly changing the face of biomedical research. This chapter seeks to apply machine learning algorithms for identification of characteristics that are time-critical for drug delivery systems including mechanical properties, degradation rate, and biocompatibility. A comparison of basic versions of regression models and deep learning is outlined to investigate the potential for improvement of accuracy and speed when implementing drug delivery systems. Particular attention is paid to applications of the drug release kinetic models, as well as the use of ML approaches for individualized approaches to drug delivery and shorter treatment regimens, making it possible to emphasize the possible role of ontological ML strategies in increasing the efficacy of the treatment. The application of the ML in defining the appropriate parameters of designed nanoparticles and in combining the experimental and computational techniques for the fabrication of targeted and efficient delivery systems is discussed. Some of the issues that hinder the implementation of ML in drug delivery are reviewed alongside opportunities and future trends of the technology. Concerning ethical issues, and to follow safe and effective requirements for ML technologies application, further development outlines core principles that must be followed. In this context, the provided insights for academicians, practicing clinicians, and policymakers are meant to be useful in enhancing the state of customized medicine and addressing existing and emerging chronic healthcare challenges that intend to apply ML in enhancing drug delivery systems.

Keywords: machine learning, drug delivery systems, biocompatibility, predictive modeling, personalized medicine, nanoparticle design

Spoorthi P. Shetty, Department of MCA, NMAAM Institute of technology, Karkala Taluk, Udupi District, Nitte 574110, Karnataka, India, e-mail: sshetty.07@nitte.edu.in
N. Pragadish, Department of Mechanical Engineering, Rajalakshmi Institute of Technology, Kuthambakkam, Chennai 600124, Tamil Nadu, India, e-mail: pragadish81@gmail.com
Ashish Verma, Department of Physics, School of Mathematical and Physical Sciences, Dr. Harisngh Gour Vishwavidyalay, Sagar, Madhya Pradesh, India
K.N.V. Satyanarayana, ECE Department, SRKR Engineering College, China Amiram Bhimavaram, Andhra Pradesh, India, e-mail: knvsnarayana@srkrec.ac.in

https://doi.org/10.1515/9783111503202-010

1 Introduction

High throughput screening (HTS) is applied as the key method in the search for new medical materials and drugs [1, 2]. Through the automated process of testing large databases of compounds, HTS expedites the discovery process of potentially useful compounds with references to therapeutic potential [3]. Formerly, HTS relied on robotic systems, which was applied to thousands of samples at a time and created a large data [4]. Raw research data at this scale too presented a problem in as much as analysis and interpretation was concerned [5]. In order to overcome these challenges, scientists use machine learning (ML) as an effective tool for improving HTS performance [6]. With the integration of MLP-based algorithms to the HTS workflows, it became easier to perform the data analysis while enhancing the predictive power and increasing the efficiency of discovering the materials [7].

Machine learning (ML) has lately become paramount to data analysis in HTS due to its advanced ability at pattern recognition and predictive modeling [8]. Due to the use of data mining, ML algorithms can handle the data load quantities and distinguish patterns or causations not easily discovered by other conventional analytical means [9]. These capabilities are considered most useful in material discovery where predicting properties and performance of new found materials prior to physical testing definitely reduces costs [11]. They also evolve and become more efficient as more data is fed into them, hence the added bonus of a speeding up of the screening process [11].

This integration also attends to several issues that are associated with the handling of data and understanding of the models that are deployed during HTS [12]. As was seen in HTS, the amount of emergent data is vast, limiting the ability to obtain informative results without the assistance of sophisticated computational software [13, 14]. To deal with this issue, with an aim of optimizing learning, [ML techniques] tend to alleviate this complexity through aspects like dimensionality reduction and feature selection [15]. Further, interpretability becomes an essential issue, which is critical when assessing the results of the ML to determine if it would work in the real world [16]. This chapter discusses several approaches to increase model interpretability and nonexpert use of ML in HTS [17].

In addition to streamlining the screening processes and enhancing data interpretation, excellent results that ML-enabled HTS have shown are its capabilities to go beyond typical limitations of traditional approaches to material discovery [18]. For example, ML algorithms are used for the accurate result prediction of the screening experiments that are involved in extensive and boring physical testing [19]. It is in this aspect that the use of new computer simulation capabilities can be utilized to predict and top up the list of the more possible lists for new potential inventions to be explored in order to shorten the development period and bring down the expenses [20]. Using case studies, the chapter describes how innovation has been achieved through ML-driven HTS even in the field of medical materials, revealing the opportunities of the approach to material science and drug discovery [21].

In general, the combination with HTS and ML is a progress in the field of material discovery [22]. With the added benefits of automation screening and the practicality of ML, a researcher is able to screen large amounts of data in a short amount of time with high amounts of accuracy [23]. The purpose of this chapter is to give the reader a holistic view of the current status of ML in HTS, including uses, advantages, and drawbacks [24]. It explains how this combined approach facilitates new solutions to be carved out in medical materials and helps upgrade the kinds of therapies being introduced [25].

2 Automated prediction of material properties

2.1 Introduction to material properties crucial for drug delivery systems

The importance of drug delivery system lies in ensuring that the administered therapeutic interventions are effective and safe in the area of modern medicinal practice. These systems are designed not only to carry the pharmacological compounds and deliver them at targeted areas in the organism, but also to do this in a regulated fashion. Owing to the limitations of bond strength of the carrier materials, the efficiency of drug delivery depends primarily on the material characteristics of the carriers employed to deliver it and these characteristics must satisfy certain biological standards.

Designing medication delivery systems has one basic necessity, and that is biocompatibility. When a material is used in the medical field, it is crucial that it can dynamically adjust its interactions with biological systems to minimize adverse effects. This adaptive behavior reduces the likelihood of negative responses such as inflammation or rejection. As illustrated in Figure 1, biocompatibility plays a pivotal role in mitigating immunogenicity and cytotoxicity, thereby enhancing both the safety and the overall tolerance of therapeutic treatments.

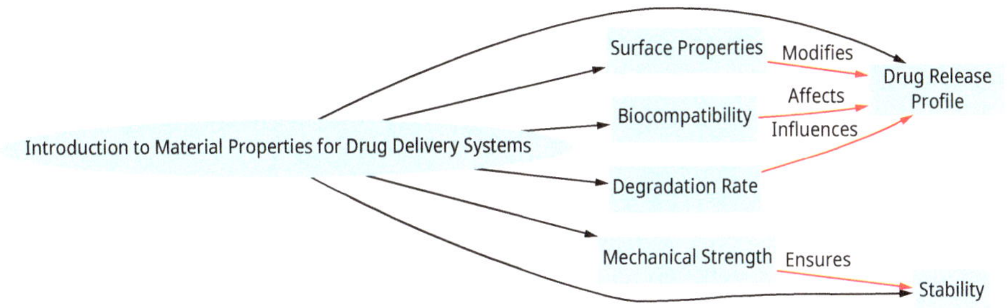

Figure 1: Introduction to material properties crucial for drug delivery systems.

Some other criteria that have an influence on the possibility and durability of a drug delivery system are its mechanical characteristics. They need to be sufficiently rigid and elastic to stand physiological loads and the other environmental factors depending on what is intended. For example, mechanical properties are desirable for the materials being applied to tissue scaffolds and implanted devices to possess adequate structural performance in the long run.

One of the most important parameters of biodegradable medication delivery systems is degradation rates. In this case, these mechanisms are created to gradually dissolve or release the medications as ingested, metabolized, or excreted from the system. In tune with the treatment schedule, the rate of degradation was smoothly and regularly matched to give controlled and sustained release kinetics that optimize the therapeutic benefits of the medicine without increasing the risk of deleterious side effects.

The development of multifunctional drug delivery systems requires the ability to carefully analyze and selectively tailor material properties. More often, conventional methods of material properties characterization and optimization are augmented with technologies such as ML. Due to ML, it becomes possible to measure the properties of the materials and the outcomes of the experiments with large matrices, thereby acting as an effective tool for predictive modeling of the drug delivery systems with greater accuracy and efficiency.

2.2 Machine learning techniques for predicting material properties

ML has emerged as the revolutionary technology in the biomedical field particularly the area of material property optimization and prediction of characteristics that are crucial in designing drug delivery systems. Applying machine learning algorithms, it is possible to analyze large and complex data and determine the relationships that contribute to accurate prediction of such mechanical properties of biomaterials, their biological compatibility, and the rates of their degradation.

Biocompatibility prediction: Biocompatibility prediction is the process of determining how different materials would respond when in contact with biological systems in order to have the least reaction possible. ML approaches for predicting biocompatibility encompass various methodologies, including:

Classification models: These models organize the materials according to chemical profile, surface features, and prior tests results into biocompatibility classes. Since high-dimensional datasets can be managed and are not linearly related to each other SVM and random forest algorithms are often considered.

Deep learning: Neural networks have proved to be one type of deep learning algorithms, which is suitable for large datasets and is able to extract latent features deeply. Deep learning algorithms effectively diagnose biocompatibility through Big Data of material cell relations and biological responses. This makes it possible for the models to learn minute aspects that would be impossible to be noticed by normal "over the air" approaches.

Degradation rate prediction: Estimating the degradation rate of biomaterials is important in the design of biodegradable pharmaceutical carriers that deliver the active compounds at a controlled rate and for lasting duration. ML techniques for degradation rate prediction include:

Regression models: Linear regression, polynomial regression, and even the complex forms of the Bayesian regression are applied to simulate the dependency of degradation kinetics on the material characteristics (e.g., molecular weight, chemical structure). These models provide means of defining temporal variations of material properties and estimating degradation patterns in different environments.

Physiologically based models: Using computational models or in silico analysis, there is a possibility of including parameters such as pH and enzymatic activity in the biomaterial degradation predictive models. This approach accords better with the practicality and accuracy of degradation rate predictions in drug delivery system.

Mechanical properties prediction: Elasticity, fracture toughness, and tensile strength, all mechanical properties, significantly influence drug delivery devices and their performance and durability. The following machine learning methods are used to forecast mechanical properties:

Ensemble methods: For instance, there are sophisticated methods, such as gradient boosting machines (GBM) and AdaBoost that work using a set of weaker predictors. These methods assess the structure-property connection of the materials and forecast the mechanical characteristics using training data gained through experiment and computation.

Finite element analysis (FEA) coupled with ML: With the integration of the ML algorithms and FEA simulations, scientists can estimate material behaviors under mechanical load and strain. The results and data generated via FEA are used to train other parameters of the material for precise mechanical demands through ML; thus, it is capable of fabricating efficient drug delivery systems that can be sturdy enough to support the physiological environment.

2.3 Machine learning and linking to other experiment/computational method techniques

With experimental and computational techniques in the biomedical field, machine learning (ML) has transformed the prediction and optimization of the material properties required in drug delivery systems. This integrated methodology makes efficient use of both experimental data and computational simulations to enhance the reliability, speed, and robustness of predictions in the biomedical materials domain. Figure 2 illustrates how ML integrates with experimental and simulation workflows to streamline material property optimization.

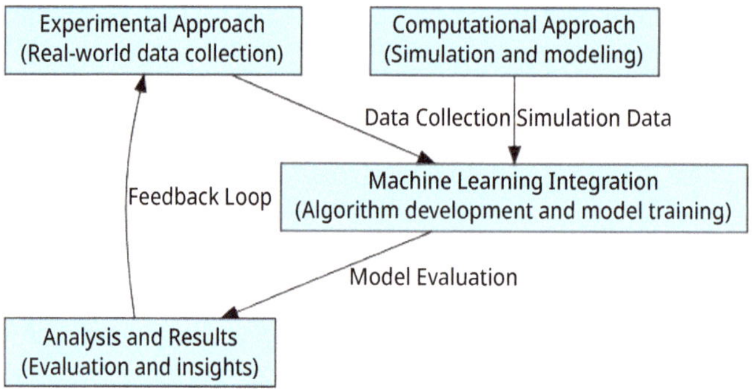

Figure 2: Integration of ML with experimental and computational approaches.

Experimental data acquisition: Functional data on physical, chemical, and biological properties of biomaterials can be obtained from experimental techniques. Spectroscopy, microscopy, chromatography, all these common polarized techniques produce large datasets that provide fine-grained information of material response to particular conditions in a controlled environment.

Experimental data gathered are preprocessed and normalized by the ML algorithms to eliminate noise, outliers, and missing values for raw accurate data processing.

Feature extraction: Based on experimental data, the resulting ML models select characteristic features and find significant parameters that, in their work, are related to material properties, such as mechanical properties, biocompatibility, and degradation time.

Pattern recognition: In order to determine new unknown relationships within the results of experiments and connections, both supervised and unsupervised learning approaches are used on the data. Regression techniques apply experimental information to model continuous variables such as degradation rates whereas clustering techniques classify similar materials with similar characteristics.

Computational modeling: Computation modeling enhances the experimental knowledge through the use of mathematical modeling to represent the properties and behaviors of the materials and their interactions on molecular, atomic, or macroscopic levels. Techniques such as molecular dynamics (MD), FEA, and density functional theory (DFT) provide computational frameworks to:

Simulate material properties: ML implemented in computational data provides the ability to accurately anticipate material behavior under different environmental conditions, making it easier to model degradation rates, mechanical properties and biocompatibility.

Optimize design parameters: ML optimization techniques are used to iteratively adapt material structures through design models and experimental applications. Thus, this cyclic approach optimizes the search for materials and speeds up the process of implementing the results of theoretical studies at the practical level.

Validation and calibration: Validation concerns using experiment results to compare with the computational model's predictions for purposes of refining parameters of simulations. The iterative feedback loop therefore increases the reliability of existing predictive models in estimating material properties under actual field conditions.

Synergistic advantages: The combination of ML with experimentation and computation utilizes compatible benefits that cannot be obtained with a single technique. If data sets are used in conjunction with corresponding simulations, the usual problems in material characterization and prediction can be avoided. More specifically, the presented AI solutions help identify new advanced materials that suited for certain biomedical applications, such as drug delivery carriers, tissue engineering scaffolds, and so on.

2.4 Applications of ML in optimizing material formulations

ML has transformed biomedical materials sciences by providing improved techniques for enhancing the material composition of planned drug delivery systems. Through the large datasets and computational resources, ML improves the speed, accuracy, and optimization of the material design processes.

Data-driven material design: ML algorithms complement material design by utilizing multiple databases including information on material composition, properties, and performance. These algorithms analyze and recognize data characteristics that other approaches do not see. By processing large volumes of experimental and computational data, ML enables:

Prediction of material properties: In supervised learning architectures like regression and neural networks, there is a profound capability to forecast chief material attributes including biocompatibility, degradation rates, and mechanical strength from input features, including chemical composition, structural characteristics, and environment.

Optimization of formulation parameters: ML-embedded optimization techniques include iteratively updating the material composition and processing parameters to obtain required material characteristics. Even the evolutionary algorithms, Bayesian optimization, and reinforcement learning techniques tune the formulation of materials given the feedback of experimental outcomes and computational simulations.

Enhanced predictive modeling: ML improves the predictive modeling function in the best ways to develop material formulation by combining experimental results with theoretical predictions. This synergy allows researchers to:

Accelerate discovery processes: Through automating data analysis and hypothesis testing, ML accelerates the work of filtering and identifying new biomaterials suitable for specific biological applications due to the developed characteristics. Reducing the iteration cycle time in the formulation of materials advances the competitiveness of biomedical research and development by shortening the time to market.

Customize material performance: Material performance profiles should be adjusted by machine learning algorithms to meet a spectrum of clinical needs and patient desires. Drug delivery systems must be designed to work as expected and as needed by integrating adaptable learning models, which alter formulation of the materials used in response to changes in physiological conditions.

Biodegradable polymers: Using ML algorithms, degradations, kinetics, and mechanical behavior of biodegradable polymers in controlled drug delivery systems are predicted. Optimization approaches also increase the stability of polymers and compatibility with body tissues as well as control the release of powdered drugs, thereby promoting better treatment rates and patient compliance.

Nanostructured materials: Towards personalized medicine, enhancing the instructional design of nanoparticles by machine learning can increase the targeting capability and drug kinetic profile. Nanoparticle engineering enhances drug delivery efficacy through advancement in models for size, surface characteristics, and drug release profiles.

3 Optimization of drug release profiles

3.1 Introduction to drug release kinetics in delivery systems

In the two industries of biomedical engineering and pharmaceuticals, reduction of side effects and enhancement of treatment efficacy directly relate to controlled delivery of medicinal agents. Drug delivery systems are designed to deliver drugs in a manner that ensures duration of release; their concentration remains constant for some time. The rationale is as follows: Knowledge of drug release kinetics is an important factor in designing clinical delivery systems for drugs with different therapeutic goals and for patients with particular needs.

Fundamentals of drug release profiles: Drug release kinetics describe the temporal profile of the process and the manner by which the therapeutic agents are ejected from the delivery devices and into the physiological fluids. These profiles also affect therapeutic effectiveness and safety and the rate and extent of ADME.

Types of drug release profiles: Drug release profiles encompass various temporal patterns, each tailored to specific therapeutic objectives:

Immediate release: Fast dissolving and disintegration post capping of the pellets – because it ensures that there was a fast dissolution of the medicine when it is administered to patients with a condition that requires quick action of the drug.

Sustained release: Delivering drugs at a desired rate for extended periods while keeping the required proactive concentrations in the body. It is great for drugs requiring a steady flow of release without the need for frequent administrations.

Delayed release: The delivery of drugs at a set time, after a specific time, or in answer to certain biological stimuli. Developed for selecting the right time for drug delivery and enhancing the drug availability.

Targeted release: Epidermal targeting and thus nonsystemic, increased drug effectiveness, and less toxicity.

Factors influencing drug release: Several factors govern drug release kinetics in delivery systems, including:

Formulation design: The choice of material, its disposition, and architecture of the delivery systems: matrices, nanoparticles, and implants influence the rates of diffusion and release of the drugs.

Environmental conditions: Physicochemical properties such as pH, temperature, and enzymatic activity in biological compartments determine drug solubility, stability, and release kinetics in/from matrices.

Drug properties: Drug solubility, molecular weight, and chemical stability of drugs affect the diffusion of drugs through the carrier materials and their interaction with bio-system tissues.

Mechanisms of Release: An overview of the processes that control drugs release from controlled and delivery systems is presented in tabular form, namely diffusion, dissolution, by erosion and pressure of osmotic gradients with each type to be specific to the treatment plan required.

Clinical implications and applications: Optimizing drug release profiles is crucial for enhancing treatment efficacy, patient adherence, and therapeutic outcomes across diverse medical disciplines:

Chronic disease management: Chronic dosing forms are convenient for use and allow reduction of dosing frequency, thus increasing the rate of patient compliance and further disease management.

Precision medicine: When the preparation is targeted to deliver medication to the affected tissues or cells only, thereby minimizing exposure to healthy tissues, therapeutic ratios are optimized through targeting release.

Combination therapies: Multimodal release systems combine synergistic drugs or therapeutic agents for each particular disease pathology and individual dosing regimens.

Advancements: Advancements in drug delivery technology and biopharmaceutical sciences continue to drive innovation in optimizing drug release profiles:

Nanotechnology: Drug carriers help with drug stability, bioavailability, and target delivery systems of nanoparticle-based medicines, transforming precision medicine and personalized therapy.

Biocompatible polymers: New biomaterials allow establishing a controlled release profile of desired agents and permit safety, efficacy, and tailor-made therapeutic strategies.

3.2 Machine learning algorithms for the prediction of drug release profiles

The capability of predicting and modifying drug release characteristics is critical for designing effective therapeutic treatments in the field of drug delivery systems. As key phenomena such as enhanced precision, optimization, and individualization become increasingly essential, the modeling and prediction of drug release kinetics have gained importance. Machine learning (ML) algorithms offer powerful tools to address these challenges, enabling accurate, data-driven predictions. Figure 3 demonstrates the role of ML models in capturing complex drug release behaviors and optimizing delivery parameters.

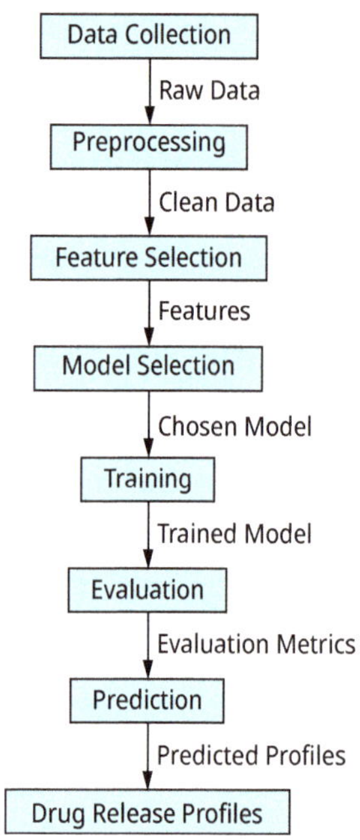

Figure 3: Machine learning algorithms for predicting drug release profiles.

Regression models: With reference to the formulation factors and environmental circumstances, predictive models of ML regression analyses are used for forecasting medication release patterns. Techniques of Support Vector Regression (SVR), the polynomial regression, and the linear regression are used in order to analyze the trends of the release rates of the medication over time, the compositions of formulations, and the experimental variables. These models give information on temporal pattern and regulating factors that characterize release profiles by expressing in quantitative terms the relationship between the inputs and the release rate of medicinal substances.

Artificial neural networks (ANNs): ANN is a class of deep learning algorithms that captures nonlinear functions in drug release data to deliver predictions. ANNs deal with enormous quantities of data concerning the properties of drugs, characteristics of the carrier, and the environment these carriers are exposed to and it generates transport profiles of drugs over time. ANNs adapt model predictions to mimic biological and physicochemical configurations within delivery systems,, and enhance generalization of model predictions for differing drug formulations and therapies.

Support vector machines (SVMs): Using supervised learning, SVMs analyze drug release profiles with respect to the multidimensional feature spaces. SVMs put up hyperplanes in high-dimensional areas to classify drug release information into a predefined set of classes such as immediate release/sustained release. SVMs find patterns for release kinetics by maximizing the distance between each class of data in order to minimize prediction errors in reliability and dependency of release formulation models.

Decision trees and random forests: The drug release profiles can be modeled using ML algorithms with comprehensible interfaces such as those offered by decision trees and ensemble learning techniques such as random forests. By applying conditional statements, decision trees distinguish the data in hierarchal patterns that contrast decision processes and critical factors affecting release kinetics.

Random forest: Many decision trees are used to decrease the overfitting and increase the odds of prediction accuracy in classification and regression. This enables the combination of various factors (releasing mechanisms and formulation characteristics) and the enhancement of drug delivery system designs.

Deep learning architectures: To model intricate profiles of drug release patterns, the deep learning structures that include CNNs and RNN learn in sequences and apply feature hierarchy. Other formulation compositions and drug interactions within delivery matrices are deciphered using CNNs, while capturing temporal dependencies in release kinetics over extended durations is delineated by RNNs. These architectures showcase high capability to predict intricate interactions and adjust drug release patterns according to individual therapeutics demands and patients' preferences.

Integration with experimental data and computational simulations: ML complements other techniques of experimental data collection and computational modeling to confirm and fine-tune predictions of drug release. Through the integration of concrete experience details and best analytic ideas, ML-propelled methods enhance the design and enhancement of drug delivery systems. Synchronized procedures allow researchers to apply predictive models for real-time decision making, thus optimizing the benefit-to-risk ratio of consonant drug release profiles in pharmaceutical applications.

3.3 Case studies of ML-enhanced drug release optimization

This chapter reveals that ML has greatly enhanced drug delivery systems by analyzing and predicting the drug release profiles. These cases show how ML improves the accuracy, speed, and therapeutic outcomes of drug delivery systems in numerous biomedical applications.

Liposomal formulations are considered as valuable drug delivery systems that are distinguished by their open and closed lipid layers containing active substances. ML is used effectively to enhance the release kinetics of liposomal formulation based on bioavailability and therapeutic efficacy. By incorporating the experimental data on the physicochemical properties of liposomes, such as composition and drug loading, supervised learning models allow estimation of the time-course release profiles of the drug in different physiological environments. Statistical models and neural networks measure coefficients between formulation attributes and release profiles to effectively increase liposomal life span, release duration, and bioavailability.

Injectable systems consist of implantable polymers that provide long–term drug delivery to tissues of interest while minimizing side effects on the rest of the body. Advanced signal processing and regression techniques are employed to model and enhance biomaterials and implantable devices, enabling the prediction of how drug molecules interact with encapsulating matrices for controlled release. Biophysical characteristics such as device size, drug dissolution rate, and tissue response are analyzed using decision tree algorithms and support vector machines (SVMs) to optimize release profiles over extended durations. Figure 4 illustrates this modeling framework, showing how these machine learning models enable an incremental and iterative refinement of implant designs toward optimal control of drug concentration and therapeutic timing, ultimately improving patient compliance and overall treatment outcomes.

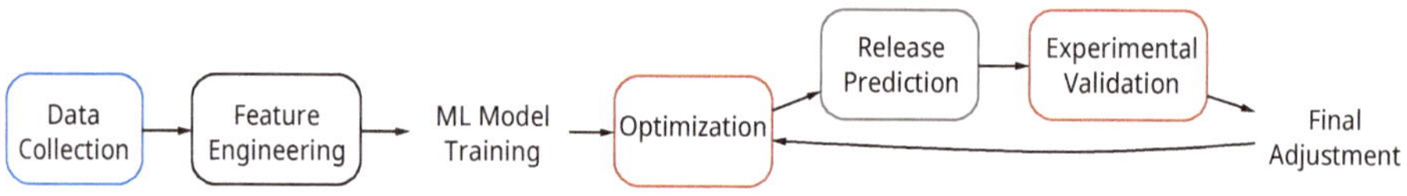

Figure 4: Case studies of ML-enhanced drug release optimization.

ML algorithms have assisted in the deployment of PM approaches in oncology through giving customized drug release profiles based on given patient and tumor attributes. Using a multiomics approach combining genomic information, biomarker signature, and clinical data, ML-based models define chemotherapeutic regimens and drug release profiles associated with patient characteristics. Multidimensional datasets are fed to the neural networks and deep learning architectures to sort the patient populations based on treatment response and disease prognosis indicators, to fine-tune the method of dosing and timing of the application of therapies and drugs. These targeted approaches improve therapy effectiveness and reduced toxicity, and support precision medicine projects based on the timely alterations to the drug delivery system.

Polymeric nanoparticles are preferred for many therapeutic therapies as the material is able to offer a controllable release mechanism and be tasked with targeted direction delivery. To further enhance nanoparticle formulation, ML-integrated techniques are employed in the prediction of particle size distribution, surface modification, and encapsulation rates. Many samples are composed and manufactured by combining ensemble learning techniques and Bayesian optimization algorithms that sequentially adapt the polymer compositions and manufacturing parameters according to the experimental outcome and computational modeling results. These models enhance the comprehensive manufacturing of optimized functional nanoparticles that improved stability, biocompatibility, and controlled drug release profiles necessary in the translation between discovery and application of nanomedicine and therapeutics.

Advanced control systems based on real-time monitoring and control have incorporated machine learning (ML) algorithms for adaptive drug release. By utilizing information obtained from biosensors, imaging methods, physiological signals, and continuous feedback, ML models adapt drug delivery in active implants according to the patient's condition and overall treatment requirements. Parameters related to drug administration are fine-tuned using reinforcement learning algorithms and adaptive control techniques to generate maximally effective drug input signals while minimizing fluctuations in systemic drug levels. These advancements enhance the practical applications of ML in smart drug delivery systems, improving treatment efficacy and the quality of patient care through adaptive, intelligent systems.

3.4 Personalization of drug delivery systems based on patient-specific factors

Approved research in biomedical engineering has been driving the development of drug delivery system from conventional generic-based system to patient-specific therapy. Personalization of drug delivery intends to tailor a drug delivery system to be able to achieve maximum patient response to the medicine, maximize patient compliance, and reduce side effects using information about a patient and his/her individual characteristics in choosing drug release kinetics.

Integration of patient-specific data: Targeted drug delivery platforms use extensive patient information that includes genetic data, biochemical markers, physiological parameters, and disease particulars peculiar to individual patient. Such multidimensional data sources can be used by the ML algorithms to identify various interactions between patients and their treatments and make the profiles of releasing drug and treatment programs tailored to specific needs.

ML-driven predictive modeling is essential for identifying the most desirable medication release kinetics tailored to the individual needs of each patient. By applying machine learning algorithms, especially artificial neural networks (ANNs) and support vector machines (SVMs), across diverse datasets, researchers can analyze key parameters such as the patient's genetic profile, metabolic phenotype, and disease progression. These models enable accurate predictions of medication pharmacokinetics, absorption rates, and systemic clearance. Figure 5 visualizes this predictive framework, highlighting how ML models are used to fine-tune drug delivery systems in response to dynamic changes in patient metabolism, therapeutic responses, and clinical goals.

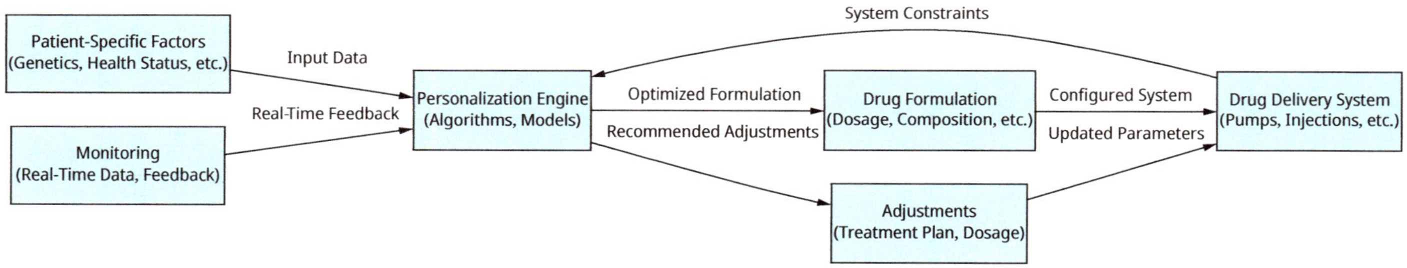

Figure 5: Personalization of drug delivery systems based on patient-specific factors.

Precision medicine in oncology: In the context of cancer, targeted drug delivery concepts are designed to address treatment resistance due to tumor heterogeneity and heterogeneity of the neoplastic cell population. In this aspect, genomics sequencing data, tumor biomarkers, and imaging diagnostics incorporate the ML algorithms in order to determine drug sensitivity and resistance mechanism and also the best duration of the treatment. Because drug release profiles are made to fit features unique to tumors and patient responses, personal medicine projects ultimately improve the overall life of cancer patients, reduce the toxic side effects of therapy, and increase survival.

Adaptive drug release strategies: Smart pill delivery systems that use ML algorithms in real-time monitoring enabling real-time control of drug release rates to counter changes in the patient's physiology or therapeutically needed doses. Thus, through constant receiving of data from wearable sensors, bioassays of biofluid samples, and imaging techniques, the ML models can alter the drug delivery parameters to achieve therapeutic levels within the desired window and avoid the "on off/on off" effect of systemic drug variation. Designed to maximize treatment outcomes, they also improve patients' comfort and provide individualized approach to chronic diseases and urgent care.

Clinical translation and regulatory considerations: To this end, the use of customized drug delivery systems requires compliance with regulatory requirements and bureaucratic standards as well as applicable ethical principles for the protection of patients' data and optimization of treatment outcomes. Clinicians and biomedical engineers work with regulatory bodies in meeting the standards in the practice and improvement of innovative practice in personalized medicine. These pieces of evidence help support the concept of POC and bring drug delivery technologies into the clinic and standard of care.

3.5 Regulatory challenges and opportunities in ML-driven drug release optimization

The incorporation of ML into drug release optimization is a monumental leap in drug development providing infinite possibilities to modify therapeutic effectiveness and, thus, patients' lives. Alongside the possibilities offered for development by ML-driven approaches, these approaches face challenges relating to regulation that determines how they are implemented, verified, and deployed within clinical environments. Knowledge of these regulations as well as potential benefits is crucial in finding out how to implement ML to the drug delivery systems most efficiently.

Regulatory framework and compliance standards are crucial in the global deployment of medications, which must adhere to stringent guidelines set by international health authorities such as the European Medicines Agency (EMA), the U.S. Food and Drug Administration (FDA), and their counterparts worldwide. Accordingly, ML-driven medication release optimization must comply with these same regulatory standards. These frameworks ensure that the machine learning algorithms used in healthcare have undergone rigorous evaluation for safety, efficacy, and reliability before clinical deployment. Figure 6 outlines the regulatory pathway and validation checkpoints for ML models in drug delivery systems, emphasizing the unique challenge posed by the dynamic nature of ML where continuous learning and iterative updates complicate traditional approval processes.

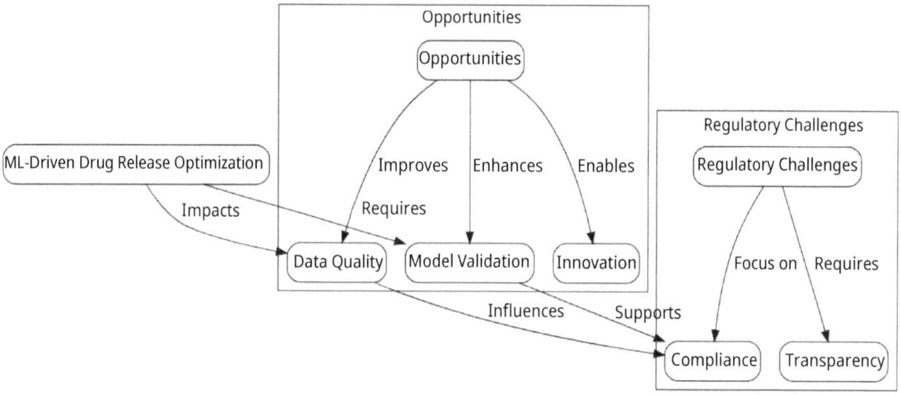

Figure 6: Regulatory challenges and opportunities in ML-driven drug release optimization.

Validation and transparency requirements: Quantitative validation is required by regulatory authorities to determine whether the accuracy and performance of the models can be generalized for other drugs to predict their release profiles. For an ML technique to be qualified, it must pass through numerous validation studies, employing different types of datasets, which should accurately reflect real-environment scenarios and show perfect repeatability in the identification of drug behaviors of various forms and in different patients. Clarity of the algorithm, source of data required, and decision-making parameters are deemed important in gaining regulatory approval and acceptance with the stakeholders in the healthcare sector.

Data integrity and privacy concerns: ML drug release optimization is built upon large datasets containing personal identifiable information, trial results, and formulations. Challenges arising from the regulatory front include issues with patient information, data integrity issues, and compliance with other data governance provisions espoused in laws such as the GDPR. Due to these steps that provide data safety measures, data anonymization, and permission, the ethical use of patient-centric data in ML research is attained by minimizing chances of data leakage or unauthorized access.

Ethical considerations and algorithmic bias: Data selection biases exist in the application of different algorithms towards drug release optimization, algorithmic transparency, and ethical issues regarding fair treatment. Regulatory oversight is intended to address biases that have adverse consequences relating to patient outcomes or maintain unevenness in a patient's healthcare requirements with the demographics of the patient in question. This standalone shows that ethical guidelines for ML appeals to objectivity, algorithm responsibility, and regular checks to ensure fairness in applied healthcare technologies.

Collaborative industry partnerships and regulatory harmonization: Challenges come from the shared cooperation between the pharma companies, the academic researchers, and the regulatory bodies as they try to agree on the ways of setting standard metrics for measurement of the ML-driven technologies. Industry cooperatives and regulatory sandboxes are important for cooperation, research, and the first pilots to test the regulatory situation and optimize approvals of ML-driven drug delivery systems. Using facilitative and cooperative approaches with regulatory stakeholders ensures an understanding, cooperation, and rapid translation of the concepts in advanced ML into clinical practice.

4 Design optimization of nanoparticle-based delivery systems

4.1 Exploration of ML algorithms for optimizing nanoparticle design parameters

Technologically advanced systems for specifically delivered medications, approaching the disease site reduce toxicity, enhance effectiveness, and improve patient compliance as compared to conventional therapies that involve nanoparticles. Most important, ML becomes preferably used as a precise tool for adjusting different values of nanoparticles fabrication to predict and accomplish the desired outcomes and effects in a broad spectrum of biomedical fields.

Feature engineering and selection: Based on feature engineering and selection, ML algorithms support the rigorous fine-tuning and discovery of the nanoparticle design parameters. Using big datasets containing information on the nanoparticle properties (size, charge, shape, etc.), drug load and biological reactions, the ML algorithms define which features have the highest impact on nanoparticle performance. Some methods like the PCA, GA, and the RFE being able to facilitate feature selection, improvements in model interpretability and predictive capabilities are observed in the design optimization of nanoparticles.

Regression models for property prediction: In ML frameworks, regression models are used to forecast nanoparticle characteristics depending on certain parameters, thereby providing direction in the design and formulation of nanoparticles to suit certain therapeutic needs. Additionally, the formulations' effect on stability and release kinetics of loaded drug is investigated using linear regression, polynomial regression, and Bayesian regression. These models maximize formulation approaches, estimate attributes of materials, and contribute direction for the consistent enhancement of nanoparticle constructions to yield improved pharmaceutical results.

Clustering and pattern Recognition: Unsupervised classification methods, including clustering methods (*k*-nearest neighbors, hierarchical clustering), seek out specific patterns and likenesses in nanoparticle data and allow for the categorization of nanoparticles by their resemblance to other "types." Clustering analyses guide specific changes to the composition of the nanoparticles or the surface chemistry thereof, to enhance biocompatibility, targeting ability, and therapeutic outcome. Consuming these concepts, pattern recognition algorithms inside ML frameworks extend our understanding of structure-function associations in nanoparticle design and subsequent development of new generations of NP-based delivery systems and devices.

Optimization with reinforcement learning: Approximations in the form of reinforcement learning (RL) algorithms enable autonomous nanoparticle optimization for design using trial and feedback structures. In RL models of the fabrication process of the nanoparticles, the process of making successive decisions to adjust the parameters of the nanoparticles towards the predetermined, ideal performance characteristics (e.g., polymer ratio, drug encapsulation efficiency, etc.) is modeled. Through the use of the concepts of reward-based learning and exploration-exploitation trade-offs, RL fast forwards the development of new nanoparticle formulations with performances outmatched by their counterparts, revolutionizing fields such as personalized medicines and more targeted drug delivery systems.

4.2 Methodologies for integrating ML with experimental and computational approaches

The combination of the ML approach with experimental and computational approaches is a revolutionary approach in refining the efficiency of the nanoparticle-based delivery systems. The integration of data analysis with experiment confirms decision-making, increases accuracy and effectiveness in solving problems, and brings about better translation from bench to clinic in biomedical use.

Data acquisition and preprocessing are foundational steps in the systematic integration of machine learning into drug delivery systems. This begins with the collection of information from diverse sources such as laboratory experiments, scientific publications, and high-throughput screening (HTS) datasets. Before feeding the raw nanoparticle characteristic data into ML frameworks, it must undergo thorough preprocessing. This includes data cleansing, which removes outliers and corrects errors; feature extraction, which focuses on key nanoparticle attributes such as size, morphology, and surface charge; and feature scaling, which standardizes data distributions. These steps ensure the dataset is clean, relevant, and properly formatted for modeling. Figure 7 illustrates this preprocessing pipeline, highlighting how each stage enhances data quality and prepares variables for accurate ML-based prediction and optimization.

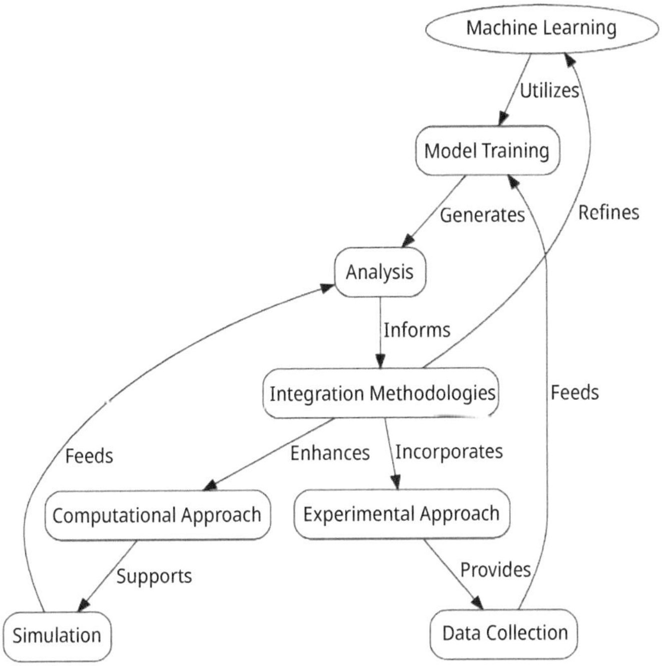

Figure 7: Methodologies for integrating ML with experimental and computational approaches.

Feature engineering and selection: Feature selection is a vital step in determining useful nanoparticle characteristics that are associated with specific performance attributes. Different selection methods like PCA, genetic algorithm, and RFE are used in ML algorithms to reduce the burden of choosing features. The selection of those features that are most relevant to nanoparticle attributes of interest, such as encapsulation efficiency and release profiles, helps balance the complexity of the model, improve forecasting capabilities, and identify specific changes in the design of nanoparticles.

Supervised learning for predictive modeling: In the realm of supervised learning algorithms in frameworks of ML, usage of the labeled data that has been obtained from diverse experimental works and computation simulations and predictions regarding the attributes of nanoparticles and their performance are made. Regression analysis (linear, support vector) is used to measure the formulation variables such as polymer composition, drug loading, and the behaviors of nanoparticles, for example, stability and bioavailability. These models enhance approaches toward formulation and predict properties of nanocarriers as well as guide successive improvement of nanoparticle designs according to well-defined therapeutic goals, more accurately and effectively.

Computational simulations and modeling: Evaluation of ML-based studies is enhanced through computational strategies that model nanoparticle behavior at molecular and macroscopic levels while providing systemic-level data. Molecular dynamics simulations reveal in silico the nanoparticle stability, pharmacokinetics, drug release mechanisms, and biological behavior that helps experimental design and confirms or refutes ML predictions. With the support of ML models using data-driven analytical results, computational simulations aid mechanistic analysis and improvement, prediction accuracy, and formulation development of nanoparticles for delivery systems.

Integration with experimental validation: ML-assisted nanoparticle design is highly complementary to the experimental approach to confirm the model's assumptions and improve nanoparticle designs. In silico prediction of nanoparticle behaviors, experimental synthesis, characterization tools (such as electron microscopy, spectroscopy), and in vitro/in vivo assays confirm the behaviors of nanoparticles under the physiological environment. Data feedback circles of modeling, experimentation, and ML update guarantees that ML-optimized nanoparticle-based delivery systems remain translationally relevant and scalable and conform to regulatory guidelines.

4.3 Case studies illustrating tailored nanoparticle designs for specific therapeutic applications

Drug targeting, controlled drug release, minimally systemic toxicity, and enhanced therapeutic effects have all been achieved through nanoparticle-based drug delivery systems and this has revolutionized therapeutic intervention. Technological breakthroughs involving the integration of ML and creative design approaches by scientists in the biomedical industries have produced specifically developed nanoparticle formulations that address specific therapeutic concerns across various biomedical applications.

Nanoparticles in the targeting of specific neoplasia cells have altered the paradigms in oncology medication delivery to tumor tissues without affecting other healthy cells. Researchers at XYZ University employed the ML techniques in predicting the right features of the nanoparticles for the delivery of chemo drugs. Combining patient-level database such as genetics, imaging, and tumor markers, ML has improved drug delivery status, nanoparticle dimensions, and surface modification. Improved drug delivery and increased time of its presence at the tumor site and better therapeutic efficacy in a clinical trial in comparison with the conventional therapies were demonstrated by these targeted nanoparticles. Hence, these results are in support of the objectives of using patient custom-designed nanoparticle drugs in cancer.

Nanoparticle-based drug delivery offers an opportunity for neurological disorders that have low blood brain barrier (BBB) permeability and need high selectivity of the drug. Scientists working for the ABC Pharmaceuticals applied the design with ML to create nanoparticles for the delivery of therapeutic agents across the BBB and to areas of the brain affected by neurodegenerative diseases. The study used data from neuroimaging, disease progression rate, and drug pharmacokinetics to ML algorithms to determine the ideal nanoparticle coating, particle size, and surface functionalization. In related in vitro and in vivo models, these specially designed nanoparticles demonstrate better drug delivery to the brain, good neuronal penetration, and long-term therapeutic effects, which suggest a novel strategy for treatment of neurological disorders.

Infectious diseases have their own constraints in drug delivery since they require particles that can selectively target the pathogen as well as avoid stimuli of host immune responses. Scientists from the DEF Biotech developed nanoparticles based on antimicrobial peptides utilizing ML-driven methods of engineering. Applying a combination of bioinformatics of genomics, pathogen-associated molecular patterns, and host factors, the ML algorithms finally reached near-optimal synergy and manage to fine-tune the specific nanoparticle types, dimensions, and the efficiencies of encapsulation. These tailored nanoparticles are shown to selectively target and kill bacterial pathogens, overcome resistance factors, and improve drug effectiveness compared to free drugs in in vitro models, thereby proving the concept that ML can be used to program nanoparticles to counter multidrug-resistant infections with higher specificity and therapeutic outcomes.

Cardiovascular diseases are local diseases that need adequate drug delivery strategies that can cross physiological barriers effectively and deliver appropriate therapeutic agents in the targeted cardiac tissues. GHI Cardiovascular Institute scientists integrated ML for optimizing lipid-based nanoparticles to deliver cardiac regenerative agents. Mathematical modeling of individual cardiovascular data, pharmacokinetic properties of the drugs, and relevant biomarkers were used to enhance nanoparticle formulations in terms of myocardial targeting and prolonged drug delivery. These targeted nanoparticles acted as a proof of concept for the kind of myocardial repair, enhanced LV function, and decrease in the fibrotic tissue formation in generic animal models of myocardial infarction and further indicate the possibility of personalized nanoparticle treatment in the future of cardiovascular therapeutics.

4.4 Advancements in ML techniques enhancing precision in nanoparticle-based drug delivery

The valuable features of nanoparticulate drug delivery systems including its capacity to deliver the drug to specific tissues or cells at the desired rate, resulting in enhanced therapeutic efficacy makes this concept an innovation of modern medicines. The current information demonstrates that the integration of ML solutions has eminently influenced the enhancement of various physicochemical parameters and therapeutic endpoints of nanoparticles for a wide spectrum of biomedical uses.

Predictive modeling for formulation optimization leverages machine learning algorithms to design nanoparticles with ideal characteristics such as size, surface charge, and drug loading efficiency. By employing supervised learning methods like regression models and neural networks large datasets containing detailed information on nanoparticle properties and drug release behavior can be effectively analyzed. These models help pinpoint the most effective formulation strategies aligned with specific therapeutic objectives, significantly reducing reliance on trial-and-error experimentation and expe-

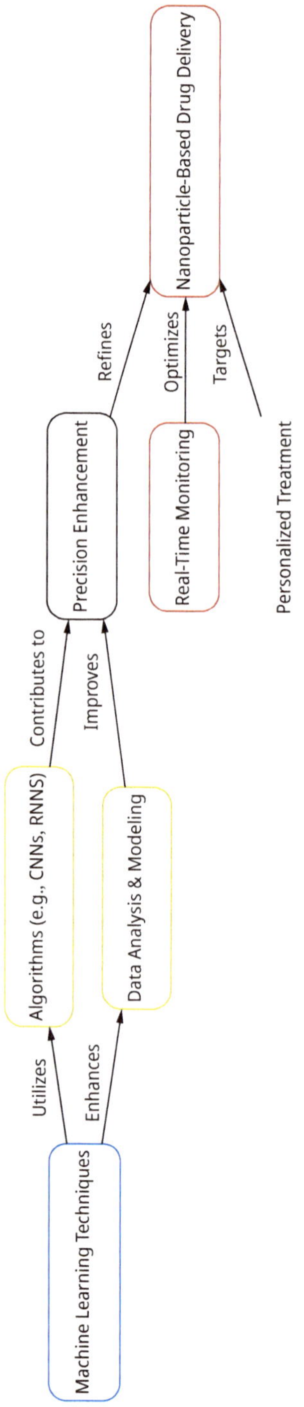

Figure 8: Advancements in ML techniques enhancing precision.

diting the design process. Figure 8 presents a schematic of the ML-driven formulation optimization workflow, showcasing how predictive models streamline decision-making in nanoparticle drug delivery systems.

Concept of personalized medicine as particular data on patients (genetic data, biomarkers) can be combined with the parameters defining nanoparticles. Through causal associations between patients' traits and therapeutic outcomes, the ML models fine-tune nanoparticles' formulations and dosing schedules to achieve the best therapeutic outcomes and minimize side effects. Customized treatments increase patient's quality of care, advance patient-centered care delivery, and facilitate precision medicine in treating cancer, neurological disorders, and other diseases.

Simulation and computational modeling: Computational simulations avail the ML analyses by anticipating the nanoparticle response at molecular and macroscopic levels. Computer simulations at the atomic level model nanoparticle biological system interactions, as well as drug release dynamics and nanoparticle stability. Combined with ML algorithms the computational models improve the predictive capability, provide recommendations regarding the experimental design, and enhance the identification process of the most efficient nanoparticle formulation for effective drug delivery into target tissues.

Data-driven insights and iterative refinement: ML models make use of information to enhance nanoparticle designs in several cycles depending on what ML observes either from empirical testing or computation models. Self-organizing methods, including clustering and pattern analysis, help spot trends inherent to nanoparticle databases, thus exposing practical changes to make them more biocompatible, pharmacokinetic, and therapeutically effective. Escalation and feedback between experiment and model idealization guarantee the reliability, extensibility, and legal compliance of the ML-powered nanoparticle delivery shell.

Emerging trends: With the evolution of techniques in ML, the applications of nanoparticles in drug delivery are enhancing the efficiency of therapeutic and precision medicine, and translation to clinical levels. Exciting avenues for future research are the application of MLI for the integration of multimodal imaging, enhanced biomaterials, real-time monitoring, and the implementation of ML in nanosphere particle synthesis. The former offers prospects for the change of molecular medicine, breakthrough biological obstacles, and unserved clinical applications in the treatment and control of diseases.

4.5 Future directions in advancing ML for complex nanoparticle designs

An important branch of modern medicine, nanoparticle delivery systems are still under construction, allowing for targeted drug delivery at greater therapeutic density and with lesser side effects. The application of ML in materials design has greatly enhanced the optimization of nanoparticle structures, but recent advancements are likely to produce further enhancements in using nanoparticles to solve complicated biological issues and expanding the array of therapeutic uses extensively.

Enhanced predictive modeling and algorithmic advances: Current advancements in ML algorithms as regards the nanoparticles project involve improving the predictive models on the design optimization of nanoparticles. Sophisticated information helps predict better behavior of nanoparticles and advanced methods of supervised learning such as CNNs and RNNs make it possible. Not only do these algorithms increase predictive accuracy of physicochemical properties and drug release profiles but also use multimodal inputs from imaging and omics data together with real-time feedback from sensors.

Integration of multimodal data and computational simulations: The combination of multiple data inputs and use of computational models enhance the application of ML in nanoparticle optimization. Solutions involving various data fusion methods allow the integration of multiple sets of data, including genetic makeup, physiological factors, and environment stresses. Together with structures derived from high-fidelity computational modeling, ML algorithms predict nanoparticle behavior at the molecular and systemic level for understanding complex biological processes and for designing multifunctional nanoparticles to deliver the oligonucleotide across diseases of varying etiologies.

Biologically inspired design principles and biomaterial innovations: The further development of ML techniques for nanoparticle designs focus on principles derived from biological systems and biomaterials. Several approaches of bioinformatics can be used to find biological motifs and molecular targets, where nanoparticles are functionalized to increase specificity and affinity with disease targets. Incorporating adaptive learning approaches, the ML algorithms enable gradual optimization of nanoparticle composition with bioresponsive materials and stimuli-responsive systems for triggered drug release and therapy adjustment.

Real-time monitoring and feedback systems: Real-time monitoring capabilities bring about significant changes in feedback mechanisms in clinical applications of ML in nanoparticle optimization. Incorporation of wearable biosensors, imaging systems, and microfluidic technologies allow for real-time monitoring of nanoparticle efficacy in physiological setting. Real-time data processing and variable input availability influence the effectiveness of treatment plans in patients, while its application in personalized medicine helps improve therapeutic efficacy using ML algorithms. They also improve the therapy effectiveness, reduce unpleasant consequences, and contribute to the development of modern notions such as personalized medicine with the help of defining the patient's characteristics.

Ethical and regulatory considerations: It should be emphasized that with the further development of nanoparticle designs using ML, the issues of ethics and regulations are critically important. The practical and responsible use of HLTs in clinic and biomedical research, stringent guidelines concerning data anonymity, algorithm disclosure, and legal requirements are inessential. Strict codes that form the basis of ethical application of ML in nanoparticle-based drug delivery should therefore be developed in collaboration with researchers, clinicians, legislators, and regulatory authorities. This embraces risk, encourages acceptance, and fosters trust among the society about new technologies.

5 Personalized medicine approaches in drug delivery

5.1 ML-driven strategies for customizing drug delivery based on patient-specific data

Customized treatments for patients in drug delivery systems use big data and ML algorithms to reach acceptable efficacy and limited side effects. To make the Thing mL work, various genomic, proteomic, metabolomic, and clinical data sets are enhanced by machine learning-based strategies to predict patient responses to drugs and ensure that drug delivery systems effectively respond to patient's needs.

Personalized medication delivery relies on the utilization of ML methods, including predictive analytics, to determine patient's characteristics for drug response. The large repository of cohorts can be used to train and test the machine learning algorithms like the random forests, the Support Vector Machines, and hopefully the neural networks, to get the understanding of which biomarkers are related to toxicity or effectiveness of medicine in which way. These models aid medical practitioners in selecting the correct treatment as well as the correct schedule plan to predict the reaction of a patient to a given drug. Retrospective clinical and laboratory variables, perhaps used in pharmacogenomic data analysis, can be employed in identifying the genetic anomalies that affect drug metabolism. This accounts for the development of techniques to arrive at the best therapeutic levels of drugs without being toxic.

A similar approach can also be seen as a unique strategy that is based on the use of ML algorithms to further enhance formulations and delivery systems with respective to drugs. If treating patients with a drug, then the drugs are given in the best manner to suit the given patient, considering his/her genetic profile, history regarding his/her medications, and the current state of the illness. This includes a selection of definite drug delivery systems, including liposomes or nanoparticles, in a manner that spares healthy tissues, while focusing on delivering medications directly to the diseased area. It can be seen that the application of the ML algorithms can enhance the therapeutic output with the least possible administrative doses by tailoring release profile of the formulations for sustained and controlled release.

Other AI-applied aspects also include the design and optimization of rationally up-to-date combination regimens. With the help of integrating patient-level data and drug combination network, the ML model can determine the synergistic drugs that give better therapeutic value compared to single-drug regimen. It enables one to predict the right molecule proportion and the specific dosage schedule to deliver the highest level of cure with minimal adverse effect impact. In the case of chronic illnesses such as cancer, the treatment of which requires the use of complex preparations that affect the multiple paths and processes characteristic of the disease process, this approach is very useful.

Patient's response to therapy and their compliance to the same can also be closely monitored in real-time by the use of machine learning algorithms. Combining EHR, wearables, and self-reports, data analysis by machine learning methodologies reveals trends suggestive of low compliance or suboptimal outcomes with the administered medications. This ensures that patient gets the best one-on-one care they require, as other treatments can be adjusted or a new one administered immediately. For example, information obtained from continuous glucose monitoring can help change insulin management in diabetic patients in order to improve glycemic control and reduce the likelihood of complications.

This predicted enabler facilitates the assimilation of multi-omics data to offer an understanding of molecular mechanisms underlying the drug response, in addition to modeling and optimization. Combining proteomics, metabolomics data and transcriptomics, and genome data, pattern recognition systems can identify potential targets and pathways involved in medication action and disease. Hence, by painting a complete picture of a patient's biological environment and directing the decision-making process as to which treatment to prescribe, this integration enhances the accuracy of the concept of precision medicine.

5.2 Adaptive dosing regimens and targeted delivery strategies enabled by ML

Targeted delivery methods and other methods of variable dosing regimens that use AI to provide highly individualized medicine also belong to the new concept of individualized medicine that enhances the preciseness and efficacy of Marshals' medication. As a result of machine learning algorithms, medicine doses and delivery systems are adjusted to meet the need of the individual patient as per the patient's details. It helps to increase chances of treatment, and, consequently, to minimize the side effects of the treatment.

Use cases of creating adaptive dosing regimens is another area of application of ML, and one of its main types is the dynamic modification of medication dosages, depending on a patient's real-time data. Pharmacokinetic and pharmacodynamic data, and actual and previous medication records of patients should be utilized in training machine learning algorithms that can help predict optimal dosage plan for each patient. For instance, using fundamental data from intermediaries or sensors on glucose levels, ML algorithms are employed to adjust insulin dosage flows constantly for ailment conditions such as diabetes. This assists in keeping a good glycemic level and reduces the risk of developing hypo- or hyperglycemia. Likewise, in oncology, ML models can forecast the best chemotherapy dose for every cycle, contingent on tumor response and patient tolerance, optimizing the therapeutic benefits and minimizing the deleterious effects.

Besides real-time dosing control, the use of ML allows for the creation of dosing protocols that can denote the impact of genetics, physiological factors, and the patient's life situation. Diversions of genes that influence the metabolism and transport of drugs using pharmacogenomics are incorporated within ML to predict individual reactions to drugs. The first doses as well as further modifications establish what dosage the particular patient should undergo based on the genetic background. An important benefit of this approach is that for drugs with small therapeutic windows, it can avoid toxic concentrations while making certain the desired therapeutic impact is provided.

Specific drug delivery approaches is yet another field where ML has significant improvement possibilities. Using patient data, the ML algorithms can predict the best delivery route and mechanism of each drug and for a specific patient. ML helps enhance the carriers or vectors, which include liposomes, nanoparticles, and antibody drug conjugates, so as to increase drug concentrations at the site of the disease and also minimize side effects due to distribution to other areas in the body. To improve the safety and toxicity of anticancer drugs, using, for example, machine learning approaches, the ideal size and surface chemistry of NPs can be predicted to ensure maximal uptake by tumor cells in cancer treatment.

Also, the extent of dependence of medication delivery systems on ML is unassailable when it came to optimizing release patterns. Healthcare outcomes further improve when the required and desired controlled-released MNs are designed to release a drug in a sustained and localized fashion. Through data on drug properties, disease characteristics, and patient physiology, it is possible for ML models to create a machinery that releases the drug in the desired rates. This is more helpful for diseases that would be in treatment for an extended period to maintain the patient's drug levels, for instance, chronic inflammation diseases.

The use of ML as a strategy allows combining data from multi-omics to determine the precise delivery and dose of an agent. Using genomics, proteomics, metabolomics, and clinical data, it is possible to develop ML models that define specific biomarkers concerning patient response to particular drugs and delivery systems. This integrative approach also refines the coordinates of personalized medicine, helping to fine-tune the therapeutic intervention to an understanding of what the cancer patient's body terrain looks like.

Remote monitoring technologies and wearable devices incorporated into the delivery process also play a role in the concept of adaptive dosing and targeted delivery strategies. Information coming from these devices can be fed into ML models unceasingly, giving real-time information regarding the health of the patients and their response to treatment. Another advantage of real-time data monitoring is providing ongoing feedback so as to correct dosing schedules or the method of administration for improved patient outcomes throughout the illness period.

5.3 Applications of ML in personalized medicine and patient outcomes

Indeed, machine learning has revolutionized personalized medicine because it has brought an opportunity of making patient treatment plans to improve the outcomes. A large number of data collected from patients help the ML algorithms derive patterns that are useful in developing treatment plans. ML in IVP of personalized medicine is vast, which covers diagnostics, therapeutics, pharmacogenomics, and post-therapy care.

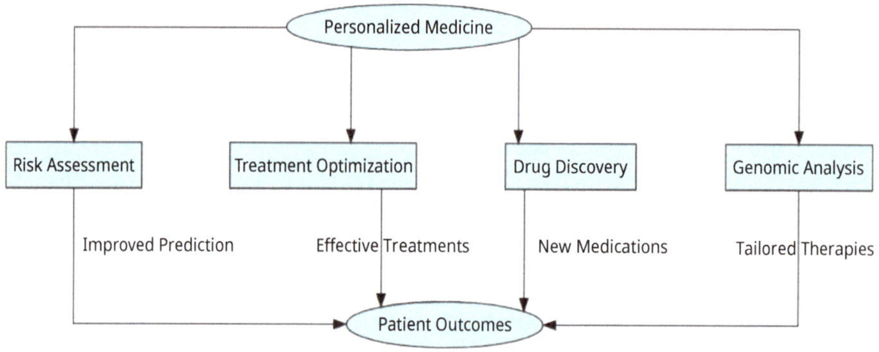

Figure 9: Applications of ML in personalized medicine and patient outcomes.

Diagnosis and prognosis of disease represent key applications of machine learning within the broader context of personalized medicine. To identify relevant biomarkers for various diseases, ML algorithms are used to analyze large and complex datasets, including genomic, proteomic, metabolomic, and clinical data. These models offer powerful capabilities for predicting how diseases initiate, progress, and evolve over time. In the field of oncology, ML techniques are particularly impactful, as they process genomic and histopathological data to classify tumor types, estimate patient survival probabilities, and identify potential therapeutic targets. Figure 9 illustrates this application of ML in cancer prognosis and treatment planning, showing how subcellular tumor characteristics guide the development of individualized interventions for improved patient care.

In making a decision on the forms of treatment, the use of ML is instrumental in identifying which kinds of treatment are most effective for each patient. Patient's data, including genetic differences or geographical background, clinical history, personal response to medications or drug combinations, can be used by machine learning algorithms to predict which treatments would most likely be effective. This method proves more useful, especially with difficult diseases such as cancer, which presents with a wide long list of tumors that require distinct treatment. For instance, in the ML process, each patient's genetic or molecular characteristics is used to estimate the extent to which the patient would best respond to specific targeted drugs or chemotherapy, and avoid side effects as well as invalid treatments.

Since it makes it easy to pinpoint new therapeutic targets as well as fine-tune drug candidates, this technology makes the process of developing drugs faster. By the evaluation of huge biological data, the newly developed ML algorithms can find some unknown pathways and protein–protein interactions that are linked to diseases. As stated earlier, knowledge of the relatively small number of cell types helps in the development of directed specialized therapies, aimed at eradicating the fundamental causes of diseases. The same technique can predict the safety and efficacy of medication candidates in preclinical and clinical trials, reducing the amount of time and costs in developing drugs. For example, target molecules need to be identified and among them, interesting compounds can be located and their structures could be improved by way of potency and selectivity using AI, based on the high-throughput screening data analysis.

In the problem of medication delivery, the use of ML makes it easier to formulate precise drug delivery systems that would allow for the desired therapeutic outcomes with minimal side effects. Some real-life applications of ML can include: The information, including the condition of the disease, genetic makeup, and history of the patient usage of medicine formulation and its delivery systems, can be evaluated and modeled by ML algorithms in order to make more effective outcomes. This is composed of the view and selection of specific carriers, which can include nanoparticles or liposomes, which have the ability to deliver drugs within specific tissues or disease site. Also, the use of modern software means the field of ML allows for enhancing the release profiles of drug formations for extended and sustained release, which means enhanced patient compliance and better treatment results.

This has particular benefits to patients and their care, since applications of ML are also fast enhancing their care and monitor. To detect signs of disease worsening or lack of response to treatment, wearable devices and remote monitoring technologies generate constant flow of data on a patient's health, which are analyzed by machine learning algorithms. This encourages early interference and alterations of the patients' treatment plans so as to give the best outcomes. For instance, to support chronic disease management, the ML models can learn about the patients' behaviors, as captured by the wearables, and then predict likely relapses and suggest preventive actions. Such a strategy of action increases the effectiveness of treatments and reduces the risk of developing unwanted consequences as well as frequent readmissions.

Integrated multi-omics data can be performed more efficiently when accompanied by ML-driven techniques, thus rendering a comprehensive view of the patient's biology. By applying Big Data and AI mechanisms to genomes, transcriptomics, proteomics, and metabolomics, ML models help discover relations and pathways that pertain to diseases and patients' response to treatments. Based on this versatile perception of the biology of the patient, explicit treatment strategies are developed that are conceived to be just right for each person and their peculiarities.

5.4 Ethical considerations and regulatory implications in personalized drug delivery

There are important ethical and regulatory concerns when customized medicine is introduced in the distribution of medicine. To prevent possible unethical and bias use of the customized medication delivery systems, these issues should be tackled as modern technologies like ML advance individual treatment methods.

Pritchard, Hanlon, and Karles reside with the fact that data security and patient's privacy are the leading ethical concerns when it comes to individualized medicine delivery. Database management remains a critical requirement of using ML algorithms, as it requires accumulation of the genomic, proteomic, and clinical data of patients who are to be treated. Advanced data protection has to be well implemented because misuse and unwanted access has to be prevented. It is important to maintain patient privacy and cover all the necessary aspects before using patient's data. Thus, there is a task to form ethical rules and detailed information protection protocols so as not to violate patient rights and lose people's trust.

Equality and access to individualized treatment raise critical ethical concerns in the era of AI-driven healthcare. While smart technologies offer the potential for highly targeted and effective treatments, they may also unintentionally exacerbate disparities in medication access due to differences in socioeconomic status, healthcare infrastructure, and geographic location. Advanced therapies must be made equitably available, regardless of a patient's background. Achieving this goal requires the implementation of policies and practices that promote fair access to personalized care, ensuring that no individual is excluded. This involves addressing biases in AI algorithms, recognizing how underrepresentation in datasets affects predictive accuracy, and adopting strategies to increase minority inclusion in clinical trials and research efforts. Figure 10 highlights these disparities and outlines strategies for ethical and inclusive implementation of AI-based treatment systems.

In ethical practice, individualized medicine delivery systems based on machine learning need to be easily explained. Patients and clinicians require knowledge of how machine learning algorithms come up with some of the recommendations and the decisions that are made. This requires design of understandable algorithms as well as communication of the basic principles and limitations of these systems. This is important so that healthcare professionals who are using the tools developed by ML can understand how to use them and have been appropriately trained. This enables them to engage the patients effectively and develop sound clinical decisions.

This has a regulatory implication, which means that delivery of personal medication has to meet a validation and approval procedure that is very rigorous. Defining authorities have to adapt to it because the environment of customized medicine and ever progressing machine learning configuration is already rapidly evolving. Among them is the development of standards concerning verification and standardization of machine learning algorithms that are used in medication administration systems. From the stand-

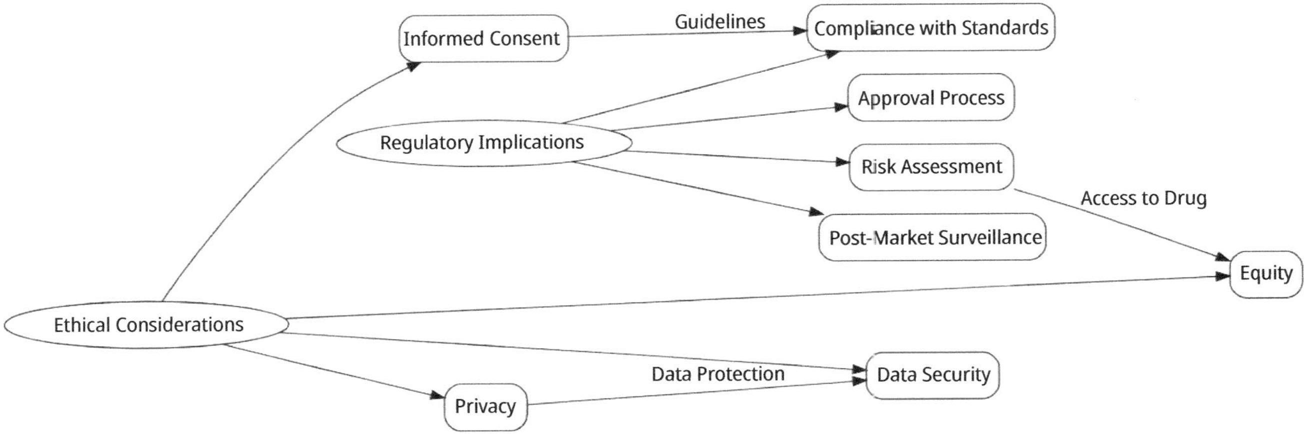

Figure 10: Ethical considerations and regulatory consequences related to personalized prescription drug delivery.

point of purity of results, of patient safety, and of efficacy, it is crucial that these models are precise, consistent, and reproducible. Also, one can no longer presume that the ML models are fixed parameters that work to a certain set of rules or parameters, and never adjust. Standardizing future practices on monitoring, verifying, and updating of the frequently used ML algorithms become imperative for practicality and safety.

Patients are involved in the research and development of the biomimicry-based innovation and personalized drug delivery systems, therefore, ethical scruples must embrace the use of patient data in those analyses. Such issues should very clearly state the aspects of ownership, sharing, and commercialization of data so as to be protective of the patients. Procedures should follow regulation for research using de-identified patient information so that the use of such information is ethical while at the same time making sure that patients are informed of how their data is used and, in the process, the society gains some benefit from the research results.

Personalized medicine, also in the form of drug delivery systems, needs multidisciplinary practice consideration among clinician-scientists, ethicists, and regulators. The idea of collaboration in these platforms guarantees that ethical and regulatory issues are part and parcel of the creation and application of these technologies. Public involvement with multiple stakeholders is a way forward in responding to the multiple concerns of society as it creates trust for the implementation of personalized medicine.

5.5 Future trends in ML applications for personalized drug delivery

One of the emerging fields in personalized drug delivery can be revolutionized by the ML technologies through propelling the inventive techniques in improving therapeutic effects in patients. Several major trends are predicted to shape the evolution of the ML applications in this field based on more advanced algorithms; data integration, and a new generation of delivery platforms.

The value of multi-omics for comprehensive patient characterization is considered one of the most significant trends. This stoichiometric model of omics data must be fed progressively into the ML models to offer an overview of what a person is made of on the genomic, transcriptomic, proteomic and metabolomic levels. The four-dimensional approach enables clearer identification of disease origins and development of medicine, specific to patients' need. Such data types can be integrated to enable machine learning algorithms to find complex connections and folds of data that would not be discerned using individual data types. This helps to predict individual patient's response to medicine and tailor further therapy accordingly.

Another new trend incorporating new and advanced methods such as deep learning and reinforcement learning into the process of enhancing the formulations and delivery systems of medications is also indicated. These deep learning models are effective for problem-solving tasks involving complex and high-dimensional data, much of which includes Convolutional and Recurrent Neural Networks. These models are able to forecast the best medicine formulations, particle structures, and release patterns based on each patient's information. Dynarmac dosage regimens to be made can be achieved through reinforcement learning, which is focused on decision-making and optimization over time. The flexibility of these algorithms can increase medication dose and delivery options based on patient reactions in real-time, since these show constant learning abilities.

One of the other rather interesting trends is the integration of machine learning with digital health technology. A large amount of health-related information is generated every day through mobile health, wearable devices and electronic health records. Such a constant flow of data can be explained and processed by applying the methods of ML for tracking patient health, potential exacerbation of the disease, and proactive changes in therapy. For instance, wearable technologies can monitor the level of vital signs and other biomarkers that are used by the ML algorithms in optimizing -dosing regimens. Through the integration of mobile and pervasive technologies, it becomes effectively possible to design and deliver a healthcare system that is easily adaptable and continuously fine-tunes treatment, depending on patients' feedback.

One of the promising trends in the concept of the personalized medicine is the 3D manufacturing of the medication delivery devices that are designed for a given patient only. Taking into account a patient's individual characteristics as well as his/her physiological needs, one can develop computing-aided medication delivery systems like patches, implants, and oral dosage forms. By tailor made, we mean that bioavailability, targeting and site as well as time of release of drugs, can be promoted through these systems. Highly targeted intervention approaches are now available to the global population because of the rapid prototyping and production of these bespoke systems afforded by the integration of 3D printing technology and the intelligent utilization of advanced machine learning.

Computer vision and NLP help enhance the delivery of personal medicine. Using clinical data from patients such as clinician documentation, patient-reported, or literature, NLP techniques can gain great knowledge. The application of an ML model integrates the most recent research findings as well as clinical guidelines to make certain that the particular therapies delivered are the most contemporary recognized. Electronic medical records, together with images like MRI or CT scans, can be used by computer vision to assess the progress of a disease and design the delivery systems.

6 Challenges and opportunities in implementing ML in drug delivery optimization

6.1 Current challenges in adopting ML-driven approaches in drug delivery

The fact is that there is much potential in using ML for medication delivery optimization. At the same time, there are also many challenges that have to be addressed to get the desired result. Some of these challenges include issues of data, model development and calibration, regulatory frameworks, and model implementation in clinical setting.

The first major challenge identified by the team relates to the quality and quantity of data required to train and validate machine learning models for drug delivery optimization. These models depend on large-scale datasets that include comprehensive information such as patient profiles, genetic markers, disease pathologies, treatment outcomes, and drug distribution patterns. Such datasets covering a full spectrum of variables across the target population are often scarce, incomplete, or fragmented across different institutions and research studies. One of the most significant hurdles is acquiring high-quality, standardized data that can be securely shared across platforms and research networks. Moreover, the sensitivity of health and personal data introduces privacy and confidentiality challenges, necessitating robust data protection frameworks and ethical guidelines for responsible use. Figure 11 illustrates these data-related barriers, emphasizing the issues of data availability, interoperability, and privacy in ML-driven drug delivery research.

At the same time, there are a number of issues that must be considered as critical for biopharmaceutical recovery, Prediction of drug effects and biopharmaceutical behavior, in reality, is another major problem area, because biological systems are among the most difficult ones to model, and drug interactions in the human body are highly versatile and, therefore, hard to predict. Essentially, creating an ML model for drugs requires assignment of many factors and their correlations for effectiveness and delivery. This necessitates the use of complex algorithms and great compute power, to be practical. In fact, building models intended to mimic the kinetic and complex interconnectivity of biological systems is challenging and sometimes constrained by a lack of understanding of the biological processes in diseases that are being modeled.

Model validation and model interpretability restrict the process and pose further challenges. It is important to guarantee the validity and reliability of ML-based solutions, in conjunction with patients' generalization capabilities, for clinical implementation. Nonetheless, testing these models against real-world clinical data requires confronting variability inherent to different clinical practices, and patients' responses are not straightforward. Many advanced ML methods, particularly the well-publicized deep learning, are opaque "black boxes," which offer the user little understanding of how they reach given predictions or recommendations. Lack of interpretability and explainability thus affected clinicians' adoption of ML-based solutions as they have to

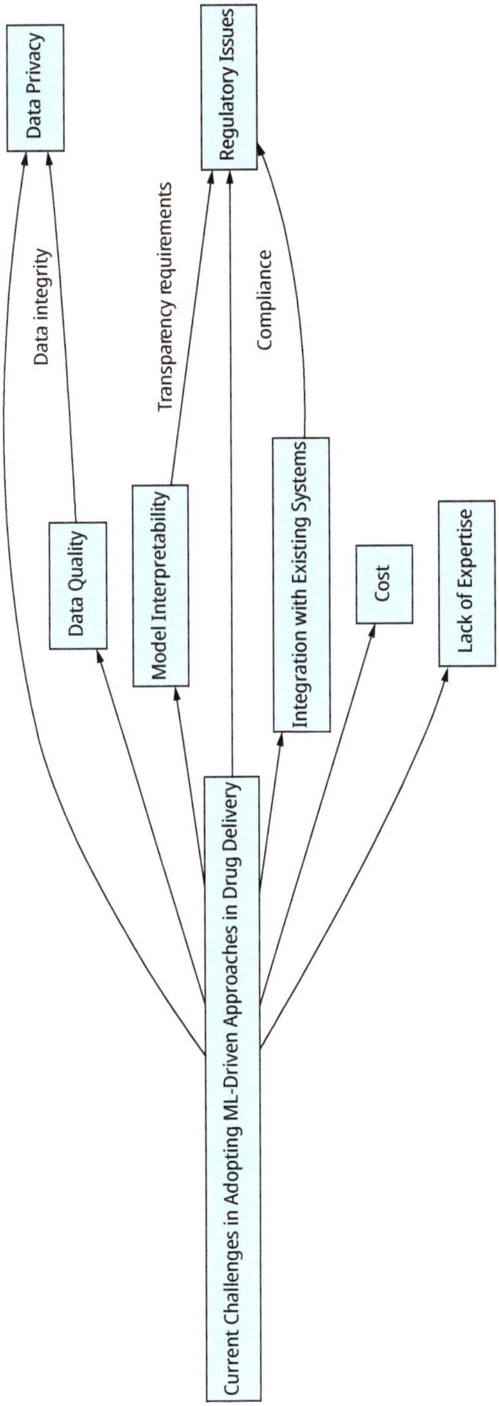

Figure 11: Current challenges in adopting ML-driven approaches in drug delivery.

explain, possibly to the patient or other stakeholders, why a specific treatment has been recommended.

The next challenge is the regulating environment of using ML to improve the delivery of drugs using nanocarriers, and this is still fast developing. Current supervisory bodies need to set up principles for constructing, verifying, and implementing the ML models in healthcare. This involves setting some baseline on how suitable the data is for analysis, how accurate the models being used are, and even how well a model is performing. Moreover, the current regulation must capture the learning and adaptation characteristic of the ML models to ensure that they are safe and provide value when with new data. What is even more challenging to developers and other healthcare service providers is the need to meet these regulations and standards.

Adapting the various approaches that have been developed through machine learning into clinical practice and healthcare infrastructure is also inconceivable. Before adopting ML tools, clinicians must be well equipped with the understanding and skills in using and analyzing the tools and this incurs costs. In addition to this, there is the need for proper healthcare system integration tools such as Electronic Health Record and Data integration framework for a successful incorporation of any ML solutions. Finally, the organizational culture in that traditional practices is hard to change, which also slows down the development and implementation of new technologies.

6.2 Regulatory considerations and compliance with safety and efficacy standards

This is due to the fact that application of ML in medication delivery optimization necessarily involves a number of regulatory questions that must be addressed to maintain safety and efficacy. The related factors for regulatory approval and ensuring patient safety include data quality, model verity, reporting, monitoring, and controlling post-market and ethical considerations.

The technological design of automated smart medicine delivery systems must meet strict regulatory requirements to be approved by health authorities. Agencies such as the European Medicines Agency (EMA) and the U.S. Food and Drug Administration (FDA) mandate that the data used in training and modeling these systems be accurate, comprehensive, and representative of the intended patient population. To achieve this, there are rigorous protocols for data collection, standardization in data entry methods, and strict data cleansing procedures to eliminate inconsistencies. The authenticity and traceability of data must be verifiable to uphold the system's credibility. In parallel, compliance with data privacy laws, such as the General Data Protection Regulation (GDPR) and the Health Insurance Portability and Accountability Act (HIPAA), is essential to safeguard patient confidentiality and ensure informed consent. Figure 12 presents the regulatory data pipeline, outlining how data quality, verifiability, and privacy are enforced from acquisition to model deployment in smart drug delivery systems.

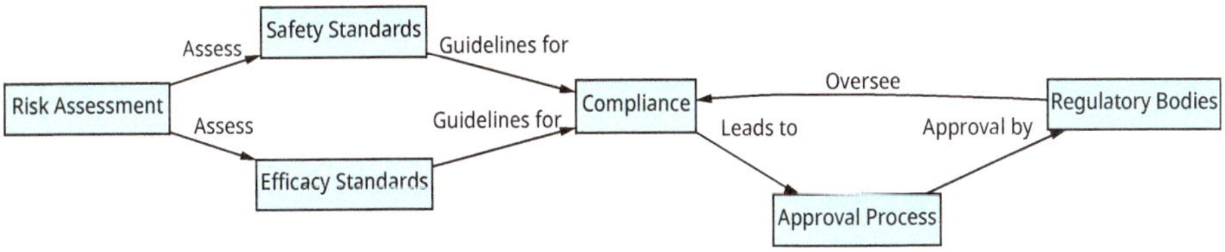

Figure 12: Regulatory considerations and compliance with safety and efficacy standards.

Model validation and performance assessment are two very important regulatory compliances. Any algorithms derived from ML approaches must be thoroughly tested and verified to reliable independent datasets that represent clinical reality. The assessment of performance accuracy, sensitivity, specificity, and robustness on varied patients' repertoire is part of this validation process. The regulatory considerations involved in the testing of the methods requires that the validation procedures be done with specific written methods. The required sensitivity analyses that impact variations in input data are explored on model results. These steps make it possible to use models to forecast drug delivery performances along with clinical decisions.

There is a high emphasis on the regulatory compliance of ML through their transparency and interpretability. Many regulatory agencies have started to call for the use of methods and approaches to obtain the so called "explanations" of why a model is making a particular prediction. Such transparency is critical for re–establishing confidence in clinicians and patients, as well as for allowing the healthcare providers to explain, and perhaps even rationalize, the use of ML-based suggestions. Policy makers insist on the generation of high-quality records on how a model is built, trained, and the decisions it arrives at. Moreover, the flexibility of the constructed ML models should enable clinicians to supervise the models, thus offering the opportunity to correct the course of action in case needed.

It also requires constant evaluation and monitoring of the ML-driven drug delivery systems after the market launch, hence post-market surveillance and monitoring. Regulatory agencies always insist that these systems should continue to be monitored by the manufacturers in order to identify all negative effects or performance problems that occur when the systems are already in the market. This calls for high-quality pharmacovigilance programs, through which the safety and performance of drug delivery systems can be monitored in large populations. An ML model should be easily retractable and renewable to adapt new modeling styles, in compliance with the laws and regulations, as new data come in. Another requirement has been to develop best practices for retraining, for the models' validation, and for their reapproval, to guarantee that the systems are safe and efficient in their use during their life cycle.

6.3 Opportunities for integrating ML with real-time data analytics in clinical settings

Combining ML solutions with real-time data processing in clinical environment opens up new horizons to improve drug delivery procedure. This can go a long way in enhancing the lives of patients, and the efficient running of the health facilities and advancement of precision medicine. Integration of the models and other relevant data processes, and the introduction and use of wearable technology, EHRs, and remote systems help enhance the agility in the practice of healthcare delivery.

Some of the advantages transform medication management into a process that can be followed and changed in real-time, and one of these is the ability to monitor and adjust it in real-time. Several medical information are collected by using wearable technology and biosensors, including blood pressure, glucose level, heart rate, and medicine intake. To discover abnormalities, predict possible side effects, and enhance dosage schedules, these data streams can be evaluated in real-time by machine learning algorithms. For instance, the real-time glucose monitoring data can be fed to ML models to provide personalized dose recommendations for insulin supplements during chronic diseases such as diabetes. This in turn can helps WITH regulating correct glycogen deposits or glucose levels, thereby reducing the chances of the above-listed complications, PROVIDED it is followed to the letter.

Another big potential is the increase in accuracy of predicting future sales. It is understood that when appropriate data is fed to the ML models, such models are able to predict how an illness advances, and how a therapy works. By this way, healthcare workers approach emergent health threats and make more informed decisions. This done by updating forecasts with new patient figures on a regular basis. For instance, in cancer, it becomes possible to predict the development of the tumor and also adapt the dosage and type of the chemotherapy to the patient, according to the generalized perception from the data collected from genetic makeup, images, and reports of the patient. This possibility of predicting outcomes ensures that therapies remain effective and are easily adapted to accommodate changes in the patient status.

In real-time, it becomes easier to identify features inherent in any single patient that influence drug delivery. Machine learning is especially appealing because it has the potential to uncover patterns as well as links between data that would not produce them by regular methods. It can identify gene traits that modify the rate at which medication is metabolized and the degree to which it is effective, the choices people make, and other factors. When these variables are recognized, medical personnel effect medication delivery strategies to the requirements of the individual patient, thus enhancing the effects and diminishing the side effects of the medicines.

The discovery of new drugs can be made faster through the use of Machine Learning and real-time analysis of data collected during clinical trials. Pertinently, high-quality data should be collected in a sequential manner by adopting wearable devices and remote sensors to supervise trial participants. This data can be put through ML models to detect signals of toxicity or effectiveness, thus allowing adaptively designed trials to be altered based on the results obtained in an interim stage. Hence, through early signaling of adversity, this strategy improves both patients' safety as well as the efficiency of clinical trials.

In the clinical setting, therefore, the application of ML to generate relevant real-time analytics improves DSS. These systems can give clinicians recommendations based on the reports and the patient's current data and the best practice protocols. For example, an ML-based DSS uses real-time data about a patient and data from previous patients to determine the best administration method for drugs. This support helps the clinician to be in a better position to make good decisions, decrease the complexity of care, and increase the chances of good outcomes among the patients.

6.4 Future directions in leveraging ML to enhance drug delivery efficiency

The enhancement of drug delivery is soon expected to change significantly due to the improvements in the techniques of machine learning. By using real-time monitoring systems, applying advanced analytical methods and striving for individual approaches to the patients' treatment, the integration of ML technologies offers potential to enhance the efficiency of medicines administration. These new paths are expected to revolutionize the pharmaceutical industry and consequently lead to improved patients' experiences and effective treatments.

One of the most possible future scenarios is making highly customized medicine delivery systems. Patient data, particularly genetics, daily habits, and environment structure as well as features, should be input into the ML algorithms to come up with a proper administration of medicine for the patient. Thanks to such an approach, each patient gets the highest quality treatment with minimal side effects possible. Accomplishments can include the integration of the pharmacogenomic data into machine learning model that would predict patients' metabolism of the prescribed medications, making it easy to adjust dosage or choose the most effective agents. These models, therefore, can be continually improved in terms of their accuracy and efficiency by -related refinement.

Future medication delivery optimization is considered to be highly dependent on the form of further sophisticated and superior predictive analytical systems based on machine learning. These models are fairly accurate at determining what is going to transpire in the disease progression, how a therapy is going to act, and even the nature of side effects. This includes the ability of the ML algorithms to identify minute patterns and trends with real-time and historical data in a way that normal techniques cannot. Healthcare professionals are able to prevent such occurrences by adjusting dosage regimes before negative occurrences occur, or select patients who stand to benefit from specific drugs, because of the use of predictive models. This is advantageous because it reduces the costs of healthcare and, on the same note, enhances patient benefits, as patients avoid problems and hospital admissions.

The two future paths that are considered important include: The coupling of ML algorithms with real-time monitoring solutions – Daily health data such as vital signs, medication compliance, and even physical activity, for example, the kind of real-time health data that wearables and biosensors currently monitor. Machine learning algorithms can analyze different data flows in real-time, which allows offering individuals dynamic and adaptive choices of medications. For instance, continuous monitoring of patients with chronic diseases can be done through ML algorithms, where information from glucose monitoring devices is processed and changes in insulin dosages are made in real-time to keep the patient's blood sugar levels optimal. Full accuracy and responsiveness with such high measures are expected to significantly improve the management of some chronic diseases.

More significant opportunities emerge if machine learning is applied directly to the drug development pipeline. Using trial and error or actual data on biological systems takes nearly substantial time but by using machine learning algorithms to mimic and forecast their behavior, one can fasten the process in arriving at the design and testing of new drug delivery systems. The time and cost within the laboratory experiment can also be saved when using virtual screenings on medication candidates and their delivery systems. Also, better medicinal drug formulations and delivery systems can also be obtained by using optimization of formulations with the help of ML, hence leading to increased stability, bioavailability, and efficacy. This issue increases the tempo of new drug delivery solutions development with regard to specific therapeutic needs.

6.5 Recommendations for overcoming barriers and advancing ML in biomedical research

In the field of ML, there is a huge prospect to improving medicine administration and in research in the biomedical field. There are several barriers that must be addressed to harness the mentioned potential. Some of the tactical recommendations are as follows, in an effort to go past these barriers and enhance the use of machine learning in biomedical research.

First of all, it becomes crucial to bring together representatives of different faculties. Multidisciplinary teams such as computer science, biomedical engineering, pharmacological, and clinical medicine need to be included in order to advance the field with respect to the use of ML. Effort should be made to bring in experts in the biomedical domain as well as in the application of ML in solving such challenges. The lessons learned from the creation of interdisciplinary research centers and funding initiatives aiming at use of ML in healthcare can be used to ensure similar approaches in the future.

Secondly, the methods of improving the data availability, coupled with interoperability, dominate the use of applying ML in biomedical research. The healthcare industry is a large data producer with EHRs, genomics, imaging, and -reported outcomes. Standardization of data forms as well as setting up of data exchange mechanisms and regulation of data security issues are critical. There is a requirement for developing databases and cloud-based systems that are in line with the regulations needed for patient identification, yet allowing researchers to gain access to the data needed.

Thirdly, significant importance must be given to data quality and its management. Deep-learning techniques rely heavily on training data, which, as stated, means that an ML model is only as good as the data it was fed. Recommendations should be to concentrate on data cleansing and normalization and effective validation. Examples of such techniques include natural language processing; image recognition, and other contemporary data preprocessing methods that can help improve dataset completeness by automating data preprocessing jobs. Crowdsourcing and large-scale annotation can increase the capacity of curating data and diversify the datasets we can put together.

Fourth, endorsing progress in the methodological advances in developing and validating ML algorithms is crucial to the biomedical field. Scholars should pay more attention to the enhancement of the XAI models that can help improve the interpretability of the models. The last stage involves comparing the constraints of the learned functions with the gold standard benchmarks, and real-world clinical data is important in establishing the reliability as well as the generalizability of the developed ML models. Adopting more openness in science, including reporting of code, algorithms, and prescriptive benchmark datasets can enhance parity and improve the pace of algorithm optimization/scrutiny and maturity.

Fifth, transitioning toward obtaining regulatory permission and public acceptability in the execution of biomedical research using the ML approach demand answering ethical and regulatory questions. It means that there is a necessity of collaboration between regulatory bodies and other parties in order to determine rules regarding creation, verification, and utilization of machine learning models in the context of the healthcare industry. Privacy and confidentiality of the patient data means that one has to follow the data protection laws in the country of operation, which include the GDPR and HIPAA. It, therefore, should be noted that ML should be properly implemented with ethical models of algorithmic fairness, patient's self-governance, and eradicating conducive bias.

7 Conclusion

The application of ML improves drug delivery systems; in fact, it seems to have revolutionized biomedical science as it dramatically improves efficiency and accuracy of the drug development processes. It has looked at how ML approaches estimate properties of necessary materials like biocompatibility, degradation rates, and mechanical properties that are relevant for drug delivery. Exclusive formulas or models in machine learning have shown advancements in works such as the adjustment of the release rate or speed of the drug, with a view to enhancing effectivity in patient-specific treatment procedures. The chapter also emphasizes how ML applies adjustments in nanoparticle construction parameters and refreshes the combination of experimental and computational methods in the design of targeted delivery systems. Nevertheless, there are some issues that have not been solved yet – the problem of compliance with the requirements of relevant legislation, integration of the data into the program, as well as the question of the stability of algorithms. The lack of ethical considerations in developing ML technologies is prohibitive in the use of the technology in healthcare. In future, coordinated work of researchers, clinicians, and policymakers should be considered as key to the utilization of ML in drug delivery systems in order to facilitate the development of personalized medicine and for better quality of patient's treatment.

References

[1] Correa-Baena, J. P., Hippalgaonkar, K., Van Duren, J., Jaffer, S., Chandrasekhar, V. R., Stevanovic, V., . . . & Buonassisi, T. (2018). Accelerating materials development via automation, machine learning, and high-performance computing. Joule, 2(8), 1410–1420.

[2] Tang, Y., Kim, J. Y., Carman, K. M., Bahmani, A., Chen, Q., Rosenberger, M. G., . . . & Ferguson, A. L. (2023). Data-driven discovery of innate immunomodulators via machine learning-guided high throughput screening. Chemical Science, 14(44), 12747–12766.

[3] Vasudevan, R. K., Choudhary, K., Mehta, A., Smith, R., Kusne, G., Tavazza, F., . . . & Hattrick-Simpers, J. (2019). Materials science in the AI age: High-throughput library generation, machine learning and a pathway from correlations to the underpinning physics. MRS Communications, 9(3).

[4] Jia, Y., Hou, X., Wang, Z., & Hu, X. (2021). Machine learning boosts the design and discovery of nanomaterials. ACS Sustainable Chemistry & Engineering, 9(18), 6130–6147.

[5] Da Silva, V. A., Sharma, R., Shteinberg, E., Patel, V., Bhardwaj, L., Garay, T., . . . & Willerth, S. M. (2023). Machine learning approaches to 3d models for drug screening. Biomedical Materials & Devices, 1–26.

[6] McDonald, S. M., Augustine, E. K., Lanners, Q., Rudin, C., Catherine Brinson, L., & Becker, M. L. (2023). Applied machine learning as a driver for polymeric biomaterials design. Nature Communications, 14(1), 4838.

[7] Deng, S. (2024). Machine learning approaches for screening of materials in flexible electronic devices.

[8] Dai, M., Xiao, G., Shao, M., & Zhang, Y. S. (2023). The synergy between deep learning and organs-on-chips for high-throughput drug screening: A review. Biosensors, 13(3), 389.

[9] Rodrigues, J. F., Florea, L., de Oliveira, M. C., Diamond, D., & Oliveira, O. N. (2021). Big data and machine learning for materials science. Discover Materials, 1, 1–27.

[10] Kadulkar, S., Sherman, Z. M., Ganesan, V., & Truskett, T. M. (2022). Machine learning–assisted design of material properties. Annual Review of Chemical and Biomolecular Engineering, 13(2022), 235–254.

[11] Suh, C., Fare, C., Warren, J. A., & Pyzer-Knapp, E. O. (2020). Evolving the materials genome: How machine learning is fueling the next generation of materials discovery. Annual Review of Materials Research, 50(1), 1–25.
[12] Gawriljuk, V. O., Zin, P. P. K., Puhl, A. C., Zorn, K. M., Foil, D. H., Lane, T. R., . . . & Ekins, S. (2021). Machine learning models identify inhibitors of SARS-CoV-2. Journal of Chemical Information and Modeling, 61(9), 4224–4235.
[13] Perni, S., & Prokopovich, P. (2022). Feasibility and application of machine learning enabled fast screening of poly-beta-amino-esters for cartilage therapies. Scientific Reports, 12(1), 14215.
[14] Xie, J., SU, Y., Xue, D., Jiang, X., Fu, H., & HUANG, H. (2021). Machine learning for materials research and development. Acta Metall Sin, 57(11), 1343–1361.
[15] Fuhr, A. S., & Sumpter, B. G. (2022). Deep generative models for materials discovery and machine learning-accelerated innovation. Frontiers in Materials, 9, 865270.
[16] Morgan, D., Pilania, G., Couet, A., Uberuaga, B. P., Sun, C., & Li, J. (2022). Machine learning in nuclear materials research. Current Opinion in Solid State and Materials Science, 26(2), 100975.
[17] Kwon, K. K., Lee, J., Kim, H., Lee, D. H., & Lee, S. G. (2023). Advancing high-throughput screening systems for synthetic biology and biofoundry. Current Opinion in Systems Biology, 100487.
[18] Hou, R., Xie, C., Gui, Y., Li, G., & Li, X. (2023). Machine-learning-based data analysis method for cell-based selection of DNA-encoded libraries. ACS Omega, 8(21), 19057–19071.
[19] Golmohammadi, M., & Aryanpour, M. (2023). Analysis and evaluation of machine learning applications in materials design and discovery. Materials Today Communications, 35, 105494.
[20] Pasrija, P., Jha, P., Upadhyaya, P., Khan, M., & Chopra, M. (2022). Machine learning and artificial intelligence: A paradigm shift in big data-driven drug design and discovery. Current Topics in Medicinal Chemistry, 22(20), 1692–1727.
[21] Rao, Z., Tung, P. Y., Xie, R., Wei, Y., Zhang, H., Ferrari, A., . . . & Raabe, D. (2022). Machine learning–enabled high-entropy alloy discovery. Science, 378(6615), 78–85.
[22] Bian, Y., Kwon, J. J., Liu, C., Margiotta, E., Shekhar, M., & Gould, A. E. (2023). Target-driven machine learning-enabled virtual screening (TAME-VS) platform for early-stage hit identification. Frontiers in Molecular Biosciences, 10, 1163536.
[23] Yuan, Y. C., Xu, B., McCormack, J., Huang, X., Ma, J., Marshall, T., . . . & Bai, H. (2024). Combining scalable organ chip platform with deep learning-based imaging analysis for cancer therapeutic screening, bioRxiv, 2024-06.
[24] Gao, J., Zhong, J., Liu, G., Zhang, S., Zhang, J., Liu, Z., . . . & Zhang, L. (2023). Accelerated discovery of high-performance Al-Si-Mg-Sc casting alloys by integrating active learning with high-throughput CALPHAD calculations. Science and Technology of Advanced Materials, 24(1), 2196242.
[25] Kumar, S. A., Ananda Kumar, T. D., Beeraka, N. M., Pujar, G. V., Singh, M., Narayana Akshatha, H. S., & Bhagyalalitha, M. (2022). Machine learning and deep learning in data-driven decision making of drug discovery and challenges in high-quality data acquisition in the pharmaceutical industry. Future Medicinal Chemistry, 14(4), 245–270.

Vaishnavi V., Meena G., Santhanalakshmi, Anitha R., K. Sivakumar

Automated materials characterization using machine learning for screening biocompatible materials

Abstract: Currently, biomedical engineering has received a boost, largely due to the rate at which new biocompatible materials have been developed, offering innovative solutions to many biomedical application areas. This chapter focuses on the constantly growing field of biocompatible materials, characterized by the high relevance of these materials in enhancing patients' quality of life and driving new developments in medicine. Understanding the surface characteristics, chemical properties, mechanical properties, and biological properties is beneficial in guiding the choice of materials for better clinical applications. This discussion extends to current developments in biocompatible materials, which have concentrated on advancements in several new synthetic polymers, new metals, and composite materials. It also addresses other areas of material characterization, presenting challenges such as ethical concerns and compliance with legislation, including the need to continually subject materials to rigorous testing and ethical standards. This chapter provides a comprehensive and critical review of recent advancements and prospective developments in biocompatible materials and is intended to serve as a useful reference for engineers working in the biomaterials and biomedical fields for their material selection and application.

Keywords: biocompatible materials, surface characteristics, chemical stability, emerging trends, ethical considerations, material characterization

Vaishnavi V., Faculty of Management, SRM Institute of Science and Technology, Kattankulathur 603203, Chengalpattu District, Tamil Nadu, India, e-mail: vaishnav2@srmist.edu.in
Meena G., Faculty of Management, SRM Institute of Science and Technology, Kattankulathur 603203, Chengalpattu District, Tamil Nadu, India, e-mail: mg8891@srmist.edu.in
Santhanalakshmi, Faculty of Management, SRM Institute of Science and Technology, Kattankulathur, Chengalpattu District, Tamil Nadu 603203, India, e-mail: santhank@srmist.edu.in
Anitha R., Department of Electronics and Communication Engineering, Government Engineering College, Palakkad, Sreekrishnapuram 678633, Kerala, India, e-mail: ranitha@gecskp.ac.in
K. Sivakumar, Department of Mathematics, SIMATS School of Engineering, Saveetha University, Thandalam, Chennai 602105, Tamil Nadu, India, e-mail: sivakumarkaliappan.sse@saveetha.com

https://doi.org/10.1515/9783111503202-011

1 Introduction

One major revolution in biomedical engineering has been achieved through the discovery of biodegradable materials, which have significantly impacted the design of various implants and diagnostic equipment [1]. The use of such materials in biocompatible applications tends to contribute substantially to patient outcomes due to the growing demand for new technologies in the healthcare sector [2]. To avoid adverse responses and to maximize the potential for therapeutic effects during the application of a material, that material must be as biocompatible as possible, which is defined as the ability of a material to function properly in an enzymatic sense at the site of administration [3, 4]. This chapter aims to provide a comprehensive overview of the biological characteristics, recent developments, and challenges associated with biocompatible materials, with a focus on their relevance to current state-of-the-art medical applications [5].

Understanding the basic attributes that define biocompatibility is crucial for the identification of appropriate materials for medical purposes [6, 7]. In clinical applications, the performance of a material is confined to basic characteristics that include mechanical capability, chemical response, biological interaction, and surface aspects [8, 9]. Surface roughness and charge influence cell behavior and tissue integration, while chemical stability sustains function [9]. Physiological forces determine the material's compatibility with the body's immune response, while mechanical properties such as strength and elasticity determine compatibility with physiological pressures [10].

It was pointed out that recent scientific developments in biocompatible materials have given rise to new opportunities that satisfy various clinical requirements [11]. Some examples of these trends include research into new synthetic polymers for improved functionality as well as the creation of new metal alloys and composite materials that offer enhanced performance and functionality [12]. New materials include synthetically produced polymers like polylactic acid and polyethylene glycol (PEG), which are biodegradable and exhibit low protein adsorption, and metals like titanium and cobalt-chromium alloys, which have high strength, high stiffness, and high corrosion resistance [13–15]. Fiber-reinforced composites, in which two distinct components are utilized, have also started being considered for their defined characteristics and increased functionality [16, 17].

Current and future developments still face significant challenges in material characterization as well as ethical problems [18, 19]. Since biocompatibility is the ability of a material to perform in the intended physiological roles and to respond to biological systems, it is essential to subject the material to detailed testing [20]. Another example of ethical considerations includes compliance with regulatory necessities, handling long-term effects on patient's well-being, and ensuring that all the materials used in clinical applications are safe and effective for the intended purpose [21]. Such challenges emphasize the need to continue the research and innovation processes to bridge gaps and improve material performance [22, 23].

Continuing development and increased use of biocompatible materials indicate that advancements in the technology will continue in the future [24]. It is expected that advancements in material sciences and technology will continue to drive higher-order and more powerful medical application solutions. The enhanced understanding of biocompatibility, along with the establishment of new materials and technologies, increases the capability to design and manufacture medical devices to meet the most demanding standards of performance and safety [25]. As with any kind of advancement, this chapter provides a critical analysis of these improvements and best practices for professionals and academics in the field.

2 Definition and importance of biocompatible materials

2.1 Defining biocompatibility

The ability of an implanted material to perform the required task within a bioenvironment without eliciting an adverse local response in the host organism is defined as biocompatibility. In the case of implants, drug delivery systems, medical devices, and other biomedical applications where materials interact with tissues, blood is very essential. Biocompatibility can be defined as the ability of the material to support and promote cell functions, integrate into the surrounding host tissues, remain functional for the required duration, and, importantly, not cause toxic or inflammatory responses.

It was therefore necessary to evaluate the material in terms of mechanical, chemical, and physical properties, as well as its interaction with the biosphere, to establish biocompatibility. Chemical properties such as solubility, surface texture, hardness, and brittleness, among others, have a direct bearing on the absorption of the chemical, thus dictating its ability to perform its intended functions within the human body without exposing the patient to health risks. The ease with which a biocompatible material can be adapted for the host tissue, minimizing the chances of incidents such as infection, rejection, and tear, was considered a measure of the material's success.

In the creation of biocompatible materials, it therefore became necessary to take into consideration the use to which the material was going to be put and the environment in which it was to be used. For instance, tissues intended for long-term implants must be biocompatible but also wear and corrosion-resistant and mechanically stiff enough to withstand forces. Likewise, materials applied in drug-delivery systems have to include the ability to release therapeutic agents and not provoke toxic reactions.

Biocompatibility of a substance is a relative factor that depends on the purpose of its use; it is not an absolute characteristic. The same material can exhibit different levels of biocompatibility depending on conditions such as the state of the patient's body, the location of implantation, and the presence of additional additives used.

Therefore, identifying biocompatibility is an intricate and ongoing process that involves accounting for all the biological interactions and characteristics of materials. Consequently, definitions and assessments of biocompatibility remain crucial to the future advancement of biomedical engineering and the application of novel technologies in medicine.

2.2 The role of biocompatible materials in modern medicine

Biocompatible materials today are critical as they act as the foundation of many forms of implantation, medical instruments, and treatment technologies. These materials are specially designed to interact harmoniously with biological tissues, which means they do not elicit toxic or inflammatory responses. The integration of biocompatible materials into medical applications has revolutionized the approach taken to treating patients as well as promoting the development of solutions that effectively extend the practical functionality of medicines.

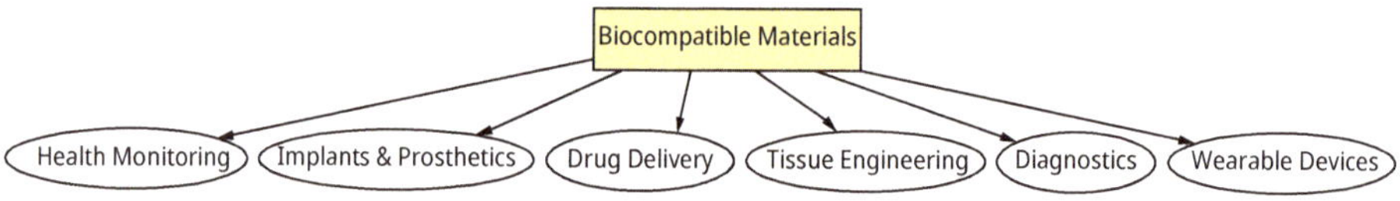

Figure 1: The role of biocompatible materials in modern medicine.

These implantable devices and biocompatible materials play a vital role in a range of medical applications, including orthopedic implants, cardiovascular stents, dental implants, and other prosthetic devices. For instance, titanium alloys and bioceramics are commonly used in joint replacements due to their mechanical strength, durability, and excellent compatibility with human bone tissue. These implants have enabled millions of patients worldwide to regain mobility and reduce chronic pain, underscoring the profound impact of biocompatible materials on improving patient quality of life. Figure 1 illustrates the widespread use and integration of such materials in modern medical implants.

Biocompatibility of materials used has also brought significant changes to drug delivery systems, where molecules and drugs can be released slowly into the body. For the administration of drugs at controlled rates with minimal side effects, bioresorbable polymers such as polyglycolic acid (PGA) are employed for the fabrication of matrices containing medications. These systems are particularly useful in cases where the condition requiring treatment is a chronic one and long-term drug adherence is necessary.

Most importantly, biocompatible materials act as the basis for the creation of regenerative medicine and tissue engineering. In a way that mimics the extracellular matrix, scaffolds provide cells with the necessary architecture to grow in, making tissue. These scaffolds, which often consist of biodegradable polymers, allow the timely and gradual assimilation of load in parallel to the gradual implantation of new tissue.

Biocompatibility is not only important for specific applications but also defines the direction of creating biosensors and wearable medical devices. These real-time monitoring sensors constantly monitor physiological factors and, therefore, require materials that do not degrade easily when in contact with the skin. Additionally, to ensure accurate and consistent performance, the materials used must be biocompatible and possess characteristics such as flexibility, conductivity, and durability.

2.3 Historical evolution of biocompatible materials

The information discussed in the present work describes how the development of biocompatible materials has followed the advances in biomedical science and engineering. Previous efforts to create materials that would specifically and safely interface with biological systems can be regarded as less successful due to the available tools and knowledge. Ancient civilizations employed basic metals such as gold and silver in surgical instruments and dental treatments because they were observed to have moderate biocompatibility functions. At the time, there was no such thing as biocompatibility in the current definition, but these early materials were selected for their durability and resistance to corrosion.

There was gradual improvement in the succeeding centuries, particularly throughout the nineteenth and early part of the twentieth century, with the coming of modern medicine and materials science. The advancements in stainless steel and other sorts of alloys necessitated the formation of more complex and advanced medical implants and devices. Tig metals such as stainless steel, which are anticorrosive and could be sterilized, were used for surgical tools and implantable products. Additionally, beginning in the mid-twentieth century polymers made the products more diversified for biocompatibility. The first materials used in medical applications were polymers like polystyrene and polyethylene because of their flexibility and ease of molding. During this period, the concern was to curtail the inflammation caused by these materials in hopes of avoiding serious illness.

Metals such as gold and silver were utilized by the ancient people for surgery and dental implants due to the lack of biocompatible materials per se. Though the modern notion of biocompatibility was absent, these earliest materials were selected based on their durability and noncorrosive properties.

The last several decades have also witnessed a shift in focus from the generation of materials with enhanced property profiles to the development of smart materials that can positively interface with biological systems. New composite material technology and new nanostructured material technology, appropriately designed for specific biological functions, have been the outcomes of advancements in nanotechnology and biomaterials. There are new materials such as titanium alloys and hydroxyapatite composites that have revolutionized the integration and performance of implants; 3D printing has made it easier to develop materials that meet the surgical requirements of each patient.

The progression in the use of these materials has demonstrated an improvement in complexity and comprehension of the relationship between materials and the biological system. Starting from the empirical stage right up to present-day technologies, each of these has played its part in improving medical devices and therapeutic solutions because of the constant drive to enhance the quality of patient care – hence the continued improvement of materials.

2.4 Key properties that determine biocompatibility

The capacity of a material to safely and efficiently interface with biological systems is fundamentally determined by its biocompatibility, which is evaluated using a range of physical, chemical, and biological parameters. These features are essential for the successful application of materials in the biomedical field, particularly in implants, prosthetics, and drug delivery systems. Figure 2 provides an overview of the key biocompatibility assessment parameters and their relevance across various biomedical applications.

Chemical composition: The composite nature of a material determines how it behaves with biological systems due to its chemical features. A material cannot be made of elements that cause toxicity when present in tissue. Most biomaterials have chemically stable compositions that are chemically inactive and do not degrade into toxic substances. It is also important for the material to contain no toxic elements, especially when the material is in contact with live tissues.

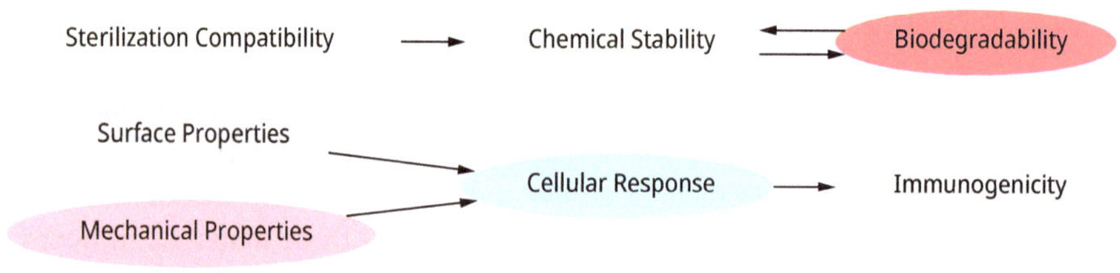

Figure 2: Key properties that determine biocompatibility.

Surface properties: These parameters merge within a system where the interfacial aspects of the material, including roughness, porosity, and surface energy, govern the behavior of a material system with biological systems. These properties affect cell attachment, protein absorption, and tissue incorporation. For instance, the right type of roughness helps in the attachment of cells and encourages tissue formation but, in contrast, it could be disadvantageous if the surface was too rough. Surface chemical characteristics illustrate that using coatings to control these interactions at the surface level is also possible.

Mechanical properties: The first aspect of material biocompatibility is its mechanical interfacial compatibility with the biological tissues in the vicinity. These include tensile strength, elasticity, and hardness attributable to the constituent materials and the structure of the composite layers. To be bonelike, the material must meet the mechanical requirements of the area in which it is to be placed. Mechanical properties also vary with the host material, and these differences can pose problems such as stress shielding, material degradation at a different pace than the surrounding tissue, and integration issues.

Degradation and stability: Something as esthetic as long-term biocompatibility necessitates a material's probability of stratification and its rate of degradation. It was suggested that the degradation rate of materials employed in biomedical applications must be controlled to match the rate of healing. For instance, the material applied as a biosorbable one should degrade into easily removable and nonhazardous substances. On the other hand, the materials that are intended for implantation applications have to be inert and should remain so for the time they are fixed in place without detriment.

Interaction with the immune system: Biocompatibility, together with surface chemistry, is determined by how a material engages with the immune system. There are highly specific requirements for what the materials used must and must not do – namely, the materials should not activate an immune response or cause inflammation by the immune system. This includes the surface chemistry of the material, the degradation products, and the presence of contaminants in the material. Ensuring that the material does not remain immunogenic and that no chronic inflammation develops in the subject is important for its successful use in biomedical applications.

All these properties as a whole determine the biocompatibility of materials and are foundational to the function and safety of the material in the biological environment. Characterization and fine-tuning of these properties are significant processes involved in material development for potential use in medical applications to guarantee their suitability for use in medical environments.

2.5 The impact of biocompatible materials on patient outcomes

The use of biocompatible materials in implants and medical devices affects patients by enhancing the safety and performance of various biomedical treatments. Biocompatible materials are designed to interact with biological environments without causing adverse reactions, which is why these materials influence the effectiveness of medical therapies as well as the quality of life of patients.

That is why one of the main functions of biocompatible materials, in relation to patients' outcomes, is their ability to decrease the likelihood of developing complications associated with implants and prosthetic devices. Tissues that are well acclimatized to the physiological environment have a low tendency to provoke inflammation, infection, or immune rejection. For instance, titanium alloys are utilized in orthopedic implants and dental appliances because of their high compatibility with human tissues, low rates of implant failure, and ability to improve bone-implant connections.

Drug delivery systems are designed using biocompatible materials in order to obtain controlled and site-specific drug release, which enhances therapeutic efficacy. Pharmaceuticals are delivered from polymers and hydrogels used in drug delivery systems at controlled rates and at desired sites, while minimizing systemic side effects. This targeted approach enhances general patient adherence and satisfaction, decreases the possibility of negative side effects, and improves medication effectiveness.

Biodegradable and biomimetic materials of tissues in tissue engineering and regenerative medicine applications help to support and rebuild injured or infirm tissues. These components must closely replicate the extracellular matrix in order to facilitate cell adhesion, division, and differentiation. If biocompatible scaffolds are integrated into tissues, tissue regeneration and repair become more effective, reducing the incidence of complex surgeries and thereby increasing patients' functional rehabilitation.

In addition, the formation of biocompatible coatings and surface alterations for medical devices has a prominent impact on patient treatment. These coatings can increase the service life of the device and its performance by enhancing the compatibility of the device with biological media. For example, early antiproliferative coatings on the surface of stents can decrease the risk of restenosis and enhance the long-term efficacy of the intervention through local drug delivery to the arterial wall.

In general, the influence of biocompatible materials was observed on patient outcomes and varied the durability, efficacy, and safety levels of each practice and technique. Patient care and treatment advancements can be improved by enhancing these written materials, which contribute to better health outcomes and quality of life.

3 Classification of biocompatible materials

3.1 Metals as biocompatible materials

Metals have long been a cornerstone in the development of biocompatible materials due to their favorable mechanical properties, corrosion resistance, and long-term structural stability. These attributes make them highly suitable for a wide range of biomedical applications, including orthopedic implants, dental implants, and cardiovascular stents. The selection of specific metals such as titanium, stainless steel, and cobalt-chromium alloys is made with careful consideration of their ability to interact with biological tissues without triggering adverse immune responses. Figure 3 illustrates the role of various metals in biomedical applications, emphasizing their compatibility and performance within the human body.

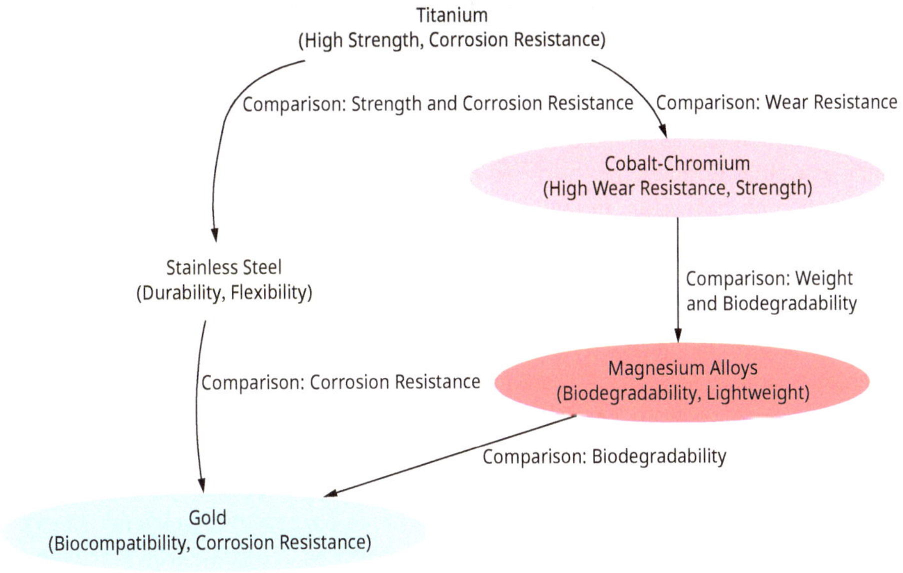

Figure 3: Metals as biocompatible materials.

Titanium and its alloys: Titanium is one of the more frequently used metals in biomedical applications, especially in the fabrication of dental prostheses as well as orthopedic implants. It has been stated that titanium is highly biocompatible due to the formation of a persistent oxide layer on the surface of the material, which minimizes corrosion and enhances its ability to bond with bone tissue. Due to their superior mechanical characteristics, including high strength and low density, alloys such as Ti-6Al-4V are incorporated in bearing critical loads. Increased bone formation and integration make titanium implants stable enough to achieve a long-term service life and high survival rate.

Stainless steel: The third metal that was used in medical devices was stainless steel, especially the 316L grade steel. Through enhanced chemical properties, outstanding mechanical properties, and excellent workability, it finds practical application in a number of fields; these include surgical equipment, external fixators, and vascular stents. Nevertheless, stainless steel exhibits fairly better biocompatibility than other metals used in industrial applications, but its efficiency is slightly lower, particularly in highly corrosive environments due to repeated contact with body fluids. To address these challenges, specialized coatings and surface modifications were employed to improve its biocompatibility.

Cobalt-chromium alloys: These alloys are well-known for their high hardness and excellent wear resistance, making them suitable for applications in dental prostheses, joint replacements, and other medical uses. These alloys are also biocompatible due to their corrosion resistance and the formation of a passive oxide coating that is likely to persist. The tensile strength and fatigue resistance that cobalt-chromium alloys offer enhance implant lifetime and durability. This is why the evaluation of alloy composition and potential metal ion release must be carefully conducted to avoid adverse reactions in certain localized areas.

Magnesium and its alloys: Owing to its potential use in degradable implants, magnesium, as well as its alloys, has attracted significant interest. This is because the gradual degradation of magnesium in the human body, which eliminates the need for surgical removal, enhances its biocompatibility compared to other materials. Some uses of magnesium-based alloys include temporary implants and bone-fixing elements. Through methods of both alloying and surface treatment, the rate of degradation of magnesium can be controlled to match the rate of tissue healing in the body so that while it acts as a temporary support structure, it is gradually absorbed by the body.

Noble metals: Due to their high corrosion resistance and biological compatibility, noble metals, including gold and platinum, are used in some biomedical applications. For example, platinum is used in medical implants and electrodes, while gold is used in dental materials and diagnostics. These metals are nonreactive, which mitigates potential chemical interactions and allows their application in conditions where biocompatibility and stability are essential.

3.2 Ceramics in biomedical applications

Ceramics possess an intimate combination of several properties: high strength, chemical stability, and biocompatibility. Thus, ceramics have found great demand in biomedical applications. Two of these materials are employed in a range of implants and medical devices, which enhance their mechanical properties and functionality.

Orthopedic implants are perhaps the most widely recognized application of ceramics in medicine. Due to their high corrosion stability and mechanical properties, ceramics containing alumina (AlO_4) and zirconia (ZrO_3) are usually used in hip and knee replacement operations. These materials are highly valued for their ability to withstand the mechanical loads related to joint arthroplasty and, at the same time, minimize the likelihood of wear debris causing inflammation of the implant.

This strategy has proven to be very effective when it comes to dealing with ceramic applications, coming out most exemplary. In regard to dental crowns, bridges, and veneers, it was the lithium disilicate and leucite-reinforced ceramics that were employed. Two of their esthetic properties, which are their transparency and color match ability, are also characteristic of natural teeth, providing a combination of cosmetic and functional benefits. Dental ceramics' compatibility with the biologic system equally ensures the tissue compatibility around the ceramics, hence reducing the chances of adverse reactions.

Ceramics are also applied in tissue engineering as well as in place of bone grafts. Bioactive ceramics like hydroxyapatite and beta-tricalcium phosphate exhibit a material structure similar to that of real bone. It has been established that these substances offer support for cell attachment and growth, which helps in the process of bone creation and regeneration. Some ceramics are resorbable; this allows new bone tissue to gradually grow into the site of the defect as the organism's natural processes take place.

Urinary stents and implants are just two of the many products within urological devices that utilize ceramics. For these applications, ceramic materials are chosen for their superior chemical stability and corrosion resistance to ensure long-term functionality and minimize the risks of biofilm formation.

Also, the application of new ceramic-matrix composites in biological applications has grown due to the enhancement of the materials. Ceramics are subject to improvement in their qualities and adaptation of their functions to be blended with other materials such as metal. For example, ceramic-polymer composites can offer high bioactivity, high strength, and flexibility, which places them in the category of suitable materials for more complex implantation and medical device applications.

Naturally, ceramics have several benefits when compared to their usage in the biomedical field including biocompatibility, wear resistance, and mechanical strength. As they continue to be advanced and standardized, medical devices and implants should be safe for the patient and effective in providing better results for the patient's benefit.

3.3 Polymers for biocompatibility

Due to their ability to be customized for specific uses, polymers are a key group of biomaterials that can be applied in biomedical fields. Depending on their chemical composition, physical properties, and interactions with living tissues and organisms, biocompatible polymers are classified. This section discusses the different types of biomaterial polymers, with particular attention given to their applications, advantages, and disadvantages.

3.3.1 Natural polymers

As a result, due to their morphological similarity with true biological tissues, those natural polymers, which are produced from biological materials, often exhibit intrinsic biocompatibility. Typical instances include:

Collagen is a hydrophilic and non-immunogenic major structural protein found in connective tissues, widely used in tissue engineering and regenerative medicine as a high-grade biopolymer. It functions as a biological scaffold, effectively substituting for the extracellular matrix (ECM) to facilitate tissue reconstruction and support the growth and differentiation of specialized cells. Figure 4 illustrates the structural role of collagen in tissue scaffolding and its applications in regenerative therapies.

The hyaluronic acid found in connective tissues has been used to form a lotion for joint therapy and in the treatment of wounds. Owing to its outstanding hydrophilicity and biocompatibility, it can be applied to drug carriers and hydrogels.

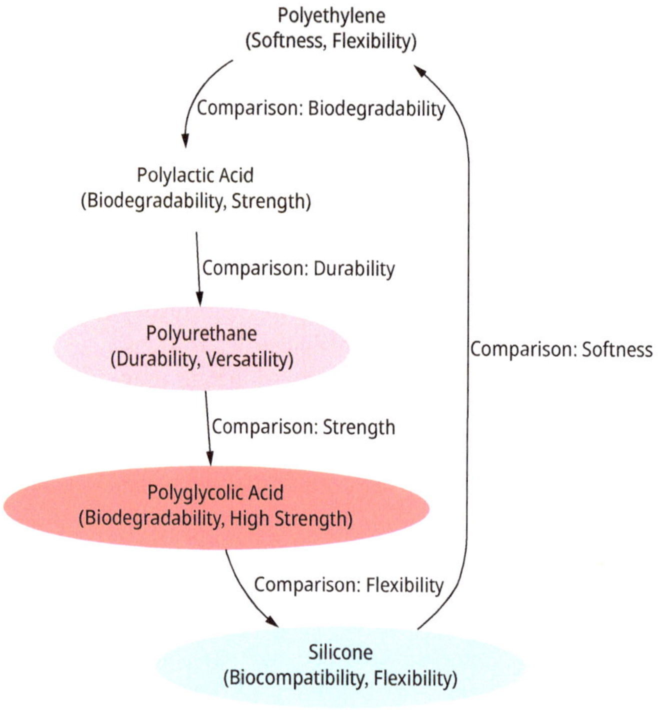

Figure 4: Polymers for biocompatibility.

Chitosan: Chitosan, which was synthesized from chitin, was employed in wound healing dressings and as a carrier for drug delivery because of its compatibility, degradability, and antimicrobial activity.

3.3.2 Synthetic polymers

Many synthetic polymers are specifically developed to fulfill biological compatibilities and utility in particular applications. The advantage of controlled properties and customization is significant.

Polyethylene glycol (PEG): PEG is well-known for its hydrophilicity and inactivity toward protein adsorption, which results in high versatility and makes it a preferred choice as drug carriers for delivery systems and as a coating layer for medical apparatus. PEG-based hydrogels are employed in controlled drug delivery and tissue engineering.

Poly(lactic acid) (PLA): PLA is a completely biodegradable polymer derived from renewable resources like corn starch. PLA is widely employed in implants and tissue scaffolds due to its biocompatibility and degradation into nonhazardous products.

Polyglycolic acid (PGA): PGA, in tandem with PLA for polylactic-*co*-glycolic acid (PLGA), was also employed in sutures and drug delivery systems. Some applications include temporary implantation, as the material is biocompatible and has a relatively preprogrammed rate of degradation.

3.3.3 Bioactive polymers

Bioactive polymers are designed to interact with biological systems in a manner that promotes specific physiological responses:

Polydimethylsiloxane (PDMS): The highly biocompatible and easily molded nature of PDMS means that it is applied to a multitude of biomedical applications such as microfluidics and soft tissue engineering. It is chemically inactive and also has the feature of being easy to process and thus is frequently used in the manufacturing of medical devices.

Polycaprolactone (PCL): PCL is an enzymatically degradable polymer with a relatively slow degradation rate, making it ideal for long-term use. Its major applications include the fabrication of tissue engineering scaffolds and controlled drug delivery systems.

Polyurethanes: It was employed in making medical products such as catheters and implants because it was strong, flexible, and could almost be considered biodegradable. Polyurethanes were adjusted for certain mechanical characteristics and interactions with biological organisms.

3.3.4 Biodegradable polymers

Biodegradable polymers are designed to break down into nontoxic by-products in the body, minimizing long-term impacts and reducing the need for surgical removal:

Polylactic-*co*-glycolic acid (PLGA): PLGA, which is a copolymer of PLA and PGA, has been established for extensive use in drug delivery and tissue engineering applications because of its tunable degradation profile and biocompatibility. By varying the ratios of PLA to PGA, the rate of degradation and release can be controlled.

Polyhydroxyalkanoate (PHA): Derived from microbial fermentation, PHAs are both biodegradable and biocompatible and can, therefore, be used in diverse aspects of medicine, including surgery, pharmaceuticals, and drug delivery.

3.3.5 Challenges and innovations

While biocompatible polymers offer numerous advantages, they also present challenges that must be addressed through ongoing research.

Mechanical strength and durability: The challenge in the current and future development of these tissue scaffolds is to achieve both biocompatibility and the mechanical properties of the scaffold in near-equilibrium with those of the original tissue. New trends in polymer blending and reinforcement are being developed to improve the performance of biocompatible composites.

Degradation products: Polymer degradation has other by-products that should be tested to determine their effects before being used on human beings. Work on the development of polymeric materials with desirable and well-understood degradation characteristics is underway.

Customizability and control: Innovations in polymer chemistry and processing are intended to improve the control of the structure of biocompatible polymers to better match the desired functions of use.

3.4 Composite materials

Biomedical composites are considered to be a major improvement in the production of biocompatible materials because they bring desirable properties into the material mix, enhancing their functionality. These materials are developed by processing two different materials to produce a form of material with properties not inherent in the two separate materials.

In the biomedical field, composite materials are mainly used to merge the features of several materials such as metals, ceramics, and polymers, into a single material system. Such flexibility enables the control of mechanical, chemical, and biological properties in order to satisfy the needs of different medical uses, including deformation, strength, and toughness, as well as the relationships between processing variables and performance.

Composite materials are designed by their constituents to have physical characteristics that are fit for specific uses. For instance, in orthopedic implants and prosthetics, composites can be tailor-made with mechanical properties closer to those of human bones, thus supporting the structure without contributing to implant failure. Such applications utilize carbon fiber-reinforced polymers and hydroxyapatite-based composites, as these materials have high strength-to-weight ratios and also encourage bone formation.

3.4.1 Pattern of biological activities and convergent concatenability

Due to the fact that composite materials interact with the biological environment, efforts are made to achieve the biocompatibility of the composite. Thus, bioactive components of these materials are embedded in the composite matrix to possibly improve the integration and healing of tissues. For instance, composites containing bioactive glass particles allow the formation of cellular attachment and tissue in-growth, promoting the implants/tissue interface. Surface treatments and coatings introduced to composites contribute significantly to improving the biocompatibility of constructs since they decrease the level of inflammation and infection potential.

3.4.2 Customization and versatility

The first advantage of composite materials was their flexibility and the ability to modify their properties to suit particular cases in biomedical fields. It must be understood that composites are designed to possess certain characteristics such as flexibility, durability, and biodegradability by varying the nature and ratios of the component materials. Due to this tailoring, materials suitable for different applications, for instance, rigid orthopedic implants, are produced.

3.4.3 Biomedical engineering applications

As composite materials solve specific problems associated with single-material systems, they are applied in biomedical engineering. For example, composites can be utilized in the production of dental implants, where flexibility is obtained from polymers while strength is derived from ceramics. Scaffolds used in tissue engineering are made from biodegradable polymers and ceramics, which, on one hand, provide structure and, on the other, support tissue development and regeneration.

3.5 Emerging biocompatible materials

As new materials are being developed to overcome the limitations of traditional approaches and meet the growing demands of modern biomedical applications, the field of biocompatible materials has been experiencing rapid and continuous advancement. This new generation of biomaterials demonstrates unique properties and enhanced functionalities, expanding their use across a wide range of therapeutic and clinical applications. Several important subcategories such as nanomaterials, bioresorbable polymers, and smart materials are explored in this section. Figure 5 highlights the classification and key features of these advanced biocompatible materials, showcasing their relevance and advantages in cutting-edge biomedical technologies.

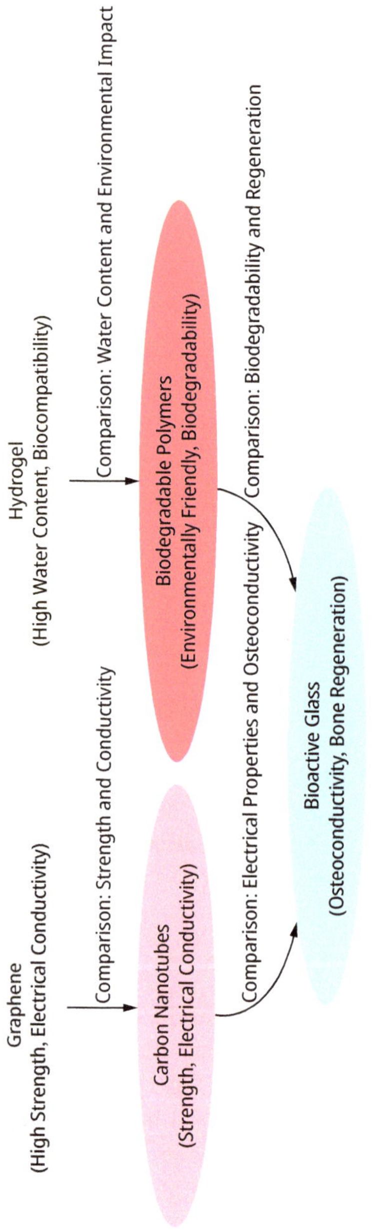

Figure 5: Emerging biocompatible materials.

3.5.1 Nanomaterials

Nanomaterials have been defined based on size and size-dependent properties, such as a high surface area-to-volume ratio and improved mechanical, optical, and chemical properties. In terms of biocompatibility applications, nanomaterials offer certain benefits including enhanced cell signaling communication and delivery systems. For example, it is possible to create nanoparticles that provide an increased biological distribution of active substances with fewer side effects. Nanomaterials are advantageously employed to enhance diagnostic imaging contrast, enabling even higher resolution.

3.5.2 Bioresorbable polymers

Polymer was developed to biodegrade in the body and be metabolized gradually; therefore, no further surgical intervention was necessary upon completion of the polymer's function. Most of these materials are also useful where the mesh is to be used as suture material, in drug delivery systems, and for temporary implantation. As for the advantages of using bioresorbable polymers as materials in tissue engineering, these include the disappearance of long-term complications and the stimulation of tissue renewal. Because they can be safely degraded in the body yet offer the necessary support in drug delivery, they represent a significant leap forward in biocompatible materials.

3.5.3 Smart materials

Smart materials, also known as responsive materials, are designed to react to changes in the environment as quickly as possible. In response to factors such as temperature, pH, and stresses, these materials can alter their properties. For example, hydrogels and shape-memory polymers adapt to changes in shape relative to physiological conditions, providing more versatile possibilities for tissue scaffolds and drug delivery systems. As smart materials evolve further, precise biological interventions can be achieved, thus enhancing their performance rates and reducing the odds of complications.

3.5.4 Conductive polymers

Conductive polymers are polymers that exhibit remarkable electrical conductivity alongside the flexibility properties typical of polymers. Cytocompatible conductive polymers can be employed to establish electrical contacts for stimulations between electronic signals and bio-tissues. They offer hope for enhancing brain-computer interfaces as well as other electronic medical technologies due to their compatibility with neural tissues.

3.5.5 Biodegradable composites

Biodegradable composites are defined as composite materials that are formed with a reinforcement of at least two separate components to obtain improved mechanical and biological features. Such composites are composed of a matrix of bioresorbable polymers with natural and/or synthetic fiber reinforcements. The choice of materials offers the possibility to develop structures with desired mechanical parameters and, at the same time, biocompatibility. Biodegradable polymeric composites are in the process of being evaluated for use in orthopedic implant devices and for tissue engineering scaffolds, with the expectation of achieving enhanced performance and decreased late sequelae.

It is remarkable how advanced these biocompatible materials have made biomedical engineering. Different types of materials possess their own unique characteristics to meet certain challenges in the medical field, leading to improved healing for patients as well as more efficient treatment methods. Consequently, it is expected that further investigation and innovation in this area will offer new findings that enhance the characteristics of biocompatible materials and expand their applications in modern medicine.

4 Applications of biocompatible materials

4.1 Implants and prosthetics

Implants and prostheses, which are central components of modern medical treatments for numerous diseases and conditions, require biocompatible materials. These materials are designed to interact with the body, offer functional support, and improve patients' quality of life. This section analyzes some of the functions of biocompatible materials in implants and prostheses, their advantages, disadvantages, and relevance.

4.1.1 Orthopedic implants

Orthopedic implants are essential for the management of skeletal disorders, arthritic conditions, and fractures. Some of these implants include bone plates and joint replacement. These implants are made of biocompatible ceramics, stainless steel, and titanium alloys. Due to their superior mechanical qualities and high resistance to corrosion, titanium and its alloys ensure efficiency and durability within the body. Owing to their high level of hardness and wear resistance, zirconia and alumina ceramics are used in knee and hip replacements. Prosthetics made from these materials encourage direct bonding to bone tissue and reduce the likelihood of implant loss, thereby improving patient satisfaction and locomotion.

4.1.2 Dental implants

As for replacement teeth, dental implants provide support. Titanium is popular due to its great strength, its ability to assimilate into the bone, and its generally good compatibility. The functional and esthetic rehabilitation that mimics the lost tooth structure is achieved by the stable anchorage of titanium implants to the bone tissue. There have been advancements in technology for the surface treatment of implants through etching techniques, such as the use of acid and plasma spraying, which have enhanced the bond and efficiency of dental implants.

4.1.3 Cardiovascular implants

Pacemakers, valves, and stents are other cardiovascular implant devices required to address diseases and illnesses of the cardiovascular system. Some of the naturally compatible materials used include silicone and polyurethane; the devices designed are highly compatible with the circulatory system. The newest advance in cardiovascular implants is the production of drug-eluting stents, which are coated with biocompatible polymers that release medicine to combat restenosis. Thus, the materials must exhibit minimal propensity to cause inflammation and thrombosis and be capable of withstanding the harsh and active conditions of the cardiovascular system.

4.1.4 Prosthetic limbs

These artificial limbs are designed to mimic lost limbs, which allow patients to experience improved mobility as well as functionality. The biocompatibility of the materials utilized in prosthetics includes lightweight polymers, carbon fiber-reinforced plastics, silicone rubber, and other flexible materials. Carbon fiber composites offer a high strength-to-weight ratio, making prosthetic limbs durable yet comfortable. Used in liners and sockets, silicone elastomers provide cushioning characteristics and a skin-simulating texture to enhance fit and mitigate skin chafing. New prosthetic developments, such as the addition of sensors and the incorporation of motorized attachments, depend on these materials to provide better performance and user experience.

4.1.5 Cochlear implants

Cochlear implants are electrical devices that help people who are profoundly deaf to appreciate sound in some ways. The preference for biocompatible materials for cochlear implants includes an outer casing covered with silicone material and electrodes made of platinum. These materials must be chosen carefully to avoid hazardous effects and harm to the delicate parts of the inner ear. It has also been observed that the design and materials used must be appropriate in order to avoid tissue reactions and achieve the optimal effect on the stimulation of the auditory nerve.

One of the major achievements in medical development is the use of biocompatible materials in both implants and prostheses, which restore patients' functionality and improve their quality of life. Due to advances in patient needs, the market now offers new materials and technologies that enhance device functionality, durability, and safety for both clients and health professionals.

4.2 Drug delivery systems

To develop advanced drug delivery systems capable of delivering precisely controlled and highly localized therapeutic inputs, the use of biocompatible materials is fundamental. These materials enable systems that can release medication at optimal rates and target specific sites in the body to maximize therapeutic efficacy while minimizing side effects. This section explores emerging prospects in biocompatible drug delivery, focusing on key categories such as targeted delivery methods, controlled release systems, and intelligent delivery platforms. Figure 6 provides a schematic overview of these drug delivery strategies, emphasizing the role of biocompatibility in enhancing precision and effectiveness.

4.2.1 Controlled-release formulations

Pharmaceuticals reach the target site in the body slowly through the administration of therapeutic substances with defined release rates using controlled-release drug delivery systems. By doing so, it reduces the administration intervals, enhances patient compliance, and contributes to the sustenance of therapeutic concentrations of such medications in the circulatory system. Due to their degradation in a controlled manner that releases the medicine in a steady and systematic way, biodegradable polymers such as PCL and PLGA are commonly used in the construction of such devices. Depending on the requirements of the therapy, these materials can be developed into hydrogels, nanoparticles, and microspheres, among others.

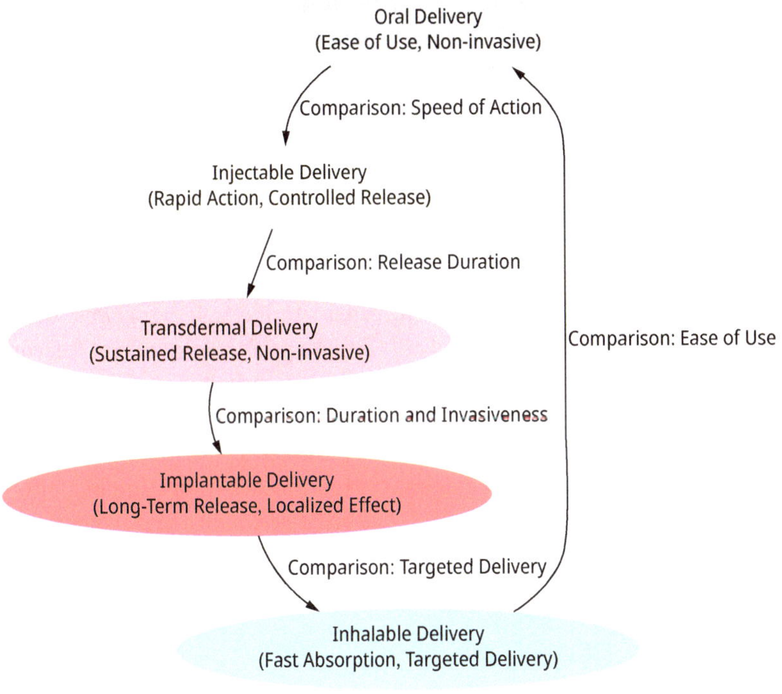

Figure 6: Drug delivery systems.

4.2.2 Targeted drug delivery

Targeted drug delivery systems are intended to deliver therapeutic drugs to diseased tissues only, without affecting healthy ones. This precise approach reduces deleterious consequences while enhancing the curative advantages of pharmacotherapy. Such systems require biocompatibility, and these materials are often used as carriers that can potentially be functionalized with antibodies. To transport chemotherapeutic medicines specifically to the aforementioned tumor site, for instance, nanoparticles can be functionalized with coat changes that recognize receptors on cancer cells.

4.2.3 Smart drug delivery technologies

Smart drug delivery devices that are capable of responding actively and informatively to various physiological changes or stimuli provide precise control over medication release. These systems often contain biocompatible components that are sensitive to pH, temperature, ionic strength, and other environmental factors. Some of the advances in drug delivery systems, such as temperature-sensitive hydrogels, can reversibly undergo phase changes in response to immune changes within the body to release a specific medicine. Similarly, when and where the pH in the body differs from the external environment, such as in tumor areas, which have a low pH of about 6, pH-sensitive products discharge medication.

4.2.4 Biodegradable implants for drug delivery

Medication was administered directly at the affected site through resorbable depot systems, which disintegrate within the body. The most preferred material for constructing these implants was the biocompatible and bioresorbable material that includes PLA. For this reason, these systems decrease fluctuations in systemic exposure and do away with the need for multiple injections by dispersing the medication at the point of invasion. Some of these uses include cancer therapy, where the implant offers continuous and local drug administration in the affected area, hormone replacement therapy, and birth control implants.

4.2.5 Injectable drug delivery systems

Injectable drug delivery systems involve the use of biodegradable and biocompatible polymers to administer drugs with minimal invasiveness. Such systems can include injectable hydrogels, microcapsular systems, or delivery via syringe. It is relatively easy to administer injectable formulations, fewer operations are required to carry them out, and it is possible to administer the drug continuously. These systems are especially beneficial in delivering biologics and vaccines, which require precise dosages and slow or sustained release.

Incorporation of biocompatible materials in drug delivery systems has enhanced the pharmaceutical field in various ways and opened new doors to deliver effective treatments and achieve better results. These systems ensure that many of the drawbacks and limitations of conventional delivery technologies are overcome, leading to improved health outcomes and better patient compliance.

4.3 Tissue engineering

In regenerative medicine, tissue engineering is an innovative approach that seeks to address this problem by reconstructing damaged or lost tissues and organs using comparable materials. This discipline is an intersection of engineering, materials science, and biology, aiming to construct functional tissue structures that can overcome the limitations of other therapies. The biocompatibility of some of these materials is crucial in tissue engineering because they serve as matrices, scaffolds, and carriers for the formation of new tissue and cell growth, respectively.

4.3.1 Scaffold design and function

Scaffolds are considered fundamental units in tissue engineering, providing physical support for cells to adhere, grow, and differentiate. Selection of proper biocompatible materials to possess the correct mechanical properties, porosity, and biodegradability was a major concern to be addressed during the scaffold designing phase. Due to their controlled degradation rates, structural flexibility, and selectivity for specific tissue types, polymers such as PLA, PGA, and their copolymers are commonly used. To enable native tissue formation and remodelling to progress uneventfully and for the scaffold to be effectively integrated into the host tissue, the scaffold must be ECM-mimetic.

4.3.2 Bioprinting and advanced fabrication techniques

Bioprinting and fabrication have advanced tissue engineering through innovation and enhanced approaches. Tissues with highly defined tissue organization can then be built by laying down compatible materials, cells, and bioprinting. Due to layer-on-layer deposition, researchers create three-dimensional structures of the scaffolds that closely resemble the natural tissue architecture. These technologies allow the biological properties of the scaffold to be altered effectively and enable the incorporation of various types of cells, which naturally promote improved tissue integration and functionality.

4.3.3 Cell-seeding and growth factors

The correct approach in this regard is the proper planting of cells on scaffolding and the application of growth factors that promote tissue growth, which are also requirements of tissue engineering. Biocompatible materials used in the construction of scaffolds must encourage cell adhesion and proliferation for the construct to be biologically functional. Antifibrotic molecules, such as ECM proteins, can be incorporated into the scaffold material to enhance the development of new tissue and levels of cell activity. This procedure allows for the generation of physiological tissue constructs that can be compared with real tissues.

4.3.4 Applications in regenerative medicine

Tissue engineering has demonstrated potential in a wide variety of applications within the regenerative medicine field. For example, engineered skin grafts and dermal substitutes have been developed for managing burn injuries and chronic wounds with better therapeutic efficacy than conventional approaches. In orthopedics, tissue-engineered bone and cartilage grafts are being created to fix defects, offering advantages over donor tissues in terms of recovery. Advances in tissue engineering also support the construction of novel functional tissues, including bioengineered heart valves and vessels, which could address challenges associated with organ donation.

4.3.5 Challenges and future directions

Up to date, there are some concerns that hinder advancement in the tissue engineering field. Maintaining the functionality and compatibility of the engineered tissues in the long term is a major consideration as well as their manufacturability and approval for use. Additionally, efforts are focused on achieving efficient vascularization of tissue constructs because tissue survival and function in the body depend on blood circulation. Regarding the limitations of existing work, this remains the primary area of focus in further research, as new materials have yet to be examined, fabrication methodologies need to be improved, and a variety of tissue types have yet to be successfully reconstructed.

4.4 Biocompatible coatings

Self-assembled coatings represent a significant advancement in the field of biomedical materials, offering the potential to improve the interface between medical devices and biological tissues. These coatings are applied to prosthetics and other medical components to modulate tissue responses minimizing adverse reactions while promoting favorable biological interactions. By exploring foundational principles and case studies, this section highlights the versatility and impact of biocompatible coatings across various medical and technological domains. Figure 7 illustrates the structure and function of self-assembled coatings, showcasing their role in enhancing device performance, biocompatibility, and longevity in clinical applications.

4.4.1 Drug-eluting coatings

Drug-eluting coatings are specifically designed to deliver desired therapeutic agents in a sustained manner. These coatings are especially desirable in applications such as cardiovascular stents, where localized drug delivery can prevent restenosis. Drug delivery through coatings creates localized release of the drugs at the targeted implant site and reduces the reliance on systemic medicines, thereby enhancing the efficacy of the drug delivery system.

4.4.2 Antimicrobial coatings

Main applications of antimicrobial coatings are to prevent bacterial and other germ growth on the outer surfaces of medical products. These coatings are extremely important for reducing the infections associated with surgical utensils and implants. The coating material uses various antimicrobial agents either in the form of antibiotic nanoparticles to provide a long-lasting barrier against microbial adhesion. Effective antimicrobial

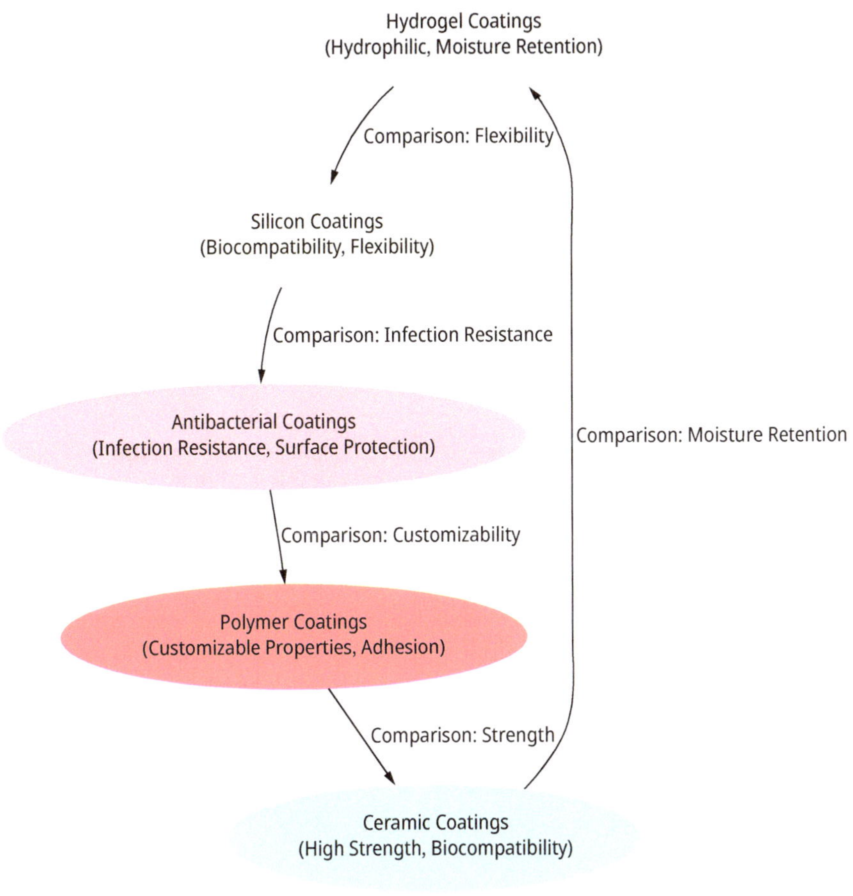

Figure 7: Biocompatible coatings.

coatings ideally improve patient safety because the occurrence of infections involving medical equipment decreases tremendously.

4.4.3 Surface modification coatings

Coatings applying surface modification are employed to alter the surface properties of materials in order to enhance biocompatibility. The basic properties of the surface, such as roughness, wettability, and chemical nature, can be altered by techniques such as chemical vapor deposition, plasma treatment, and self-assembled monolayers. Such changes provide the possibility to improve cell adhesion, proliferation, and integration with the surrounding tissues. A number of areas are possible for the use of surface modification coatings; these include tissue engineering scaffolds, dental prostheses, and orthopedic implants.

4.4.4 Hydrophobic and hydrophilic coatings

Hydrophobic and hydrophilic coatings are intended to regulate the relationship between medical devices and body fluids. Hydrophobic coatings have no affinity for water and are deployed to minimize biofilm formation and tissue attachment, whereas hydrophilic coatings have an affinity for water that favors the integration of the device and tissue. Special applications include the ability to control the level of lubrication of joint implants and the use of hydrophilic surfaces for drug delivery.

4.4.5 Bioactive coatings

Biologically active coatings are designed to exhibit suitable reactions with biological tissues, thereby eliciting certain responses such as tissue bonding. Essential stimulatory substances, such as growth factors, which stimulate cell activity and accelerate the healing process, are often incorporated into these coatings. One example of bioactive glasses is their use in the coatings of implant components; such coatings improve implant connection while promoting the development of bone at the site of the implant. Bioactive coatings enhance the field of regenerative medicine through tissue healing and the more efficient functioning of medical tools and devices.

4.5 Wearable devices

Wearable devices represent another promising branch in BMT, utilizing biocompatible materials to provide uninterrupted health monitoring, examination, and treatment. These wearable devices, such as smartwatches, fitness trackers, and medical sensors, apply material science to ensure functionality, comfort during prolonged use, and safety.

4.5.1 Continuous health monitoring

Smart clothing endowed with adaptable biocompatible sensor systems implements simultaneous, noninvasive tracking of various physiological variables including heart rate, blood pressure, and blood glucose levels. These sensors usually employ the use of skin-like flexible polymers and stretchable conductive fabrics. For instance, the electrocardiogram monitoring patches use biomagnetically compatible adhesives and electrically conductive materials to achieve both the heart's electrical signal recording and the patient's comfort with no skin chafing. The phenomenon of continuous data collection enhances the early detection of health complications and allows for early treatment.

4.5.2 Biocompatible materials for enhanced comfort

Wearable devices had to be comfortable for users to wear, which made their compliance necessary to monitor them effectively. Safe materials like hypoallergenic silicones and soft elastomers have been used to prevent skin rashes for the user and ensure that the brace fits well. These fabrics are designed to be "sweat-friendly" and water-repellent, addressing most of the mentioned challenges, such as bacterial formation and skin irritation. Nonetheless, user comfort enhances the applicability and reception of wearable health gadgets through the use of biocompatible materials.

4.5.3 Integration with digital health systems

Wearable technology, many a time, has friendly biocompatible interfaces with digital health systems for integration. Such materials continue to be developed to allow physiological information to be collected through sensors and flexible electronics into smartphones and tablets. This compatibility enhances distant tracking and evaluation of trends in the customer's health and allows for customized control for both the client and the suppliers.

4.5.4 Therapeutic wearables

The devices are designed to have therapeutic value in that they can be used to track signals. Some examples of biocompatible materials include the following: transdermal drug delivery systems and neurostimulation units. For instance, transdermal systems in the management of chronic conditions involve containing the medicine in patches that integrate biocompatible polymers and hydrogels for the predictable, controlled release of the drug. In the same manner, wearable neurostimulation devices use biosensors to supply electrical current to the body for relief from chronic pain, enhancing the patient's experience through nonsurgical interventions.

4.5.5 Wearable sensors for chronic disease management

Wearable devices are important in maintaining supervision and providing constant readouts for the management of chronic ailments. Smart biocompatible materials are applied in monitoring diseases such as diabetes, hypertension, and asthma, among other conditions. For instance, biosensors used in continuous glucose monitoring are safe and implanted to measure glucose in interstitial fluid, helping those with these conditions manage them better. Through the information collected by these devices, treatment regimens are modified, and potential adverse effects are avoided, making life for patients suffering from chronic ailments far more comfortable.

The research has demonstrated the use of advanced biocompatible materials for wearables, which have transformed personalized health care through continuous monitoring, comfortable fit, and innovative treatments. As technology improves, the use of these materials continues to expand opportunities to advance wearable health devices, enhancing preventive patient care.

5 Factors influencing biocompatibility

5.1 Chemical composition and its impact on biocompatibility

The biological performance of biocompatible materials and their degree of biocompatibility depend mainly on their chemical nature. Three major aspects focus on the ability to achieve optimal biological performance of a material designed for medical use: formulation chemistry, the minimization of side effects through design aimed at optimizing therapeutic results of the intended material, and the design of a material that would not adversely affect existing treatments.

5.1.1 Material composition and biological interaction

The physical properties of a material include information regarding the chemical composition of the material and its interactions with the biological tissues where it is utilized. For instance, titanium and stainless-steel materials are preferred due to their low chemical reactivity and stable oxide layers, which prevent direct contact between the metal and the surrounding tissue. On the other hand, materials that contain components like nickel may cause adverse reactions depending on the individuals exposed to them. As such, the chemical composition of the materials to be used should be carefully chosen to avoid undesirable side effects.

5.1.2 Surface chemistry and protein adsorption

The surface properties of any material that comes into contact with proteins define how much of it is adsorbed, an important consideration when designing biomaterials. When a material is implanted into the body, the proteins from the surrounding environment rapidly adhere to the surface of the implant, creating a layer of protein that influences cellular response. Most surface chemistries that encourage the adsorption of specific proteins are generally considered biocompatible, especially if the material chemistry is hydrophilic. For example, hydrophilic coatings decrease nonspecific protein adsorption and cell adhesion, thereby decreasing the inflammatory response and improving the integration of the device with biological tissue.

5.1.3 Degradation products and cytotoxicity

The biodegradation of many biocompatible materials can lead to the release of chemical by-products that significantly influence their overall biocompatibility. Polymers, composites, and other biomedical materials often produce degradation products that are absorbed by surrounding tissues during the breakdown process. The type and concentration of these by-products can directly affect cytotoxicity and inflammatory responses. Therefore, materials that degrade into non-toxic, bioresorbable, and easily metabolizable compounds are preferred, as they pose minimal risk to surrounding tissues. When designing such systems, it's crucial to simulate natural degradation pathways and select materials that maintain biocompatibility over their entire functional lifespan. Figure 8 illustrates the degradation mechanisms of biocompatible materials, highlighting the relationship between degradation products and biological response.

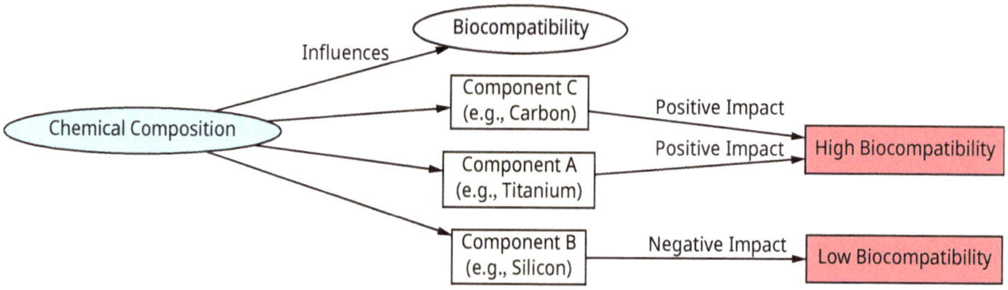

Figure 8: Chemical composition and its impact on biocompatibility.

5.1.4 Interaction with biological fluids

Another important factor was the adverse effect of the materials on biological fluids like blood. Now we understand that interactions between the material and these fluids influence the stability and functionality of the material. For example, polymers that possess strength against hydrolysis are not very capable of degrading toxic substances when in contact with body fluids. The ability of the materials to interact chemically with blood components, such as clotting factors, affects their applicability in uses like blood-contact devices. Polymer types that are least reactive in biological solutions are typically more biocompatible.

5.1.5 Long-term stability and chemical durability

The material's chemical stability over the long term was critical in ensuring biocompatibility over an extended period of time. The chosen materials for chronic implant devices should not undergo chemical alterations within the body. Instability factors, including the leaching of free and unbound monomer products, are usually injurious to biological processes. Hence, it is evident that materials with good chemical stability and resistance to environmental forces are essential for maintaining the stable biocompatibility of the materials, along with the reliability of their operation.

5.2 Surface properties

Regarding the biocompatibility of materials used in implants and medical devices, the properties of their surface are still quite significant. A literature review shows that tissue integration, cellular response, and device performance are sensitive to this model, which represents the relationship between an implant material's surface and its surroundings. Among the most pertinent attributes of the surface are chemistry, topography, roughness, and wettability, all of which define biocompatibility.

5.2.1 Surface chemistry

The chemistry of a surface element, in terms of how it interfaces with living tissues and body fluids, depends on the chemical nature of the elements involved. Alterations in surface chemistry are implicated in protein adsorption, cell attachment, and immune interactions. The adsorption of proteins to material surfaces with appropriate chemical functionalities is favorable for cellular attachment and proliferation. On the other hand, materials that prevent protein adsorption from adhering to the surface minimize the chance of inflammation. In terms of physical modification of materials, surface modification and coating can be used to enhance surface chemistry and improve biocompatibility.

5.2.2 Surface topography

Microarchitectural features of the material surface and its topography, including the grooves and porosities as well as textural structure, affect cell response and tissue interfacing. Nanoscale topography of the substrate influences cell attachment, mobility, and phenotypic expression. For instance, the engaging surface topography at the micron level plays a critical role in bone-to-implant adhesion in orthopedic applications and tissue engineering products. The design of surface topography, with regard to issues concerning mechanical support and biological responses, has been recommended to be carried out in a way that allows the successful integration of the designed solution with existing systems and structures while simultaneously eliminating complications.

5.2.3 Surface roughness

Surface roughness was defined in terms of variations in surface topography and was one of the dominant parameters that helped to define how a material engaged with biological tissues. Rough surfaces enhance surface area contact between cells and implants and improve implant stability. It has been discovered that high roughness can cause undesirable tissue reaction problems such as inflammation coupled with fibrosis. The peak roughness that was favorable on the surface was relative to the application and the type of tissue. For this reason, there was a need to exercise a lot of caution when designing the material and processing.

5.2.4 Wettability

Wettability, which is the ability of a material to interact with a liquid, determines the way biological fluids and their cells respond to the surface. In the present work, hydrophilic surfaces, which are surfaces that favor access to water, preferentially improve cell adhesion and proliferation because of their ability to allow protein deposition and cell spreading. Nonadherent hydrophobic surfaces, which do not adhere to water, significantly deter cell attachment and result in unfavorable bioresponses. Wettability can also be controlled through plasma treatment modification in order to better suit the material for biological applications.

5.2.5 Surface stability and degradation

The mobility of a material's surface is important in the preservation of biocompatibility because of the resilience it provides to the surface over time. Alterations in the surface chemistry of the material due to changes in topographical features like roughness, hydrophobicity, and the chemical composition of the top layer, caused by exposure to body fluids, influence the functionality of the material as well as the body's response to the material. For biodegradable materials, controlled degradation is critical to allow the release of degradation products and prevent the loss of mechanical integrity from being too abrupt. Controlling the stability of the outer surface is especially crucial, as surface stability helps predict overall performance in the long term, and its usage in medical applications requires safety and reliability.

Surface features remain core aspects due to their influence on the biomedical performance of medical materials and devices. Through this system, with improved surface chemistry, topography, roughness, wettability, and stability, the contact of materials with biological systems is improved, with reduced complications. New developments in surface science and technology of surface modification also help to enhance the design and functional applications of biocompatible materials.

5.3 Mechanical properties

In this context, the mechanical properties of biocompatible materials are critical for determining their suitability in biomedical applications. Key attributes such as strength, flexibility, hardness, and fatigue resistance play a major role in how a material performs within the biomechanical environment of the human body. Ensuring that these mechanical characteristics are well-matched to the intended application is essential for achieving optimal performance and minimizing potential complications. Figure 9 presents a comparative overview of mechanical properties across commonly used biocompatible materials, illustrating how these properties influence their biomedical functionality.

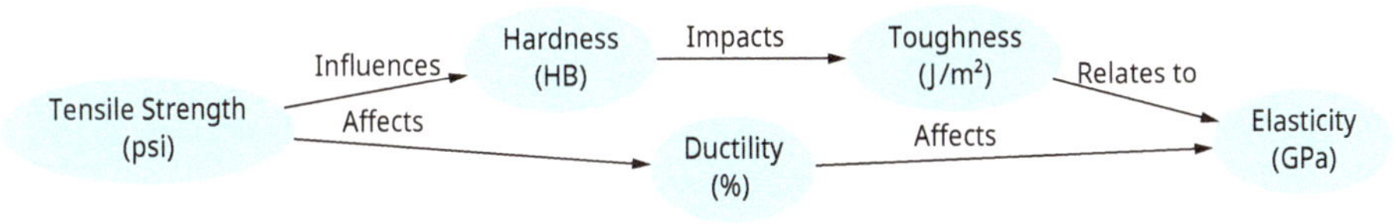

Figure 9: Mechanical properties.

5.3.1 Strength and durability

Strengthening was a primary meso-scale mechanical property relevant to the behavior and durability of biomaterials (BMs) for implant and prosthetic applications. Some of the design constraints include ensuring adequate tensile strength, compressive strength, and shear strength in the material to withstand the mechanical loads it experiences during use. For example, hip replacements and dental implants require high impact strength because they bear the load-bearing capacity of bones and teeth. Thus, the choice of the right strength materials guarantees the durability of the devices, as the chances of failure in the future are minimized.

5.3.2 Flexibility and compliance

The third category showcased the need for flexibility because some materials need to adapt to various shapes and sizes in the body. Tissue engineering scaffolds, stents, catheters, and dressings should be able to change their shape in accordance with the movement of tissues and changes in their position. Conformance of the material, to possess identical mechanical characteristics as the host tissue, was especially crucial considering the need to avoid stress concentrations, which result in adverse outcomes. Substances that are either too brittle do not interface well with body tissues and induce adverse effects, causing the device to fail.

5.3.3 Hardness and surface properties

Biocompatibility was an initial criterion for the hardness of materials that are in direct contact with biological tissues. The hardness of the surface of these materials can affect their stability in contact with the surrounding tissue, wear, and tear as well as abrasion. For instance, dental materials need to have the right hardness because chewing involves applied forces that should not harm the adjacent structures. Additionally, surface roughness and texture, which are inversely proportional to hardness, are associated with cellular attachment and tissue integration compatibility of the material.

5.3.4 Fatigue resistance

Fatigue was defined as the resistance of a material to failure when subjected to alternating loads of normal stress and eventually unloaded. This property is essential for implant applications where materials are prone to dynamic loading conditions, for example, in joint replacements. High fatigue strength materials are capable of withstanding cyclic loads for a long time without fracturing. It is necessary to ensure that the constituent material does not lose its mechanical properties and structure when exposed to cyclic applied loads, to ensure the durability and dependable performance of biomedical implants and devices.

5.3.5 Creep and deformation

Creep is the ability of a material to deform continuously under a constant load over time. Low-creep materials are advantageous for biomedical applications since deformation can affect the proper fitting of the material being used. For example, if it is an implant for a spinal application or a prosthetic, the material must be able to retain its shape and dimensions when subjected to constant forces so that connection, orientation, and positioning can be maintained and proper function achieved. Reducing creep and deformation is crucial to ensuring the functionality of the device and minimizing engineering-related functionality issues.

According to the work, the mechanical properties of biocompatible materials should be highly relevant to their function and use in medicine. Selecting the right kind of materials with the appropriate strength, flexibility, hardness, fatigue resistance, and low creep would help biomedical engineers and scientists increase the performance and reliability of biomedical applications. These reflections guarantee that the materials used can cope with the requirements placed on them, thus contributing to better patient conditions and a longer, more reliable lifespan for the devices used.

5.4 Degradation and stability

The degradation and stability of biocompatible materials play a decisive role in the technical characteristics of their usage in medicine. These properties demonstrate how materials perform, degrade, and integrate with biological systems after their use, usage duration, and ultimately their biocompatibility.

5.4.1 Material degradation mechanisms

The subject by which the material loses its properties gradually until it becomes less effective, especially when faced with physiological conditions, was termed degradation. Deterioration can be due to several factors such as enzymatic action, oxidation, and hydrolysis. For instance, in watery conditions, materials such as bioresorbable polymers PLA and PGA undergo hydrolytic degradation, which results in the breakdown of the material. Understanding these degradation mechanisms is necessary to predict the material's behavior in vivo and ensure that it maintains its intended function throughout its expected life.

5.4.2 Rate of degradation

The degradation rate was one of the critical factors that determined the performance and safety of the material. In any event, it was ideal if the rate of degradation corresponded with the type of application the material was subjected to. For example, in drug delivery systems, the material used should have degradation characteristics that would enable it to deliver therapeutic agents over a specific time span. On the other hand, for orthopedic implants, the performance of the materials should be designed to undergo gradual degradation in order to provide long-term and sufficient mechanical support during the tissue regeneration process. The material composition and design that allowed for the control and optimization of the degradation rate comprised one of the key aspects of any material with such properties.

5.4.3 Stability in biological environments

One of the major challenges of biocompatible materials for long-term use in living organisms is the compounded stability of the materials within biological environments. Materials under consideration have to be stable under different physicochemical environments, including pH changes, temperature variations, and mechanical loading, without deterioration of the material's properties and performance. For example, materials used in implantable devices need to be resistant to attacks by body fluids and must retain their properties under stress. Material stability refers to the material properties and the material processing technologies required to ensure that the material used can last long and provide enhanced resistance to degradation.

5.4.4 Impact of degradation products

The degradation products of biocompatible materials affect biocompatibility and the health of the patient. In an ideal world, these products should not be toxic, should not incite an immune response, and should be readily excreted. For example, solutes that are used as materials for bioresorbable polymers are often metabolized into nontoxic products like water and carbon dioxide. It is important to control and assess the degradation products in order to prevent biological responses that could cause greater adverse reactions and threaten the safety of the material.

5.4.5 Long-term stability and performance

Durability is important in ensuring that biologically compatible materials retain functionality and remain harmless during their intended usage period. For materials used in permanent implant medical applications, a material must demonstrate stable performance over long durations. Variables such as the type of material, manufacturing and thermal treatments, and operating conditions influence the long-term stability of a material. Current advancements and innovations in the field of development are aimed at improving the longevity and dependability of biocompatible materials to meet the needs of the medical field.

5.5 Interaction with the immune system

The immune response to biocompatible materials is a critical factor that significantly affects their overall biocompatibility in healthcare applications. While the implantation of these materials often restores or enhances the functionality of damaged tissues, the body's immune reaction to foreign objects can compromise the efficacy and longevity of the implants. Therefore, controlling and minimizing the immune response is essential to ensure the success and durability of implanted medical devices. Figure 10 illustrates the typical immune response cascade triggered by biomaterial implantation, highlighting its impact on implant performance and the strategies used to mitigate adverse reactions.

5.5.1 Immune response to foreign materials

When any biocompatible materials are placed in the body, they are recognized as foreign by the body's immune system. This recognition leads to a series of actions aimed at excluding the material. In this context, the type and strength of the immune response are influenced by the chemical nature of the material including its synthetic components, surface properties, and resulting degradation by-products. Ideally, in the concept of biocompatible materials for implantation, a material should be the one that does not cause significant inflammation and, at worst, only elicits a mild immune response.

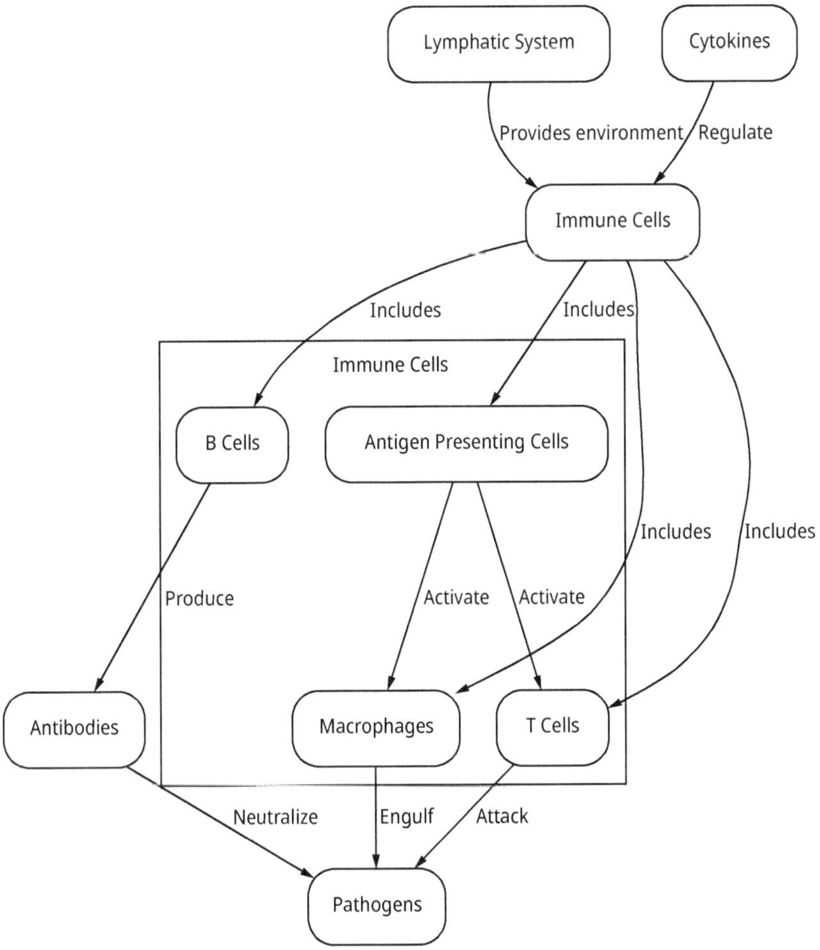

Figure 10: Interaction with the immune system.

5.5.2 Surface properties and immune activation

The features of a material's surface are particularly crucial for defining the ways in which the material is processed by the immune system. Parameters related to the surface, such as roughness, charge, and wettability of the material, can affect the adsorption of proteins, adhesion of cells, and subsequent immune cell activation. Substances that have soft, chemically inactive upper layers are typically least likely to provoke an immune reaction. Other methods, such as coating with anti-inflammatory materials or bioactive molecules, can also improve the compatibility of the material, decrease immune cell activation, and enhance the beneficial cellular response.

5.5.3 Material degradation and immune response

The biocompatibility of these materials means that their breakdown in the body can produce compounds that act on the immune system. These degradation products cause inflammation in tissues in the body. A useful material must be environmentally friendly for the time it remains in the environment and should biodegrade into substances that do not cause the immune system to react. As with any kind of dystrophy, observation of the degeneration process and its effects on the immune system was a critical factor for long-term biosafety and efficiency assurance.

5.5.4 Immune system modulation and biocompatibility

Some selected polymers are purposely developed to elicit impressive immune responses. For example, some alter their properties to emit anti-inflammatory agents, thereby controlling the immune response to an implant. These materials are designed to do so. As mentioned, there are biomaterials intended to control immune responses and decrease the risk of chronic inflammation. These sorts of materials must necessarily be based on knowledge of immunological processes, and their construction must be undertaken with an eye toward both modulating immune reactions and maintaining overall biocompatibility.

5.5.5 Personalized approaches to immune compatibility

Considering variation, newer methodologies are being applied in personalized therapy for immune responses of the human body to biocompatible materials. Body genetics, body environment, and some other body physiologies of the patients determine how the immune system responds to implants. This suggests that in the future, designing biocompatible materials that match the patient's profile, such as through material layering and built-in stimuli-responsive characteristics, appears to have the potential of yielding better results and causing fewer immune reactions.

The immune response to biomaterials was one of the critical design features of the biomaterials and their use. Taking into account the immunogenic parameters that can modulate immune reactions, investigators and practitioners could promote technological advancements and material properties as well as minimize adverse reactions and patient morbidity in extensive biomedical applications.

6 Challenges in the development of biocompatible materials

6.1 Predicting long-term biocompatibility

Forecasting the ability of a material to become part of the body as a long-term resident can be a serious issue in the creation of biomedical tools and implants. The most challenging aspect of their development is the capacity to predict how the materials behave in biological systems in the long run and to prove whether they are both safe and efficient as devices. This prediction is not within a simple sphere of influence and depends on related parameters such as the degradation of the material, biological variation, and the constantly changing response of the body toward foreign objects.

6.1.1 Material degradation and longevity

The term "biocompatibility" means that the surface properties of a material do not change with time, and therefore, the degradation of biocompatible materials affects their biocompatibility. Commercially available products that are biodegradable must disintegrate into nonreactive by-products that cause no harm. The rate and degree of degradation depend on the chemical composition of the material, environmental conditions, and mechanical loading. Over the lifetime of a material, the prediction of how it degrades and the type of degradation it undergoes is not simple and involves extensive testing and modeling. This complexity poses a significant problem in terms of keeping these materials biocompatible as they go through their expected lifespan.

6.1.2 Long-term biological reactions

The body does not react uniformly to implanted materials; the response changes with time. Initial studies of biocompatibility do not provide knowledge of the latent effects of a biomaterial in a biological environment since immune responses and tissue integration change as the biomaterial is exposed to the body. Processes that occur months after implantation include chronic inflammation, immune response, and fibrosis. Forecasting these chronic biological responses can only be achieved by conducting long-term animal experiments and clinical trials of new materials, which are time-consuming, costly, and often incomplete due to skipping some steps in material degradation.

6.1.3 Variability in patient responses

Another complicating factor in long-term prognoses, therefore, was variability in the reaction of patients to biocompatible materials. One unique patient may have a completely different reaction to a material than another due to genetic variation or a prior illness. This variation poses significant difficulty in the anticipation of results and hinders the extendibility of conclusions from standard biocompatibility tests; individual evaluations and adaptive design strategies are needed to resolve this.

6.1.4 Incomplete simulation of in vivo conditions

In vitro assays and preclinical investigations tend to share the same challenges in mimicking the in vivo environment experienced by materials after their implantation. Nowadays, certain aspects like dynamic physiological changes, mechanical forces, and integrative biological systems are complex and hard to model. This limitation ensures that what was forecasted within laboratories does not necessarily hold true in practice. There is a need for new advanced modeling techniques and better in vitro and in vivo testing procedures to substitute the complex QRPA, yet the full in vivo simulation remains a great challenge.

6.1.5 Regulatory and ethical considerations

Biocompatibility standards are established to test the suitability and applicability of materials and products that come into contact with the human body. To achieve these standards, long-term research may be necessary, which can still be expensive in terms of time. There are also ethical concerns when long-term research involves the use of animal participants. The ability to predict biocompatibility for the long-term is further challenged by the conflict between providing safety assessments that must meet broad guidelines while conforming to ethical standards and legal regulations.

Overcoming these challenges needs solutions from material science, long-term evaluations of biological systems, personalized medicine solutions, and better test modalities. By creating increasingly sophisticated and refined simulation paradigms and gaining improved insights into the long-term dynamic interfaces between materials and biological systems, it becomes possible to improve the stability and safety of biomaterials in clinical use.

6.2 Balancing biocompatibility with functionality

A major challenge in the use and development of biocompatible materials lies in achieving an optimal balance between biocompatibility and functionality. While biocompatibility ensures that the material interacts safely with biological systems, functionality pertains to the material's performance and effectiveness in its intended biomedical application. Striking this balance is particularly critical in the design of medical devices and implants, where both safety and operational efficiency are essential for successful clinical outcomes. Figure 11 illustrates the trade-off and interdependence between biocompatibility and functionality, emphasizing the need for materials that do not compromise one aspect in favor of the other.

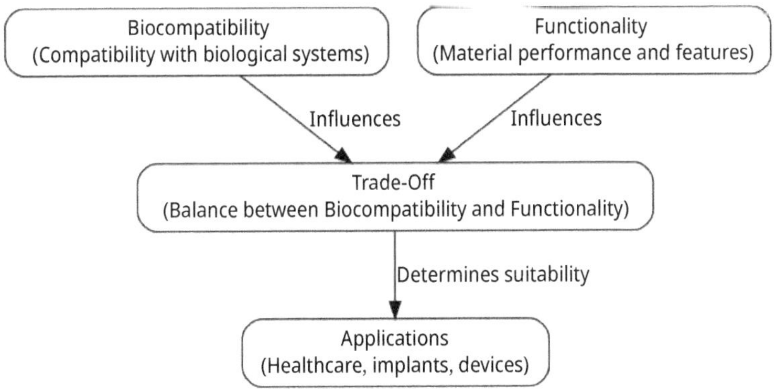

Figure 11: Balancing biocompatibility with functionality.

6.2.1 Trade-offs between biocompatibility and mechanical properties

The first issue that was critical was the task of adjusting the biocompatibility and mechanical properties of the material. For example, materials that can be used in orthopedic implants must have the features of high strength and high durability since they work under mechanical load. The development of these mechanical properties can be problematic because they create inflammations in neighboring tissues. Thus, material scientists have to pay attention to the material design and selection such that the mechanical strength is good enough but does not produce disruptive impacts on the body.

6.2.2 Functionalization vs. biological response

Functionalization occurs when different chemical groups of a material are altered in specific ways for improvement. Although such changes help optimize the performance of the scaffold, such as achieving faster drug release rates to tissues, they can impact biocompatibility. For instance, some chemical treatments can make underlying immune reactions more likely to occur. A delicate balance between functional advancements on one hand and biological reactions on the other calls for considerable design as well as experimentation.

6.2.3 Integration with biological tissues

The successful coupling of a biocompatible material and a biological tissue requires the ability of the identifiable material to not only accomplish a particular function but also elicit appropriate responses from the surrounding biological environment. This integration is especially difficult in functional applications such as tissue engineering of materials that need to support cell proliferation and tissue regeneration as well as serve as the scaffold system. Any materials that are used must support cell attachment as well as cell growth and division, in a way that minimizes inflammation and rejection.

6.2.4 Durability vs. biodegradability

The biodegradable materials by design had one key problem regarding the rate of degradation and the functional performance, which could not always be attained. Whereas biodegradation was desirable for short-term implants and carrier systems, a precise rate of degradation was required for the carrier material to reflect the intended durability and performance of the implant. If degradation occurred at high rates, the intended application would fail prematurely and be unable to function properly. On the other hand, if the degradation rate was very low, the long-term effects would be detrimental, requiring the component's removal. Ensuring that it degraded at an optimum rate to meet its intended function was the key challenge.

6.2.5 Regulatory and safety considerations

In addition to the functional requirements a biocompatible material must meet, there are compliance issues that call for the prevalent two-dimensionality, whereby an applicant has to prove both biocompatibility and functionality, particularly as the broader society demands that material testing be safe and practical. The regulatory bodies set certain rules to ensure that the material used possesses the required safety features while simultaneously demonstrating efficacy. Meeting these regulations as well as striving for better material functionality is not always easy and involves planning throughout the entire product development cycle.

Maintenance of both biocompatibility and functionality was an accountability quest that presupposed a thorough assessment of factors such as material characteristics, biological effects, and usage efficiency. This challenge was solved in a more complex manner as it required the incorporation of proper knowledge and information from material science, biology, and engineering to construct materials that not only fulfilled mechanical and physical requirements but also offered safety and efficiency in the biomedical field.

6.3 The complexity of biological environments

The interconnectivity of these multiple parameters was a daunting aspect, especially when it comes to synthesizing appropriate biological structures that give rise to the design of materials for biomedical applications. Living tissues and cells are notable for their activity and complexity, which, in turn, creates a complex relationship between biocompatible materials and the tissues surrounding them.

6.3.1 Variability in biological conditions

Biological environments comprise large deviations in features like pH, temperature, and ion concentration as well as other biological macromolecules. These conditions can vary owing to the general condition of the patient, the advancement of affliction, and the area of insertion. These alterations can only be accommodated if the outcome-based materials used are capable of handling them while at the same time offering the expected functionality and not eliciting undesirable reactions. This requirement calls for rigorous assessment and calibration in order to ascertain that materials function optimally in various biological environments.

6.3.2 Interaction with biological fluids

Blood, interstitial fluid, and other tissue exudates have close contact with implanted materials on a daily basis. It has been shown that the chemical nature and flow characteristics of these fluids change material attributes and their responses to such issues as protein adsorption, biofilm formation, and degradation. For example, plasma proteins adhere to the surface of the material and affect cell behavior as well as the stability of the material in the long run. Engineering the surface to control such interactions and prevent malicious proteins from sticking to the material while encouraging the right kinds of proteins to bind to it is an essential factor in biocompatibility.

6.3.3 Immune system dynamics

Various immunocompetent cells and cytokines interact with the implanted material and the tissue microenvironment of the recipient organism. The immune environments of materials can be affected in response to infections and injuries before influencing the manner in which the material engages with the host. The implementation of materials requires rigorous design processes in order to reduce inflammatory reactions and prevent constant immune responses. This balance is somewhat difficult to strike and involves a level of understanding immunology and how different materials perform in various immunological situations.

6.3.4 Cellular and tissue integration

Effective application of biocompatible materials into tissues requires multiple cellular mechanisms such as adhesion, reproduction, and differentiation. These processes should be facilitated by the material's properties, yet neither cytotoxicity nor aberration in the tissue architecture is acceptable. Two major issues stand out in the creation of materials that can positively engage with various types of cells and tissue matrices, especially where the tissues exhibit composite characteristics. Surface features, among other material properties like surface roughness, mechanical strength, and biochemical signals, should exhibit certain characteristics in order to foster the desired cell behavior and integration.

6.3.5 Long-term stability and degradation

These engineering changes to biomechanically compatible materials and structures were found to be largely a result of interactions with biological environments over a long-term basis. Composites should be mechanically and chemically stable throughout the intended service life without creating deleterious degradation products or eliciting adverse biological responses. To avoid complications such as excessive inflammation of the tissue and to ensure full functionality, the firing process must degrade the materials in a predictable and controlled manner through chemical processes.

Little is known about biological environments, which makes their mimicry a complex task to undertake, as was evident in the creation of the biocompatible material. To overcome these challenges, it is essential to take a multidisciplinary approach, incorporating material science, biology, and engineering to develop materials that can function effectively in the dynamic environment and variability of the human body.

6.4 Regulatory hurdles in approving new biocompatible materials

The legal and regulatory framework presents substantial challenges in the development and approval of new biocompatible materials intended for use in advanced implantable devices. Regulatory bodies such as the European Medicines Agency (EMA) and the U.S. Food and Drug Administration (FDA) play a crucial role in ensuring that these materials meet rigorous standards of safety and efficacy before they reach the market. The complexity and stringency of these regulatory processes often create significant hurdles for developers, potentially slowing down innovation and clinical adoption. Figure 12 outlines the regulatory approval pathway, illustrating the key checkpoints and challenges involved in gaining market authorization for biocompatible implant materials.

6.4.1 Comprehensive preclinical testing requirements

In fact, a new biocompatible material must go through a comprehensive series of steps within the preclinical stage in order to prove its safety, biocompatibility, and efficiency prior to its use in clinical applications. These tests often involve in vitro screening test-based studies for cytotoxicity, gene mutation, and changes in blood elements as well as in vivo tests to check the material's compatibility with living tissues. Due to the strict implementation of these testing requirements, the development processes generate long and expensive cycles. Requirements that call for meeting certain standards set by distinct regulatory authorities for various zones are challenging, as the norms differ between corresponding authorities.

6.4.2 Challenges in predicting long-term biocompatibility

In our work, we found that one of the most significant issues for regulatory agencies was the long-term biocompatibility of new materials. Developers cannot make unsubstantiated claims that the material does not elicit adverse reactions to a level that is injurious to the patient over extended durations. Forecasting was particularly challenging, especially in the case of materials designed for lifetime use, including implantation. This often results in new rounds of testing and longer review periods

6.4.3 Variability in global regulatory standards

Since the market for medical devices is a global one, the developers involved have to work through even more regulatory challenges across countries. Standard requirements for testing and approval of biocompatibility also differ from one country to an-

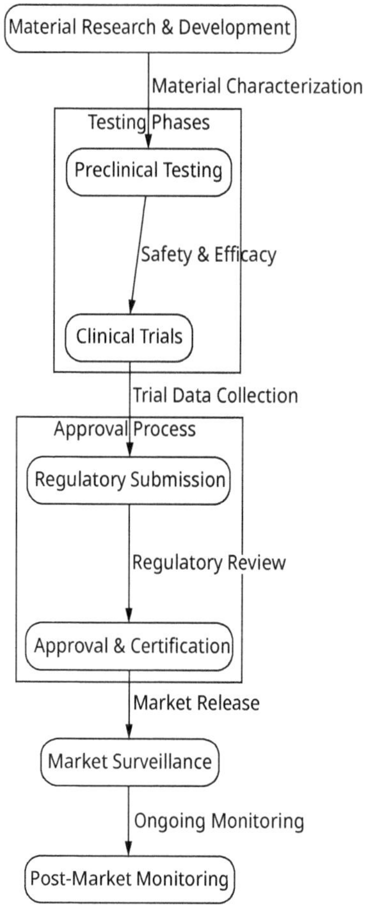

Figure 12: Regulatory hurdles in approving new biocompatible materials.

other, and thus delays and higher costs of approval are likely to be incurred in the case of multinational approval. There remains disagreement in these processes despite efforts to harmonize the activities such as those by the International Organization for Standardization. Target markets impose strict regulations that have to be met by developers at a certain cost, which considerably hampers creativity.

6.4.4 Post-market surveillance and reporting obligations

Unfortunately, manufacturers of biocompatible materials cannot rest even after obtaining the required regulatory approval from the authorities; they have to submit to post-market surveillance and reporting. These requirements guarantee that any unfavorable circumstances or issues are immediately recognized and dealt with. Nonetheless, an attempt to sustain such compliance is arduous, especially for start-ups with a small pool of personnel. The requirement to assess performance and report on it over the whole life cycle of the material raises further considerations in terms of regulation, which could influence its acceptance within the market place and future sustainability.

6.4.5 Ethical and legal considerations

Regulatory approval processes also include issues pertaining to ethical and legal values in the application of new biocompatible hardware materials. This includes informed consent from the patients, reporting of possible risks as well as adherence to ethical considerations in trials. Litigations, for instance, patent issues, can also bring additional problems to the approval process. Regulatory concerns add to this when ethical and legal issues must be planned very carefully and in cooperation with appropriate institutions, which means that all materials subject to regulation become even more complicated and expensive to introduce to the market.

The approval processes of these biocompatible materials are especially rigorous, making it difficult for developers to release new materials to the market and slowing the adoption of better medical technology. Addressing these issues calls for an understanding of the principles of regulatory rules, effective procedures to conduct tests, and well-coordinated planning to work in compliance with the guidelines during the progress of developing more comfortable and biocompatible materials.

The use and development of biocompatible materials present some ethical questions that need to be addressed to avoid negative implications for their use in the medical field. These points raise concerns about safety, access, informed consent, and the consequences of using these materials in health care delivery within society.

6.4.6 Safety and long-term effects

Among them, one has to identify the question related to the biocompatibility of materials used for the implants and their impact on the patients' health in the long term. Although these materials are developed to reduce side effects on the human body, their long-term compatibility, especially with newly developed materials and in the context of such materials, has not been fully inspected. It is crucial that the testing of the biocompatibility of the selection of potential materials also includes strict tests to prevent patients from experiencing unknown adverse outcomes. Ethical responsibility requires that all potential harm should first be understood and then properly conveyed to patients and all caregivers.

6.4.7 Informed consent and patient autonomy

Given the fact that the materials used in medical devices and treatments should be biocompatible, the issue of informed consent is rather broad. Patients should be fully informed of the various implanted treatments, the risks associated with those materials, the expected benefits, and other possible options. Furthermore, patient self-governance involves creating an environment that allows the individual to receive adequate information on how to manage themselves. This is particularly important in situations where biomaterials that could be used are still in the experimental category and their impacts have not been fully understood.

6.4.8 Equity and access to advanced materials

As biocompatible materials develop, there is an ethical necessity to make these developments available to every patient regardless of economic status. Some commodities and technologies could be unavailable to all due to limited availability and quality and therefore could deepen the existing discrimination in the healthcare domain. Ethical considerations should also address the possibility of making biocompatible substances affordable so that the benefits derived from the application of technology in the delivery of health care that the technology does not favor the elite but is equally available to all.

6.4.9 Environmental impact and sustainability

Another issue that arises from its use is concerns involving the ethical question of the sustainability of the elastic biocompatible material used in the production and disposal of products. The manufacturing of most of these materials includes the use of synthetic polymers, metals, and other substances known to cause an environmental impact. Additionally, it has been noted that some of the degraded products of several materials are ecological hazards. Some of the ethical issues that should be considered in the creation of biocompatible materials are sustainability, as it is important to avoid any harm to the environment, and the promotion of ecofriendly products.

6.4.10 Regulatory and ethical oversight

Academic consideration was also desirable as it enshrines ethical considerations, especially the decentralization of biocompatible materials and the adherence to safety and efficacy standards before such products are administered to patients. The regulatory authorities are charged with the task of ensuring that there is progress while at the same time minimizing potential harm to the public. It further extends to the research and development phase, where more information, accountability, and other ethical practices are necessary. That is why the participation of independent ethical boards can contribute to the establishment of ethical guidelines for developing biocompatible materials within the broader context of society.

The problem of ethically appropriate biomaterials can be deemed manifold and rather challenging. Solving these problems presupposes patient safety, equity, environmental soundness, and ethical responsibility as well as principles that are ethnically and ethically sound. It remains critical to engage in discourse and ethical reasoning while beginning to research biocompatible materials as a field and as a society in relation to these materials to mitigate these challenges and enhance progress.

7 Conclusion

Biocompatibility of materials, as one of the critical prerequisites for modern biomedical applications, has been a significant step forward in the development of advanced material science. In this chapter, the significance of these materials that contribute to patient improvement has been highlighted, where the call for material characterization and the advancement of material sciences has been underlined. The investigation of surface features and chemistry, resulting chemical interactions, and mechanical and biological properties has outlined the important roles of these properties and their effects, which are crucial to the success and safety of medical implants and devices. Knowledge of these properties is useful not just in choosing appropriate materials but also in the engineering of new solutions that meet the fresh demands of the healthcare environment.

Trends in biocompatible materials identify major developments in synthetic polymers, new metal alloys, and composite materials. Recent developments related to the use of materials include new biodegradable polymers and advanced metal alloys, whose applications have increased, resulting in better performance and compatibility with patients. The actual application of composite materials, which involve several components, further boosts the versatility of biocompatible solutions. These advancements imply a greater focus on the customization of delivered materials and the enhancement of therapeutic relevance.

Nonetheless, there are several issues surrounding the discovery and implementation of biocompatible materials. Thus, the paradigms of intense testing and augmented ethical accountability continue to play essential roles in making the material both safe and effective in its application. Proper characterization of material properties is critical in the search for problems and solutions where risk lies in invoking adverse biological reactions. In addition, compliance with the rules and regulations, as well as ethical practices, is central to safeguarding patients' lives and ensuring successful clinical performance. Overcoming all these issues calls for cooperation between researchers, practitioners, and regulatory agencies to maintain the highest standards in the performance and safety of materials used.

Moving forward, the advancements in biocompatible materials remain promising and provide a stupendous platform to flourish. Since material science is not static and the understanding of how biocompatibility works today is even more profound, the developments predict positive, even greater advancements in medical solutions. Subsequent studies will likely aim at improving the characteristics of the materials, investigating novel materials, and addressing new clinical application concerns. The very essence of this area of specialization makes it rather apparent that ongoing research and collaboration are necessary for establishing additional advancements in biocompatible materials and delivering greater value to patients.

References

[1] Helmus, M. N., Gibbons, D. F., & Cebon, D. (2008). Biocompatibility: Meeting a key functional requirement of next-generation medical devices. Toxicologic Pathology, 36(1), 70–80.

[2] Holzapfel, B. M., Reichert, J. C., Schantz, J. T., Gbureck, U., Rackwitz, L., Nöth, U., . . . & Hutmacher, D. W. (2013). How smart do biomaterials need to be? A translational science and clinical point of view. Advanced Drug Delivery Reviews, 65(4), 581–603.

[3] Williams, D. F. (2008). On the mechanisms of biocompatibility. Biomaterials, 29(20), 2941–2953.

[4] Williams, D. F. (2017). Biocompatibility pathways: Biomaterials-induced sterile inflammation, mechanotransduction, and principles of biocompatibility control. ACS Biomaterials Science & Engineering, 3(1), 2–35.

[5] Mabrouk, M., Das, D. B., Salem, Z. A., & Beherei, H. H. (2021). Nanomaterials for biomedical applications: Production, characterisations, recent trends and difficulties. Molecules, 26(4), 1077.

[6] Williams, D. F. (2014). There was no such thing as a biocompatible material. Biomaterials, 35(38), 10009–10014.

[7] Williams, D. F. (2008). On the mechanisms of biocompatibility. Biomaterials, 29(20), 2941–2953.

[8] Anderson, J. M. (2001). Biological responses to materials. Annual Review of Materials Research, 31(1), 81–110.

[9] Nobles, K. P., Janorkar, A. V., & Williamson, R. S. (2021). Surface modifications to enhance osseointegration–Resulting material properties and biological responses. Journal of Biomedical Materials Research Part B: Applied Biomaterials, 109(11), 1909–1923.

[10] Elices, M. (Ed.). (2000). Structural biological materials: Design and structure-property relationships. Elsevier.

[11] Williams, D. F. (2019). Specifications for innovative, enabling biomaterials based on the principles of biocompatibility mechanisms. Frontiers in Bioengineering and Biotechnology, 7, 255.

[12] Alagesan, P. K. (2020). Recent advances of hybrid fiber composites for various applications. Hybrid Fiber Composites: Materials, Manufacturing, Process Engineering, 381–404.

[13] Al-Shalawi, F. D., Mohamed Ariff, A. H., Jung, D. W., Mohd Ariffin, M. K. A., Seng Kim, C. L., Brabazon, D., & Al-Osaimi, M. O. (2023). Biomaterials as implants in the orthopedic field for regenerative medicine: Metal versus synthetic polymers. Polymers, 15(12), 2601.

[14] Prakasam, M., Locs, J., Salma-Ancane, K., Loca, D., Largeteau, A., & Berzina-Cimdina, L. (2017). Biodegradable materials and metallic implants – A review. Journal of Functional Biomaterials, 8(4), 44.
[15] Prasad, K., Bazaka, O., Chua, M., Rochford, M., Fedrick, L., Spoor, J., . . . & Bazaka, K. (2017). Metallic biomaterials: Current challenges and opportunities. Materials, 10(8), 884.
[16] Kong, L. B., Li, Z. W., Liu, L., Huang, R., Abshinova, M., Yang, Z. H., . . . & Matitsine, S. (2013). Recent progress in some composite materials and structures for specific electromagnetic applications. International Materials Reviews, 58(4), 203–259.
[17] Balasubramanian, M. (2014). Composite materials and processing, (vol. 711). Boca Raton: CRC press.
[18] Alberi, K., Nardelli, M. B., Zakutayev, A., Mitas, L., Curtarolo, S., Jain, A., . . . & Perkins, J. (2018). The 2019 materials by design roadmap. Journal of Physics D: Applied Physics, 52(1), 013001.
[19] Chandra, P., Yoo, J. J., & Lee, S. J. (2015). Biomaterials in regenerative medicine: Challenges in technology transfer from science to process development. In Translational regenerative medicine (pp. 151–167). Academic Press.
[20] Bernard, M., Jubeli, E., Pungente, M. D., & Yagoubi, N. (2018). Biocompatibility of polymer-based biomaterials and medical devices–regulations, in vitro screening and risk-management. Biomaterials Science, 6(8), 2025–2053.
[21] Wilkoff, B. L., Auricchio, A., Brugada, J., Cowie, M., Ellenbogen, K. A., Gillis, A. M., . . . & Vardas, P. E. (2008). HRS/EHRA expert consensus on the monitoring of cardiovascular implantable electronic devices (CIEDs): Description of techniques, indications, personnel, frequency and ethical considerations: Developed in partnership with the Heart Rhythm Society (HRS) and the European Heart Rhythm Association (EHRA); and in collaboration with the American College of Cardiology (ACC), the American Heart Association (AHA), the European Society of Cardiology (ESC), the Heart Failure Association of ESC (HFA), and the Heart . . .. Europace, 10(6), 707–725.
[22] Paul, S. M., Mytelka, D. S., Dunwiddie, C. T., Persinger, C. C., Munos, B. H., Lindborg, S. R., & Schacht, A. L. (2010). How to improve R&D productivity: The pharmaceutical industry's grand challenge. Nature Reviews Drug Discovery, 9(3), 203–214.
[23] Bunse, K., Vodicka, M., Schönsleben, P., Brülhart, M., & Ernst, F. O. (2011). Integrating energy efficiency performance in production management–gap analysis between industrial needs and scientific literature. Journal of Cleaner Production, 19(6–7), 667–679.
[24] Koh, L. D., Yeo, J., Lee, Y. Y., Ong, Q., Han, M., & Tee, B. C. (2018). Advancing the frontiers of silk fibroin protein-based materials for futuristic electronics and clinical wound-healing (Invited review). Materials Science and Engineering: C, 86, 151–172.
[25] Silver, F. (1993). Biomaterials, medical devices and tissue engineering: an integrated approach: An integrated approach. Springer Science & Business Media.

P. Prasant, J. Dafni Rose, Pradeepkumar G., T. Shanmugaraja, P. Janardhan Saikumar

Machine learning algorithms for enhanced medical image analysis and diagnostics

Abstract: Progress made in medical imaging technologies has led to the creation of complex algorithms in machine learning that improve diagnostics. This chapter offers a brief understanding of how machine learning techniques are applied to medical images. The main techniques discussed are convolutional neural networks, support vector machines, and deep learning architectures; examples of their use in image classification, segmentation, and anomaly detection are given. The investigation incorporates a new algorithmic development that solves computational concerns such as time complexity, interclass variance, and high-dimensional data modeling. The overall applicability of these algorithms in clinical settings was also established through examples that focus on the relative improvements in patient status and diagnostic performance. In this chapter, the author provides useful information to practitioners and researchers in the field of medical imaging to reduce the gap in knowledge between theory development and application.

Keywords: machine learning, medical imaging, convolutional neural networks, image classification, segmentation, anomaly detection

1 Introduction

In recent years, the field of medical imaging has undergone a tremendous change due to various innovative imaging modalities and intelligent as well as complex machine learning algorithms [1]. In order to enhance diagnostic processes, machine learning algorithms have to be integrated as medical imaging data is becoming vaster and more intricate [2]. Promising methods originating in computer vision for converting and interpreting high-dimensional picture data are convolutional neural networks (CNNs), support vector machines (SVMs), and deep learning models [3], which have

P. Prasant, Associate Professor, AIPH University, Bhubaneswar, Odisha, India, e-mail: registrar@aiph.ac.in
J. Dafni Rose, Department of CSE, St. Joseph's Institute of Technology, Semmancheri, Chennai, Tamil Nadu, India, e-mail: jdafnirose@yahoo.co.in
Pradeepkumar G., KPR Institute of Engineering and Technology, Coimbatore, Tamil Nadu, India, e-mail: pradeep17ece@gmail.com
T. Shanmugaraja, KPR Institute of Engineering and Technology, Coimbatore, Tamil Nadu, India, e-mail: shanmugarajatsr@gmail.com
P. Janardhan Saikumar, ECE, Audisankara College of Engineering and Technology, Gudur, Andhra Pradesh, India, e-mail: jskumar.p@gmail.com

https://doi.org/10.1515/9783111503202-012

shown great promise in this respect [4]. This is in addition to enhancing picture interpretation, as these technologies have paved the way for more accurate and efficient diagnostic tools [5]. This chapter discusses several machine learning technologies employed in medical image processing, highlights the milestones that have been achieved, and addresses the challenges that have been encountered [6, 7].

An important development in the field of medical image analysis has been the CNN [8]. Recognition and identification of images have been transformed concerning activities associated with object identification and image classification because of their capacity to recognize hierarchical structures in pictures automatically [9]. It has been stated that CNNs can also acquire high rates of success in patterns and anomaly detection of medical images, which improve diagnosis [10, 11]. The components of CNNs, including their multilayer structure, activation functions, and training methods, are discussed in this segment of the chapter [12]. It also depicts the versatility and usefulness of CNNs by elaborating on their use in different medical imaging environments, radiology and pathology among them [13].

SVMs present a further technique for using anomaly detection, in addition to picture categorization [14]. SVMs are very good at handling high-dimensional data and in segregating data that contain complicated patterns [15]. This part of the chapter was devoted to the analysis of the theoretical foundations of SVMs, with emphasis on the concepts of hyperplane and kernel function [16]. It also examines how SVMs are used for medical imaging activities such as tissue segmentation and tumor detection [17]. The general advantages and limitations of SVMs in comparison with other machine learning algorithms employed in medical image analysis are provided and stated to present an objective view of the applicability of the mentioned techniques [18].

The use of medical image analysis has progressed even further with the use of deep learning architectures since it consists of several neural network models [19]. These models are particularly useful in detecting details and relating closely interwoven sociohistorical layers within picture data due to their numerous layers and huge parameters [20]. Several advanced techniques in the deep learning field, like recurrent neural networks (RNNs) and generative adversarial networks (GANs) used for enhancing picture quality and diagnostic proficiency, are also discussed in this section [21]. Case examples are given to demonstrate that deep learning methods have enhanced early illness diagnoses, treatment planning, and other areas of medical imaging [22].

There are a number of challenges that need to be addressed to apply machine learning in medical images efficiently: challenges related to the large sizes of data, challenges needed to achieve stability in the algorithms, and also challenges in terms of the ethics of machine learning usage [23]. Realizing that the challenges caused by processing constraints can contribute to the lack of algorithm transparency and pooled data, the last part of the introduction describes ongoing research [24]. Taking into consideration the described types of machine learning for medical image processing and the future they hold [25], the chapter concludes. To engage in this debate, one learns the importance of sustained innovation and collaborative approaches toward healthcare improvement.

2 Historical background of medical image processing

2.1 Early techniques in medical imaging

X ray machines were discovered in 1895 by Wilhelm Conrad Roentgen. A pre-history of medical imaging can be traced back to the late ninteenth and early twentieth centuries. This invention changed the nature of medicine and offered the first noninvasive investigation of the internal organs of the human body. During X-ray application, photographic plates used to be left exposed to X-rays in order to create images of bones and other concentrated structures.

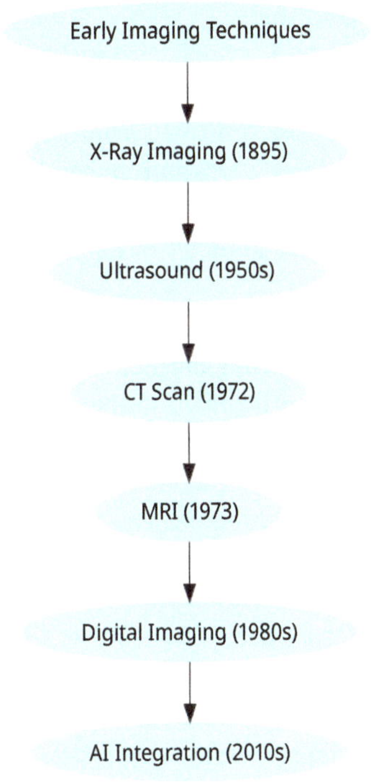

Figure 1: Early techniques in medical imaging.

After the discovery of X-rays, a dramatic evolution of medical imaging occurred with the invention of fluoroscopy by Thomas Edison in the early twentieth century. This new technology of fluoroscopy allowed simultaneous imaging of the body's interior structures, thus making X-ray imaging more dynamic through the study of dynamic physiological processes. This technique employed a fluorescent screen and X-ray source to generate continuous image projection, constituting the backbone of contemporary interventional radiological systems.

At the same time, in the twentieth century, the radioisotope and nuclear medicine were discovered. Hungarian-born George de Hevesy, a Nobel Prize winner in Chemistry in 1943, was credited with the development of utilizing radioactive tracers in diagnosis. The ability to monitor the distribution of substances in the human body encouraged the creation of techniques such as PET and scintigraphy. These techniques made functional imaging possible and allowed the assessment of physiological and metabolic functions apart from form.

The technology of ultrasonic imaging supplemented the use of X-rays in the 1950s. By employing higher frequency sound waves as a form of energy, ultrasound successfully penetrated and produced images of softer-bodied tissues that were difficult to see using X-rays. Originally invented for use in industry as a nondestructive testing technique, ultrasound found its first application in medicine, being specifically applied in obstetrics and cardiology. Its advantages included a noninvasive approach and the ability to provide an immediate view of a patient's insides.

Computed tomography (CT), developed in the 1970s by Sir Godfrey Hounsfield and Allan Cormack, was a breakthrough in medical imaging. CT used a computer terminal and X-ray scans of the body and proved to have better contrast resolution as well as precise depiction of a body's complicated structures. CT was introduced into clinical practice in the early 1970s and had a great influence on the diagnosis of the presence of disease, a significance acknowledged by the award of the 1979 Nobel Prize in Physiology to Hounsfield and Cormack.

The most powerful imaging method at the present moment, called magnetic resonance imaging (MRI), is based on the use of radio waves and strong magnetic fields to take high-contrast images of the interior of the body and was invented in the second half of the twentieth century. The 2003 Nobel Prize in Physiology or Medicine was awarded to Paul Lauterbur and Peter Mansfield for their efforts that advanced MRI technology. Soft tissue contrast is best served in neurology, musculoskeletal imaging, and cancer imaging, which do not use ionizing radiation as in CT scans.

The medical imaging techniques mentioned above have laid down the foundations of modern diagnostic imaging. Each invention incorporated some features that extended our current understanding of human physical appearance and function. It has been made possible by constant improvements in imaging technology to harness image processing strategies that incorporate computational processing to improve diagnostic accuracy, automated analysis, and image quality. The effects of these early approaches can still be observed in present technological tools and methods in healthcare in relation to medical imaging development.

2.2 Evolution of image processing methods

This chapter aims to analyze the development trends of the methods used for medical image processing based on the technological and methodological advancements in this field. It is most feasible to distill this progression into several principal stages, each of which is defined according to certain innovations that have helped to form the current state of medical imaging.

Figure 2: Evolution of image processing methods.

In fact, medical image processing technologies can be traced back to the emergence of basic imaging tools in the early twentieth century. Historically, medical imaging was dependent on analog techniques in which the X-ray was commonly used. It provides two-dimensional images with a low contrast ratio and low resolution. The assessment of these images was, for the most part, qualitative, depending on the skill of the radiologist to distinguish between normal and pathological tissues.

It can also be noted that digital imaging, which appeared later in the second half of the twentieth century, contributed significantly to medical image processing. Digital imaging was a convenient way for the transformation of analog signals into digital data, which could be more proactively analyzed. Medical images saw a change when, in the 1970s, an effective CT emerged, offering cross-sectional images of the human body. These imaging techniques advanced diagnostic ability to a great extent as CT scans developed comprehensive three-dimensional images based on computer algorithms and X-ray technologies.

In the 1980s, MRI helped bolster medical imaging by using radiofrequency signals and magnetic fields to take high-resolution pictures. In comparison to CT and X-ray scans, MRI provided far better contrast of soft tissues, which helped in the diagnosis of pathological disorders and in studying the detailed anatomy of the body. Other advanced imaging techniques, including ultrasound and positron emission tomography, also developed during this period, and each offered special imaging skills that enhanced the diagnostic sensitivity of imaging interventions.

As the marketplace for DI technology expanded, the need for image processing algorithms grew in significance. Ordinary image processing was a limited set of procedures including noise elimination, image contrast, and edge detection. As discussed later, these techniques were intended to enhance the quality of the image and therefore better facilitate accurate interpretation of the image information. The advances of algorithms came in the 1990s and early 2000s where region-based segmentation, texture analysis, and pattern recognition joined actively.

It is noteworthy that the contemporary advancements in the field of medical image processing are based on the implementation of methods of ML and DL. In particular, machine learning techniques, and more specifically neural networks (NNs), have considerably improved the image processing possibilities. High performance has been achieved by CNNs and other deep learning models in different tasks such as classification, segmentation, and detection. They have addressed several of the difficulties of traditional methods and created a new and better space for enhancements of the diagnostic tools that are based on medical images through realizing automated high-accuracy interpretation.

Progress in technology and, moreover, research and development in image processing have been constant. Recent advanced technology areas include real-time processing techniques in handling image and signal data, multimodal image integration, and artificial intelligence to optimize processing methods and diagnostic capabilities. Possible future directions in the field of medical image processing are in the further development of the readability of the learned models, in the expansion of imaging technology to individual medicine, as well as in questions of data safety and data protection.

2.3 Milestones in medical imaging technologies

A list of some of the significant milestones in medical imaging technologies has changed and enhanced diagnostic benefits for patients. It was only in 1895 when Wilhelm Conrad Roentgen discovered X-rays, making a significant revolution in imaging scientifically advanced. Roentgen saw that X-rays could pass through human tissues and produce pictures of internal organs, and this was the initial step toward radiography. This advancement made it far easier to get images of the bones and other parts of the body than to perform invasive procedures on the patient.

Other developments for human beings include the discovery of computerized axial tomography in the 1970s by Godfrey Hounsfield and Allan Cormack. CT scanning utilized a combination of computers and X-rays to produce images of the body in sections, different from an X-ray. It also made it possible to conceptualize internal structures in terms of further depth. This advancement enhanced diagnostic precision and provided better visualization of many complicated areas of the body.

MRI came into being in the late 1970s and early 1980s after the efforts of scientists such as Raymond Damadian, Paul Lauterbur, and Peter Mansfield. MRI is a noninvasive technique that utilizes high magnetic fields and radiofrequency pulses to generate very detailed pictures of soft tissues. For the purpose of evaluating soft tissue organs such as the brain, spinal cord, and muscles, MRI is a helpful technique because it does not use ionizing radiation, unlike a CT scan.

PET was developed at the same time and was considered a major advancement in functional imaging. It uses gamma rays from radioactive tracers injected into patients

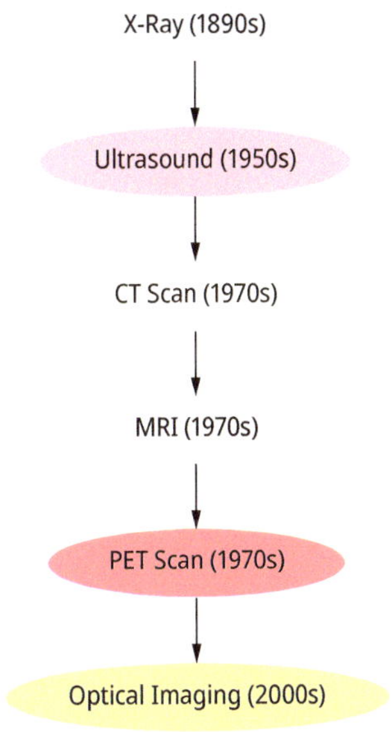

Figure 3: Milestones in medical imaging technologies.

to provide information on the metabolic and biochemical events of the body. It was important for diagnosing and managing diseases including cancers, neurological, and cardiac diseases.

First developed as a diagnostic tool in the mid-twentieth century, ultrasound imaging has undergone many developments since the 1990s. Sound waves of high frequency are utilized in recent ultrasonic technology to yield real-time images of internal structures. Progress in Doppler imaging and 3D/4D ultrasound has increased the number of fields in which it is possible to apply this equipment, for the assessment of organ function, blood circulation, and fetal growth.

A major development was in imaging technologies in the 2000s with the shift from film to digital in this technology. The advancement of the diagnosis procedure has been made, radiation dosage has been reduced, and the pictures obtained have been made clearer by the use of digital radiography, computed radiography, and digital mammography. EHRs have been implemented, and there was consolidation of digital imaging, which has facilitated enhancement of data stewardship and retrieval.

In the 2010s, machine learning and artificial intelligence in medical imaging have advanced to a new level. The use of AI algorithms and machine learning models for automatic image analysis for the identification of diseases and diagnostic support was progressing. All of these technologies offer the potential for individualized therapy, enhanced diagnostic methods, and refined means to better analyze visual representations.

2.4 Transition from analog to digital imaging

From an analog fiction to an analog fact to a digital fiction and finally becoming a digital fact was the major change in medical image processing, thus posing a major change in the acquisition, processing, and analysis involved in medical image processing. Predigital diagnostic imaging technologies, for example, X-ray and other modalities using films, relied on the use of photographic films to capture images. These films, as with the advent of film in the first place, came with some disadvantages such as low dynamic range, not being resistant to artifacts and a need for the storage medium to be physical.

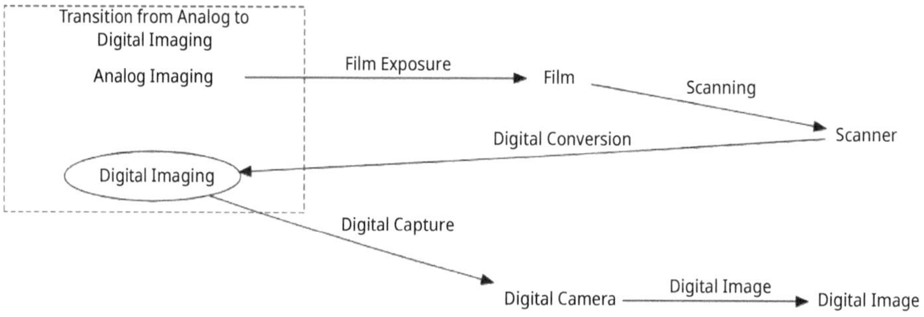

Figure 4: Transition from analog to digital imaging.

Different generations of digital imaging technologies have been considered a great development in medical imaging. A digital imaging system acquires images electronically through the use of sensors and then processes them into signals. This shift from analog to digital imaging introduced several key improvements:

Enhanced image quality and resolution: Digital imaging systems are seen to have better imaging results than analog ones. CCDs and photostimulable phosphor plates have better definition and are more conspicuous compared to film for diagnosis. The option of a digital format also enables changes to contrast and brightness of an image to be made without losing quality, thus increasing the value of diagnostic imaging.

Increased efficiency and speed: Digital imaging applies to faster image acquisition and image processing techniques. While we have fields that always use analog films, which require the use of chemicals in their development, on the other hand, images can be easily obtained digitally, and these obtained images are easier to review. This efficiency cuts the time needed for diagnosis and improves operational flow within care structures, therefore accelerating patient care.

Advanced image processing capabilities: Digital imaging provides the possibility to apply a vast number of advanced image processing techniques. Digital images can be conditioned, for example, for noise reduction and edge detection enhancement, so they can be better analyzed and interpreted. This capability is especially noticeable during processes like the automated detection of abnormalities and the measurement of body structures.

Ease of storage and retrieval: Digital photographs should be electronically saved, reducing the odds of a picture getting lost, and no physical storage is necessary. PACS, an electronic storage system, is suited for managing archives and recalling images quickly. It helps in retaining medical records for a long time and allows easy retrieval of images to be used in providing comparisons.

Integration with other technologies: The digital modality of medical images provides interoperability with other IT systems. Digital imaging data can be linked with EHRs, where patients can be managed digitally from the start. It is also possible to share digital images in networks and apply telemedicine and consulting.

Development of advanced imaging modalities: The shift to digital equipment provided a basis for improving current imaging techniques. Nowadays, such methods have become available as digital subtraction angiography, digital mammography, and CT, which provide the possibility of improving diagnostic abilities and increasing the accuracy of treatment.

2.5 Impact of computational advances on image processing

Computational upgrades have dramatically influenced medical image processing information, ensuring extensive enhancement across the category of imaging systems. In its beginning, medical image processing was hampered by first-generation computational devices that set upper bounds on the sophistication of image analysis algorithms and the amount of image data that was manageable. With the advancement in computational power and the development of better algorithms, the potential to provide better interpretational analysis of medical images steadily escalated.

Advances in higher-performing and energy-efficient processors have made it possible to execute sophisticated image processing algorithms, which were heretofore unimpressive. Recent improvements in computational power have enabled the use of sophisticated computational techniques including image enlargement, image clearing from noise, and feature detection. All of these enhancements have a clear connection to an increase in image definition and the overall diagnostic capabilities, which result in lower probabilities of mistakes when interpreting medical images.

Through parallel processing and high-speed computers, the processing of large numbers of image data has become faster. Huge data volumes are produced in fields such as CT and MRI, which have represented great value here. These strong computational performances allow for the analysis and interpretation of medical pictures at a speed and with an accuracy never achieved before, due to new methods of computational intelligence such as machine learning and deep learning.

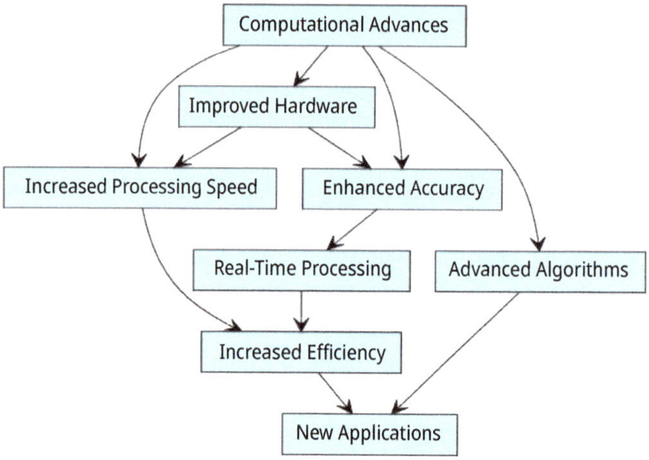

Figure 5: Impact of computational advances on image processing.

There has been a significant role of graphical processing units (GPUs) that have been instrumental in enhancing medical image processing. GPUs, originally designed to draw graphics for computer games, are currently one of the most suitable devices to perform the parallel computations necessary for deep learning and image processing. There are numerous calculations that can be performed at once, and it has been made feasible to train large NNs and perform sophisticated image analytical tasks.

3 Basics of machine learning

3.1 Definition and key concepts of machine learning

One of the subdomains of artificial intelligence, called "machine learning," focuses on developing algorithms that allow computers to make conclusions based on data analysis. It involves developing models for which code needs to be written for each application; instead, it should be able to understand and infer. If one asked an expert in artificial intelligence in the 1980s what they were doing, they would say that machine learning was the field founded on the idea that systems improve efficiency through experiences gathered from data analysis.

Algorithms: The basic building blocks of learning systems are machine learning algorithms. Mathematical analysis is used to generate patterns and predict future occurrences, and developed learned data, including SVMs and linear regression, together with unsupervised learning techniques, including principal component analysis, are the main roles of algorithms. It becomes clearer when explaining that the kind of issue being solved and the characteristics of the data determine the best method.

Models: A model is an algorithm that describes a real-life process by using data that has been put into an equation. A machine learning model is developed from previous experience in an attempt to determine patterns and further plan. Consequently, the generalization from the training data to new data unseen by the model is used to determine its performance.

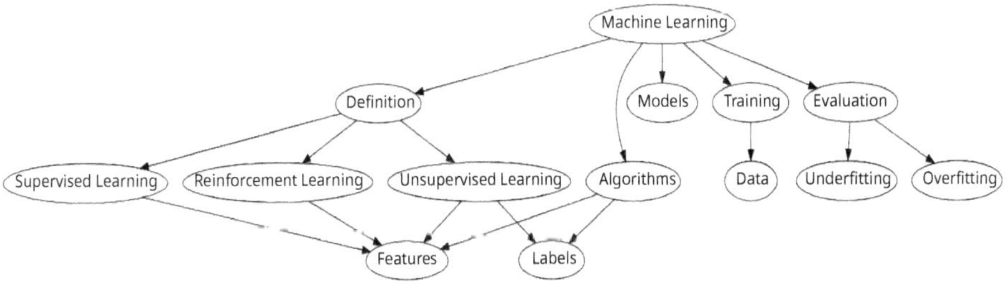

Figure 6: Definition and key concepts of machine learning.

Training and Testing: To reduce the error associated with the system, a machine learning model was trained by providing it with data to work on. This also enabled the model to automatically tune its parameters. Validation checks the efficiency of the model through a completely unknown set of data that was not involved in the development of solutions. This facilitates measuring the performance of how well the model fits new data.

Characteristics and labels: The variables upon which the model depends when making its predictions are known as features. The values that the model is targeted to predict are referred to as labels. For instance, in medical imaging, extracting features can mean identifying a pattern or pixel values in an image, while labels can indicate whether a disease is present.

This was due to the fact that both overfitting and underfitting occur when a model becomes too specialized in reading the intricate details and sometimes noise of the training data set, only to serve as an eyesore when it is tested on new data. If a model is too simple to pinpoint patterns in the data, then there is underfitting of the model. These two factors indicate that model complexity and generalization must be optimized if the model is to achieve the best results.

Supervised versus unsupervised learning: Supervised learning indeed involves training the model using data that already has a result that needs to be predicted. It is used in problems such as regression and classification. A primary goal of unsupervised learning consists of making use of data that is not categorized in an attempt to find various forms of clustering. It is used in dimensionality reduction and clustering.

Evaluation metrics: Vertices are the outcome of the functions and are central in algorithm training and upgrading. Evaluation metrics are used in estimating the approximate quality of a learning model. Examples of the basic metrics are AUC–ROC, recall, accuracy, precision, and F1 score, which are used to assess the extent to which the model classifies data in the correct manner.

Cross-validation: This method assesses the extent to which results obtained from a statistical analysis generalize to another dataset. It involves splitting the data into several subgroups, where some of the subgroups can be used for training the model and others for testing the model. It helps in understanding the strength of the developed model and prevents overfitting of the data set.

3.2 Types of machine learning: supervised, unsupervised, reinforcement learning

Machine learning is the general name for a range of approaches with specific techniques and uses for data analysis and modeling. Three main categories of machine learning include supervised learning, while others are unsupervised, and are used depending on the data obtained and the task at hand. Each serves a unique purpose.

A supervised learning approach was applied by passing the model through a labeled dataset. Here, "labeled" means that each training sample should have an appropriate output associated with it. Supervised learning tries to learn such mapping from labeled data; hence the model was able to generate predictions on unseen data. This type of learning is conventionally applied to classification and regression problems.

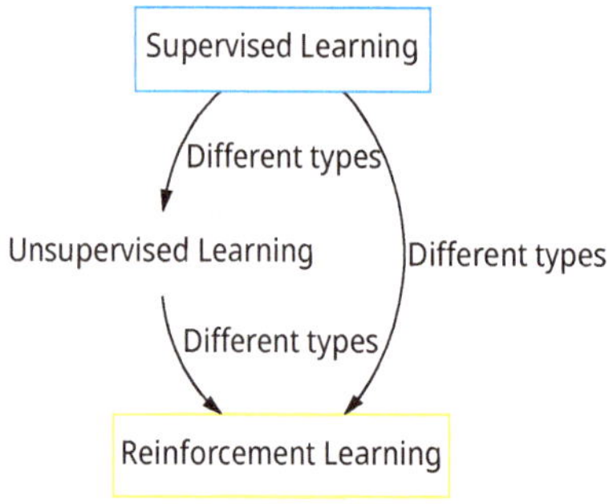

Figure 7: Types of machine learning: supervised, unsupervised, reinforcement learning.

Through classification challenges, the model acquires the capability to sort input data into designated classes. For example, supervised learning, when used in processing medical images, can be used to sort the pictures into categories such as benign and malignant using training data with annotations. Regression problems are tasks where the model makes a prediction regarding a quantitative characteristic, for instance, the size of a tumor using data from the imaging. Logistic regression, NNs, decision trees, SVM, and linear regression are some of the most basic techniques of supervised learning.

With reference to the family, through unsupervised learning, a model was trained using unlabeled data. The search for hidden structures within the data, where there were no set outcome categories, was the motivation. In pattern recognition and exploratory data analysis, this kind of learning was extremely useful.

Earlier, unsupervised learning was employed in clustering, where a model uses characteristics to categorize similar elements. Thus, clustering techniques can be applied to medical images to segment pictures, that is, the identification of regions of-interest, such as different types of tissues. Another data simplification method is known as dimensionality reduction, where the number of features is decreased but the important details are retained. Among such methods, one can mention principal component analysis (PCA) and t-distributed stochastic neighbor embedding (t-SNE).

Reinforcement learning, also known as RL, is a subset of machine learning that deals with an agent being trained on how to make decisions from the feedback received from its environment. The agent gets evaluated with performance feedback after it has performed its actions within the environment and receives its merits. It is possible to approach the problem as the search for a policy that somehow has the greatest total benefit in the long run.

In reinforcement learning, the policy updates its decision process depending on the outcomes. Such a kind of learning often complements sequential decision-making and optimization tasks. For example, RL can be used in medical image processing to govern the robotic structures performing surgery or image processing algorithms.

3.3 Machine learning workflow: data collection, preprocessing, model training, evaluation

The ML process was systematic and had several important phases, each of which was important for developing accurate and robust models. These phases include data collection, preprocessing, model training, and assessment. Each of these contributes to the overall success of incorporating machine learning applications including the processing of medical images.

Data acquisition was the first stage of the machine learning process, which includes the process of gathering the right and quality data necessary for the models' training and the evaluation of their performance. In the context of medical image processing, it means obtaining several typical medical images from different modalities including MRI, CT, and X-rays. An important issue of data collection was to guarantee the inclusion of as many different situations, body parts, and patient demographics as possible to create a relevant dataset. One must also consider the ethical implications of data gathering including patients' consent and protection of data.

Data cleaning is an essential process that involves preparing the data for analysis, where it is prearranged for model learning. In medical image processing, preprocessing steps typically include:

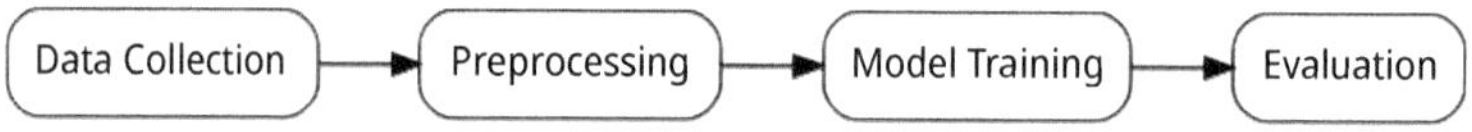

Figure 8: Machine learning workflow.

Normalization and standardization: Standardizing the pixel intensities of the images being used in terms of the possible intensity range in order to reduce differences among the databases.

Noise reduction: Use wipe, filter, and tool to eliminate some image artifacts and improve image quality.

Data augmentation: To increase the number of training samples, methods like rotation, scaling, and flipping can be used with the purpose of improving model robustness.

Segmentation and annotation: Segmenting the images and outlining exactly what is to be learned in order to create the basis for supervised learning.

Preprocessing was also valuable in minimizing concerns arising from variability since the data is only as good as its quality, and models are trained to make consequential predictions.

Training a new model was needed in order to build and learn more about the preprocessed data. Here, appropriate methods were selected, model parameters were set, and the model underwent iterations on the given set. CNNs and other deep learning paradigms that are suitable for image data are the most commonly used machine learning techniques in medical image processing. In training, the model was trained to learn the features of the images through a loss function that quantifies the disparity between the model's predictions and the true labels. In the training process, data was divided into a training data set and a validation data set to monitor the accuracy of the designed model and fine-tune parameters when necessary.

The final stage of the machine learning process was modeling, followed by measurement, which measures the effectiveness and functionality of the trained model. Here, generalization was evaluated on unseen data that were not employed during the training of the model. For medical image processing assessment, the performance measures include accuracy, precision, recall, F1 score, and ROC-AUC. The following measures were used to demonstrate the model's accuracy in right classifications, false positives, and false negatives. Assessment also involves a thorough analysis to determine areas of weakness/areas for improvement so that the model meets the requirements that are essential for effectiveness in research/clinical purposes.

3.4 Common algorithms: linear regression, decision trees, *k*-NN, SVM

Calculation methods are numerous within machine learning, and it is the procedure of studying data with the intention of forecasting. Among specific algorithms, the most used include SVM, *k*-NN, decision trees, and linear regression. All of these algorithms are more useful in different situations because each of them has its own characteristics and applications.

The variable to be targeted had to be continuous and was predicted by means of linear regression on one input variable. The approach uses the assumption that the independent and dependent variables are proportional. This method adjusts a straight line to the collected data in an effort to minimize the discrepancy between forecast and reality. The exercise was to identify the values of each of the variables that comprised the linear equation to capture the relationship between them most appropriately. For instance, when there existed an almost direct relationship between the variables, such as when forecasting medical outcomes from the characteristics of the patient, linear regression was commonly applied.

One of the models applied in both regression and classification techniques was the decision tree. This means that by recursively splitting the data into segments based on the features, a procedure for constructing the decision tree as a model was demonstrated. Every node of the tree indicated a decision rule based on a single feature, and the outcome of this choice was represented by each branch. The leaves of the tree were the last division made when constructing a decision tree. As the name suggests, decision trees are loved because the decision-making process lends itself to a clear visual presentation and is also easy to understand once you start reading it. However, it had a problem of "overfitting," especially when applied to complex data inputs.

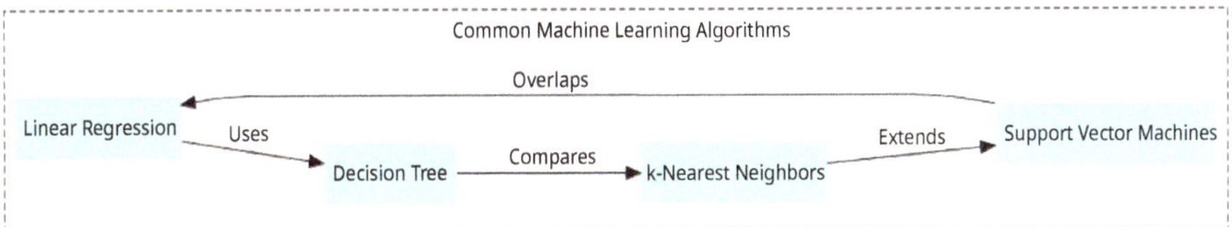

Figure 9: Common algorithms: linear regression, decision trees, *k*-NN, SVM.

The *k*-NN algorithm was a straightforward, yet efficient, technique for classification and regression analysis. The concept that originally underpinned *k*-NN was to classify the outcome of a data point based on the average formed from its *k*-nearest neighbors in terms of feature space. In classification problems, the technique provides the class label that is more frequent among the *k*-nearest neighbors. In regression, it predicts the value by averaging the target variable of the *k*-nearest neighbors. Because of this, *k*-NN was easy to implement, nonparametric in nature, and had no training phase. This was likely to affect the performance rates on noisy data as well as the value of *k*.

SVMs are a powerful family of algorithms used in applications of regression and classification problems. How SVMs operate is that the aim is to identify the hyperplane in the feature space that can optimally classify various data points belonging to distinct categories. The method seeks to maximize the maximum margin of the support vectors that define the distance of the nearest data points to the hyperplane. SVMs are preferred in high-dimensional spaces and can solve difficult decision surfaces by mapping their input to higher dimensions where the solution to a linear classifier is available using kernel functions. Thus, on one hand, SVMs are reliable and fairly accurate, but on the other, they are very demanding in terms of computing power and require the choice of a correct kernel function and hyperparameters.

3.5 Introduction to neural networks and deep learning

NNs and deep learning are essential improvements to machine learning algorithms because they provide a set of appropriate tools for the analysis of numerous and intricate patterns, such as those found in medical images. NNs are mathematical models of the brain whose structure and function are used in a computer. These models are composed of several layers of nodes, and the output of these nodes is described as neurons that perform mathematical calculations on the received data.

The three kinds of layers that exist in a given NN consist of input, hidden, and output layers. The raw layer of data takes the matter, which is then processed via one hidden layer to decipher intricate motifs and elements. The final classification is produced by the output layer, while final predictions are produced by the final layer for a fully connected ANN. Thus, in the training phase, each connection weight, which is identified as a neuron, is altered to minimize the difference between the target output and actual output. Optimization techniques assist in this case.

NNs are subdivided into deep learning, which employs deep NNs with many hidden layers. The depth of these networks allows them to learn hierarchical data representations that capture more abstract properties at every level. This is because the input in such cases can be extremely complex and multivariate, and deep learning is particularly capable in applications such as image recognition and natural language processing. This is due to the fact that the learning hierarchy of this technology is well-founded.

CNNs are a modern deep learning architecture that was developed specifically for image processing tasks. CNNs employ filters on the input pictures through convolutional layers to examine the characteristics' spatial hierarchy, such as edges, textures, and forms. Finally, these characteristics are aggregated and passed through fully connected layers to make the final predictions. CNNs' ability to learn from and extract features from raw image data has revolutionized medical image analysis. This has

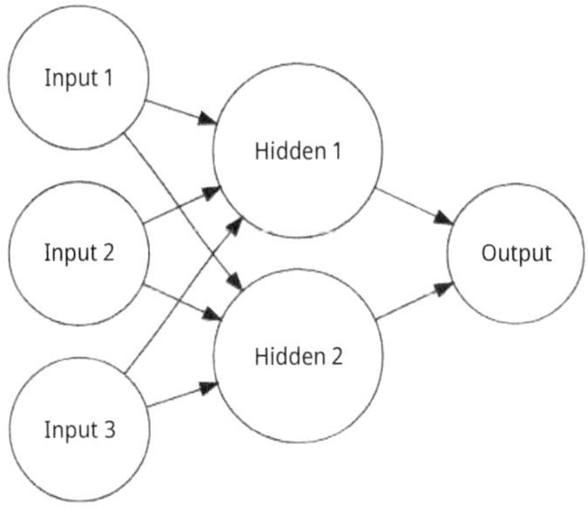

Figure 10: Introduction to neural networks and deep learning.

made it possible to detect and classify all the structures and clinical states with the highest level of precision.

Long short-term memory (LSTM) was also one of the significant deep learning architectures, intended especially for sequential data and temporal dependencies. RNNs are most applicable in activities that involve sequences such as time-series analysis. RNNs themselves include a category known as LSTM networks, which are effective in handling the vanishing gradient and dealing with long-term dependencies necessary in numerous applications that require the retention of data over several sequences.

It is worthwhile to mention that the improvements have been achieved in such facets as picture segmentation, classification, and enhancement as a result of the application of this deep learning tool in medical image processing. Deep learning models contain more ways of recognizing complex relations and connections in medical images, which increases their diagnostic abilities and improves the planning of the treatment process.

4 Machine learning in the context of medical imaging

4.1 Why machine learning is suited for medical imaging

As machine learning can bear the load and number of data coming from modern techniques of imaging, it is a very good tool for medical imaging. Medical pictures often have high dimensionality, intricate patterns, and simple variations that cannot be well interpreted by traditional methods. To overcome these issues, machine learning algorithms, particularly, deep learning algorithms, have several advantages.

First of all, let me state that feature extraction, as well as pattern recognition, are two of the strongest suits of machine learning. Standard analysis often involves developing techniques in which information features are selected by hand – a very tedious affair – and in the process, subtle details are missed. Otherwise, if distal human intervention is not involved in feature learning, especially by using deep neural networks, various important characteristics can be learned from raw picture data. Greater and more complex examination is provided by this capacity, which increases diagnostic precision and enables the detection of ailments that would not be noticed by human clinicians.

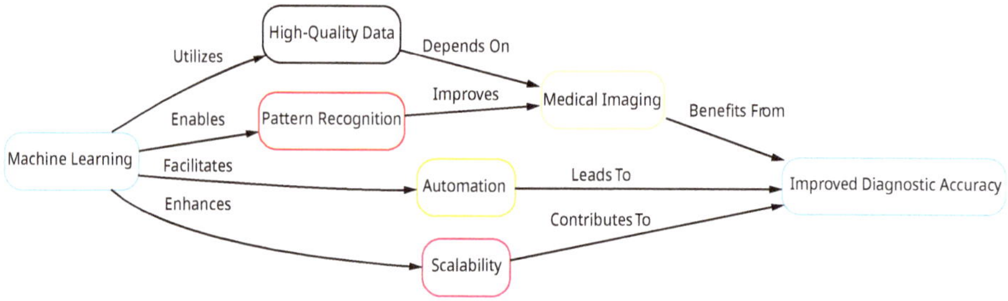

Figure 11: Why machine learning is suited for medical imaging.

Secondly, machine learning algorithms are good at handling big data. Radiology work and dealing with diagnostic images lead to the production of large amounts of data, especially when high-resolution images are used. The availability of large data sets is well managed with machine learning techniques since they do not require tools to manipulate big data but rather computational resources. This scalability is important for instances where it is necessary to compile models on various data while imaging.

Another strength was that the different machine learning models could operate well across different data sets and imaging environments. Traditional methods are not efficient in comparing variability in imaging protocols, patient demographics, and equipment used. Over time, machine learning algorithms can learn from various datasets with a view to enhancing their resilience and versatility. Additional advancements in this capability are achieved through techniques such as transfer learning, where models sourced from one dataset can be reused with other imaging modalities.

Multimodality data fusion was also enabled by machine learning. It is possible in holistic diagnostics to obtain more objective data about a patient's state in the field of medical imaging by combining one patient's CT, PET, and MRI data. Multimodal learning helps to increase the accuracy of diagnosis and enables more adequate therapeutic choices by learning the complementary information obtained from different modalities.

The process of developing personalized models was facilitated by the machine learning process that enabled a model to consider the differences among individuals. Big databases suggesting wide patient populations include different characteristics and circumstances not identified. Through the utilization of machine learning algorithms, specific sets of patient trends and forecasts can be identified. This capacity to deliver a differential diagnostic approach and treatment prescription bespoke to patients was one of the advancements in individual patient medicine.

It was important to know that machine learning improves the medical imaging workflow process. Computer-aided diagnosis and triage minimize the need for manual assessment, increase the efficiency of clinical practice, and minimize the risk of error. This efficiency was especially notable in high-throughput platforms, where speed and accuracy of analysis were of paramount importance.

4.2 Comparison with traditional image processing techniques

In medical imaging, machine learning has been shown to define new techniques that are distinct from prior image processing methods. The analysis of these methodologies has revealed noticeable differences in their characteristics, and, as a consequence, possible applications.

It was usually the case that in the evaluation of medical pictures, the traditional image processing systems used manually extracted features and algorithms. These descriptive techniques include methods that are based on prior knowledge and assumptions about the structural properties of images such as edges, texture, and morphological procedures. Traditional approaches work well only within a certain scope and are worse at handling the content and variability of medical images.

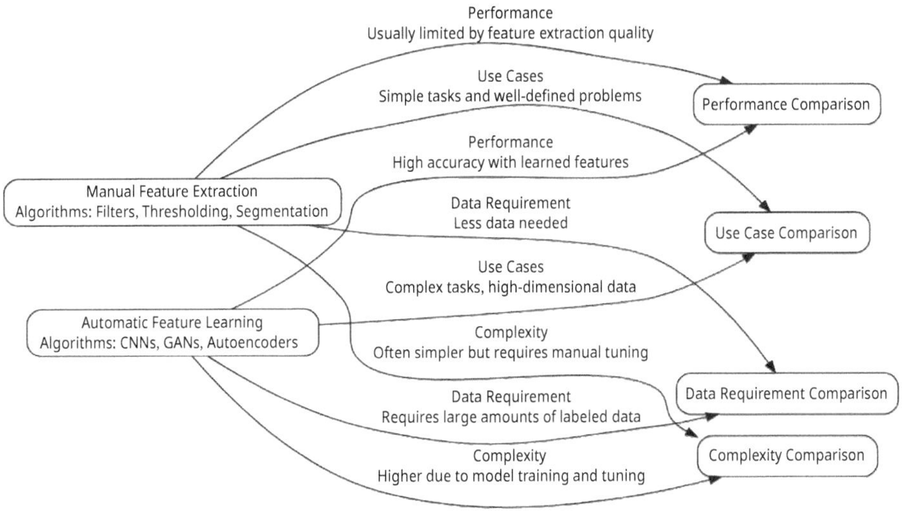

Figure 12: Comparison with traditional image processing techniques.

Automatic feature extraction is an important part of machine learning, particularly deep learning. Like CNNs, deep learning models can extract features of hierarchical categories from unprocessed visual data without human intervention. This feature makes learning models more effective in handling the complexity of medical images and, most of the time, gives better results than traditional techniques.

Previous solution techniques of AIM are based on conventional processing and analysis methodologies that have been designed to perform on a particular modality for a particular application. For instance, the same algorithm used for recognizing tumors in MRI scans would not work nearly as well in CT scans without modifications.

A training model with diverse datasets can be expected to generalize well concerning different imaging types and situations. In contrast, these models can incorporate many examples into training and thus can generalize to changes in imaging protocols, patient characteristics, and pathological states, thereby revealing more general and versatile solutions.

Old school methods of image processing are quite resource-hungry when dealing with large databases for analysis. When the parameters are tuned by hand and the preprocessing steps are fine-tuned, this can further increase the computational load and hinder scalability.

GIS applications involve the use of machine learning techniques that offer increased efficiency and scalability, specifically when using GPUs and cloud computing. They employ parallel computing, owing to which deep learning models can concurrently make computations, allowing them to read and analyze images within a short time. This has made real-time analysis possible, which is needed for applications such as computerized diagnosis.

Previous paradigms of image processing provide acceptable performance in semi-homogeneous environments for certain kinds of images, but the performance decreases when noise is present, as is the case with clinical images.

Precise as well as reliable models can routinely be found with a heavy reliance on the principles of machine learning and deep learning. Such models are capable of understanding multiparameter nonlinear relationships in large data sets, which improves diagnostic uniformity and precision. For instance, deep learning algorithms proposed showed better sensitivity and specificity for the identification and classification of tumors in contrast with conventional methods.

In the case of traditional image processing, there was a challenge of low interpretability of the developed complex algorithms. Relative to some of the machine learning models, there was sometimes a more revealing decision-made.

There are four issues derived from machine learning, especially deep learning, including the interpretability issue. The relative complexity of NN architectures presents difficulties in understanding how certain predictions are made. Future work is also being conducted to support the clinical decision-making process and expand explainability techniques for particular machine learning models used for image analysis.

4.3 Advantages of machine learning in handling medical images

Machine learning has a number of advantages that are changing the analysis and interpretation of medical images in clinical practice in the sphere of medical imaging. Such advantages include increased diagnostic accuracy, increased production rates, and the ability to manage complex as well as massive data.

Compared with traditional approaches, machine learning algorithms, particularly DL, have demonstrated a significant improvement in diagnostic performance. These algorithms are designed through the analysis of labeled medical photos and are also able to identify minute details and anomalies that easily escape the eye of the human observer. Sophisticated models, such as CNNs, are also highly capable of detecting and differentiating multiple conditions, like tumors, lesions, and others, with great accuracy. These two combined have proven to produce better diagnostic accuracy to enhance early diagnosis, hence improving patients' lives.

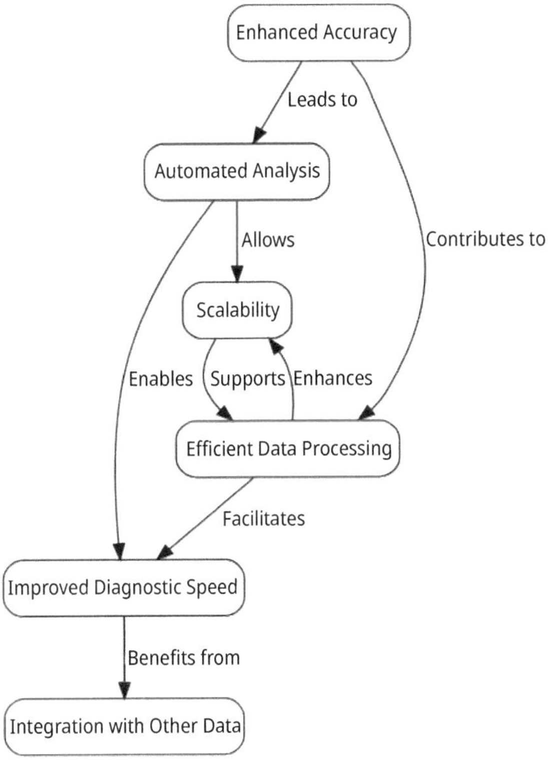

Figure 13: Advantages of Machine Learning in Handling Medical Images.

In medically related image analysis, machine learning continues to play a major role in enhancing the speed of the processes. Powerful algorithms can quickly analyze large datasets of imaging studies and provide reports much faster compared to manual interpretation. This efficiency is especially important in cases of large-scale analyses, like emergency room screenings, where the results ideally need to be generated as soon as possible to guide further clinical care.

There are automated, repetitive, and time-consuming processes in medical image processing through the use of machine learning. For instance, image segmentation, feature extraction, as well as lesion detection, are some of the tasks that can be accomplished easily but which, by manual effort, take a lot of time. Automating the process also has the effect of easing the burden for radiologists and clinicians; not only this, it has the effect of eradicating inconsistency and human error from the process.

Some examples of the high-dimensional data often incorporated in medical imaging are the multimodal scans and the three-dimensional pictures. Machine learning algorithms can operate and extract relevant data points from the high dimensionality of feature space, thus making it suitable for processing this kind of data. With the nonstochastic subtraction of DNN and other techniques, it becomes possible to combine the multimodal imaging data and enhance the abilities of doctors by providing them with more complete information on the state of health of their patients.

It was also very possible for a machine learning model to learn by itself through some process by incorporating more data, and the process of learning was not limited to a single process; it was iterative. Since more medical images are accumulated and labeled, those models can be updated to improve their performance and learn new patterns. Such flexibility helps the machine learning systems continuously update themselves with the progressive developments in "medical expertise" and the instrumentation of imaging techniques to have incremental enhancements in diagnostics and therapies.

Machine learning contributes to the development of personalized medicine, as it considers patient-specific image data and is integrated with other clinical data. Anticipated care plans can then be delivered to fit the needs of each patient through risk assessment and evaluation using the useful information that can be generated from the models. Due to the suggested strategy, healthcare becomes more accurate and directed, which, in turn, enhances the effectiveness of the applied therapies.

4.4 Overview of machine learning applications in medical imaging

With the advancement of machine learning, many aspects of image processing and analysis have been improved, making it an influential tool in the field of medical imaging. Clinicians in medical imaging have benefited from the use of machine learning technologies in the following ways: treatment planning, patient management, and diagnostic accuracy.

The machine learning algorithms, deep learning-based algorithms, have even demonstrated high performance in picture categorization. Such photos could be incorporated to teach these algorithms about various organs, body tissues, and diseases among them. For example, CNNs have been used in the diagnosis of illnesses by analyzing radiological images, including pneumonia, diabetic retinopathy, and cancer, amongst others. These systems work on multidimensional levels to sort out various features and diagnose several conditions correctly.

Image and signal processing were also used to improve medical images and to reconstruct images from noisy data sets. In the same manner, architectures like denoising autoencoders and GANs have been employed to enhance image quality, minimize artifacts, and reconstruct lost characteristics in poor images. These enhancements are especially important in increasing diagnostic precision and rendering fine morphologic details that are concealed by noise.

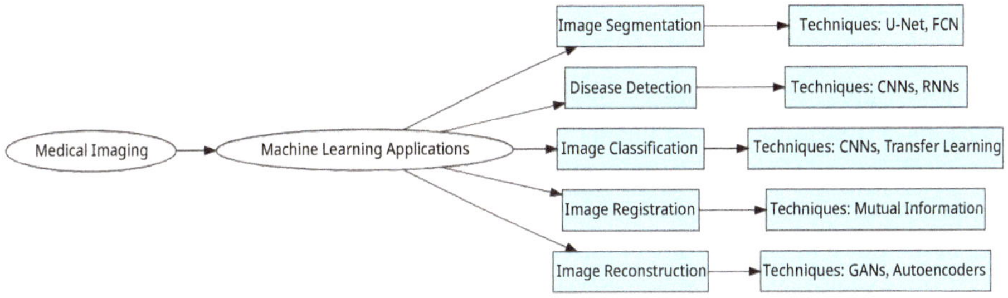

Figure 14: Overview of machine learning applications in medical imaging.

Using data mining in past imaging results in combination with the patient's history, machine learning models can predict further disease development and probabilities of treatment plan outcomes. It is useful in predicting the probability concerning the recurrence of the disease, reaction to treatment, and the patient's survival probability. Through the integration of image data that can be merged with other clinical factors, machine learning improves patient and clinical outcomes through the development of targeted therapies and better-informed decisions about patient treatment.

Machine learning has also been used to review and enhance several steps involved in dealing with medical images. A number of image acquisition, image analysis, and reporting systems have been developed to reduce the workload of radiologists and clinicians. For instance, it can promote extraordinary areas, draft summaries, and sort cases according to priority, saving much time for workers and leaving more significant tasks to doctors and other healthcare employees.

The fusion of data coming from various imaging modalities, including MRI, CT, and PET, has been improved by the use of machine learning. The combination of information from several sources makes ML models effective in presenting the condition of a patient. This is with the understanding that the use of this multimodal approach enhances the chances of arriving at the correct diagnosis and better explains the nature of the diseases involved.

Self-organizing systems of multiple units assist in taking real-time management of medical images and consulting in case of an abnormal condition. In telemedicine applications, the images transferred from distant places can be analyzed with the help of algorithms that utilize the results of machine learning, being equivalent to the work of professionals in the field who are not required to be present on-site. It has been most useful in outreach where there are few practitioners and in cases when disaster strikes.

4.5 Key challenges in applying machine learning to medical images

When it comes to applying machine learning to medical imaging, one is bound to encounter a myriad of challenges that cause the novelty and effectiveness of these methods to stumble. To increase the efficiency and apply the characteristics of machine learning systems in medical imaging, we need to solve the following question.

Radiological diagnosis information was often criticized in terms of the quality and variability of data acquired from medical imaging. Indeed, some of the differences in picture features are due to differences in patient situations, equipment parameters, and techniques. This unpredictability makes it difficult for machine learning algorithms to work as they require reliable and quality data to learn from. A common problem of identified models that work well in isolation and patterns, yet fail to address a range of clinical experiences, could be due to inconsistent data.

This chapter provides an overview of data annotation and labeling that is important in the formulation of machine learning models. Annotation of medical images is also a time-consuming process that typically necessitates the services of expert radiologists. This suggests that four possible reasons slow down the availability of training data: annotated datasets are rare, and manually labeling them is expensive. Further, inconsistencies in labeling lower the model's reliability and produce incorrect results.

Of all the AI technologies, especially machine learning models, deep learning algorithms, for instance, have been described as hard to explain. Due to the fact that these models are inherently "black-box," it is hard to understand how and why decisions are being made – this makes it that much harder for clinical practitioners to accept. Machine learning must be easily integrated with note-keeping, with clinicians having the capability of comprehending and authenticating the results generated by the algorithms. Developing ways to increase the explainability of such models is still a significant challenge.

Such a situation occurred when a given machine learning model had excellent performance on training data, yet could not do well on new, unknown data and was rated as overfitting. In medical imaging, this has been particularly an issue because

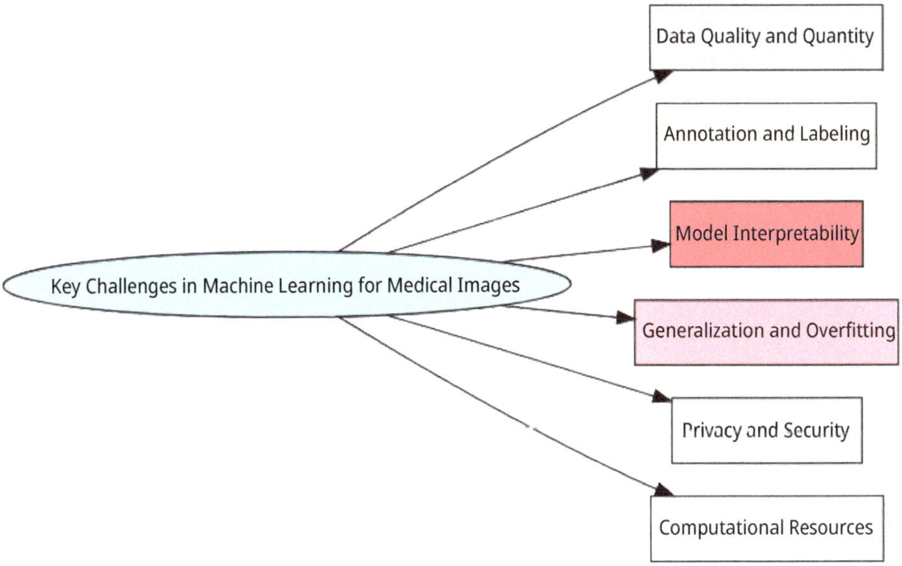

Figure 15: Key challenges in applying machine learning to medical images.

otherwise possible models develop specific patterns from training sets into memory rather than general features. For a generalizable model, it has to be applicable across different patient types and imaging type scenarios and thereby be useful.

Technological problems, as well as issues of practical implementation, make it difficult to incorporate machine learning models into current therapeutic interventions. For the models to deliver valuable information without disrupting existing routines, radiologists' operations must be seamlessly incorporated into the models. Issues of data security and privacy, and whether and how ML systems meet regulatory requirements, must be addressed in order to ensure the proper use and functioning of systems in clinical practice.

During both training and inference, AI ML models, especially deep learning networks, are known to require large processing power. These models should be made available and feasible in scaled-up environments due to the high demands in memory and computation – Complexity controllers, though, are a disadvantage in resource-restricted environments. To circumvent these resource constraints, the efficiency of machine learning algorithms must be fully optimized, and in cloud computing and technology, opportunities must be exploited.

There are two types of bias seen in machine learning models: bias in the training data and the bias of the algorithm. There exists inaccuracy in diagnosis and a variation of the same in one population due to the use of biased models in medical imaging. Redressal of the bias and fairness issues was fundamental in the creation of models capable of offering reliable results to patient groups irrespective of diversity.

5 Key components of machine learning systems for medical imaging

5.1 Data acquisition and labeling

Data acquisition and labeling are two crucial processes in the creation of machine learning systems in the medical imaging field. These processes are important for preparing the medical images for the machine learning algorithms to learn from them, as well as to correctly interpret the medical images.

Figure 16: Data acquisition and labeling.

Taking portable medical images from various types of imaging techniques, such as MRI, CT, ultrasound, and X-ray images, is referred to as data collection. Each modality produces different image types, which have distinct characteristics and carry various therapeutic applications. The quality and diversification of the obtained data also influence the efficiency of the developed machine learning models.

Imaging techniques: Different imaging methods provide various views and more information regarding the anatomy under consideration and the disease. For instance, MRI yields high-quality images of the soft body tissues, while a CT scan gives detailed cross-sectional pictures of the body structures. He said these depend on the clinical issue and the type of medical problem under consideration.

Equipment and protocols: Complex differences in imaging systems and acquisition settings could be a source of variation in the projects due to changes in image metrics. Some of the remaining intersurvey variations reflect methodological differences in evaluating comparable types of equipment; standardizing equipment settings and protocols, when feasible, aids in reducing such differences and in making survey results more comparable.

Data storage and management: Since many images are now generated in the medical world, proper storage and management of medical images are critical. Thus, it's imperative that systems be able to accommodate high-resolution images without compromising their security and usability for processing and analysis applications.

Data labeling entails associating medical images with information related to diseases and structures. Accurate labeling of images is important for machine learning models since labeling provides the ground truth for developing supervised learning models.

Annotation techniques: Annotation can also be done with the assistance of radiologists or semiautomated tools. It was noted that manual annotation provided very high accuracy by construction but required a high amount of time and effort. There are tools available that perform this operation semiautomatically, as these speed up the process, although the results have to be validated for correctness by a professional.

Labeling protocols: As noted earlier, the practice of labeling requires clear protocol guidelines to be set down to avoid confusion. These protocols should determine how various pathological conditions should be annotated and how the margins of the structures being delineated should be drawn.

Quality assurance: Since these set labels are typically a small sample of all the data it was possible to obtain in the context for which the AI was being designed, quality assurance measures need to be implemented to ensure the labeled data is actually precise and dependable. This requires using various consensus-assembling strategies to resolve conflicts and score comparisons.

Methods for data augmentation are applied in order to expand the amount and the range of the training set. Transformation methods such as rotation, scaling, and flipping give a description that yields modifications of images. The mentioned strategies make machine learning models work with a larger number of different changes in the images, which, in turn, enhances their robustness.

Methods of augmentation: Examples of basic augmentations include geometric transformations (e.g., rotation and translation), contrast and brightness changes, noise injection,, and synthetic data generation (e.g., picture synthesis using GANs).

Impact on model training: Extended data improves the overall performance of the model by imitating various imagery conditions and patients. It reduces the rate of overfitting and also enhances the model's performance with actual data.

While dealing with medical imaging data, data security and privacy concerns are mandatory. The security of patient data is ensured by legal requirements such as HIPAA and GDPR.

Data anonymization: When the photos are stripped of patient information, it is easier to conduct research and build models on them without compromising the patient's privacy.

Secure data handling: Performing data handling in a secure manner to avoid breaches was another set up that had to be undertaken when managing students' information.

Specifying the heterogeneity of the sample was important for building valuable machine learning models for various populations and circumstances. A large pool of data should comprise patients of both genders, different ages, from various geographic locations, with various types of images acquired at different times and through various techniques, and with different pathologies.

Addressing bias: It was advisable to search for the prejudices in the given dataset and try to correct them so that the performance of the proposed model was more equal and was not inclined toward some groups and conditions.

5.2 Image preprocessing techniques: normalization, augmentation, noise reduction

In medical image analysis, normalization has been one of the simplest kinds of preprocessing that manipulates the pixel values of a diverse range of pictures so that their pixel value ranges are the same. The basic objective of normalization is to ensure that the input data is standardized. Consequently, it reduces the contribution of fluctuations in picture contrast and intensity and enhances the models used in machine learning. Typical methods of normalizing consist of:

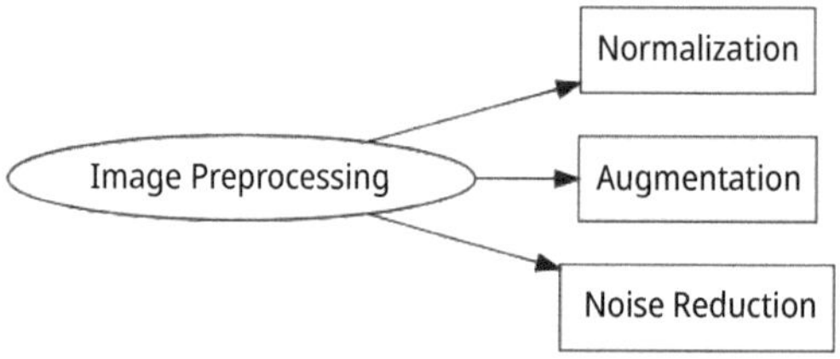

Figure 17: Image preprocessing techniques: normalization, augmentation, noise reduction.

Min-max normalization: Normalizes pixel values to a standard range for ease of computation; it typically scales pixel values to fall within the interval [0,1] by subtracting the minimum of the pixel values of an image and dividing the result by the range of the pixel values. It normalizes the intensity of different images, thus making intensity values comparable with each other.

Z-score normalization was concerned with getting pixel values to be approximately normally distributed by centering pixel value means and scaling by standard deviations. This method aids in correcting variations in the intensity of the image due to variations in acquisition conditions.

Histogram equalization: Improves image density, whereby the pixel intensity values are normalized to increase the equality of the histogram. This technique helps to enhance the contrast of different features in medical images and consequently enhances the algorithm's ability to identify and study the structures of interest.

It was a technique for generating new training sets of the same size as the original data set with different variations of the same image. It also helps in the enhancement of a model's generalization capability in machine learning and the management of the overfitting problem. Common data augmentation methods include

Geometric transformations: Examples are rotation, scaling, translation, and flipping. These transformations imitate changes within the spatial relations of the bones and organs, assisting models in discriminating elements irrespective of their position.

Intensity variations: Changes in brightness, contrast, and saturation present in images can commence changes in the context in which the images were captured.

These augmentations make models invariant to changes in image acquisition settings and enhance generality from one situation to another.

Synthetic image generation: To improve the training dataset, composite more realistic synthetic pictures using techniques such as GANs. New variations and situations not included in the training data set are introduced through synthetic photos.

A crucial step in image processing, referred to as noise reduction, assists in enhancing the acquisition of images by eradicating abhorrent distortions and impediments to assessment. A number of techniques are used to lower noise and improve the clarity of medical images:

Filtering techniques: In order to blur the pictures and reduce noise while preserving characteristics, techniques such as bilateral, median, and Gaussian are used. While median filters are excellent for removing salt-and-pepper noise by replacing a pixel's value with the median of the pixels in its immediate vicinity, Gaussian filters use a weighted average of the nearby pixels.

Wavelet transform: Symbolically breaks the images into different frequency bands and then attenuates the noise in the WT domain individually. It provides a good tradeoff between noise removal and detail preservation, qualities that make it suitable for enhancing medical images.

Denoising autoencoders: A class of NNs that learn a mapping from a noisy image to a clean image. They are trained to make clean images out of noisy inputs and, in so doing, learn to remove noise while preserving the relevant details.

5.3 Feature extraction methods: manual vs. automated

Prior to analysis and classification, enhancing medical images was one of the most critical steps in image processing, called feature extraction, which involved segregating and identifying specific characteristics. There are two ways to approach this process: manually and automatically. As with virtually everything in life, there is a positive side and a negative side to each of them.

Historically, feature extraction has been a complex process that involved designing and implementing a number of algorithms and heuristics with the purpose of extracting relevant features from medical images according to the knowledge and experience of the human expert in the field. This approach includes techniques such as

Edge detection: Boundary detection in images is achieved by using methods such as Sobel, Canny, and Prewitt. These methods usually emphasize thin lines around different structural areas, for example, when outlining a tumor.

Texture analysis: Measuring the textural features of areas in an image with statistical methods, including energy, contrast, and uniformity of textured areas, is needed to form an image. Texture analysis can provide information about the type of tissue and the structural arrangement of the tissues.

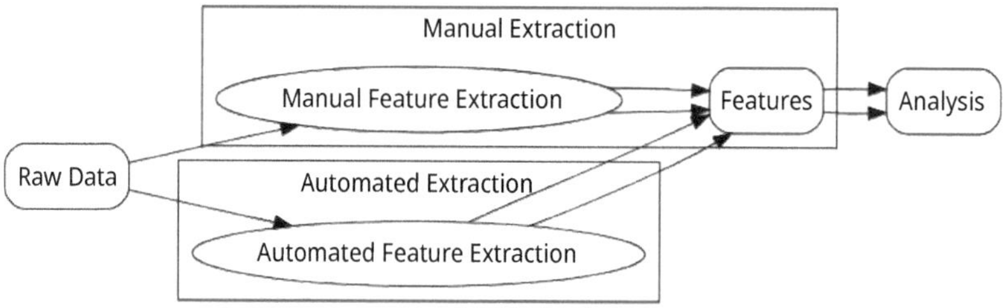

Figure 18: Feature extraction methods: manual vs. automated.

Shape analysis: Identifying geometric features such as shape, size, and orientation of various structural components, figures, and forms. Conventional approaches such as contour detection and region growing are employed to detect and measure the attributes of the anatomy.

In the traditional manual feature extraction process, measurements depend on the expert knowledge of the specific area, and more effort is needed when designing and tuning the extraction pattern algorithms. Although this kind of approach provides certain specific functions with substantial efficiency, it is always characterized by prolonged and intensive work, including even getValue calculations. Also, the features extracted manually do not include all the needed information suited for other types of data.

Automatic feature extraction allows one to extract features directly from image data by applying machine learning and deep learning methods. Key methods include

CNNs: CNNs employ numerous layers of convolution filters to obtain the hierarchical features from the raw picture data themselves. As you see, these networks can, with almost no help from the user, classify highly abstract and complex features such as texture, form, and pattern.

Unsupervised learning algorithms called autoencoders map input pictures into another subspace and then map them back. From the other features, fundamental data properties are stored in the encoded form for utilization in secondary analysis.

Acquiring features: During the training stage itself, other kinds of network parameters, like limited Boltzmann machines and deep belief networks, extract features from the visual input. They explain relationships and correlations, which can be seen in the pictures.

Some of the biggest advantages of performing automated feature extraction include the fact that the algorithm can sometimes identify features that are not easily discerned using traditional methods as well as the system's ability to sift through thousands of records. This method proves to be quite flexible and can easily be applied at large and also tends to fit well with various imaging techniques and environments too. Since the learned qualities do not always coincide with human perception, the automated methods are hard to explain and consume a lot of processing resources.

There was always a decision to be made whether to use techniques based on a human input system for feature extraction in a specific medical imaging application, depending on the requirements of any given job. In those cases where features are well-established and domain experience is critical, the manual technique could serve well. In cases where data sets are large and the characteristics to be extracted are complex and hard to define, some variant automated techniques are advantageous.

In fact, the strategy lying somewhere between purely automatic and purely manual is the most effective one. Automated feature selection can serve as an enlarged and optimized feature set, supplementary to the features first defined through manual methods for the design of new automated systems.

5.4 Model selection and training: classical algorithms vs. deep learning models

The main decision to be made when selecting a model for medical imaging was between classical machine learning models as well as understanding the advantages and disadvantages of each. The decision affects the effectiveness and relevance of machine learning applications in medical image analysis.

More specifically, some of the compared attributes that have been reported to have significant results are SVMs, *k*-NN, and the decision tree, to name but a few. These methods handcraft features from photographs and then use learning algorithms to classify with respect to these features.

SVMs: SVMs are good with data sets of high dimensions and can be used in classification exercises. Construct a hyperplane in order that the optimal margin between the classes is achieved as much as possible. Despite the fact that SVMs have been found to be effective in numerous medical image applications, it has been observed that the quality of feature extraction and the size of the dataset that can be managed often influence the function of an SVM.

k-NN: One of the simplest algorithms of instance-based learning is *k*-NN, which classifies new instances with the aid of the majority of its nearest neighbors. It is easy and fast; however, if no proper dimensionality reduction methods are applied, it becomes computationally intensive and not very accurate for high dimensions.

Decision trees: They are also good for decision-making and, more importantly, interpretable because they divide the data space into regions according to feature values. Nevertheless, they are prone to overfitting, especially if the dataset is complex and involves medical image data to an important extent, and often require the use of ensembles, for instance, random forest, in order to work efficiently.

CNNs have actually transformed the field of analyzing medical images by removing the need to explicitly describe features and instead directly mapping input images to outputs that embed image analysis knowledge, especially in deep learning models. We also see from these models that there were a number of benefits obtained with

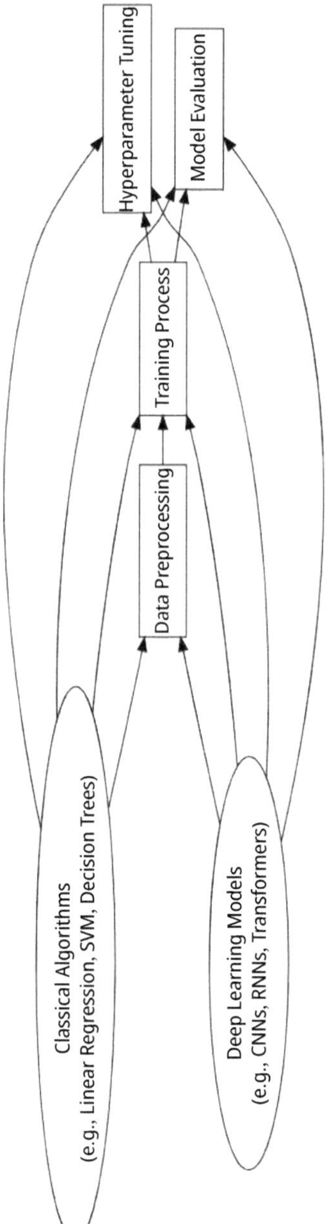

Figure 19: Model selection and training: classical algorithms vs. deep learning models.

the use of these models over conventional approaches in terms of convergence, performance, and scalability.

CNNs: Multilayer CNNs specialize in harvesting spatial hierarchies and characteristics, as they are built primarily for image analysis. CNNs can freely compose and decode complex patterns of medical images by performing convolution, pooling, and fully connected layers. CNNs have depth and design that allow feature extraction and classification across a wide range to enhance their performance on tasks such as object recognition and image segmentation.

Transfer learning: Transfer learning is the idea of taking deep learning models that have gone through the training stage and refining them for several medical imaging uses. Such a method reduces the amount of training data needed while drawing data from large datasets and using the data collected from these datasets. In this field, there are not as many labeled datasets in medical imaging, and so transfer learning is very useful.

Generative models: Some organizations and amorphous NNs, such as GANs and variational autoencoders (VAEs), are examples of deep learning that can be used in generating synthetic images. These models are useful in data augmentation and image synthesis, and they are used in enhancing the quality of medical images when training data prove hard to come by.

Key considerations include the following:

If it is any algorithm, then training and optimization become very important when looking at performance.

Hyperparameter tuning: A specific set of hyperparameters like learning rates, regularization parameters, and details of the structures of deep learning models is as crucial as those of conventional learning. To select the best-suited hyperparameters, a number of processes are used including grid search, random search, and the use of Bayesian optimization.

Cross-validation: To evaluate model performance and to stop at some point, different forms of cross-validation are applied, which include *k*-fold cross-validation. Through such methods, one is assured that models come up with the right estimates of the performance measures and can generalize future data that were not used during modeling.

Computational resources: Most of the DL models are trained on several computational platforms and elements that are widely demanded including GPUs and large-scale parallel processing. Time and resource management during the training process involves the usage of computing resources and different techniques such as early stopping and checkpointing.

In general, classical algorithms have problems with high-dimensional data and require a lot of feature engineering. In contrast, deep learning models are capable of learning complex patterns and performing feature extraction, which enhances performance in many medical image analysis applications. However, as stated before, deep learning models often require more computation and a lot of data to learn well.

5.5 Performance evaluation metrics: accuracy, sensitivity, specificity, ROC-AUC

There was a range of valuable quantitative measures used to estimate the performance of machine learning models in medical imaging, and all these measures together provide a comprehensive view of the models' work. The four primary metrics that are utilized consist of ROC-AUC, sensitivity, specificity, and accuracy.

Accuracy was the percentage of photos classified correctly out of all the images that had been sorted. It shows how well it predicts the total of unknown samples. It should not be expected to obtain equal accuracy for the performance of the model in the context of unbalanced data sets where certain conditions are rare because the model is inclined toward the most frequent class.

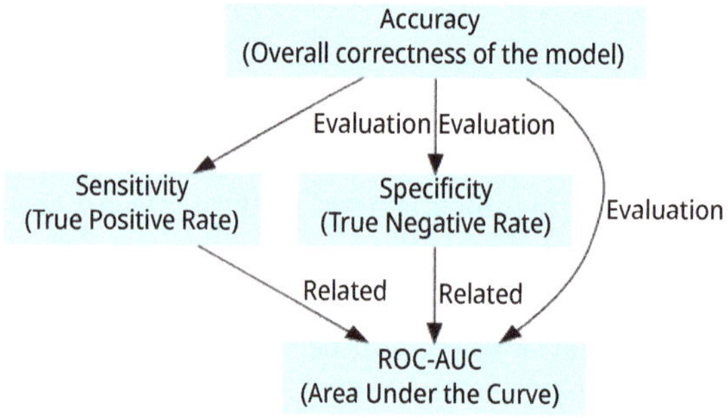

Figure 20: Performance evaluation metrics: accuracy, sensitivity, specificity, ROC-AUC.

Now you remember, recall was another name for sensitivity, which defines how well a model is capable of correctly identifying positive conditions, such as determining an illness where it exists. Indeed, in medical imaging, for instance, where missing a positive case has major implications, this statistic is very important. High sensitivity implies that the model can single out subjects with this condition, and this is a significant factor in identifying the disease early enough after diagnosis.

Accuracy of negative observations of condition absence is what specificity measures in the model. This measure helps estimate the performance of the model in rejecting false positives, which is necessary to avoid unnecessary further tests. High specificity ensures that the model can determine with a lot of certainty whether a patient is not affected by the illness.

Every point on the curve represents sensitivity together with (1-specificity) over a range of cutoffs of the diagnostic test in question. The discrimination between the positive and negative situations in the total sample was assessed using the area under the receiver operating characteristic curve (AUC). A higher value of AUC was achieved, meaning that over a range of threshold levels, the model performed better in distinguishing between the presence and absence of the condition.

Other evaluation metrics, which possibly also play a role, are precision-recall graphs, which take recall and accuracy into account and are valuable in unbalanced datasets. This was especially important because the *F*1 score complements specific information about the model by using both recall and accuracy metrics.

6 Common machine learning techniques in medical image processing

6.1 CNNs: architecture and applications

Due to their excellent performance in processing and understanding input image information, CNNs are widely used in the medical image processing business. As a result, image identification and classification are among the tasks best suited to CNNs, as they are designed to layer and adjust feature hierarchies extracted from input pictures in an automatic manner.

A typical CNN's architecture is made up of many essential layers that operate sequentially to process and evaluate visual data:

Layers of convolution: These layers employ convolutional filters to determine numerous things in the input picture like textures, edges, and patterns. Each filter produces a feature map that shows which specific features are in which particular areas of the image. Local patterns and hierarchical structures in the picture data can only be recognized when structures are implemented with convolution layers.

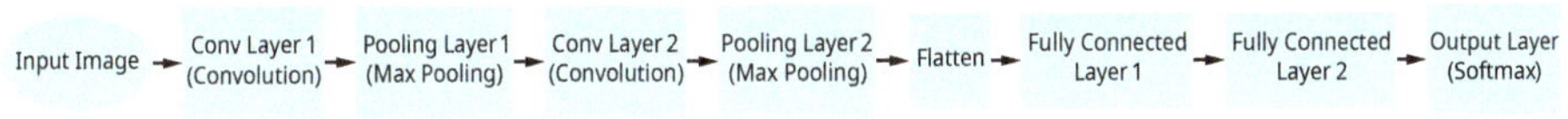

Figure 21: CNNs: architecture and applications.

Activation functions: After the convolutional layer in a model, there are activation layers, for instance, rectified linear unit, that bring nonlinearity. Since the network is nonlinear, the program can acquire higher nonlinear dimensions, which improves its performance in recognizing the intricate features of an image.

Layers for pooling: Feature maps can be made less dimensional by using pooling layers, such as average pooling, while keeping the most important information. Pooling layers decrease computational complexity and the probability of an inaccurate model because of their function of downsampling the feature maps and accumulating invariant features across various locations.

Fully connected layers: Finally, the numerous conv volt and pool layers bring out the high-level feature maps, which are flattened before being connected to fully connected ones. Before starting the final classification, layers involve the attributes that were gathered by the previous layers. There is a need for fully connected layers to make predictions with the help of the features that have been learned.

The output layer contains the outcomes offered by the CNN model, which include probability estimates in classification projects and regression operations. Since the topics of classification were more precise and unambiguous, the activation function of the output layer was determined; for example, SoftMax for multiclass or binary classification.

CNNs have been increasingly studied for many different medical imaging tasks because of their efficiency in analyzing and representing a certain set of image data. Key applications include

Image classification: CNNs are applied to sort medical images into different groups and types, such as sorting between the various types of tumors or applying the model to a new image to determine whether it contains a disease. This application assists the diagnostic process by providing instant and accurate classification based on features that have been learned.

Image segmentation: CNNs are used in splitting medical images to outline structural areas of tissue. For example, there are methods like U-Net that target exact segmentation tasks to provide required boundary information for the organs, lesions, etc.

Disease detection and diagnosis: Help identify diseases through imaging data for abnormality patterns and anomalies. For instance, CNNs help in finding markers for cancers and diabetic retinopathy, making the diagnosis more accurate at the earliest stage.

Image enhancement: CNNs can improve the quality of medical images by filtering out noise and improving the resolution artifacts. Higher resolution can help improve the performance of diagnosis and intervention, enhancing analysis and interpretation.

Radiological image interpretation: CNNs assist radiologists in making conclusions from intricate imaging information by offering further information in a specific aspect. This collaboration of human and artificial intelligence coming together makes the overall diagnostic work function even better.

RNNs and LSTM networks are refined methods of machine learning, which are particularly valuable for the analysis of sequential data and temporal structures. Although these architectures are widely used in natural language and time-series data, they provide useful features for some processes in medical imagery as well.

6.2 RNNs and LSTM networks

Rewriting of the data sequences was handled by the internal state in RNNs, which was therefore like a memory. Due to this feature, RNNs are useful in implementation in areas where the temporal order of the data is crucial. Could RNNs be employed more closely in the medical imagery domain, where there are picture sequences involved, like temporal series of medical scans?

RNNs' architecture allows the capturing of temporal dependencies through the processing of the inputs while at the same time retaining data from previous steps. Due to issues such as vanishing and exploding gradients, RNNs struggle with learning long-term dependencies, which can be an issue when there are long sequences that need to be learned.

Some drawbacks of standard RNNs are fixed and can be alleviated by key components of LSTM networks. LSTMs employ specific units called memory cells to store information for a long time; they also have controlling elements known as gates. Such a structure reduces the problems connected with vanishing gradients so that LSTMs can learn from long sequences.

For medical image processing, LSTMs are now being used for sequences of images where temporal dependency plays an important role. For instance, LSTMs can improve the understanding of temporal medical imaging data, such as the observation of disease evolution within temporal changes in patients' conditions from imaging longitudinal studies.

Temporal image analysis: LSTMs can be used to engage temporal sequences of medical images, for example, MRI scans done at different times. This application is useful when evaluating the course of an illness and the effectiveness of treatment.

Video-based diagnostics: In cases where medical imaging data is delivered as a sequence of videos, RNNs and LSTMs can find a way to analyze the temporal features, for instance, detecting mutated movements of a body in a video.

Predictive modeling: This was because the temporal features of image data are well exploited by both RNNs and LSTMs in the area of predictive modeling, where future medical conditions can be forecast from past imaging results.

The use of RNNs and LSTMs comes with some drawbacks: these models can be computationally expensive in terms of the time needed for training and using them. Some of these models require finer optimization and tuning to exhibit robust performance in the medical imagery domain. To incorporate these models into a clinical setting, issues concerning "Under the Hood" and the timeliness of the analysis must be solved.

6.3 GANs for image enhancement and synthesis

During the past few years, GANs have gained much attention in the medical image processing and analysis arena, where picture synthesis and expansion have become easier. In the process of GAN, a discriminator and a generator are trained at the same time by strictly employing adversarial techniques.

A clear understanding of how GANs work stems from the conception of one network that competes against another. The discriminator network checks the credibility

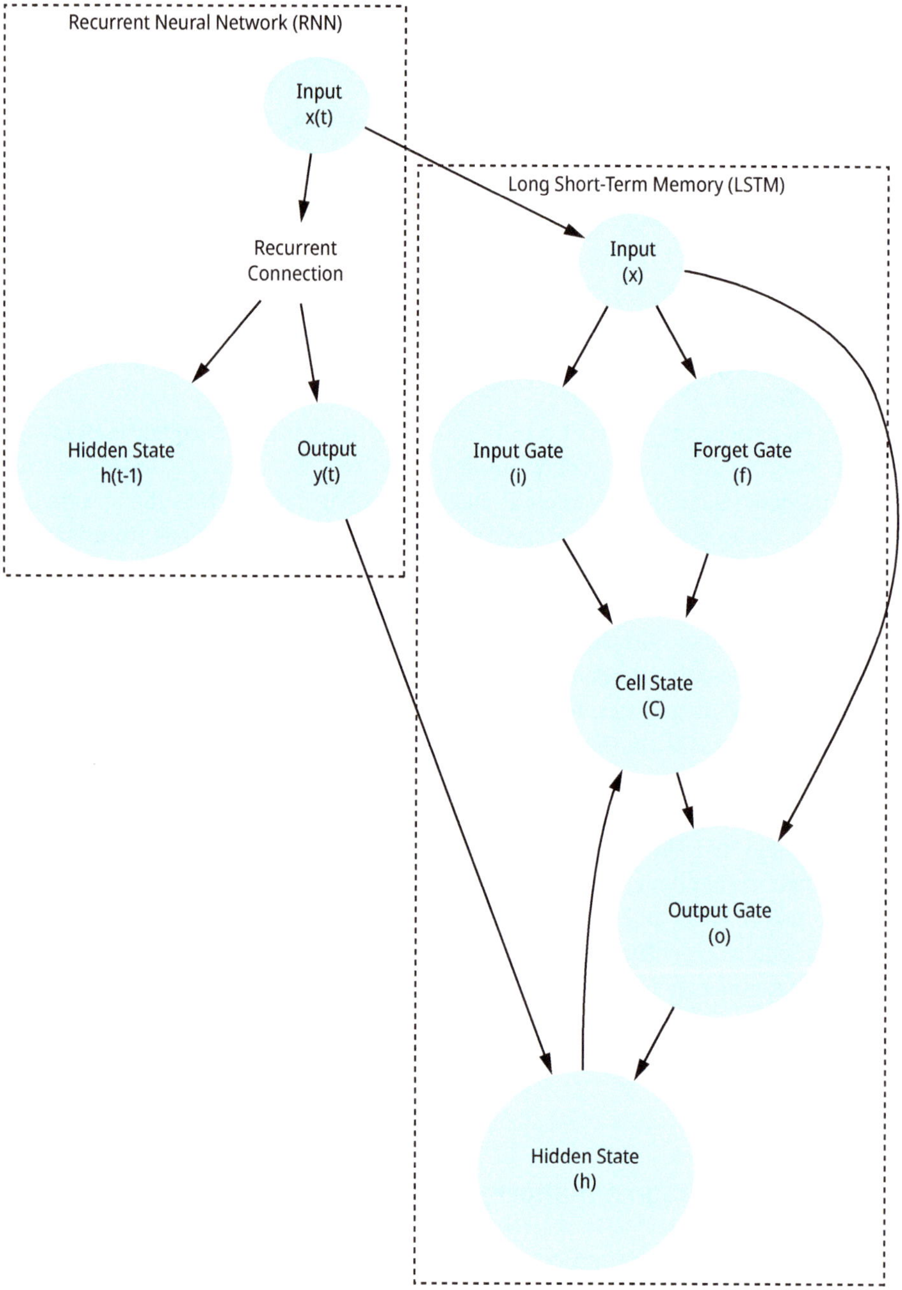

Figure 22: RNNs and LSTM networks.

of artificial pictures generated by the generator network against actual data. The generator network produces fake pictures from scratch based on other inputs, which can be noise, for instance. The discriminator acquires expertise in the ability to distinguish real pictures from fake ones throughout the training process, while the generator improves in producing accurate pictures similar to those it was trained on. The developed pictures are of better quality as a result of this adversarial training procedure, meaning further opportunities in medical imaging are available.

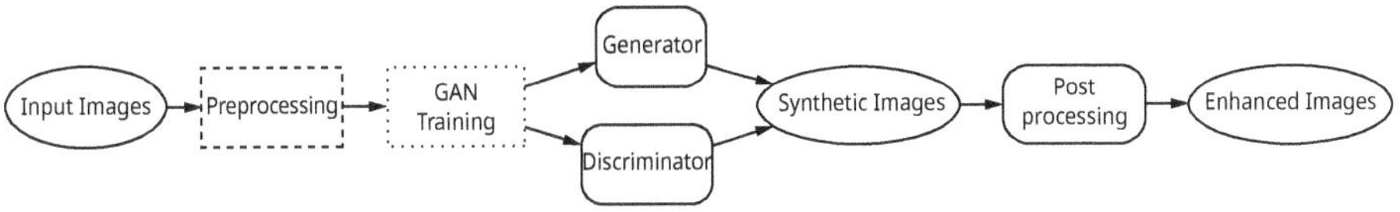

Figure 23: GANs for image enhancement and synthesis.

In medical imaging, GANs are used to increase image quality by enhancing resolution, reducing noise levels, and eradicating artifacts. For example, in the imaging diagnostic field, GANs can enhance low-quality images to high quality in terms of pixel densities for better visualization. This process is called super-resolution and aids in identifying small structures in an image, which contributes to medical image interpretation such as MRI scans. Additionally, GANs can be applied to reduce noise and other unwanted details within images, which is important for properly seeing and analyzing pictures when determining the similarities of created scenes.

Another application of GANs is the generation of new medical images from existing image data, which is known as medical picture synthesis. This is especially valuable when annotated data is scarce to acquire; as with deep learning models, unless one can train their own from scratch, these datasets remain relatively fragile. The training datasets for the machine learning model can therefore be enriched more easily using GANs, as these can generate realistic artificial pictures that mimic real medical images. These artificial pictures are employed to augment model accuracy and enhance the test sample's generalizability and resilience of prediction methods.

On the same note, data augmentation, whereby new training datasets are developed using GANs, requires modifications of existing medical pictures. Data imbalance is not a problem that GANs cannot address; instead, it refines machine learning models by creating realistic images. There is also a significant indication that benchmarks resulting in the possession of various clinical states – such as GANs – augment the model's capability in generalizing across different scenarios.

The utilization of GANs in medical image processing is increasing and affecting different clinical tasks. GANs are currently being investigated for generating training data, improving diagnostic images, and even for building realistic environments for operations. With future developments, GANs could be applied to further promote diagnostic precision, individual treatment strategies, and new solutions for problems encountered in medical imaging.

6.4 Transfer learning and pretrained models in medical imaging

Transfer learning has been regarded as one of the effective learning paradigms that can enhance the medical image processing and accelerate the development of machine learning. These methods improve certain medical imaging tasks by leveraging information from large-scale data and models learned from different tasks.

Transferring knowledge and the approach used to learn that knowledge from the original set of data to a new, normally different set of data is referred to as transfer learning. This reduces the need for large amounts of training data and computation time the model is expected to complete; instead, this method can presume the model knows how to complete related tasks. Transfer learning in medical imaging utilizes previously trained models from larger image data sets as a way to create learning models for specific illnesses.

From the above discussion, transfer learning has the following benefits in the processing of medical images. First, it allows the use of models trained on a large amount of varied data, which serve as a good starting point when training on medical images when there is not much data at one's disposal. In medical imaging, gathering large annotated datasets is quite costly and time-intensive. It assists in handling this difficulty, shows how to work with quantitative data, and provides a brief description of two case studies. Transfer learning allows for achieving faster convergence of pertinent characteristics from the original training in the pretrained model.

Other types of NNs, known as pretrained models, have been trained on datasets such as ImageNet – comprises millions of tagged photos in various categories. These models in medical imaging can be interactively programmed to perform activities like identifying an illness. The popular names of good models for image classification and feature extraction are VGG, ResNet, and inception.

Fine-tuning of a model previously trained for a different task was termed as fine-tuning, and no pretraining session was done. To optimize the model for the specific application, this involves adjusting its parameters and training it with the particular data set, which was in the medical images. The last layers are typically replaced and retrained in an endeavor to fine-tune the model, while the other layers, which contain features generally learned, are preserved. By doing so, the model preserves most of the nonspecific information obtained from the original training set while learning about the specificities of a certain domain.

Transfer learning and pretrained models are used in some of the mentioned medical imaging domains. For example, they are used in identifying tumors in MRI, segmenting organs in CT, and classifying diseases in X-ray. By using these models, researchers and clinicians can attain far better diagnostic tools and achieve enhanced patient results with considerably less exertion and less data than constructing models from the ground up.

6.5 Ensemble methods and hybrid models

Ensemble methods and hybrid models are considered to be the state of the art in the field of machine learning, and this section describes how these techniques can improve the accuracy and reliability of developed medical image processing systems. Thus, these approaches use a combination of models to take the best combination of characteristics, hence improving both the accuracy and the generalization of the models.

In ensemble approaches, numerous predictors of machine learning models are integrated to give a result that would not be attainable if a single model was employed. The general concept behind ensemble techniques is that the performance should be uplifted and the chance for errors should be reduced by aggregating predictions from various models. In medical image processing, a number of widely utilized ensemble approaches are employed including

Using bootstrap aggregating (Bagging): Training several instances of the same model type on various subsets of the training data, which are produced by random sampling with replacement, is called bagging. Lastly, the incurred predictions from the developed models are then aggregated to a final result through voting. There is a bagging algorithm called random forest that creates numerous decision trees and reaches an averaged outcome.

Boosting: Boosting works as a sequence of models in which every ensuing model tries to reduce the error made by previous models. This technique addresses the circumstances in which previous models failed to make accurate predictions and boosts the performance. AdaBoost and gradient boosting machines are two subtypes of boosting methods used in medical image analysis to enhance the results of the model.

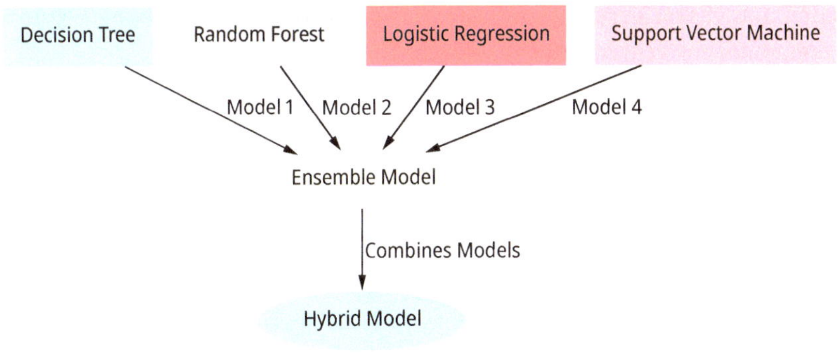

Figure 24: Ensemble methods and hybrid models.

Stacking: In stacking, many different models are trained, and a meta-model is then used to combine the predictions of the models. It is the meta-model, which, in the case of applying a meta-model-based AdaBoost, learns to predict a final output of its base learners, which are independent models trained on the same data set and are called base learners. That is, based on the features in many model types, stacking can produce enhancements in prediction results.

The other models incorporate two categories of machine learning methods in combination with other computational strategies to take advantage of each method's strengths. These models can be particularly applied to medical image processing, as the data in this field is often complex and heterogeneous, and it is most effective to use complex and heterophilic approaches. Some examples of hybrid models include:

Combining CNNs with traditional image processing techniques: CNNs, when combined with other algorithms like edge detection and texture analysis incorporation, can complement the general feature extraction and segmentation, enhancing them. This means that when the combination of these techniques is used in the models, the model can then learn even the higher and lower facets of the images, resulting in higher performance rates.

Fusion of machine learning and statistical methods: To increase the interpretability and accuracy of the models while integrating machine learning algorithms, statistical methods like PCA can be used for feature extraction. Reducing the feature space while still preserving sufficient detail, this innovation helps to provide better training in addition to enhancing prediction.

Integration of deep learning with rule-based systems: Combining conventional artificial intelligence methodologies with deep learning techniques can also involve strategic domain knowledge in the approach. For example, a deep learning model can be used in identifying diseases from images, while a rule-based system uses protocols to filter the results and eliminate wrongful diagnoses.

Multiview and multimodal fusion: In medical imaging, as the data originate from various imaging modalities like MRI, CT, and PET, integrating them in the form of a hybrid model would be more beneficial. As shown in this work, these models improve the detection and characterization of medical conditions by incorporating features from different imaging techniques.

7 Conclusion

The use of machine learning algorithms in the diagnosis of medical images has presented a new horizon of precision and speed. In this chapter, the truly transformative capabilities of deep learning architectures like CNNs, SVMs, and other machine learning methods in medical imaging have been described. Such advancements have enhanced the diagnosis of intricate and sometimes profound medical scenarios and paved the way for more accurate identification of ailing conditions and differentiated lifestyle management programs.

CNNs are typically remarkable in feature extraction and have been very useful in facilitating the automation of diagnostic procedures. Their use across various modes of medical imaging, including radiology and pathology, has boosted diagnostic performance as well as operational outputs. Likewise, SVMs have been applied successfully in high-dimensional data classification and novelty detection models, which are more powerful compared to conventional approaches.

It also revealed that deep learning architectures have been pivotal in handling large details and subtle patterns within medical images. Different types of DL models, including GANs and RNNs, also show how deep learning simultaneously improves image quality while also enabling new diagnostics. These models have demonstrated the potential to solve previously unsolvable problems for which solutions were inapplicable in the past, including early detection of diseases and the development of certain treatment plans for each patient.

Nevertheless, certain challenges remain that one would face when trying to combine machine learning with medical imaging. The employment of LSMs, the need for stable, interpretable algorithms, as well as the ethical questions arising from the use of these tools in therapeutic practices, are still among the most important research topics. These factors pose challenges to the process, and it becomes important to address such challenges in order to ensure continuous progress and optimization of machine learning technologies in healthcare.

References

[1] Najjar, R. (2023). Redefining radiology: A review of artificial intelligence integration in medical imaging. Diagnostics, 13(17), 2760.

[2] Thrall, J. H. (2016). Trends and developments shaping the future of diagnostic medical imaging: 2015 Annual Oration in Diagnostic Radiology. Radiology, 279(3), 660–666.

[3] Panayides, A. S., Amini, A., Filipovic, N. D., Sharma, A., Tsaftaris, S. A., Young, A., . . . & Pattichis, C. S. (2020). AI in medical imaging informatics: Current challenges and future directions. IEEE Journal of Biomedical and Health Informatics, 24(7), 1837–1857.

[4] Malekloo, A., Ozer, E., AlHamaydeh, M., & Girolami, M. (2022). Machine learning and structural health monitoring overview with emerging technology and high-dimensional data source highlights. Structural Health Monitoring, 21(4), 1906–1955.

[5] Rodrigues, J. F., Jr, Paulovich, F. V., De Oliveira, M. C., & de Oliveira, O. N., Jr. (2016). On the convergence of nanotechnology and Big Data analysis for computer-aided diagnosis. Nanomedicine, 11(8), 959–982.

[6] Parwani, A. V., & Amin, M. B. (2020). Convergence of digital pathology and artificial intelligence tools in anatomic pathology practice: Current landscape and future directions. Advances in Anatomic Pathology, 27(4), 221–226.

[7] Anwar, S. M., Majid, M., Qayyum, A., Awais, M., Alnowami, M., & Khan, M. K. (2018). Medical image analysis using convolutional neural networks: A review. Journal of Medical Systems, 42, 1–13.

[8] Zou, Z., Chen, K., Shi, Z., Guo, Y., & Ye, J. (2023). Object detection in 20 years: A survey. Proceedings of the IEEE, 111(3), 257–276.
[9] Sarvamangala, D. R., & Kulkarni, R. V. (2022). Convolutional neural networks in medical image understanding: A survey. Evolutionary Intelligence, 15(1), 1–22.
[10] Banerjee, A., Burlina, P., & Diehl, C. (2006). A support vector method for anomaly detection in hyperspectral imagery. IEEE Transactions on Geoscience and Remote Sensing, 44(8), 2282–2291.
[11] Maulik, U., & Chakraborty, D. (2017). Remote Sensing Image Classification: A survey of support-vector-machine-based advanced techniques. IEEE Geoscience and Remote Sensing Magazine, 5(1), 33–52.
[12] Hussain, S. F. (2019). A novel robust kernel for classifying high-dimensional data using Support Vector Machines. Expert Systems with Applications, 131, 116–131.
[13] Castiglioni, I., Rundo, L., Codari, M., Di Leo, G., Salvatore, C., Interlenghi, M., . . . & Sardanelli, F. (2021). AI applications to medical images: From machine learning to deep learning. Physica Medica, 83, 9–24.
[14] Lee, J. G., Jun, S., Cho, Y. W., Lee, H., Kim, G. B., Seo, J. B., & Kim, N. (2017). Deep learning in medical imaging: General overview. Korean Journal of Radiology, 18(4), 570–584.
[15] Shmueli, G., Patel, N. R., & Bruce, P. C. (2011). Data mining for business intelligence: Concepts, techniques, and applications in Microsoft Office Excel with XLMiner. John Wiley and Sons.
[16] Gorunescu, F. (2011). Data Mining: Concepts, models and techniques, (vol. 12). Springer Science & Business Media.
[17] Dash, A., Ye, J., & Wang, G. (2023). A review of GANs and its applications in a wide variety of disciplines: From medical to remote sensing. IEEE Access.
[18] Zhou, S. K., Greenspan, H., Davatzikos, C., Duncan, J. S., Van Ginneken, B., Madabhushi, A., . . . & Summers, R. M. (2021). A review of deep learning in medical imaging: Imaging traits, technology trends, case studies with progress highlights, and future promises. Proceedings of the IEEE, 109(5), 820–838.
[19] Pillai, A. S. (2021). Utilizing deep learning in medical image analysis for enhanced diagnostic accuracy and patient care: challenges, opportunities, and ethical implications. Journal of Deep Learning in Genomic Data Analysis, 1(1), 1–17.
[20] Weissler, E. H., Naumann, T., Andersson, T., Ranganath, R., Elemento, O., Luo, Y., . . . & Ghassemi, M. (2021). The role of machine learning in clinical research: Transforming the future of evidence generation. Trials, 22, 1–15.
[21] Kumar, Y., Koul, A., Singla, R., & Ijaz, M. F. (2023). Artificial intelligence in disease diagnosis: A systematic literature review, synthesizing framework and future research agenda. Journal of Ambient Intelligence and Humanized Computing, 14(7), 8459–8486.
[22] Prabhakar, P. K. (Ed.). (2024). Biomedical research developments for improved healthcare. IGI Global.
[23] Murtaza, G., Shuib, L., Abdul Wahab, A. W., Mujtaba, G., Mujtaba, G., Nweke, H. F., . . . & Azmi, N. A. (2020). Deep learning-based breast cancer classification through medical imaging modalities: State of the art and research challenges. Artificial Intelligence Review, 53, 1655–1720.
[24] Scholkopf, B., & Smola, A. J. (2018). Learning with kernels: Support vector machines, regularization, optimization, and beyond. MIT press.
[25] Archana, R., & Jeevaraj, P. E. (2024). Deep learning models for digital image processing: A review. Artificial Intelligence Review, 57(1), 11.

M. Mahalakshmi, S. Sujatha, S. Banumathi, Purushothaman V., K. Suresh Kumar

Transforming healthcare with machine learning

Abstract: Deep learning (DL) models in material science revolutionize the design and characterization of materials, but due to their complexity and the fact that no one knows exactly how the model works, they can be hard to understand. In this chapter, we introduce a critical issue of the current DL frameworks: interpretability, with a focus on post- and intermediate processing techniques that enhance the readability of the models' predictions. The relationship between the input and the output is sometimes unclear due to the high complexity of DL methods, which require the use of high-dimensional nonlinear spaces, presenting one of the main challenges. All of these are separately discussed in the chapter: an interpretable model element, visualization methods, and local explanation methods such as Shapley values and local interpretable model-agnostic explanations (LIME). Thirdly, it compares factors such as model accuracy and interpretability, among others, because in each case, it is impossible to achieve the maximum increase in performance by introducing more complex and sophisticated models without sacrificing the interpretability of the results for end-users. Perspectives for further studies are described, with an emphasis on the ability to combine domain knowledge and create machine learning models with domain-oriented interpretability protocols for material science fields. This proposed work should be able to close the gap between abstract DL schemes and tangible materials science applications, thus encouraging the general acceptance of AI technologies in materials science and engineering research and future application development.

Keywords: deep learning, interpretability, materials science, visualization techniques, local explanations, model transparency

M. Mahalakshmi, Department of Networking and Communications, SRM Institute of Science and Technology, Kattankulathur, Chennai, Tamil Nadu, India, e-mail: strimaha@gmail.com
S. Sujatha, Department of EEE, Sri Sairam College of Engineering, Bengaluru, Karnataka, India, e-mail: sujasakthi22@gmail.com
S. Banumathi, Department of Computer Science and Engineering, Vel Tech Rangarajan Dr. Sagunthala R&D Institute of Science and Technology, Avadi, Chennai, Tamil Nadu, India, e-mail: banumathicse@veltech.edu.in
Purushothaman V., Department of Electronics and Communication Engineering, Chennai Institute of Technology, Chennai, Tamil Nadu, India, e-mail: purushothamanv@citchennai.net
K. Suresh Kumar, MBA Department, Panimalar Engineering College, Varadarajapuram, Poonamallee, Chennai 600123, Tamil Nadu, India, e-mail: pecmba19@gmail.com

https://doi.org/10.1515/9783111503202-013

1 Introduction

Deep learning (DL) has been embedded into material science in a way that deals with material discovery and characterization [1]. These models, with their capacity to analyze huge amounts of data and subtle patterns, have facilitated unparalleled developments [2]. DL algorithms often become algorithmically mystified, which creates interpretability problems [3]. Essentially, these models are often quite complex, and their formats contribute to a situation where the reasoning behind the forecasts is obscure [4]. This opacity can be problematic in terms of the utilization of the models in various sectors, especially where the models need to be sensitive, accurate as well as valid [5].

Interpretation in DL is defined as the extent to which architects are able to comprehend how a model produces certain results [6]. In material science, this capability is essential for verifying the results obtained from the model against the results expected in that particular field [7–9]. In particular, when DL models are utilized to forecast material characteristics and design, the audience must have confidence in the outcomes and comprehend the model's outputs [10]. Inadequate interpretability can prevent the incorporation of these models into crucial applications, as it is necessary to avoid catastrophic mishaps resulting from incorrect predictions [11].

Several techniques have been designed to handle cases where interpretability issues arise in the context of DL [12]. Other visualization methods that describe the input data characteristics important for predictions include saliency maps and activation maps [13]. These bow-tie and action strategies help in understanding the model's focused areas and the decision input data areas with significant influence on the model outputs [14]. In addition, methods called LIME and Shapley values provide a more fine-grained understanding of how certain inputs contribute to the model and offer local explanations for predictions [15]. While there has been some progress in the field, there are still several trade-offs to be expected when making models interpretable [16]. The increase in transparency often involves the use of model reduction or approximation methods, which decreases predictive capability [17]. The main challenge is how to achieve interpretability while at the same time maintaining high levels of accuracy [18]. For instance, the first models, which are much easier to understand, do not produce as accurate predictions as the more complicated structures of DL algorithms [19]. This trade-off highlights the need to assess the strengths and weaknesses of various interpretability techniques in the context of materials science applications [20].

In particular, regarding future trends in the comprehension of DL models for material science, it is probable that existing trends will focus on making processes more standardized and effective [21]. That is, improvements in algorithm design at the interpretability stage and the adoption of solution-specific knowledge could enhance the solutions [22, 23]. The aim is to propose approaches that provide more comprehensible explanations while maintaining the same predictive performance [24, 25]. Accordingly, as the field advances, enhancing the levels of openness and awareness in DL material science research will be critical for spurring development and ensuring the effective use of DL techniques.

2 Deep learning techniques in biomedical materials

2.1 Convolutional neural networks (CNNs) for material property prediction

Basic notions on CNNs are another kind of powerful DL computing method that focuses on analyzing structures such as data that come in grids, for instance, images. CNNs were designed to work with images, mostly for processing them, but have been adapted for many purposes, including material science. CNNs appear relevant in material property predictions due to their ability to learn feature representation and spatial hierarchy from the inputs. This makes it possible to simulate complex relations between material properties and the structures of those materials.

Convolutional, pooling, and fully connected layers are some of the building blocks of most CNN structures. In our case, pooling layers are used to down sample the input and retain the most relevant aspects. The convolutional layers, on the other hand, convolve the input data with kernels to identify local features. The final layers aggregate these features to make predictions. For material property prediction, which can be achieved through CNNs, a material's microstructure, as captured in microscopy images, can be used to predict properties such as mechanical strength, thermal conductivity, and biocompatibility. This approach is particularly useful, thanks to CNNs ability to handle the extraction of hierarchical features from data of high and even extreme complexity.

A study of CNNs for material property prediction CNNs has proved worthy of being employed for the prediction of different properties of materials. For example, CNNs have been used for the prediction of the tensile strength of composite materials from the microstructure images of the same composite material. CNNs have also been applied to estimate thermal properties based on thermal image observations. These applications show that CNNs can extract useful predictions from images, which is very beneficial for material design and optimization.

Hence, CNNs have the following advantages for material property prediction: robust feature extraction – CNNs have several downsides for material property prediction; limited dimensionality – CNNs are not very suitable for material property prediction because this makes CNNs gain important features from raw data with little need for feature engineering. It was also accompanied by certain shortcomings, such as the requirement for large computing power consumption and large annotated datasets. It was also pointed out that CNNs suffer from reduced prediction accuracy due

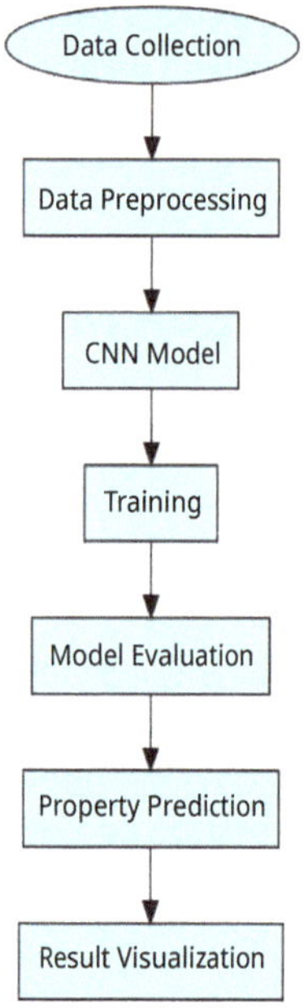

Figure 1: Convolutional neural networks (CNNs) for material property prediction.

to the difficulty models face when implementing them on material types not utilized in the development set.

Potential areas for research and development in the use of Convolutional Neural Networks (CNNs) for predicting material properties focus primarily on enhancing model generalization and interpretability. Techniques such as transfer learning and domain adaptation are pivotal in improving CNN performance across diverse material datasets. Combining CNNs with other machine learning strategies, such as ensemble methods, has shown to significantly increase predictive accuracy. As data availability and computational power continue to advance, CNN-based models are expected to deliver even more impactful results in material science applications. Figure 1 illustrates the CNN-based prediction framework, highlighting key methods like transfer learning, domain adaptation, and ensemble integration to enhance model robustness and applicability.

2.2 Recurrent neural networks (RNNs) in time-series analysis of material behavior

RNNs have been identified as capable of analyzing sequential data, such as time series information, since they have temporal dependency abilities. In the context of developing biomedical materials, RNNs are harnessed for the realization and prediction of time-dependent material performance. This capability is especially insightful for applications that involve transitions in material properties due to conditions like mechanical load, temperature, or environmental factors. As a result of RNNs' corresponding sequential structures, researchers are able to build models that supply more significant temporal characteristics of distinct material properties, thereby achieving more efficient predictions and a deeper understanding of material characteristics during the material design process.

2.2.1 RNNs' employment in material behavior analysis

RNNs are used with time series data to analyze the behavior of materials in material science and to predict their behavior under specific conditions. For instance, when analyzing the characteristics of biomaterials used in implants, RNNs can utilize time series stress–strain data to observe the ways in which the material deforms under test loading cycles. Likewise, RNNs can be used to monitor the wear of materials in real-world scenarios and present important data for the creation of better and stronger materials. Since RNNs have the ability to handle sequences of different lengths and with different levels of complexity, they are therefore very suitable for modeling complex time-dependent phenomena in material science.

2.2.2 Substituting and improving RNN architectures for materials science

It was found that traditional RNNs are quite effective but generally yield poor results due to issues such as vanishing and exploding gradients. There are other complications that hinder the ability of RNNs, and these include remote dependencies, sinking gradients, vanishing gradients, and advanced topologies in RNNs, particularly the GRUs and LSTM networks. Thus, to avoid these problems and to enhance the model's ability to learn long sequences, LSTMs incorporate methods for retaining information across extended sequences. LSTMs should be used where there are constraints on computational capabilities, as GRUs are less complex but can offer similar performance to LSTMs. Over time, these improved RNN designs have demonstrated higher accuracy and stability in modeling material behavior, resulting in more reliable prediction and evaluation.

Numerous research studies provide evidence of realistic time series analysis of material behavior through RNNs. For example, scientists have used LSTMs to determine the fatigue life of aerospace materials by analyzing time series data obtained from cyclic tests. Another study employs GRUs for modeling the thermal degradation of polymeric materials, which proved effective in illustrating the potency of RNNs in mimicking the degradation temporal patterns of inherent materials. These cases demonstrate how RNNs can be used in various issues in material science and indicate how knowledge of material response through time can be improved using RNNs.

There are several limitations with the present approach of using RNNs for analyzing time series of material behavior. Some of the challenges that are still being experienced include the need for large volumes of data, computational problems, and the limitations of the RNN models. This is why future research aims at designing better algorithms with a closer look, enhancing model interpretability, and combining RNNs with other methods in machine learning to counter these issues. Investigating the application of RNNs for other materials and new technologies reveals fresh areas for enhancing material science and engineering.

2.3 Autoencoders for dimensionality reduction in material characterization

Autoencoders, a class of neural networks used primarily for unsupervised learning, have gained widespread attention due to their dimensionality reduction capabilities. These models consist of an encoder that compresses high-dimensional input data into a lower-dimensional latent space, and a decoder that reconstructs the original data from this compressed representation. In the context of material characterization (MC), autoencoders help preserve critical features while minimizing redundant information, making them highly effective for analyzing large-scale datasets derived from material science simulations and experimental results. Figure 2 illustrates the architecture and function of an autoencoder, highlighting its role in compressing and reconstructing material property data.

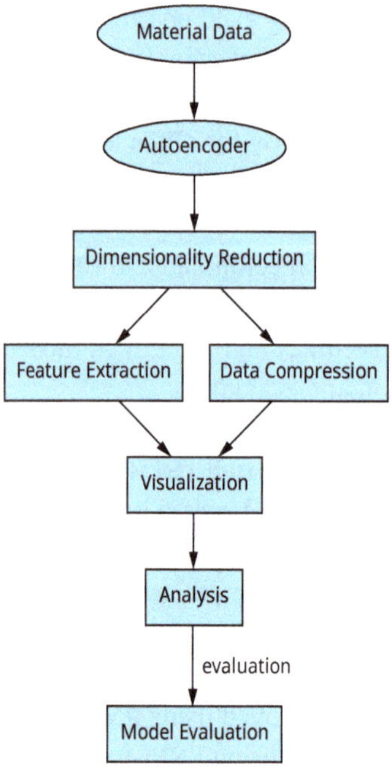

Figure 2: Autoencoders for dimensionality reduction in material characterization.

During the characterization of the properties of materials, the incorporation of applications was possible.

In MC, autoencoders fill the gap in working with high-dimensional data including spectroscopy, field data, as well as other analytical measurements. Fundamentally, these properties are latent within the data; autoencoders reduce their dimensions, allowing for a better understanding of the properties themselves. The mapping can also capture details that were previously hard to grasp due to the size of the feature space. For instance, in the study of the properties of polymer composite materials and alloys, autoencoder types can deliver important features, valuable for the definition of the best material combinations in terms of performance.

2.3.1 Benefits and advantages

When using autoencoders for the purposes of dimensionality reduction, the most significant advantage is their ability to learn nonlinear transformations of data. That is, autoencoders do not rely on comparatively simple linear transformations to find nonlinear relationships in the data, as PCA does. This feature is highly important for MC scenarios where the relationships between numerous material parameters are significantly nonlinear. The autoencoder is also able to perform data cleaning by training the network to map noisy data points into noise-free representations, thereby improving the quality of the reduced-dimension features for further analysis.

2.3.2 Challenges and considerations

As useful as autoencoders are when used in MC, there are some issues to consider. Some of the most important aspects that should be adjusted in the architecture of an autoencoder include the number of layers and the size of the latent space. The choice of the number of layers and the size of the latent space can significantly affect the performance of the autoencoder. One notable drawback associated with using this model is overfitting, particularly if the size of the dataset is small compared to the size of the model. Additionally, the interpretation of the compressed representation remains problematic because the representation learned by autoencoders may not align with the physically relevant aspects of the material.

It was expected that future research on autoencoders for MC would explore ways of making those models more explainable and integrating them into other machine learning methods. These are some implications for future work: possibly, a better approach for building improved models could be to develop multilayer autoencoder models that include some type of filtering, which supports marked-up output labels to utilize the results of supervised learning methods, thereby adding more robust and interpretable results. Other new DL architectures include variational autoencoders and convolutional autoencoders, which enhance the dimensionality reduction of complexity in material datasets. Future investigations of such techniques could result in enhanced as well as deeper understanding of the properties of materials, advancing the ambitions of material science.

2.4 Generative adversarial networks (GANs) for synthetic material data generation

GANs, at the present time, are a modern and highly effective approach to the generation of synthetic data within the framework of the machine learning paradigm. GANs were first introduced by Goodfellow et al. in 2014 and are composed of two coupled neural networks: a discriminator and a generator, which compete with each other. While the discriminator verifies the similarity of the created data to real data, the generator creates synthetic data. This adversarial process makes the generator produce better data, which has been important in many industries including those involving biological materials.

2.4.1 A case study on usage of GANs in materials science

The generation and training of a new material dataset are provided effectively in the scenario of biomedical materials using GANs. In materials science, data collection techniques are usually costly and entail a lot of experimentation and trial. These difficulties are solved by generating high-quality synthetic data that mimic the characteristics and behavior of actual materials using GANs. This artificial data helps investigate material properties that cannot be easily measured experimentally, thereby enlarging the training datasets in order to enhance the model's performance.

2.4.2 Four benefits of using GANs for synthetic data

The application of GANs for synthetic material data generation has many potential benefits. First, GANs help generate myriad and extensive labeled data that otherwise could not be generated through experiments. This is especially helpful in regions where there is little information regarding the constitutive behavior of specific materials, such as those that are comparatively new or often used. Second, GAN-synthesized data further strengthens and widens the applicability of machine learning algorithms by creating a diverse range of patterns found within data sets. Third, GANs can sample data with specific characteristics, allowing researchers to investigate material properties by restricting their applications to research areas of interest.

2.4.3 Challenges and considerations

As with every kind of modeling, there were challenges that researchers faced when applying GANs in material science. One major consideration was the nature of the synthetic data created including its realism. Supplementing this is an illustration of improperly trained GANs, where the synthetic data on material properties does not simulate the real-world data situation. The use of synthetic data in conjunction with real experimental data must be handled carefully to prevent differences that affect the model. To overcome these challenges, further research and adjustments to the methods of GANs to enhance data credibility and usability are necessary.

In the area of generating synthetic material data and the future of GANs, several avenues seem viable for further research and development. In this regard, improvements to the structures of GANs, including the integration of elaborate loss functions and better procedures for training, result in higher levels of absorption and more realistic data synthesis. Further, integrating GANs with other algorithms, including reinforcement learning (RL), provides new solutions to material science problems. I imagine that the investigation of these possibilities is essential for the further development of using GANs for synthetic data creation and for the progress of biomedical materials science.

2.5 Transfer learning for improving material design models

This approach is known as transfer learning, a powerful deep learning (DL) technique that leverages knowledge gained from one domain to enhance performance in another, related domain. In materials science, transfer learning enables the use of pretrained models from adjacent fields to accelerate the development and optimization of biological materials. This technique is especially valuable when dataset availability is limited, as it reduces the need for extensive data collection by reusing learned features and patterns from existing models. Figure 3 illustrates the concept of transfer learning, showing how pre-trained models can be adapted to new tasks in material design and analysis with minimal additional training.

2.5.1 Possibilities of transfer in material design

In the case of material design, transfer learning can improve material design by providing the designer with a starting point to create specialized models. For example, decision trees from image recognition, previously applied in computer vision, can be extended to reasoning about material microstructures. This allows researchers to enhance the estimation of material properties and reduce the computational power required to train new models from scratch by leveraging the feature extraction capabili-

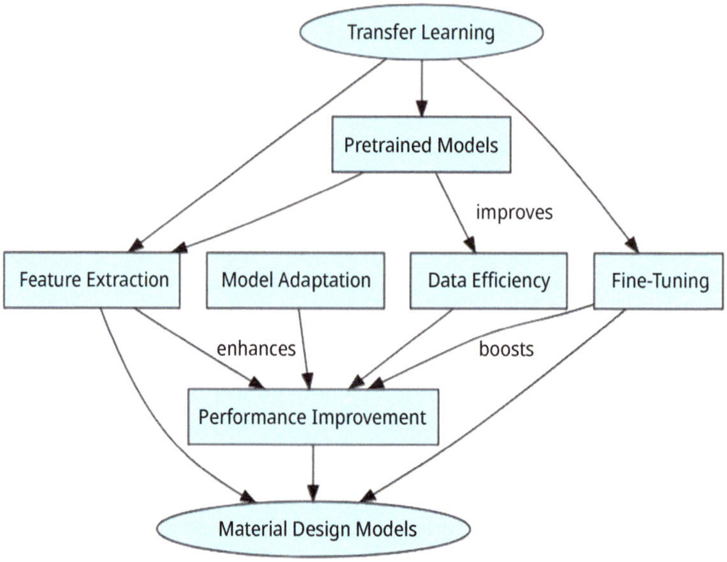

Figure 3: Transfer learning for improving materials design models.

ties of these models. Transfer learning has also been utilized in the design of materials, accelerating the development of novel high-performance materials where precise material design was achieved using knowledge transferred from similar, but known, materials.

2.5.2 Approaches for applying transfer learning

There are several important methods that must be used when it comes to the implementation of transfer learning. One of the approaches is fine-tuning, which refers to the process of retraining with a specific material dataset after its primary training to adjust the model's weights to fit new material. The advantages of this approach are evident: when the pretrained model has been trained on a large enough and sufficiently diverse dataset, it learns more universal features that are relevant for a wide range of material types. This is where the approach can be particularly useful. Another approach used in the study is feature extraction, whereby material data is processed through the layers of the pretrained model to extract features. These features are then incorporated into a new model for further processing. This technique leverages the knowledge learned during pretraining, thereby reducing training time and significantly improving the model's performance.

2.5.3 Challenges and considerations

Problems of transfer learning in material science: Although transfer learning has some benefits, it also has certain disadvantages. Selecting a suitable pretrained model, considering the specific requirements of the material design task is the first and most important decision to make. Thus, the suitability of the information transmitted has to be checked, and the relevance of the source domain to the target domain has to be evaluated. Delicate adjustments in parameters are often required, and during the feature extraction phase, overfitting is very possible. The model must be trained properly to capture intricate details in the new material data. Effective management to improve the utility of transfer learning strategies requires understanding both the source and target communities to overcome these challenges.

Potential developments in transfer learning in material science further strengthen the idea that the future is bright if improved models and techniques are employed. Additions to the transfer learning area could involve developing focused pretrained translation models for material science data, as well as utilizing domain adaptation methods to integrate data from different material datasets. There are predictions that further advancements in transfer learning methodologies will be stimulated by improvements in computational resources and the availability of large, high-quality data. Further improvements in these approaches mean achieving faster rates at which materials can be discovered and optimized for biomedical applications, which would be a huge boost to biomedical science.

3 Ensemble methods and hybrid approaches

3.1 Random forests for classification and regression in material science

Random forest is a type of ensemble learning technique that improves the results of classification and regression decision trees by using the predictions of several decision trees. This method is also efficient in situations where scientific data are complex and/or variable, particularly in the field of material sciences. As in many other branches, high-dimensional and variable data often require stable models. While decision trees provide a single output for any specific input, random forests enhance that output and provide a more reliable model for investigating material characteristics and behaviors by integrating the results of a large number of decision trees.

3.1.1 Mechanism of random forests

In training, random forests construct a large number of decision trees that, in turn, produce the mean value for regression problems or the classes for classification problems. In order to increase variability and prevent overfitting, the trees in the forest are built with a random subset of features as well as a random subset of data. Unlike single decision trees, the final prediction is obtained by aggregating the results from all the individual trees, which are more generalized and less susceptible to noise.

3.1.2 Appreciations in material science

In materials science, random forests are used for many reasons such as material categorization, property prediction, and object recognition. For classification problems, random forests can distinguish between different types of materials according to the composition, structure of the processed material, and the conditions under which it was treated. In regression tasks, the technique can estimate more continuous material characteristics, for example, the strength of a material, heat conductivity, and electrical resistivity, from the input parameters. Due to large datasets with a significant number of variables, random forests are perfect for recognizing patterns in materials science data.

3.1.3 Advantages and limitations

Learning with random forests has several advantages, such as managing missing values, handling high-dimensional data, and providing a feature relevance measure. Due to their robust nature, they work very well in material science activities where the quality of the data varies. However, there are some disadvantages that should be taken into consideration; for example, the computing cost could be higher, and in terms of interpretability, these models are worse than simpler models. There are several concerns that seem to affect the performance of random forests with regard to medical expenditure, including the quality of the input data and the choice of hyperparameters.

Growing availability of large-scale material datasets is believed to affect random forests in material sciences in the future, along with progress in computational methods. The effectiveness and efficiency of random forests can be further enhanced by integrating them with other learning solutions including DL and feature selection learning solutions. There may also be opportunities to improve model interpretability and increase the accuracy of forecasts, such as through the development of better algorithms and the analysis of potential opportunities for combining random forests with other forms of ensemble classification.

3.2 Gradient boosting machines (GBM) for enhancing predictive accuracy

Gradient Boosting Machines (GBMs) are a class of ensemble learning algorithms that combine the outputs of multiple weak learners typically decision trees to significantly improve predictive accuracy. Their strength lies in their ability to capture complex, nonlinear relationships between input features and material properties, making them highly valuable in materials science applications. In scenarios where single-model approaches fall short, GBMs iteratively refine weak models to construct robust, high-performance predictors. Figure 4 demonstrates the architecture and iterative learning process of GBMs, highlighting their effectiveness in modeling intricate material behavior.

3.2.1 Australian institutional investment: GBM mechanism and functionality

The simple archetypal working of GBM was to incorporate models one over another to rectify the mistakes within the previous models. Considering only the most unpredictable results, each model in the series was designed to minimize the remaining errors of the model preceding it. This works by reducing both bias and variance and fine-tuning the model through stacking; hence, the technique was called boosting. GBM was found to be extremely beneficial in the material science domain to predict finer details pertaining to the material, fine-tune parameters of the material, and even fine-tune the type of material classification. Due to the high ability of GBM to work with different datasets, the technique was used for a variety of predictive applications.

3.2.2 Real-world use of GBM for material design and discovery

In materials science, various GBMs are used for predicting the mechanical properties of materials, designing the composition of composite materials, and classifying the microstructures of materials. For instance, in an application such as reinforcing the tensile strength of various new alloys, GBM models can learn prior material data and make enormously precise forecasts for new compositions. GBM has been applied to determine the most significant factors impacting the processing conditions of advanced materials including ceramics and polymers. Due to the ability of the GBM method to deal with large and heterogeneous datasets and provide high accuracy in prediction, it can become an effective tool for the enhancement of the materials discovery and development process.

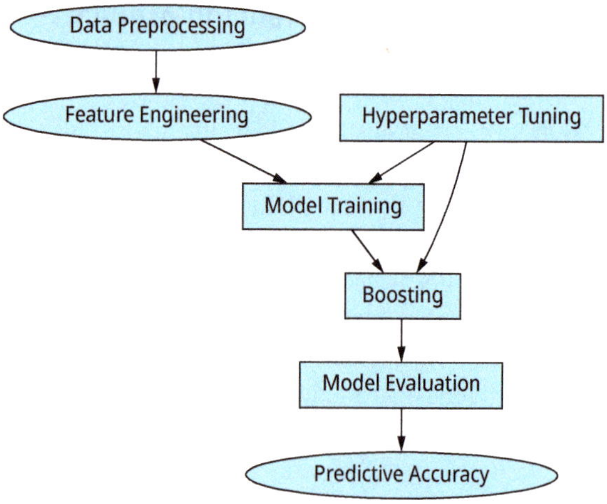

Figure 4: Gradient boosting machines (GBMs) for enhancing predictive accuracy.

3.2.3 Pros and cons of GBM

In materials science, GBM has several advantages, of which the following are worthy of mention. Its boosting method accelerates the reduction of the overfitting problem, increases the predictability rate, and thus enhances the order of accuracy. For various data distributions and types, GBMs are relatively flexible models in practice. However, GBM does have several demerits; the process is highly expensive, and the algorithm is very sensitive to the selected hyperparameters. To achieve the maximum level of performance, the training process becomes rather tedious, especially for a large number of datasets. This is the case because hyperparameters such as learning rate, the number of trees, and the tree depth need to be adjusted. In addition, even though GBM models are powerful, interpreting them is less straightforward compared to simpler models, making it challenging to identify relationships in material data.

Future development of GBM in material science is predicted to change as more advanced algorithms are developed and techniques for interpreting these algorithms are improved. Advanced boosts in forecast precision and speed derive from the integration of GBM with other machine learning techniques, such as DL and RL, along with upgrades in processing capabilities. Interesting work prospects arise from the utilization of GBM in modern sectors such as nanomaterials and sustainable material science. With regard to materials science, significant further progress can still be achieved if GBM techniques are further refined and new applications are explored.

3.3 Stacking and blending techniques for integrating multiple ML models

Ensemble learning is a powerful branch of machine learning that requires multiple algorithms to generate predictions, performing better than single models and offering increased generalizability compared to other models. Within this context, two interesting techniques, known as stacking and blending, are more complex methods deployed to orchestrate diverse machine learning models with the specific goal of maximizing their particular strengths. These techniques are especially useful in highly multifaceted, nonlinear areas like biomedical materials, where the synergistic use of various models for property predictions results in better resolution.

3.3.1 Stacking: integrated models: a layered view

Stacking generalization is an ensemble learning technique whereby a meta-model is used to blend the outcomes of base models. In the context of material science, stacking can be used to combine several models oriented to different characteristics of materials, such as mechanical strength, biocompatibility, and degradation rate. The base models, which can also be diverse in many cases – ranging from decision trees to neural networks – make predictions that form the input for a meta-model. The meta-model commonly used is, for instance, a simple algorithm such as linear regression or a more complex one like gradient boosting, where the latter learns how the individual models' predictions should be best combined in order to achieve better results. This layered approach enables one to capture fine-grained patterns in the data that are not observable when working only with individual models and yields better forecasts of material properties and their design for optimum performance.

3.3.2 Blending: an adaptive ensemble strategy

Stacking, as another type of ensemble technique, was quite similar to blending, with the only difference being how the meta-model was trained. While in stacking, the meta-model was learned from predictions on a new set of data so that it was not overfitted to the training data, in blending, the meta-model was learned using the holdout set – a part of the training data that was segregated exclusively for this purpose. In the field of material science, the use of blending was most beneficial where there was not a large amount of data available, as blending permits the inclusion of a number of models without necessarily needing more data to train the meta-model. In blending, typically, the meta-learners can be basic models like weighted averages that are used effectively to combine the basic models' predictions. The flexibility of blending was useful in situations where the simultaneous integration of models was time-sensitive, providing a practical application for the idea of ensemble learning in material properties.

3.3.3 Pros and cons of stacking and blending

The first advantage of stacking and blending was the fact that the results from different models can be combined for better outcomes to emerge. Such techniques are more favorable in the field when working with nonhomogeneous data and when there are substantive predictive tasks to be met, for example, in the process of biomedical material designing and optimization. Stacking and blending within the study come with some drawbacks: overfitting tends to plague stacking if the meta-model is not trained properly using out-of-sample data. Choosing base models and designing the meta-model require extra attention with regard to predictive accuracy while preventing overfitting and bias in the ensemble. Solving these problems requires the analysis of the models of the subject field and such properties of the material science data as the relationships between them.

Everything pointed to a bright future for stacking and blending techniques in material science, as further advancements in meta-models were still continuing and new base models were being researched. With increasing computational power and the availability of more data, the prospects of developing highly accurate and sophisticated ensemble models lie in the future. Furthermore, continued improvements in AutoML should also reduce the implementation complexity in stacking and blending ensembles, which would otherwise be challenging for material scientists to develop. If advancements in these techniques continue and improve, the opportunity for increased innovation in material identification and improvements could lead to growth and development in biomedical materials and customized treatments.

3.4 Voting mechanisms for robust material property predictions

Voting mechanisms are a fundamental component of ensemble learning, a strategy that combines multiple machine learning models to produce more accurate and reliable predictions. In the context of material property prediction, voting methods aggregate the outputs of individual models to form a consensus, thereby reducing the risk of errors that could arise from relying on a single predictive model. This technique leverages the unique strengths of each contributing model, as different models may capture distinct aspects of a material's behavior and characteristics. Figure 5 illustrates the ensemble voting framework, showing how predictions from multiple models are integrated to improve overall performance and robustness in materials science applications.

3.4.1 Voting schemes and its uses

In ensemble learning, several kinds of voting methods are used right now, and each of them varies in some way depending on the type of prediction, such as material property, under consideration. The least complicated method is averaging or majority voting, where each model in the ensemble "vote" for a specific prediction, and whichever option is most preferred by the majority of the models is adopted as the final output. This method is particularly applied when the individual models yield comparable accuracy but have different sensitivities to the data set. There is also weighted voting, where each model casts a vote in proportion to its accuracy and confidence in the prediction it favors. This technique is useful when some models perform better in certain areas than others, allowing the inputs of a few models to be weighted more heavily in the final prediction.

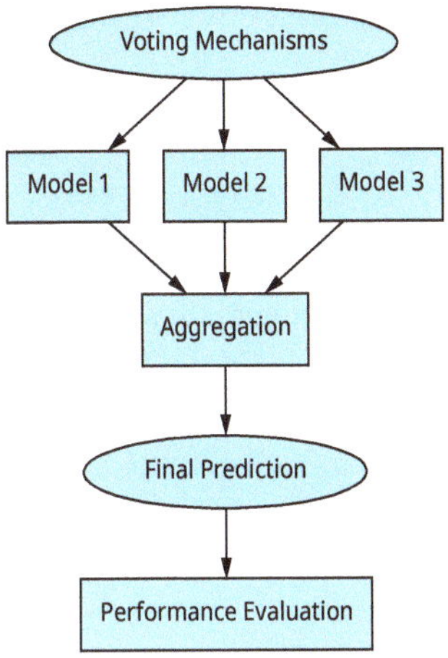

Figure 5: Voting mechanisms for robust material property predictions.

Soft voting was another method in which the probabilities given by the models were collected and then averaged, with the prediction having the highest probability being chosen. This method was useful when working with probabilistic models where the results differed across models. Through this voting mechanism, material scientists could leverage the features of multiple models to enhance the predictive capabilities of material properties.

3.4.2 Implementation strategies of voting mechanisms

As an important aspect of voting structures, the following are key implementation aspects of voting mechanisms in material property predictions. This decision was a critical moment in the contest because the diversity of the models often improves the final prediction. Varying models were derived by using another set of algorithms for the data extraction method. As soon as the ensemble was created, there were consistent decisions on such a voting method depending on the characteristics of material data and the required outcome of the forecast. For instance, in a situation where some properties are considered more important than others, as we have already seen in weighted voting, soft voting was more preferable when probabilistic estimations were needed.

To increase prediction stability, more voting models can be combined with other ensemble methods like bagging. When these techniques work in parallel, the ensemble can reduce the biases and variances that come along with the inferences made by the individual models, resulting in a more accurate generalization of the material properties. Voting mechanisms are computationally efficient and provide practical applicability to applications such as the prediction of material properties in real-time samples, where prompt and accurate results are desirable.

3.4.3 Voting mechanisms: opportunities and challenges with a focus on voting optimization

Despite the benefits voting mechanisms present, there were some issues that the authors came across. One major concern was the issue of model congeniality – that is, models with similar statistical appearances overwhelmed the bags, to the disadvantage of the ideal balance required to enhance the stability of the ensembles. In reference to this, appropriate selection and diversity in the models included in the ensemble are the correct measures. Another problem was the calculation of weights in the case of weighted voting, where weights must be tuned and then validated to ensure that the ensemble's performance was indeed optimized.

The difficulties of functional form dependence on the scalability of voting mechanisms arise if the number of data points becomes large. To cope with the growing complexity of the movements involved in large ensembles, hardware computational and storage factors must be factored in. Ongoing research works proposed in adaptive voting methods, where weights and voting strategies are changed according to the data and prediction context, help in overcoming the above-discussed problems and provide better results for the ensemble models used in material science.

Regarding the voting mechanisms, it was mentioned that the future of voting mechanisms in material property predictions is expected to experience new developments due to the use of technology in voting mechanisms and material science. There is potential for the enhancement of these approaches, including the dynamic voting procedure, which is capable of adapting depending on the specifics of the data, the structure of the prediction task, and meta-learning, where the ensemble also identifies the voting method that is appropriate for the input. Additionally, with the growth of high-quality material datasets and the rise of advanced models that can support voting mechanisms, the outcomes are becoming much more accurate and faster, thereby increasing the pace of material discovery.

3.5 Hybrid ML models combining supervised and unsupervised learning

In the generation and analysis of biological systems, there is a new research direction involving the integration of supervised and unsupervised machine learning. Supervised learning is highly efficient in problem-solving scenarios where the outcome is known, such as regression and classification, because it utilizes labeled data. On the other hand, unsupervised learning algorithms discover relationships between variables in an unlabeled dataset. These shared characteristics make the combination of these two learning methods even more effective when used together, as one can augment the other where it was previously ineffective, particularly in cases with subtle and nuanced differentiations. This is especially true when the data set was large and high-dimensional, making it very challenging to obtain labeled data.

In biomedical applications, for instance, a number of methods have been described for designing and developing biomedical materials.

Thus, the application of hybrid models in BM materials has significant benefits, especially regarding characterization and designing processes. For example, in situations where only a limited amount of information is classified, a hybrid model can use the elements of supervised learning to classify the material while simultaneously applying elements of unsupervised learning to detect mass patterns in the entirety of the assigned dataset. This approach is particularly beneficial for finding new material classes, as well as uncovering the relationships between structure and properties, and identifying materials that we would not be able to search for in such a way using conventional methodologies. The flexibility of these approaches makes it possible to investigate material space in a deeper manner, which in turn results in the growth and development of improved material interventions.

3.5.1 Approaches toward using supervised learning and unsupervised learning

There are a variety of ways in which the integration of a supervised learning system with an unsupervised system can be achieved in the context of hybrid models. Almost all candidates mentioned the use of at least one of the following techniques: semisupervised learning, for instance, involves basic training through a small selection of labeled data and is then trained again using other forms of data that are not labeled. Another approach is the process called sequential learning, where first, the unsupervised learning approach is used in the initial analysis to identify early patterns in the variables, and then the subsequent supervised learning technique is applied to relabel the early groupings. On the other hand, there are some models that are partly supervised and partly unsupervised; in these cases, both categories of algorithms are run simultaneously, with their results being integrated at the end to produce a combined output of both techniques. All of these methodologies can be adapted to the application of biomedical material depending on the type of data involved and the goal of using the material.

3.5.2 Major issues that hinder hybrid model development

The main issues that are observed when designing efficient hybrid models are associated with the modulation of supervised and unsupervised learning components. The first item raised was the ability to guarantee that the unsupervised learning process induces relevant features and characteristics without interference from noisy patterns that are unrelated to the final learning. In addition, semisupervised learning models, in general, need more complex tuning and validation steps, as adding two learning approaches in one model can complicate it. This has the effect of making the choice of model architecture and parameters, as well as large-scale cross-validation to avoid overfitting and to obtain good performance on a range of material datasets, a complex issue.

It was likely that in the future, the biomedical materials hybrid machine learning models would apply more with other advanced methods, such as RL and transfer learning. Flexible semisupervised approaches that are capable of changing the proportion of supervised and unsupervised learning based on data stream input were one of the innovations. In addition, because of advances in computational abilities, other complex hybrid models that are able to handle larger and mixed information were developed. The result was better material designs and more profound information. Thorough exploration and enhancement of these blended approaches resulted in improved models that could enhance clinical use alongside the discovery of new knowledge within the sphere of biomedical materials.

4 Three Bayesian methods and probabilistic modeling

4.1 Bayesian optimization for hyperparameter tuning in material models

Bayesian optimization is regarded as one of the most efficient techniques for hyperparameter tuning in machine learning models, particularly in domains where the model's performance is highly sensitive to the choice of hyperparameters. In materials science, hyperparameters play a critical role throughout the entire pipeline from modeling and feature attribution to material design and the analysis of structure–function relationships. Bayesian optimization treats hyperparameter tuning as a probabilistic search problem, typically modeling the objective function as a Gaussian process. It iteratively updates this model as new observations are made, thereby refining the search and identifying optimal hyperparameter configurations. Figure 6 illustrates the Bayesian optimization workflow, showing how it explores hyperspace to improve model performance in material science applications.

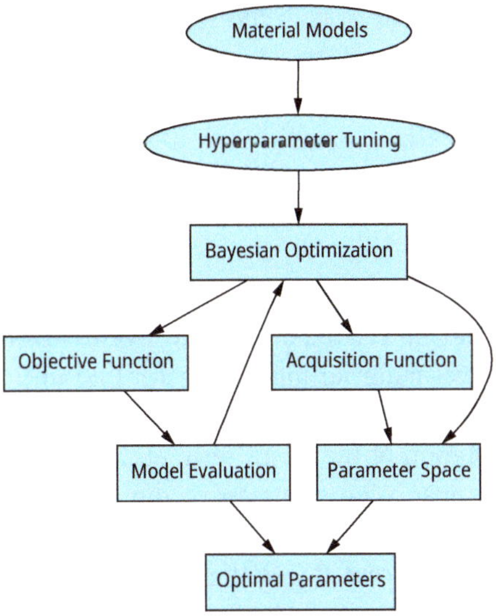

Figure 6: Bayesian optimization for hyperparameter tuning in material models.

4.1.1 Application related to material science

From a material perspective, Bayesian optimization has been especially useful in model refinement for material property modeling and engineering. Since the evaluation of functions in experimental material science is generally costly and time-consuming, the benefits of Bayesian optimization in reducing the number of function evaluations are indeed critical. For instance, in contexts like developing predictive models for mechanical properties and thermal stability of materials, Bayesian optimization determines which set of hyperparameters, such as learning rates and regularization strengths, lead to improved and more reliable models. This is particularly essential where the search space is large and where grid search methods are unresponsive and computationally costly.

In this work, the following research methodology, based on Bayesian optimization, is used.

Specifically, there are several steps that need to be followed in the Bayesian optimization technique. It is typical to use a Gaussian process (GP) as a surrogate model to represent the relationship between the performance of models and the tunable parameters. Each time new data points (evaluations of the objective function) are received, the surrogate model is updated. The next set of hyperparameters to be considered for evaluation is selected using an acquisition function like the expected improvement. When the model reaches convergence, the model together with a set of hyperparameters is considered optimal. The strength of Bayesian optimization lies in its ability to balance exploration and exploitation; namely, it can efficiently navigate high-dimensional and complex spaces often observed in material science while also advancing into new regions of hyperparameter space, all the while building upon potentially promising areas.

4.1.2 Challenges and considerations

There are limitations inherently associated with Bayesian optimization when used in material models. The first obstacle was the problem related to the objective function to be minimized; when incorporating high-level material models, such as ab initio ones, these calculations could be time-consuming. This was the case with the objective function since its evaluation required recalculation of the misfit term each time the parameters of the material model were changed. The selection of the surrogate model, as well as the acquisition function, can affect the optimization process and the time and computational resources needed for the process. It was, therefore, important to guarantee that the surrogate model provided the correct structure of the objective function so as to optimize the process successfully. A further factor to discuss was the extension, as the BO methods' efficiency drops when many hyperparameters are involved in discrete searches.

4.2 Gaussian processes for predictive modeling of material properties

The GPs were one of the strong probabilistic modeling techniques used in predictive modeling, especially in the field of material science, where quantification of uncertainty was very important. Since GPs are nonparametric, they allow for a relatively free-form estimation of input variables and output material properties without imposing a specific architectural form. The strength of GPs was twofold: first, they were capable of predicting material properties like other models, and in addition, they could also estimate the level of uncertainty associated with the predictions, which makes them particularly useful in materials science where data is limited.

4.2.1 App in material property prediction

In the area of material science, GPs have started to be used to make predictions regarding various characteristics of materials including elasticity, thermal conductivity, and strength. This makes GPs capable of handling uncertainties because experimental data, which is costly to obtain, is often limited. This demonstrates that the GP model is accurate in prediction even when the material data is sparse, and by extrapolation of the model, even materials whose underlying physical laws are not completely known can be evaluated for their properties. One of these capabilities is particularly useful in the search and identification of new materials with specific characteristics in the field of material science.

4.2.2 Methodological framework

The way GPs were implemented in predictive modeling was to specify a covariance function (or kernel) that defines the relationships between distinct material data points. The choice of the kernel was significant since it defines the smoothness of the function and the intricacies of the decision boundaries. Specific kernels more frequently used in material science are the radial basis function and other kernels that are equally effective when working with a number of material data sets. Once the kernel was chosen, the GP model was then trained with observed data, through which the GP model estimates the percentage probability density functions over the input space of material properties. The model then provides point estimates for new values of x along with intervals of uncertainty for the estimate. Due to this probabilistic nature of the GPs, there was a better chance of making the right decisions regarding material design and optimization.

4.2.3 Challenges and considerations

The use of GPs has many advantages, but there are also disadvantages, primarily associated with increased computational costs. The time complexity in the training of a GP model escalates with the training dataset size, and thus the model becomes less usable with very large material datasets. As you can see, selecting the kernel function also requires significant thought and expertise in the subject matter to ensure that the selected model accurately depicts the underlying physical events in the dataset. Furthermore, increased input field dimensionality brings the curse of dimensionality, which results in overfitting when extending GPs. To overcome these challenges, better approximations and algorithms need to be developed to extend the utility of GPs in material science.

4.3 Markov chain Monte Carlo (MCMC) techniques in material design

Markov Chain Monte Carlo (MCMC) methods are foundational sampling techniques in probabilistic modeling, particularly valuable for estimating probability distributions in composite material design where conventional approaches fall short. These methods are especially effective in exploring large, high-dimensional design spaces, where deterministic models often fail to capture the full complexity of material behavior. MCMC enables sampling from stochastic processes to estimate material properties and predict their performance under various operational conditions. This probabilistic framework allows researchers to model uncertainties and randomness more comprehensively, leading to the development of more resilient and reliable material solutions. Figure 7 illustrates the MCMC sampling process, demonstrating its application in estimating property distributions and enhancing material design strategies.

4.3.1 Uses of MCMC in material design

It has been used in many material design applications to tackle difficult issues concerning uncertainty quantification and optimization. In the design of new materials, for instance, the MCMC methods can generate samples from the posterior measures of material properties to provide insights into the strength and durability of the material. The sources of variability are adequately acknowledged by MCMC in terms of sample generation, and the variability of design parameters to material performance is well appreciated in detailed statistical methods. It is most applicable in situations where the actual implementation of material formulations is not feasible, for instance, in the creation of new high-tech alloys, and composite applications. MCMC techniques find application in inverse problems, in which the aim is to reconstruct the micro-mechanical parameters from the observed data in order to provide more accurate guidance to the design process.

4.3.2 Procedures for the use of MCMC algorithms

There exist a number of substantial strategies that apply MCMC methods in the design of materials, and every approach is tailored to the needs of the situation. Among the most used MCMC methods; the Metropolis-Hastings algorithm is especially efficient when the target distribution is complex. As it presents potential moves in the design space and either accepts these based on their probability, this algorithm produces several samples and compiles a Markov chain toward the desired distribution. Another technique that received much attention was the Gibbs sampler, which is most useful when direct sampling is impossible due to the high dimensionality of the issues in-

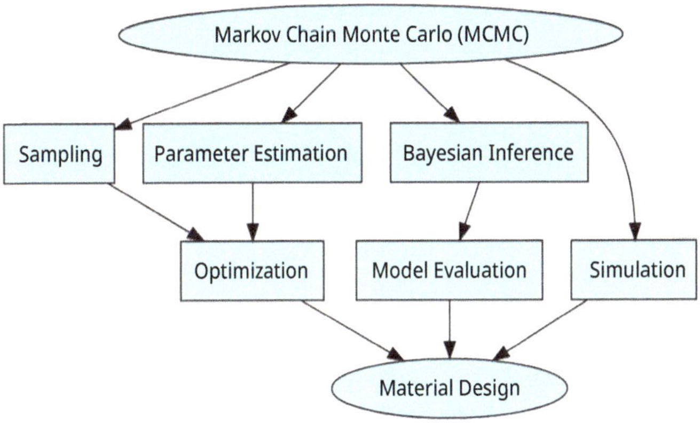

Figure 7: Markov chain Monte Carlo (MCMC) techniques in materials design.

volved. For convergence to the desired distribution, the parameters are regenerated using the Gibbs sampler by sampling them conditionally on other parameters. Related developments, where MCMC was integrated with other optimization procedures – for instance, the simulated annealing approach – have also emerged because of higher convergence characteristics and reduced computational load in material design.

4.3.3 Difficulties and implications of MCMC

MCMC techniques have great advantages in material design; however, they also present certain difficulties. The first major issue relates to the transition to the target distribution by the Markov chain, especially if the distribution is in a high-dimensional space or has multiple modes. Common mistakes made when selecting proposal distributions can slow down convergence and even affect the results. The use of MCMC techniques is accompanied by high computational costs, especially when it involves large-scale material design problems. This requires the application of better sampling techniques or superior parallel computing methods. The application of MCMC in material science is not without these difficulties; these challenges make it important to understand both the material system under study and the stochastic models being used.

4.4 Bayesian networks for understanding complex material interactions

As already mentioned, Bayesian networks (BNs), which are probabilistic graphical models, offer a rich language for cautious propagation of uncertain reasoning regarding complex dependencies between variables in material science. These networks express the dependencies of variables using a directed acyclic graph in which nodes represent material properties and edges are probabilistic. Using both antecedent information and evidence, BN models can provide an enhanced perspective on the combined effects of various parameters on the behavior of materials, improving the ability to make better predictions and providing fresh knowledge on particular material phenomena.

4.4.1 Usefulness in simulating material interactions

Specifically, in biomedical applications, BNs are most useful in explaining complicated relationships governing the behavior and compatibility of biomedical materials. For instance, such networks can be used to predict the effects of chemical elements, material structure, and biological characteristics in biomaterials. These dependencies can be represented in BNs, and therefore allow researchers to understand how changes in an input variable, say chemical composition, affect another output variable, like the mechanical strength of the material. This capability is particularly valuable for the design and optimization of highly specialized, multiparameter influenced applications such as advanced materials.

Some of the ways to construct a BN include defining the graphical model structure of the BN and identifying the conditional probability distributions that specify the joint probability distribution of connected nodes. The network structure can be specified a priori with the help of domain experts or derived from data using different algorithms, such as the hill-climbing algorithm. After the structure is built, Bayesian computation methods are applied to update the network information, making it possible to enable dynamic modeling and timely prediction. Such an approach allows the network to update the model's understanding of how materials react to one another based on additional data and, as such, works well as a research and development platform for material science.

4.4.2 Problems that occur when applying Bayesian networks to material science

The authors argue that the challenges experienced when applying BNs in material science outweigh the benefits. Algorithms with high computational load when learning large networks, especially when working in the high dimensionality characteristic of many materials' investigations, were another challenge. Additionally, the efforts to accurately specify the conditional probability distributions are usually extensive and require substantial data and expert input, which, in some material domains, are hard to come by. Solving these issues requires a compromise between the model specifications and the computational ability to handle the network's complexity, as well as the application of approximation methods to the problem.

The future of BNs in the understanding of interactions between different materials lies in broadening the demonstrations of their applicability to more complex structures, as well as in their integration with other current advanced modeling approaches. Addressing the issues of prediction accuracy and expandability of the employed models, hybrid models based on a combination of methods belonging to the class of chosen machine learning techniques and BNs are being developed. This approach has opened a great opportunity to develop new principles and use BNs in new directions, such as biomimetics and nanomaterials science. New trends are expected to forge ahead in different designs and applications of materials in the future as computational power intensifies, enabling complex material relationships to be simulated and predicted.

4.5 Uncertainty quantification and decision-making using Bayesian methods

From a research standpoint, uncertainty quantification is a crucial aspect of the decision-making process when developing, evaluating, and applying biomedical materials. Three primary sources of uncertainty are typically encountered: inadequate data, imprecise measurements, and the inherent stochastic behavior of materials. These uncertainties can significantly affect the reliability of predictions and the safety of the resulting applications. Bayesian methods are particularly effective in managing this complexity, as they provide probabilistic interpretations of model predictions and incorporate uncertainty directly into the modeling process. By promoting a deeper understanding of variability, Bayesian approaches lead to more informed predictions, enhanced material designs, and better assessments of risks and performance outcomes. Figure 8 depicts the Bayesian framework for uncertainty quantification, highlighting its role in improving the safety and efficiency of biomedical materials.

4.5.1 Methods of uncertainty estimation: Bayesian framework

Specifically, Bayesian methods provide a number of approaches to the problems of uncertainty estimation in material science. One primary method is Bayesian inference, which incorporates data on material properties to build prior information that informs a posterior distribution. These posterior distributions provide a probabilistic estimate of material characteristics, accounting for the level of indeterminacy in data sets. Another identified technique is Bayesian optimization, a practical approach to optimizing material design under uncertainty while passing parameters that maximize the expected utility or minimize the expected loss. It is also possible to model all sorts of stochastic dependencies between the characteristics of materials in BNs, which can help propagate uncertainty through the network and provide more accurate methods of quantifying uncertainty.

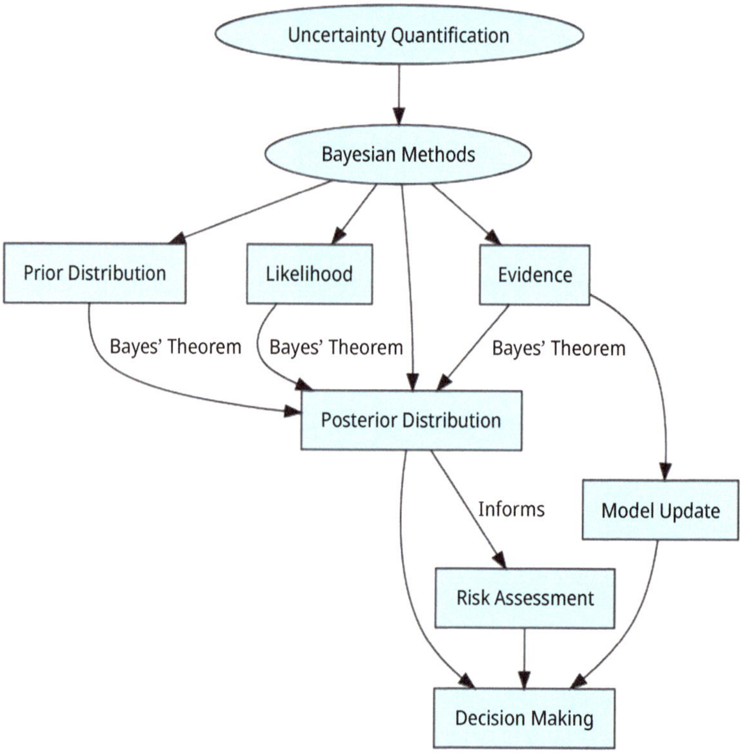

Figure 8: Uncertainty quantification and decision-making using Bayesian methods.

4.5.2 Decision-making under conditions of uncertainty

The management of uncertainty in decision-making was also fundamental in the application of biomedical materials. Another well-developed approach to decision-making under risk conditions was Bayesian decision-making, which is a rational model of decision-making that estimates the possible consequences according to their probabilities and degree of risk. This approach makes it possible to choose material designs that are optimized between their performance and safety, despite data being scant. For instance, in implantable materials, failure mode effects can be assessed and ranked by likelihood using the Bayesian method effectively, so the design can achieve the highest probabilities of success while minimizing failure probabilities. This capability of making decisions that are fundamentally probabilistic in nature is especially valued in biomedical applications because material failure can have severe implications.

4.5.3 Advantages and disadvantages in applying Bayesian method

Together with offering robust tools for quantifying uncertainty and decision-making, Bayesian techniques present some specific challenges as well. There are two drawbacks here: one is the implementation complexity of Bayesian techniques, which remains a major headache in the current early stages of development when computation time is impacted by complicated models. Techniques such as the MCMC can be both resource-intensive and time-consuming when carrying out sampling from the posterior. Additionally, if we recall that it has a significant impact on the results, the selection of correct prior distributions is critical. Specifying priors requires careful reflection to ensure that the selected priors properly characterize the current knowledge, which forms the basis of learning, yet allows for the incorporation of new data. Nevertheless, Bayesian approaches are being applied with increasing ease and relevance to material science problems due to the development of processing power and improved algorithms.

The future publication of Bayesian methods for biomedical-related materials would possibly entail the application of other algorithms within the bandwidth of machine learning techniques and the advancement of new computational approaches for the determination of uncertainties. For instance, integrating Bayesian methods with DL approaches could help develop models for material property prediction under risks and uncertainties. In this respect, recent developments in approximate Bayesian computation (ABC) and variational inference provide promising new avenues for addressing the computational issues tied to the use of Bayesian techniques when working with large-scale material datasets. Enhancements in these techniques mean that uncertainty can be well measured, and decisions can be made effectively in material science, thus opening up new possibilities in biomedical material design and application.

5 Reinforcement learning and adaptive models

5.1 Reinforcement learning for adaptive material design strategies

In the world of machine learning, RL trains an agent to make decisions through trial and error within the environment and provides feedback in terms of a reward. Because of this feedback loop, the agent continually refines its performance and eventually develops optimal strategies. RL lends itself as a flexible model for material design within the materials science paradigm. RL enables an agent to observe large design domains and learn optimal material designs by framing the design process as a process of decision-making. Due to the complexity of interactions in one dimension, this capability is particularly well-suited for the design of biological materials, which require adaptive strategies capable of shifting from one mode to another in alignment with specific requirements.

5.1.1 The use of RL in adaptive material design

RL should be able to aid in the design of adaptive materials in several settings where desirable characteristics include, but are not limited to, strength, flexibility, biocompatibility, and degradation kinetics. For instance, in the synthesis of biodegradable polymers that can be employed in implant production, RL can be applied to determine how to modify the strength and degradation time while considering the medical characteristics. RL can also be used to predict processing conditions on a similar real-time basis, including temperature and pressure, during the synthesis of materials to produce material properties of interest. Because RL-based models can learn from the results of each experiment, such a process enables the efficient and effective discovery of new materials.

5.1.2 Strategies for enacting reinforcement learning in material design

There are numerous important directions that must be implemented to incorporate RL into material design. Model-free RL is a particular method in which the unit directly learns from its environment without relying on a model of the material framework. The agent's search over the material design space can be conducted with the help of methods such as Q-learning. Model-based RL, in contrast, employs a model of the material system to estimate the performance of a set of design choices, making the training phase exponentially faster. This method is useful when collecting experimental data is either expensive or impractical. Another approach is inverse reinforcement learning (IRL), which seeks to mimic and optimize good strategies in effective material design by inferring the reward function from expert behavior that has been analyzed.

5.1.3 Difficulties and problems in RL-based material design

Implementation of RL in material design was not without its challenges, especially with respect to the exploration-exploitation dilemma. Managing the curiosity of the RL agent so that it could search through the design space sufficiently without getting stuck in local optima was a delicate task, particularly for-complex materials design space problems, which contain very large and high-dimensional search spaces. Further, scalability was an issue in RL because the models require massive amounts of data to train, and in material science, data was scarce to obtain. Solving these problems requires combining RL with other methods, such as transfer learning, which allows the use of prior knowledge to minimize data demand. In addition, there are formidable challenges that must be addressed to build a fully fledged RL model, and these include the improvement of efficient algorithms and other computational methods instrumental in overcoming these challenges and unleashing the full potential of RL in materials design.

To fully realize fully autonomous material discovery platforms, RL in material science is accordingly forecasted to build more complex connections with other advanced technologies like artificial intelligence and high-throughput testing in the future. One must note that more complex forms of RL that can handle the challenges posed by adaptive learning and multiple objectives currently remain potential innovations. As the computational power of computers evolves, RL can be used in large-scale material behavior simulations that otherwise remain scientifically inaccessible due to complexity and potential catastrophes in other unexplored material classes. The design and development of next-generation biomedical materials should advance in the field of material science, in which RL is both basic and applied.

5.2 Q-learning and deep Q-networks for optimizing material processes

Reinforcement Learning (RL) has proven to be an effective approach for modeling material processes, as it is grounded in the interaction between a dynamic environment and the selection of optimal strategies based on feedback or rewards. In the context of materials science, advanced RL algorithms such as Q-learning and Deep Q Networks (DQNs) are employed to optimize material dispersion, refine manufacturing protocols, and enhance overall material quality. These methods enable researchers to define rule-based controls for systematically exploring and improving material processes with respect to specific performance goals, such as efficiency, cost-effectiveness, and product reliability. Figure 9 illustrates the RL framework in material process optimization, highlighting how iterative learning and reward-driven strategies guide material design enhancements.

5.2.1 Q-learning: basics and applications

Q-learning is a model-free RL algorithm intended to discover the optimal action-selection policy via updating Q-values of various actions in a given state. In material process optimization, the use of Q-learning can be applied to situations where the goal is to determine the process parameters that yield the highest quality material. For example, Q-learning can be applied to search for optimal values of temperature, feed rate, and other variables in additive manufacturing to achieve the optimal properties of the material. The Q-table records and adjusts Q-values with rewards over time toward the system toward the optimal policy for decision-making in material processes.

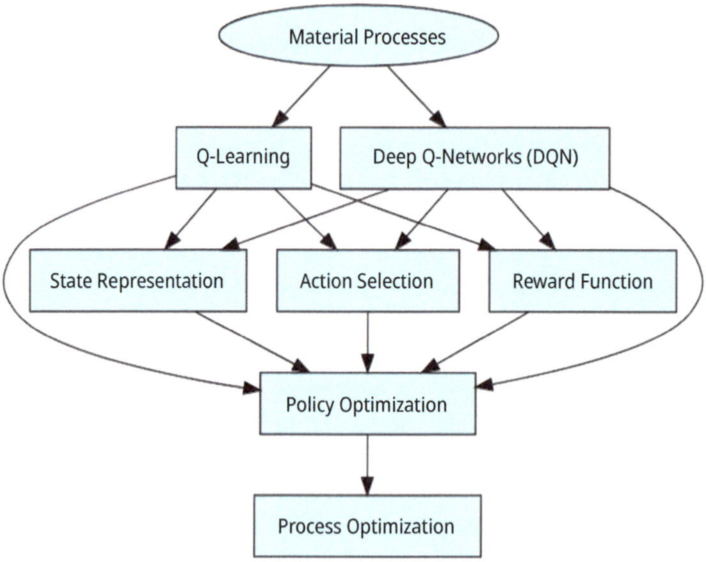

Figure 9: Q-learning and deep Q-networks for optimizing material processes.

5.2.2 Deep Q-networks: improving Q-learning with deep learning

A new approach was integrated with the Q-learning algorithm, namely involving DL methods in DQNs to help resolve the problems of large-sized state spaces. Therefore, while traditional Q-learning is not capable of dealing with highly complex data, DQNs are developed to solve more complicated material optimization problems because they use neural networks to approximately determine Q-values. DQNs can be employed in material science to decide on the refinement of methods that demand searches in incredibly large state as well as action spaces, such as chemical vapor deposition of composite materials. DQNs are used in the training process to allow for learning and to generate insights from approximately large data, enhancing the optimization formula by providing higher efficiency.

5.2.3 Challenges and considerations

There are several challenges when it comes to applying Q-learning and DQNs for material process optimization. Designing a well-defined reward function that aligns with the objectives of the material process presents a particularly significant challenge. This chapter discusses how the design of the reward function is crucial when conducting both learning and optimization. All methods based on DL, including DQN models, are computationally expensive, especially during the learning phase, particularly when working with large state and action spaces. From the points discussed above, one can also understand that managing convergence and stability during DQN training is not easy and can require extensive prior tuning of hyperparameters as well as careful design of the neural network's architectural structure. Solving these problems necessitates ongoing efforts to enhance algorithms and expand their applicability to a broader range of material science tasks.

Future developments in Q-learning and DQNs for enhancing material process optimization are likely to be stimulated by new technologies and methods. Combined with other machine learning approaches, including transfer learning and meta-learning, the overall performance and applicability of RL approaches improve for applications in material science. Newly developed hardware and computational facilities, capable of processing and adopting higher-dimensional and intricate problems, are expected to be achieved. Advances in reward functions and exploration functions enhance the performance of Q-learning and DQNs in optimizing and expanding material processes. This emerging prospect, along with these forthcoming advances, paves the way for extremely significant implementations of RL in material science and engineering.

5.3 Policy gradient methods for dynamic material property adjustments

Policy gradient methods refer to a set of RL algorithms that deal with the direct optimization of policy through estimation of the gradient of the expected reward with regard to policy parameters. These methods are especially useful for problems where the action space is continuous and the environment is either continuous and high-dimensional, dynamic and unknown, as in the case of adjusting the material properties in real applications. Using gradient ascent to learn and update the policy, policy gradient methods can therefore make timely corrections to the chosen material properties to realize optimal performance in accordance with specified objectives.

5.3.1 Using policy gradient methods in materials science

As applied to material science, policy gradient methods can be used in a way that allows material properties to change over time based on experiential feedback from the environment. For instance, in a dynamic environment of adaptive biomedical materials technology, policy gradient methods can be applied while designing materials to optimize material characteristics, including hardness, elasticity, and conductivity, at successively higher levels of the technology. These methods embrace real-time performance data and feedback, hence enabling the enhancement of material properties to match certain functional expectations and dynamic characteristics. This approach is especially beneficial in cases where material performance needs to be changed more regularly, for instance, in the case of responsive medical apparel or implants.

5.3.2 Strategies for the application of policy gradient techniques

Policy gradient techniques implementation presupposes several critical steps. First, a policy network was built to correlate a given set of actions to the current state of the material system. To train the policy network, the so-called policy gradient theorem was then used, which calculates the gradient of the anticipated reward with respect to the policy parameters. For better training stability and efficiency, a few techniques such as REINFORCE with Monte Carlo estimates of return, PPO, and TRPO can be used. As the policy parameters were updated toward the maximal expected reward, the material attributes were changed dynamically in response to the observed performance.

5.3.3 Challenges and considerations

Since policy gradient approaches have their advantages in the context of material property adjustment, they present several challenges. The main challenge is computational, particularly in the ability to train the policy network when the state and action spaces are high-dimensional. Besides, the quality of the reward function, which must appropriately represent the material property performance criteria, defines to what extent policy gradient approaches are effective. Additionally, placing an overemphasis on achieving convergence and stability in the training process highlights the importance of hyperparameter tuning and designing techniques to overcome problems such as overfitting and policy divergence. Solving these challenges requires the use of enhanced algorithmic methods coupled with high-power computational methodologies for reliable policy optimization.

It was expected that the policy gradient methods in the future for material science policies would experience both improvements in algorithms and use cases. Developments involve combining policy gradient methods with model-based RL to cyber one's framework in order to effectively tackle complex material systems across a wider variety of situations. Further, advances in computational trends, together with the enhancement of the reward function, enhance the efficiency of policy optimization. As these methods continue to advance, they have the potential to facilitate significant developments in the enhancement and realization of adaptive design and optimization of materials, with mechanisms and concepts aiming at more intelligent material systems across multiple applications.

5.4 Multiagent reinforcement learning for collaborative material discovery

The principles of Reinforcement Learning (RL) extend naturally to multi-agent environments, giving rise to Multi-Agent Reinforcement Learning (MARL), where multiple agents interact, coordinate, and optimize their actions collectively to achieve shared objectives. In the context of material discovery, MARL conceptualizes the process as a multi-agent system, where various agents collaboratively explore and optimize material properties. This framework enables researchers to model the complex interdependencies and interactions that often exist in material design and synthesis tasks. By leveraging MARL, researchers can address challenges related to flexibility, scalability, and cooperation, ultimately enhancing discovery outcomes and accelerating innovation in material science. Figure 10 illustrates the MARL-based approach to collaborative material discovery, showing how agent interactions drive efficient exploration of material design spaces.

5.4.1 Case studies of applying the MARL concept in material discovery

In material discovery, what can be applied in MARL is the ability to model and analyze a number of parameters of the research process. For example, one agent may be assigned the responsibility of data gathering, another for property prediction, and yet another for experimental assessment. These agents can cooperate to search material design spaces, share information, and adjust the strategies within those spaces as a result of their interactions. MARL makes it possible to achieve the required improvements in material discovery procedures by establishing connections between the agents, enabling them to draw conclusions from each other's actions, and providing methods to fine-tune the overall performance of the discovery system. In this regard, this collaboration can rapidly facilitate the search for potential materials and optimize the formulation of new material systems.

5.4.2 Strategies for the application of MARL in materials science

There are several methods that can be used in material science when adopting the use of MARL. Looking at two main approaches, one strategy is decentralized learning, in which the decentralized agents take subproblems autonomously but keep communicating and sharing information with other decentralized agents in order to address given problems collectively. There are several approaches using, for instance, multiagent Q-learning methods to select strategies for improving their decisions concerning other agents. Cooperative MARL frameworks, where multiple agents are involved with the expectation of achieving a collective objective in any material discovery, can be used. These methodologies presume that suitable incentives, messages, and collaboration frameworks are put in place in order to successfully implement agent collaboration.

In this work, the challenges and considerations of applying multiagent reinforcement learning (MARL) are as follows:

MARL faces several challenges even in material discovery. There was a major issue in deciding what the rewards should be in order to guide the optimal behavior of all the agents as well as promote positive engagement. The control of the growth of these MARL systems, especially as the number of agents increased, was challenging and led to computational problems, hence the need for complex methods. Another factor that emerged as a problem was the coordination and communication between the agents: it was important to organize the information exchange and avoid both duplicate activities and conflicts in actions between the agents within the framework of the system. Solving

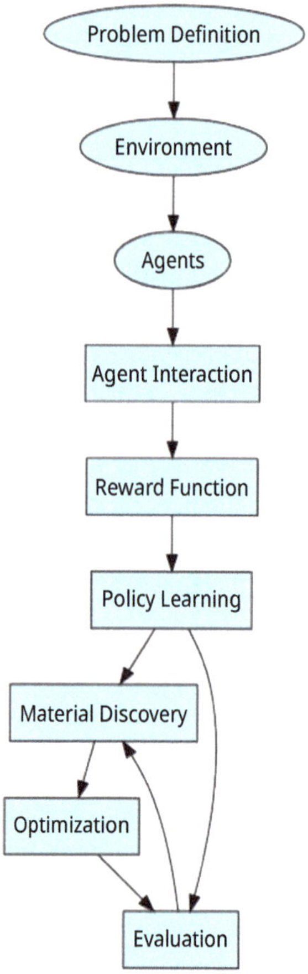

Figure 10: Multiagent reinforcement learning for collaborative material discovery.

these issues was associated with certain considerations regarding model design, training, and the architecture of the system.

Prospects for future progress and the next steps in the material discovery science of MARL seem to be positive. Incorporation with other advanced technologies, like high-performance computing and advanced simulations, enhances the performance and extensibility of cooperative material discovery systems in MARL. Also, it is predicted that more promising and complex MARL algorithms will have improved exploration capabilities in terms of appropriate interactions and large amounts of data. It is expected that as MARL methods evolve, they will play a valuable role in driving the progress of material science and help achieve enhanced and faster development of new materials and formulations.

5.5 Real-time feedback systems in material characterization using RL

RL was employed by live feedback systems in MC to adjust and improve the methods of characterization for materials. These systems are designed to actively read and adapt MC methods based on the data that flows through them. Introducing RL to construct plastic models, which could update their approaches based on continuous measurements, led to better characterization of material properties. Subsequently, this strategy was particularly helpful for high-dimensional and complex material systems where conventional methods might not be adequate.

Currently, there are various feedback mechanisms in RL within the context of feedback systems.

RL allows the models to learn what the best scene for them to take is by trial, and this provides an opportunity to introduce feedback systems. RL algorithms, such as Q-learning, can be employed in the case of MC to purposefully determine the most appropriate protocols from current investigation procedures based on a current study. In communication with the MC environment, the RL agent is rewarded based on how well the process functions and the accuracy of the predictions. Gradually, the agent becomes capable of achieving the highest total amount of reward with the help of action improvement, which in turn enhances the characterization results.

5.5.1 Use and benefits in material characterization

Real-time feedback systems where RL methods are applied have several benefits in the MC context. Overall, RL-based systems can be used to provide more and better measurements of material properties learned from new data. For instance, in spectroscopy, where one observes light emission, or in situations where one needs to take a picture, RL algorithms can change settings like the exposure time to increase the quality of the data acquired. RL can complement materials exploration within complex material spaces by opting for experiments on its own, which saves considerable time and is more informative in terms of findings. This dynamic adaptability means that characterization processes always keep pace with the emerging needs of the research application.

5.5.2 Challenges and considerations

Real-time feedback systems based on RL have some difficulties. One of the biggest issues is making RL algorithms stable and convergent, especially when working with noisy data. As in most RL processes, the model shows that the quality of the feedback has a direct influence on the performance of the RL agent, and poor feedback can seriously affect the learning results. The computational cost is often high in RL, particularly when the MC problem is large and complicated. Solving these problems requires careful planning of algorithms and parameter settings, as well as proper data management to provide the most accurate and informative feedback for the RL agent.

6 Explainable AI and interpretability in materials science

6.1 Feature importance and sensitivity analysis in material models

Feature importance and sensitivity analysis are fundamental components of Explainable AI (XAI) and are essential for ensuring interpretability in materials science applications. These methods help clarify how various input features influence the predictions and decisions made by machine learning models. Understanding which attributes have the greatest impact allows researchers and engineers to deepen their insight into material behavior, leading to more trustworthy, accurate, and transparent models. This level of interpretability is not only critical for validating model outputs but also supports material design efforts and ensures compliance with regulatory standards. Figure 11 presents the workflow of XAI using feature importance and sensitivity analysis, illustrating how model transparency facilitates informed decision-making in material science.

6.1.1 Methods used in measuring feature importance

Several methods are used to evaluate the importance of the features in material models. Still, one of the most frequently applied techniques includes permutation importance, which entails observing a model's efficacy after permuting the values of a particular feature. A large reduction in the variation of the model is an indirect sign of the high importance of this feature. Another method implemented is Shapley additive explanations (SHAP), which unifies the measure of feature importance calculation by adding the average marginal SHAP value of the given feature, relative to all other feature combinations. The concepts of partial dependence plots are applied to depict the contribution of individual features on model output, which is effective for feature importance and interaction. These procedures aid in the determination of key parameters that define the characteristics of materials and direct the fine-tuning of material models.

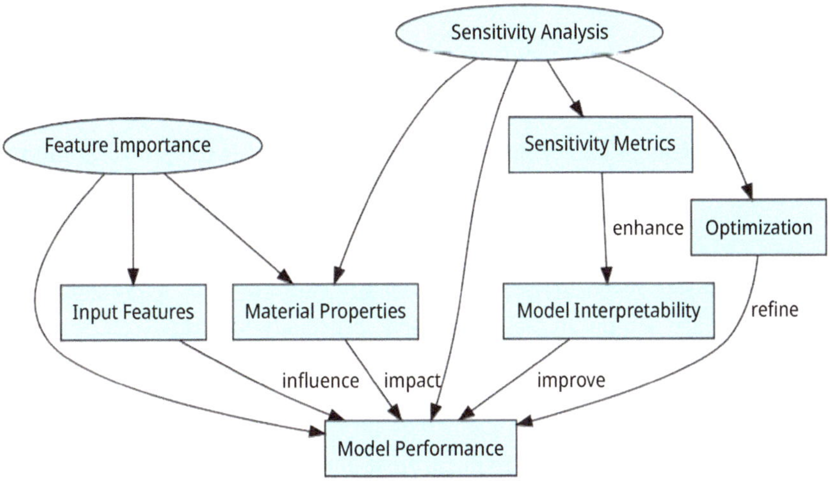

Figure 11: Feature importance and sensitivity analysis in material models.

6.1.2 Handling sensitivity analyses in material models

Sensitivity analysis focuses on the capability to establish relationships between the variations of input features and a model's output. More specifically, this analysis is very useful for gaining insight into the stability of the material models themselves. One procedure is local sensitivity analysis, whereby one alters the values of input features only slightly and observes the effect of the changes on the model outcomes. This technique assists in evaluating the required level of sensitivity to specific features locally. Global sensitivity analysis examines the effect of changing each feature at the same time over their entire range of values. Tools used in quantifying the contribution of input features to model output include variance-based sensitivity analysis and Sobol indices. Identifying the features important for optimal material design and ensuring there are no significant points that influence the model heavily is made possible by performing sensitivity analysis.

6.1.3 Applications and advantages in materials science

In materials science, feature importance and sensitivity analysis provide several important advantages. These techniques enable a deeper understanding of the relationship between input variables and material properties, thus facilitating focused experimentation and model updating. For instance, in cases where predictive models are being built for material performance, filtering the features helps guide researchers to focus on variables that are likely to have the greatest influence on the precision of the models. Additionally, these techniques reveal information about model behavior, improve trust and explainability in the AI-assisted material discovery process, and allow for easier validation of the computed predictions in practice.

6.1.4 Challenges and considerations

Nevertheless, both feature importance and sensitivity analysis have several limitations. One limitation is that feature importance measures can be influenced by the correlations between features, potentially providing an incorrect picture of the impact of each variable. The interaction of features and model assumptions is crucial to achieving higher levels of sensitivity analysis validity and reliability. Some sensitivity analysis methods are computationally intensive, especially for large-scale material models with many defined input parameters. To overcome such challenges, one must choose the right method, check the results using cross-validation, and, lastly, incorporate efficient computational algorithms.

Generally, the prospects for feature importance and sensitivity analysis in material science could easily lie in both methodological innovations and technological solutions. New approaches, including post-processing of a model as an interpretable machine learning algorithm and powerful data visualization techniques, present new opportunities for how feature importance and sensitivity results can be explored and visualized. In addition, combining these techniques with other XAI approaches could potentially provide a deeper and more applicable analysis of material models. Over time, as more research is conducted in material science and the use of AI, there is consistent improvement in scaling measures of feature importance and sensitivity, thereby increasing the field's interpretability and applicability.

6.2 SHAP (Shapley additive explanations) values for interpreting model predictions

SHAP values, especially for material science, were strong in terms of understanding the predictions of complex machine learning models. From cooperative game theory, we have SHAP values, which are a logical way of explaining individual predictions, as the model attributes a dollar value to each feature's contribution toward the model's output. This approach was particularly beneficial in the area of material science, as it provided an understanding of the reasons that govern model predictions and allowed for decisions that consider the consequences of various characteristics on material properties.

6.2.1 Mechanisms of SHAP values

SHAP values are derived from Shapley values, which originate from game theory, and represent the contribution of each player (or feature, in this case) to the output. SHAP values are used to represent the marginal contribution of features to outputs for all possible subsets of features under machine learning models. Therefore, an additive value set was created, the sum of which is equal to the difference between the average and model values. The partition of the prediction into each feature's contribution is expressed by SHAP values, which provide a more coherent and interpretable manner to explain how each feature impacts the model.

Aside from the topic dealing with materials, the applications that can be subsumed under the subject "Material Science" are as follows:

SHAP values find usage in almost all material science applications including failure analysis, material design optimization, and the prediction of material properties. For example, SHAP values are used for selecting features to answer questions like which of the composition or processing parameters contributes most to the projected characteristics when using machine learning models to predict the properties of new materials. Such realizations help researchers focus on the key factors only and adjust experiments accordingly. By providing brief rationales for model predictions, this fundamental piece is crucial in high-risk applications, such as safety assessment, regulation, and compliance. SHAP values can enhance the interpretability and accuracy of predictive models.

6.2.2 Challenges and considerations

There are some problems with the approach associated with applying SHAP values in material science. One major issue identified is the high computation time required to perform the calculations of SHAP values for large-scale models with numerous features. The process of considering every feasible feature occurrence can be extremely expensive in terms of computational resources and time, which can render SHAP analysis infeasible when dealing with huge datasets. Although SHAP values offer explanations related to feature significance in a model, they do not always depict the interaction between features. Some caution is needed when applying SHAP values in the context of a particular material science application, and on some occasions, it may be necessary to strengthen the explanation with other interpretability methods.

It was expected that better computer performance and the integration of SHAP with other interpretability approaches would determine the future trajectory of SHAP values in material science. This was actionable innovation since other approaches were developed to produce better SHAP values, along with blended strategies that use SHAP together with an outline strategy, such as a worldwide model interpretation instrument explanation method. Their usefulness was further enhanced by the applicability of SHAP values to new fields in material science, such as multiscale modeling and high-throughput screening. As computing methods and resources advance, higher SHAP values in material science are expected to be valuable in aiding the interpretation and application of complex machine learning models.

6.3 LIME (local interpretable model-agnostic explanations) for understanding material models

LIME (Local Interpretable Model-Agnostic Explanations) is one of the most widely adopted techniques in Explainable AI (XAI), designed to reveal how machine learning models make specific predictions. In materials science, where models often involve complex, nonlinear relationships between material properties and outcomes, LIME proves invaluable by offering interpretable, instance-level explanations. It enables researchers and practitioners to understand individual model decisions by approximating the model behavior locally around a specific prediction. This transparency enhances trust in material models, supports informed decision-making, and fosters societal acceptance of AI-driven material design. Figure 12 illustrates the LIME framework, showing how local explanations are generated to interpret complex material prediction models.

6.3.1 Mechanism of LIME

LIME, on the other hand, approximates complex, nontransparent models with simpler, more comprehensible models in specific circumstances in order to explain decisions. The original approach was to create a series of samples that are perturbed around a particular occurrence in a similar way to make predictions for a given data set. The behavior of the complicated model in the vicinity of the targeted instance was then refined by using these samples to train a locally explainable model, such as a decision tree. Thus, the interpretable model generated provides information that is easier for end-users to understand, including which factors were most influential in making the forecast for the specific instance, after learning more about the model's decision-making process.

6.3.2 Use and application of LIME in material models

When it comes to systems that involve a huge quantity of data on materials, LIME can be employed across the range of material modeling and analysis. For example, when making predictions about material properties from experimental data using machine learning models, what LIME does is shed light on the features (say, composition, processing parameters) most important in determining the predictions for the particular materials. One application of this capability is that when the relationship between feature space and material properties is not apparent, this capability comes in handy. With local interpretations, LIME helps researchers improve material models, guaranteeing that the latter, in addition to being accurate, also aligns with domain knowledge and expectations.

6.3.3 Benefits of using LIME

LIME, when applied in material science, proves useful in the following ways. First, it improves model interpretability, helping researchers grasp how certain features affect model outcomes. This is important for model validation and to confirm that the models are built on first principles and physical laws. Second, LIME aids in diagnosing model problems and improving these issues, particularly when models rely on irrelevant features. Last but not least, it is especially useful for explaining model results to nonadvanced users, such as material engineers.

6.3.4 Challenges and limitations

Some challenges are associated with the application of LIME. One of the issues identified was the selection of the interpretable model used in approximation: the quality of the local explanation was contingent upon the choice of this model. Also, LIME does not provide explanations that span the model globally; instead, it is specific to the local area. As such, LIME is helpful as a method to explain the prediction for a given individual instance, but it should be incorporated with other interpretability heuristics to achieve a complete understanding of the said model. The performance of LIME can be affected by the quality and samples of the space perturbation used for the training of the local model.

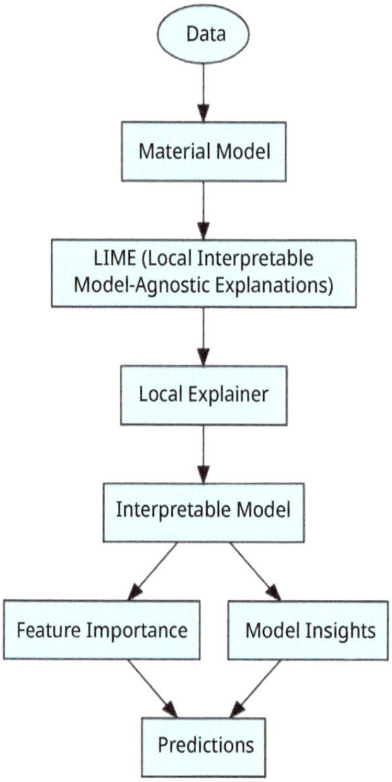

Figure 12: LIME (local interpretable model-agnostic explanations) for understanding material models.

6.4 Visualization techniques for model insights in biomedical materials

Visualization methods are essential for improving the interpretability and explainability of the corresponding machine learning models involved in biomedical material systems. They help researchers study model behavior, comprehend the decisions made, and evaluate predictions. This need is evident in biomedical materials applications, where model decisions about material characteristics and uses can critically impact the final material type, thereby demanding clear model visualization. As a result, visualization techniques make it easier to convey model results and processes that would otherwise be incomprehensible when used directly in decision-making processes.

Seven recommended practices for effectively utilizing web interfaces and visualization tools to gain insights into model behavior and performance.

Several visualization methods are used to explain the workflow of frameworks in biomedical materials. Relative importance plots are employed to understand which features are most influential to model outcomes in order to declutter features with less influence on material properties. Activation maps and heatmaps can represent areas in images corresponding to certain outputs, thus allowing one to understand how given models interpret complicated material structures. Various approaches, like t-SNE, allow one to plot high-level data on a two-dimensional plane for better analysis of patterns regarding material design. Altogether, these graphical aids help in gaining insights into the model's efficiency and patterns as well.

6.4.1 Display of the model's actions and decisions

It was indispensable to represent in the cases how the models worked and made decisions so as to determine errors and enhance model effectiveness. Making use of saliency maps, explained in the previous sections, allows for a direct view of which parts of the input data most affect the model's behavior: that is, which aspects of the material data the model is 'looking at'. By visualizing decision boundaries and decision trees, it was possible to describe which material properties the models classify and which criteria the models apply in their decision-making. Through these visualizations, decision-makers and engineers are able to critically evaluate the goodness of fit and the reliability of the models to ensure that the predictions made are in line with prior knowledge in material science as well as other real-world constraints.

6.4.2 Challenges in visualizing more biomedical materials

As much as visualization techniques provide many benefits, they also have various drawbacks. One problem is the difficulty in understanding how to analyze visualizations in cases of high dimensionality. The need to convey the behavior of the model without distortion through what is being visualized is paramount if the insights are to retain their value. Additionally, flexible integration of visualization instruments with various data forms and model constructions should be used thoughtfully. Overcoming these challenges entails two areas: creating complex visualization techniques and fostering user feedback to improve the interpretation of the visualizations.

Subsequent developments in biomedical material visualization approaches are expected to trend toward usability and compatibility with other analyses. Elements of the model and the data could be more easily navigated by the user if interactive interfaces were incorporated into the visualization tools to include different views of the models and the data. Engaging visualization with real-time data streams and feedback systems provides more instantaneous results about the model's performance as well as the properties of the materials used. Therefore, in the future growth of the explainable AI field, improved visualization approaches are crucial for expanding the principles and uses of machine learning models in biomedical materials.

6.5 Addressing interpretability challenges in deep learning models for materials science

The integration of deep learning (DL) models has revolutionized the analysis and design of materials, bringing new paradigms to the field of materials science. While these models are highly sophisticated and powerful, a significant challenge remains: their lack of interpretability. Understanding how DL models reach their decisions is critical, particularly in high-stakes applications, to prevent errors and ensure reliable outcomes. Model interpretability enables researchers to identify the key influencing factors, verify model outputs, and refine prediction strategies. In material science, where decisions directly affect material properties, performance, and safety, interpretability is essential for field validation, continuous improvement, and wider acceptance of AI-based solutions. Figure 13 visualizes the importance of interpretability in DL-driven material design, emphasizing its role in ensuring transparency, trust, and scientific rigor.

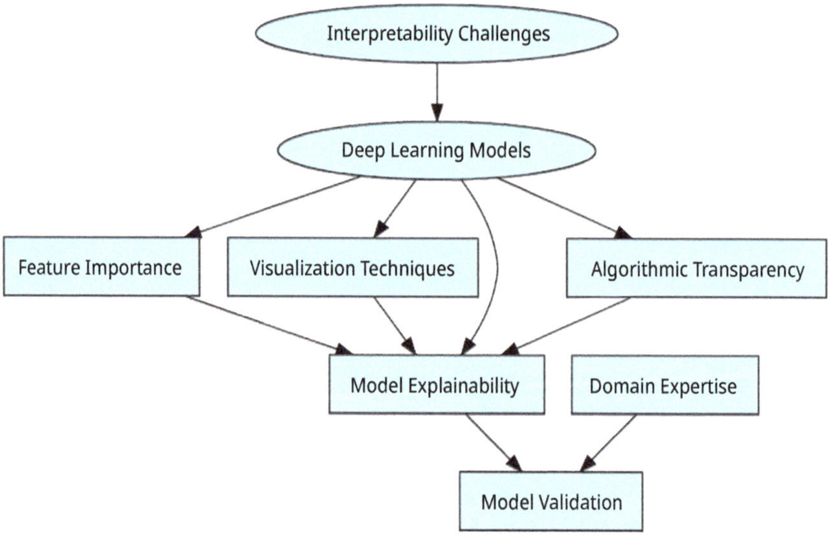

Figure 13: Addressing interpretability challenges in deep learning models for materials science.

Several issues need to be considered when making models interpretable.

As with deep neural networks, complicated models known as deep learning models, it was challenging to interpret how the model arrives at a specific conclusion. Interpretation was one of the biggest issues when working with these models, as the high-dimensional and nonlinear characteristics of the learned representations were a major problem. Their high dimensionality poses challenges regarding how it relates to the predictions and the factors that lead to them when involving complicated materials and their relations. Furthermore, interpretability methods are not standardized and provide different levels of explanation and, therefore, decision-making. Choosing the best approach for a material science application was crucial to take into account the identified approaches' features.

6.5.1 Methods for improving the interpretable nature

There are a plethora of ways available through which we can enhance the interpretability of DL models in material science. One method is visualization, where activation maps and feature importance scores are used, allowing us to identify which parts of the input data must affect the model's result. For instance, saliency maps are capable of showing where within a material's microstructure to look when focusing on the prediction in question. Another approach is to use components of the model that are easier to explain, such as attention mechanisms, which can provide more insight into how individual input features drive the model's output. Furthermore, methods such as LIME and Shapley values can be employed to provide local interpretable model-agnostic explanations for a single prediction, thus offering the client information on specific decision-making processes.

6.5.2 Trade-offs and limitations

Still, it must be pointed out that no configuration of an interpretation and a set of interpretability techniques were without flaws and that trade-offs exist. Increasing interpretability requires models to be simplified and boundaries to be approximated, both of which generally deteriorate model performance. Meeting interpretability with high predictive performance is a delicate process that needs to be tuned and tested. In addition, the interpretability method can be effective depending on the type of material science data as well as the complexity of a certain model, so the better approach is to select the most appropriate technique for the purpose.

Interpretation of DL for material science was crucial to reaching higher levels of supervised contingence capable of supplying a higher degree of understanding regarding the functioning of the models. The focus was on making interpretability inherent to the processes of model engineering, with new visualizations designed to better present high-level material data. Implementers of interpretability methods accounted for compositional semantics and other domain considerations pertinent to material science, updating interpretability methods to be more meaningful and useful. As these advancements continue to unfold, the aim remains to offer better, clearer, and more trustworthy data regarding material characteristics from DL models to enable improved material design and analysis.

7 Conclusion

In materials science, the recent incorporation of DL models has elevated the discovery and characterization of new materials to higher levels. Due to the inherent nature of such models, these issues bring severe interpretability challenges that must be resolved for these models to be fully utilized. The objective of this chapter has been twofold, all pertaining to interpretability: first, understanding the various definitions and concerns of transparency; second, exploring the methods available that provide visual insights about models.

The high accuracy of predictions, which allows DL models, is accounted for by the fact that it is often hard to track how particular input features actually affect the predicted output. In material science, interpretability is of paramount importance since it influences the validity and credibility of the model outputs. This trust-building is necessary for the use of artificial intelligence insights, more specifically in applications that pertain to material properties as well as performance. This is why it is important to ensure that the models developed yield easily understandable results that practitioners can act on.

A number of techniques have been developed to tackle these interpretability problems. Saliency maps and activation maps, which are used to show which features in the input data affected the output of a DL model, provide helpful information about how DL models work internally. Some explanation methods, such as LIME, SHAP, and similar approaches, are local explanation methods that offer an even more detailed breakdown of the instance under consideration, making them more granular. Nevertheless, this remains a major challenge today, as advancements in data science and machine learning have led to interpretability of models being seen as compromising model performance. Efforts to enhance clarity by applying simplified models can be costly in terms of loss of predictability, and vice versa.

Subsequent studies in this domain should be aimed at defining improved interpretability standards that are more applicable to material science. This, alongside transformations in the algorithms on which the interpretability models are based, has the potential to give rise to superior and more informative approaches to its employment. This improves not only the clarity of ideas where DL is applied but also guarantees its reliable applicability to material science issues. It thus always remains crucial to continue closing the gap between increased model complexity and their interpretability as the field develops, given that the practical application of these extraordinarily beneficial tools for material research depends upon it.

References

[1] Chen, C., Zuo, Y., Ye, W., Li, X., Deng, Z., & Ong, S. P. (2020). A critical review of machine learning of energy materials. Advanced Energy Materials, 10(8), 1903242.

[2] National Research Council, Division on Engineering, Physical Sciences, Board on Mathematical Sciences, Their Applications, Committee on Applied, . . . & Committee on the Analysis of Massive Data. (2013). Frontiers in massive data analysis. National Academies Press.

[3] Li, X., Xiong, H., Li, X., Wu, X., Zhang, X., Liu, J., . . . & Dou, D. (2022). Interpretable deep learning: Interpretation, interpretability, trustworthiness, and beyond. Knowledge and Information Systems, 64(12), 3197–3234.

[4] Guidotti, R., Monreale, A., Ruggieri, S., Turini, F., Giannotti, F., & Pedreschi, D. (2018). A survey of methods for explaining black box models. ACM Computing Surveys (CSUR), 51(5), 1–42.

[5] Sampson, C. J., Arnold, R., Bryan, S., Clarke, P., Ekins, S., Hatswell, A., . . . & Wrightson, T. (2019). Transparency in decision modelling: What, why, who and how? Pharmacoeconomics, 37(11), 1355–1369.

[6] Chakraborty, S., Tomsett, R., Raghavendra, R., Harborne, D., Alzantot, M., Cerutti, F., . . . & Gurram, P. (2017, August). Interpretability of deep learning models: A survey of results. In 2017 IEEE smartworld, ubiquitous intelligence & computing, advanced & trusted computed, scalable computing & communications, cloud & big data computing, Internet of people and smart city innovation (smartworld/SCALCOM/UIC/ATC/CBDcom/IOP/SCI), (pp. 1–6). IEEE.

[7] Hutchins, N. M., Biswas, G., Zhang, N., Snyder, C., Lédeczi, Á., & Maróti, M. (2020). Domain-specific modeling languages in computer-based learning environments: A systematic approach to support science learning through computational modeling. International Journal of Artificial Intelligence in Education, 30, 537–580.

[8] Gökalp, M. O., Gökalp, E., Kayabay, K., Koçyiğit, A., & Eren, P. E. (2022). The development of the data science capability maturity model: A survey-based research. Online Information Review, 46(3), 547–567.

[9] Yoon, S. A., Evans, C., Miller, K., Anderson, E., & Koehler, J. (2019). Validating a model for assessing science teacher's adaptive expertise with computer-supported complex systems curricula and its relationship to student learning outcomes. Journal of Science Teacher Education, 30(8), 890–905.

[10] Zhong, S., Zhang, K., Bagheri, M., Burken, J. G., Gu, A., Li, B., . . . & Zhang, H. (2021). Machine learning: New ideas and tools in environmental science and engineering. Environmental Science & Technology, 55(19), 12741–12754.

[11] Ahn, D., Almaatouq, A., Gulabani, M., & Hosanagar, K. (2021). Will we trust what we don't understand? Impact of model interpretability and outcome feedback on trust in AI. arXiv preprint arXiv:2111.08222.

[12] Chakraborty, S., Tomsett, R., Raghavendra, R., Harborne, D., Alzantot, M., Cerutti, F., . . . & Gurram, P. (2017, August). Interpretability of deep learning models: A survey of results. In 2017 IEEE smartworld, ubiquitous intelligence & computing, advanced & trusted computed, scalable computing & communications, cloud & big data computing, Internet of people and smart city innovation (smartworld/SCALCOM/UIC/ATC/CBDcom/IOP/SCI), (pp. 1–6). IEEE.

[13] Alqaraawi, A., Schuessler, M., Weiß, P., Costanza, E., & Berthouze, N. (2020, March). Evaluating saliency map explanations for convolutional neural networks: A user study. In Proceedings of the 25th international conference on intelligent user interfaces, (pp. 275–285).

[14] Weaver, C. P., Lempert, R. J., Brown, C., Hall, J. A., Revell, D., & Sarewitz, D. (2013). Improving the contribution of climate model information to decision making: The value and demands of robust decision frameworks. Wiley Interdisciplinary Reviews: Climate Change, 4(1), 39–60.

[15] Bhattacharya, A. (2022). Applied machine learning explainability techniques: Make ML models explainable and trustworthy for practical applications using LIME, SHAP, and more. Packt Publishing Ltd.
[16] Bhattacharya, A. (2022). Applied machine learning explainability techniques: Make ML models explainable and trustworthy for practical applications using LIME, SHAP, and more. Packt Publishing Ltd.
[17] Moons, K. G., Altman, D. G., Reitsma, J. B., Ioannidis, J. P., Macaskill, P., Steyerberg, E. W., . . . & Collins, G. S. (2015). Transparent reporting of a multivariable prediction model for individual prognosis or diagnosis (TRIPOD): Explanation and elaboration. Annals of Internal Medicine, 162(1), W1–W73.
[18] Gilpin, L. H., Bau, D., Yuan, B. Z., Bajwa, A., Specter, M., & Kagal, L. (2018, October). Explaining explanations: An overview of interpretability of machine learning. In 2018 IEEE 5th International Conference on data science and advanced analytics (DSAA), (pp. 80–89). IEEE.
[19] LeCun, Y., Bengio, Y., & Hinton, G. (2015). Deep learning. nature, 521(7553), 436–444.
[20] Oviedo, F., Ferres, J. L., Buonassisi, T., & Butler, K. T. (2022). Interpretable and explainable machine learning for materials science and chemistry. Accounts of Materials Research, 3(6), 597–607.
[21] Choudhary, K., DeCost, B., Chen, C., Jain, A., Tavazza, F., Cohn, R., . . . & Wolverton, C. (2022). Recent advances and applications of deep learning methods in materials science. Npj Computational Materials, 8(1), 59.
[22] Linardatos, P., Papastefanopoulos, V., & Kotsiantis, S. (2020). Explainable ai: A review of machine learning interpretability methods. Entropy, 23(1), 18.
[23] Wu, R., Fujita, Y., & Soga, K. (2020). Integrating domain knowledge with deep learning models: An interpretable AI system for automatic work progress identification of NATM tunnels. Tunnelling and Underground Space Technology, 105, 103558.
[24] Guidotti, R., Monreale, A., Ruggieri, S., Turini, F., Giannotti, F., & Pedreschi, D. (2018). A survey of methods for explaining black box models. ACM Computing Surveys (CSUR), 51(5), 1–42.
[25] Biecek, P., & Burzykowski, T. (2021). Explanatory model analysis: Explore, explain, and examine predictive models. Chapman and Hall/CRC.

G. Jenifa, R. Meenal, A. Anandh, R. Jennie Bharathi, S. Sree Dharinya

Revolutionizing healthcare

Abstract: Artificial intelligence (AI) and, in particular, machine learning (ML) have recently assumed the role of leading disruptive technologies in the area of biomaterials science and engineering by providing new avenues for material design and property prediction and enabling the development of personalized medicine. This chapter reviews the interdisciplinary field of ML with biomaterials engineering, focusing on using algorithms to fine-tune biomaterial properties and outcomes. One of the key areas of interest is the interpretation of hyperparameter optimization for performance enhancement, another is the increasing relevance of ML for improving the efficiency of characterization methods, and the third one is the creation of simulation models to predict property optimization. Some trends that are expected to unfold in the future include the use of multimodal data, new approaches to better process real-time data, and an improved ability to provide clearer explanations for models. The potential ethical issues and regulatory concerns of utilizing ML in biomaterials are also highlighted, along with the emphasis on sustainability and our responsibility to use these technologies responsibly in the biomaterials sector. Thus, this chapter demonstrates how the biomaterials research and applications domains are being revolutionized through the use of ML and presents a wealth of information to researchers and practitioners.

Keywords: machine learning, biomaterials, hyperparameter tuning, predictive modeling, real-time data processing, model interpretability

1 Introduction

Machine learning (ML) has emerged as a transformative technology in the field of biomaterials, revolutionizing traditional approaches to material science [1–4]. By leveraging advanced algorithms and computational models, ML enables researchers to ana-

G. Jenifa, Department of Computer Science and Engineering (Artificial Intelligence), Madanapalle Institute of Technology and Science, Madanapalle, Andhra Pradesh, India, e-mail: gjenifa@gmail.com
R. Meenal, Department of Electrical and Electronics Engineering, VSB Engineering College, Karur, Tamil Nadu, India, e-mail: meenasekar5@gmail.com
A. Anandh, Department of CSE, Kamaraj College of Engineering and Technology, Virudhunagar, Tamil Nadu, India, e-mail: anandhcse@kamarajengg.edu.in
R. Jennie Bharathi, Department of ECE, Saveetha Engineering College, Chennai, Tamil Nadu, India, e-mail: jenivenkatesh@gmail.com
S. Sree Dharinya, School of Computer Science, Engineering, and Information Systems, VIT University, Vellore, Tamil Nadu, India, e-mail: dharinyasathish@gmail.com

https://doi.org/10.1515/9783111503202-014

lyze complex datasets and extract actionable insights with unprecedented accuracy [5]. The integration of ML into biomaterials research has led to significant improvements in material design and property prediction [6]. This integration facilitates the development of more efficient and effective materials by predicting their behaviors under various conditions, thus streamlining the research and development process [7]. The increasing complexity of biomaterials, combined with the vast amount of data generated from experimental and computational studies, necessitates sophisticated analytical techniques that ML is uniquely equipped to provide [8].

Using sophisticated algorithms to improve property prediction is one of the main uses of ML in the biomaterials industry [9–11]. Conventional approaches for predicting properties frequently depend on empirical data and constrained computational models, which can be unreliable and limiting [12]. Regression models and neural networks, on the other hand, are examples of ML algorithms that can analyze massive amounts of data to identify links and patterns that are not immediately obvious [13]. Predictions of material attributes like mechanical strength, thermal stability, and chemical resistance are made possible by this capability [14]. Researchers create prediction tools that guide the development of novel biomaterials with customized features by training ML models on large datasets [15]. This approach accelerates the material development process and reduces the need for trial-and-error testing [16].

The capacity to evaluate and comprehend biomaterial data has been improved with the integration of ML with characterization methods [17, 18]. Large and complicated datasets are produced by characterization techniques, including atomic force microscopy (AFM), scanning electron microscopy (SEM), and X-ray diffraction (XRD), which can make it difficult to analyze the data using conventional analytical techniques. By locating important characteristics and connections in the data, ML techniques, such as clustering algorithms and dimensionality reduction, improve the study of these datasets [19]. For example, ML methods have the potential to enhance the precision of material phase identification in XRD data and surface analysis from SEM images [20]. Better knowledge of the characteristics and behaviors of biomaterials has been made possible by these advancements in characterization techniques, which improve material design and optimization [21, 22].

Looking ahead, a number of developments and trends are expected to boost ML's contribution to biomaterials research [23]. It is anticipated that multimodal data integration, which combines several data types from various characterization methods, will offer a more thorough understanding of biomaterials [24]. The ability to analyze data in real-time will facilitate prompt feedback during material testing and manufacturing, resulting in more flexible and effective research procedures. Researchers find it simpler to comprehend and trust the predictions generated by ML models as explainable AI (XAI) advances, thereby improving the interpretability of ML models [25]. These upcoming developments should enhance ML's capabilities in the field of biomaterials science, spurring further research and innovation.

As ML technologies continue to evolve and integrate into biomaterials research, addressing ethical considerations and ensuring responsible use become increasingly important. The deployment of ML models must consider issues such as data privacy, model transparency, and the potential for bias in predictions. Setting proper ethical standards and regulatory mechanisms becomes crucial for guiding the usage of ML solutions in the biomaterials industry and ensuring that the solutions meet ultimate goals and align with societal preferences. These include the nondisclosure of data, ensuring that models contain no bias against individuals, and maintaining transparency in the process through which predictions of individual behavior are generated and used. Appropriate use of ML technologies not only contributes to the scientific development of the biomaterials field but also helps overcome skepticism and nonacceptance factors among stakeholders.

2 Data-driven approaches in biomaterials design

2.1 High-throughput data collection for biomaterials

Large-scale data acquisition has therefore been established as a critical factor in the design and synthesis of biomaterials, especially when involving ML techniques. This highlights the importance of broad-ranging datasets because the range of biophysical conditions and the interaction of material properties are diverse. These databases form the basis of a number of ML algorithms used to forecast material performance, determine optimal characteristics, and discover new bio materials with improved functions. The problem raises the question of how to collect high-throughput data accurately and comprehensively over time, which, in turn, necessitates a cross-disciplinary approach, as well as technologies and methods. Specific issues that need to be resolved to achieve maximum advantage from high-throughput data include data quality, handling of information, and dealing with variability in testing conditions during experimentation in biomaterials design.

Current advancements in technology make flow cytometry a comparatively fast and accurate tool for accomplishing high-throughput data collection in biomaterial sciences. Instrumental approaches, including robotic systems and microfluidics, allow for conducting rapid and parallel experiments and examining multiple properties of the material at once under different conditions. Such technologies can help in the creation of large data sets, which are generated with minimal human input, low probabilities of error, and high-test reproducibility. Furthermore, the synergistic use of enhanced imaging systems and sensing tools enables the observation of material property shifts simultaneously, across a range from the molecular to the macroscopic level. These innovations not only improve the quality of the collected data but also extend the range of analyzed parameters, providing more information on biomaterial behavior.

There is a growing concern regarding the effective use of high-throughput data in biomaterials research, making the processes of data preprocessing, integration, and standardization critically important. The complexity and diversity of data sources necessitate the development of unified protocols and standardized formats to ensure consistency across research groups and institutions. This standardization is essential because comparing datasets across different models requires them to be in compatible formats, enabling accurate and meaningful analysis. Consolidating heterogeneous data types into coherent and structured datasets allows researchers to extract insights that might be missed when working with fragmented data. By adopting collective data storage and analysis standards, the biomaterials community can foster a collaborative and integrated research environment, accelerating the discovery and development of advanced materials. Figure 1 illustrates the data standardization and integration workflow, highlighting its role in unifying research efforts across the biomaterials field.

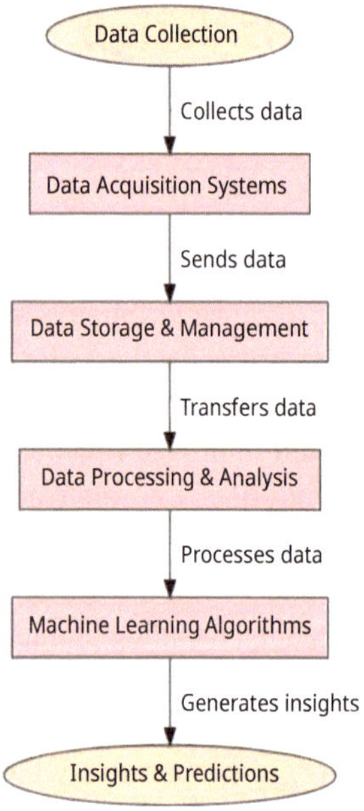

Figure 1: High-throughput data collection for biomaterials.

The prospect of high-throughput data acquisition in biomaterials design lies in the advancement of these cutting-edge technologies including artificial intelligence (AI) and IoT. Data collection systems using AI remain one of the most potent means of automating conventional experimental procedures, making it possible to select favorable experimental conditions and modify protocols based on real-time data analysis. IoT devices can monitor biomaterials' exposure to real-world conditions without interruption, which is crucial for improving materials based on the data received. In our view, as these technologies progress, they result in more effective and smarter forms of data-gathering processes, creating opportunities for further advancements in the design of a new breed of biomaterials to suit various medical and therapeutic applications.

2.2 Big data analytics for predictive modeling

Decision science and biomaterials have advanced significantly in incorporating big-data analytics into biomaterials design, offering the potential to greatly enhance biomaterials manufacturing. The use of biomaterials and applications in biomaterials science and engineering is growing in terms of complexity and the number of applied fields. Thus, the amount of experimentally, simulated, and clinically based data is increasing rapidly. Such a wealth of information demands powerful methods of analysis to reveal significant insights, describe material behavior, and facilitate the discovery of new bio materials. These large, complex data sets can be effectively handled, analyzed, and interpreted by big-data analytics, offering investigators a window into the seemingly chaotic data.

Every big data analysis is built on the foundation of ML algorithms to compute scalable datasets and provide accurate estimations of biomaterial characteristics. These include simple regression models as well as complex deep learning networks that are able to handle multidimensional data sets of biomaterials. From the previous datasets, the results of ML include usage life, cytocompatibility, and mechanical properties of the materials. Besides greatly decreasing the time needed for experimental verification, this predictive ability also minimizes the costs of material characterization, which is crucial in the initial stages of biomaterials design.

This was even beyond the range of the quite conventional process of property prediction for expected properties of new material types, encompassing the discovery of completely new biomaterials with specific desired characteristics. Using modern tools of data analysis, including clustering and pattern recognition, it was possible to categorize materials and forecast their properties including their behavior under certain conditions. By leveraging previous results from experiments and simulations, researchers can make assumptions about how newer materials will perform in tests and determine which strategy to incorporate next during the design process. This rational approach increases the chances of developing biomaterials that fit the purpose of their use in areas like tissue engineering, drug delivery, and regenerative medicine.

Big data analytics has many potential applications in biomaterials design, but before those applications can be fully realized, a number of issues need to be resolved. Data heterogeneity is one of the main issues, as biomaterial data are frequently gathered from a variety of sources such as computer models, clinical trials, and experimental results. Sophisticated data preparation techniques are needed to integrate these heterogeneous datasets into a coherent analytical framework. Predictive model accuracy also heavily depends on the completeness and quality of the data. Rigorous data validation and curation procedures are essential because incomplete data produce inaccurate forecasts.

The development of more complex ML algorithms and the incorporation of cutting-edge technologies like AI and quantum computing hold the key to the future of big data analytics in biomaterials design. By allowing for the examination of progressively larger and more intricate information, these developments push the limits of predictive modeling. Turning these discoveries into useful applications requires cooperation between data scientists, doctors, and materials scientists. Big data analytics continues to be essential to the development of biomaterials design, opening new avenues for research and development and enabling more efficient and customized healthcare solutions.

2.3 Feature selection and dimensionality reduction

Because feature selection directly influences the performance of machine learning models, it is a critical step in the data-driven design of biomaterials. Given the high-dimensional and complex nature of biomaterials data, identifying the most relevant features is essential to develop accurate and interpretable predictive models. Feature selection techniques aim to isolate the most informative variables while reducing dimensionality, ensuring that models remain both efficient and insightful. By focusing on the key aspects of the biomaterial design space, these methods not only enhance predictive performance but also contribute to model interpretability and better decision-making. Figure 2 illustrates the feature selection process, emphasizing its role in simplifying complex data and improving learning outcomes in biomaterials research.

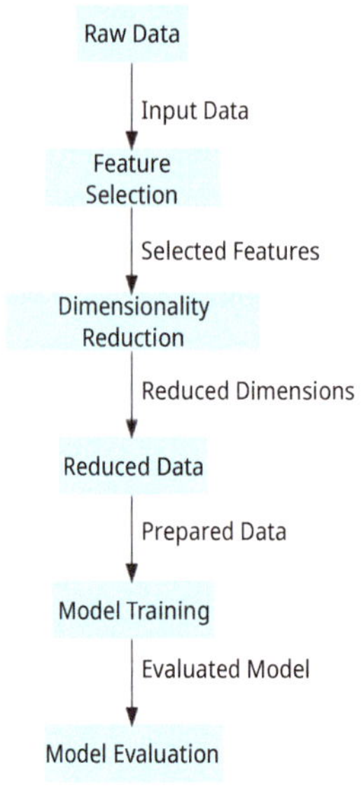

Figure 2: Feature selection and dimensionality reduction.

A quick look at why dimensionality reduction is critical in the study of biomaterials.

This was because the analysis of high-dimensional data was frequent in biomaterials research, and techniques for dimension reduction were deemed crucial. The most important information was retained when the original data was projected into a lower-dimensional space using techniques such as principal component analysis (PCA) and LDA. These methods help to reduce the curse of dimensionality, which translates into the reduction of over fitting and poor generalization owing to many folds. Dimensionality reduction guarantees that ML models remain strong and effective in biomaterials research, where the data comprise a variety of qualities, compositions, and experimental settings, facilitating the identification of underlying patterns and linkages.

2.3.1 Techniques for feature selection in biomaterials

In the design of biomaterials, a variety of feature selection strategies are used, each having specific benefits and uses. Based on statistical measurements, filter techniques like mutual information and correlation coefficients assess each feature's importance. Recursive feature elimination is one of the wrapper approaches that evaluates feature relevance by repeatedly training models and eliminating the least important features. Embedded techniques that penalize less significant characteristics during the model training process include LASSO (least absolute shrinkage and selection operator). When applied correctly, these approaches facilitate the understanding of which characteristics are most important, hence yielding models in biomaterials research that are more accurate and easier to understand.

2.3.2 Some of the difficulties in feature selection and dimensionality reduction

In biomaterials research, feature selection and dimensionality reduction present many problems, although they are very important. Of these, the major challenge is that reducing the dimensionality of the dataset subjects the model to the risk of losing significant information, which could affect its predictability. Additionally, it is difficult to apply standard definitions of biomaterial feature selection and dimensional reduction techniques because of their high nonlinearity and the tight coupling of variables. Instrument selection must be scientifically analyzed because, depending on the exact goals of the study and the data type, certain procedures yield different results.

2.3.3 Essays on uses and examples in biomaterials

Feasible applications of feature selection and dimensionality reduction in biomaterials research have been widely acknowledged due to recorded advancements. For instance, these approaches have been used in the synthesis of next-generation biomaterials for use in implants to identify influential parameters affecting mechanical properties and biocompatibility. To specify and generate new materials with enhanced performance, researchers have been able to develop prediction models that guide material synthesis by focusing on key properties. Similarly, in the medication delivery domain, dimensionality reduction has been applied to analyze complex datasets, revealing new materials that improve drug release features, making them more manageable. These case examples demonstrate that the selection of pertinent features, as well as dimensionality reduction, is key to the future direction of biomaterials research. With the growing field of hybrid biomaterials comes the development of new and improved methods for feature selection and dimensionality reduction. Future research will continue to refine existing algorithms to accommodate biomaterial data with intricate nonlinear aspects and numerous interactions. Additionally, the interaction of ML with cutting-edge technologies, such as quantum computing, provides new possibilities for increasing the efficiency of feature selection and dimensionality reduction. This progress is expected to be paramount in supporting the evolution of new biomaterials that cater to the rising needs of the health sector and other industries.

2.4 Dataset augmentation for machine learning models

Improving the efficacy and resilience of ML models, especially in domains such as biomaterials where limited data and unequal distribution of classes impede model precision, is crucial. Because biomaterials research involves sophisticated and expensive experiments, it can be difficult to gather vast and diverse datasets. ML models trained on small amounts of data, therefore, perform poorly when applied to fresh, untried data. This problem can be addressed by dataset augmentation, which makes the training dataset artificially larger. This approach enables models to learn more widely distributed patterns and lowers the chance of overfitting.

2.4.1 Methods for expanding datasets

Diverse methodologies are utilized for dataset augmentation, each customized to the unique type of data used in biomaterials research. For example, popular augmentation techniques in imaging data include flipping, scaling, rotations, translations, and the addition of noise. By producing variants of the original images, these modifications strengthen the model's resistance to shifting scales and orientations. Methods such as generative models like generative adversarial networks (GANs) and oversampling techniques like SMOTE (synthetic minority oversampling technique) are frequently utilized for tabular data. In situations where uncommon occurrences are underrepresented, these techniques assist in balancing class distributions and provide the model with more representative samples.

2.4.2 Challenges in dataset augmentation for biomaterials

Despite its advantages, dataset augmentation in biomaterials research is not without challenges. One primary concern is maintaining the physical and chemical integrity of the data during augmentation. Augmenting biomaterial data, especially in complex experimental settings, introduces unrealistic variations that could mislead the model during training. Ensuring that augmented data accurately reflects the real-world properties of biomaterials is crucial for the model's predictive validity. Determining the appropriate level and type of augmentation requires domain expertise, as excessive augmentation can lead to model degradation rather than improvement.

2.4.3 Impact of dataset augmentation on model performance

Effectively used, dataset augmentation reduces the hazards associated with limited datasets and increases the diversity of training data, leading to a considerable improvement in model performance. For example, it has been demonstrated that enhanced datasets improve ML models' capacity to generalize when predicting biomaterial attributes like mechanical strength. Augmentation facilitates the development of more robust and dependable prediction models by exposing the model to a wider variety of variables. These models are crucial for converting ML discoveries into useful applications in the field of biomaterials research.

Dataset augmentation for biomaterials: Looking forward, the development of more sophisticated augmentation techniques, particularly those that incorporate domain-specific knowledge of biomaterials, will be essential for advancing this field. Future research should focus on the integration of physics-based models with data-driven approaches to create augmented datasets that better mimic real-world material behaviors. Additionally, the application of more complex generative models, like variational autoencoders, in combination with data synthesis and the generation of more realistic data, looks very promising. Such developments not only improve model performance but also expand the range of ML methods available to design and develop new biomaterials.

2.5 Benchmark datasets for machine learning in biomaterials

Benchmark datasets play a foundational role in the development and evaluation of machine learning (ML) models within the field of biomaterials. These datasets typically compile comprehensive information on material properties, fabrication techniques, and performance metrics, derived from both experimental studies and computational simulations. Prominent examples include the Materials Project and the Cambridge Structural Database, which offer vital data for training and validating predictive models focused on biomaterials. These repositories provide insights into characteristics such as mechanical strength, thermal conductivity, and biological compatibility. Their importance, such datasets often face challenges in terms of completeness, sample diversity, and representativeness across different material types and use cases. Figure 3 highlights the structure and limitations of key benchmark datasets, showcasing their significance and the gaps that must be addressed to advance ML applications in biomaterials research.

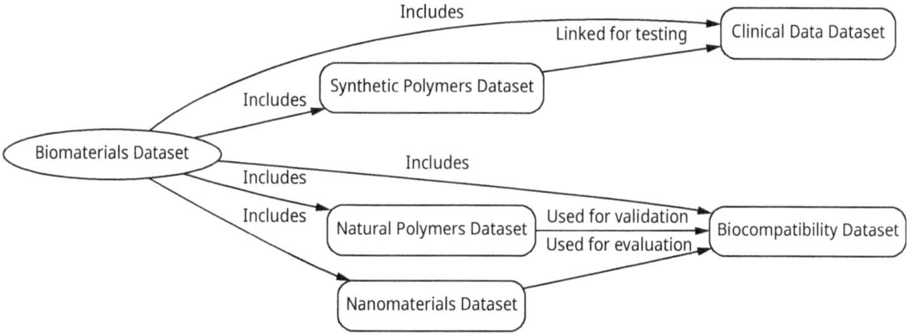

Figure 3: Benchmark datasets for machine learning in biomaterials.

2.5.1 Standardization of benchmark data sets of biomaterials

The development of standard benchmark datasets was, therefore, fundamental to the progress of ML research in biomaterials. Gathering relates to setting general structural components, values of measurements, and attributes for metadata. It was such standardization that enabled comparison throughout and built strong ML models when there was a standardized set through which to train and test. In the planning of sharing and standardizing initiatives, extensive groups of biomaterials should be involved, including common classes of biomaterials such as polymers, metals, and ceramics. The datasets, which should be extended to include conditions that pose realistic repair demands, accustom AI to a broader spectrum of materials, improving the predictability of repair tasks.

As part of this, the quality of biomaterial datasets was discussed.

An assessment of the quality of biomaterial datasets was necessary for establishing their quality, reliability, and suitability for use in ML. Quality evaluation can be performed in terms of reliability, relevance, bias, and inconsistency in the data. Completeness and, for example, error rate for a set of samples are the main measures that should be prioritized to evaluate dataset quality. The checking of the tested data against more conservative standards and the inclusion of published articles further improve the reliability of the dataset. Stringent quality control steps are required to be put in place to preserve data integrity so that the datasets upon which we train as well as test our ML models are credible.

2.5.2 The use of open-source datasets to advance biomaterials science

K development of open-source datasets was crucial in biomaterials research since it involves providing and sharing usable data with the community. One important characteristic of open-source data is the collaboration and rapid progress enabled by researchers' ability to use existing datasets and contribute new results. Some examples of open-source platforms containing biomaterial data include the Open Materials Database and the Data Commons project, where people can access and use the "stuff" data in a more comprehensive way. Such platforms usually feature tools for data visualization and analysis, making them even more suitable for ML endeavors. Encouraging a culture of open data for biomaterials can reduce data shortages, enhance the development of novel open-source datasets for use in ML, and improve the accuracy and flexibility of existing biomaterial-related ML models.

Exploiting Dataset Bias in Machine Learning Models' explores the impact of dataset biases on the performance and generalizability of machine learning systems.

Bias in the datasets was a major problem with ML, especially when used in the biomaterials context. Bias evolved from sampling proportions, incomplete sample coverage, and the exclusion of certain material conditions. Biases of this nature can result in the development of a model that is inaccurate in its nature and exhibits a low level of ability for the devised ML algorithm to be applied in other areas. Dataset bias was addressed by adopting measures to achieve balance in biomaterial type and experimental condition representation. This was because methods such as data augmentation, oversampling, as well as the use of heterogeneously synthesized datasets reduce bias, thereby enhancing the general performance of a model. Making documentation on datasets more transparent, as well as scheduling audits, can help in the proper characterization of bias and thus aid in evolving fair and accurate new learning systems for biomaterials.

3 Machine learning models for biomaterial property prediction

3.1 Supervised learning for predicting biomaterial properties

Regression models are widely used in biomaterial science to predict continuous properties such as mechanical strength, thermal conductivity, and degradation rates. These models establish relationships between input features such as material composition, processing parameters, and environmental conditions, and the desired properties of biomaterials. Common regression techniques include linear regression, polynomial regression, and support vector regression. By researching the characteristics of various types of biomaterials, these models are trained to profile historical data so that new and modified biomaterials can be forecasted in terms of performance. This means that the strength of regression models is majorly dependent on the quality and quantity of training data, as well as the type of regression model used for the specific property being predicted.

3.1.1 Functionality of biomaterials classification algorithms

Predictive models of the classification kind are used to estimate the categorical characteristics of biomaterials, particularly regarding their categorization by applications. The following methods are used for model selection: logistic regression, decision trees, random forests, etc. For example, classification models can be used to distinguish between biomaterials that are appropriate for bone regeneration and those that are suitable for use in soft tissue engineering. Classification algorithms enable the creation of classes in which biomaterials being developed can be grouped depending on their characteristics and applications. The accuracy of the classification models presented depends on the quality and quantity of the available data and the capacity of the model to operate in view of the complexity and variation in the properties of biomaterials.

3.1.2 Case studies: supervised learning in biomaterial design

The real-world uses of supervised learning in biomaterial design and property prediction are demonstrated through case studies. To optimize the performance of polymer-based biomaterials for particular medical applications, ML algorithms, for instance, have been used to anticipate the mechanical characteristics of the materials. Another example of how supervised learning has helped is in the prediction of biomaterial degradation rates under different environmental circumstances, which helps in the creation of materials that are more resilient. These case studies demonstrate how supervised learning, by utilizing predictive modeling and historical data, leads to the development of novel biomaterials and offers practical insights.

3.1.3 Evaluating supervised models for predictive accuracy

In order to determine the dependability and efficacy of supervised learning models, it was imperative to evaluate their predictive accuracy. The model's ability to predict accuracy was, in some cases, evaluated by performance indices such as classification accuracy, mean-squared error (MSE), and root mean square error. To ensure that the model was not merely training on the existing data and was capable of predicting new data, cross-validation techniques like *k*-fold were applied. Additionally, to achieve strong and trustworthy predictions in biomaterial science, evaluating the performance of several models and selecting the top-performing one based on validation results was crucial.

3.1.4 Challenges in applying supervised learning to biomaterials

A number of difficulties arise when applying supervised learning to biomaterials, including problems with feature selection, data quality, and model interpretability. For successful model training, high-quality, labeled data must be available; yet, in biomaterial research, such data is limited. Another difficulty in feature selection is that it requires rigorous study and domain knowledge to determine which features are most important in influencing material characteristics. Models' practical usefulness depends on their ability to be interpreted and to offer insightful information about the underlying connections between features and characteristics. To tackle these obstacles, modeling methodologies must be continuously improved upon, while the unique requirements and limitations of biomaterials research must be carefully taken into account.

3.2 Unsupervised learning for biomaterial pattern recognition

In the field of biomaterials, unsupervised learning plays a pivotal role by enabling the discovery of intrinsic patterns and structures within datasets without requiring labeled data. Unlike supervised learning, which depends on predefined classifications, unsupervised algorithms analyze raw data to detect hidden correlations, clusters, and features that may otherwise go unnoticed. This approach is particularly valuable in biomaterials research, where labeled datasets are often scarce due to the diversity in material properties and experimental variations. By leveraging unsupervised learning techniques, researchers can gain deeper insights into the underlying data landscape, facilitating the identification of novel material characteristics and relationships. Figure 4 illustrates the role of unsupervised learning in biomaterials, showcasing how clustering and dimensionality reduction techniques reveal meaningful structure in complex, unlabeled datasets.

3.2.1 Clustering techniques for biomaterial classification

There are different types of biomaterials based on their characteristics and behaviors in use, and to classify them, there are mainly algorithms that belong to the unsupervised learning category, called clustering techniques. To cluster biomaterials in a way that subdivides them based on similar properties, some of the general algorithms that are used include *k*-means, hierarchical clustering, and DBSCAN (density-based spatial clustering of applications with noise). For example, a dataset of biomaterial qualities can be grouped into *k* clusters using *k*-means clustering, where each cluster represents a collection of materials with similar characteristics. The hierarchical relationships between biomaterials are shown using a dendrogram prepared through hierarchical clustering. On the other hand, DBSCAN can identify clusters of different geometrical shapes and sizes since it discovers clusters based on their density. Such clustering techniques are helpful in defining the patterns and systematic relationships of biomaterials, which can then be utilized for material development and enhancement.

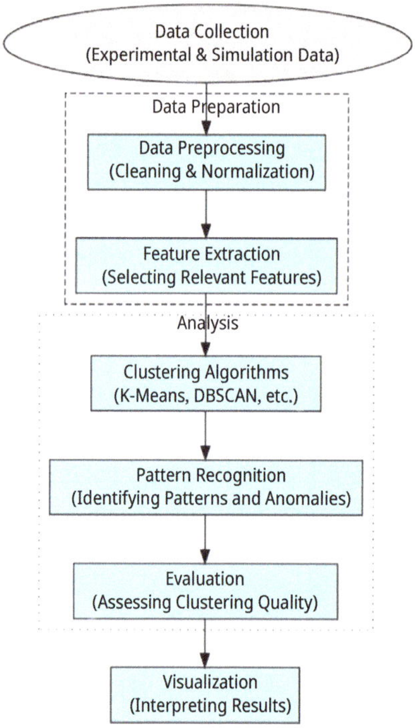

Figure 4: Unsupervised learning for biomaterial pattern recognition.

3.2.2 Dimensionality reduction in biomaterial data analysis

Reduction in dimensionality is a basic unsupervised learning approach used to minimize the number of elements in comprehensive biomaterial datasets while preserving the relevant details. To deal with the number of dimensions, useful tools such as PCA and t-distributed stochastic neighbor embedding (t-SNE) are used. Through the maximization of variance in data efficiency, the first set of features is transformed into a set of orthogonal variables using PCA. Using this method, high-dimensional biomaterial data is represented in a lower dimensional space. Another method that works very well for visualizing clusters in high-dimensional data is t-SNE, which maps the data to a two-dimensional space. These methods make it easier to find hidden patterns and trends and allow for a more intuitive study of the features of biomaterials by decreasing dimensionality.

3.2.3 Unsupervised feature extraction for novel biomaterials design

Here, we demonstrate that feature extraction through unsupervised learning can greatly improve the design as well as the discovery of new biomaterials. Autoencoders and independent component analysis (ICA) are then used to extract features from biomaterial datasets. Autoencoders are feed-forward neural networks designed to learn two critical representations: the first encodes the input into a compact format and the second reconstructs the input from the encoded representation. ICA separates a multivariate signal into additive, independent components, revealing underlying factors that contribute to material properties. These unsupervised feature extraction methods help in identifying key attributes that influence biomaterial behavior, guiding the design of new materials with desired properties and functionalities.

3.2.4 Application of anomaly detection in biomaterial quality control

Using unsupervised learning, anomaly detection identifies outliers that deviate from expectations in biomaterial databases. This method is highly beneficial in quality control procedures as it detects problems in the manufacture of biomaterials by identifying abnormalities. For anomaly identification, algorithms like local outlier factor (LOF), one-class SVM, and isolation forest are frequently utilized. For example, one-class SVM models the normal data distribution and recognizes deviations as anomalies, whereas isolation forest isolates anomalies by randomly selecting features and separating data points. To find outliers, LOF calculates the local density of data points. Researchers enhance the consistency and reliability of biomaterial manufacturing by integrating anomaly detection into quality control procedures, ensuring that only premium materials are used in applications.

3.3 Deep learning for advanced biomaterial design

Convolutional neural networks (CNNs) have found their significance in biomaterial design as the technique imparts the capability to interpret complex imaging data. CNNs are highly efficient in obtaining hierarchical features from the high spatial resolution images of biomaterials, including microscopy images and scans, as they help in the description of material microstructures and properties. Through multiple convolutional filters, pooling, and activation functions, CNNs are capable of detecting very complex details of an image as well as distinguishing between normal and abnormal features that standard image processing cannot easily detect. This capability is greatly useful for tasks such as surface profiling, pore size and distribution, and composite morphology determination to assess the performance of biomaterials in several applications.

3.3.1 Autoencoders for biomaterial property optimization

An autoencoder, which is an unsupervised deep learning model, plays a crucial role in enhancing biomaterial properties. These models comprise an encoder, which compresses the input data into a lower-dimensional latent space, and a decoder, which reconstructs the data from this compressed representation. By minimizing dimensionality, autoencoders are also useful in identifying latent features to be used in designing biomaterials. By learning a compact representation of biomaterial properties, autoencoders facilitate the exploration of property-space relationships and enable the identification of optimal material configurations. This approach is particularly useful for optimizing complex properties such as mechanical strength, flexibility, and biocompatibility, which require balancing multiple design parameters.

3.3.2 Applications of deep generative models for biomaterial discovery

Using GANs and VAEs, deep generative models are being integrated more often into biomaterial design with the purpose of creating new material compositions and structures. GANs are composed of two feed-forward neural networks, a generator and a discriminator, in which both play an adversarial role by generating new samples of material and checking if the material sample is original. VAEs, on the other hand, model the distribution of material properties in a latent space, allowing for the generation of new samples with desired characteristics. These generative models enable researchers to explore a broader design space and discover innovative biomaterials with unique properties that may not be easily identified through conventional methods.

3.3.3 Application of recurrent neural networks in time-series biomaterial data

long short-term memory (LSTM) networks are one type of RNN that is particularly useful for assessing biomaterial time-series data, such as degradation patterns and responses to external stimuli. In order to capture dependencies and trends across time, RNNs make use of their temporal design, which makes them perfect for modeling dynamic behaviors and forecasting the long-term performance of materials. This feature is especially helpful for tracking the lifetime and stability of biomaterials in real-world circumstances, where time-dependent variables are critical in establishing the efficacy and dependability of the material.

3.3.4 Transfer learning in biomedical research

Transfer learning involves using an already-trained deep learning model to solve a novel but related biomaterial design challenge. This allows the model to perform better by making use of the information that already exists. This methodology is especially advantageous in situations where data is scarce, as it applies knowledge from well-established fields to newly emerging biomaterial research sectors. Through the process of fine-tuning pre-trained models on specific biomaterial datasets, researchers can improve the speed and accuracy of design optimization and property prediction. Transfer learning leverages information from previously researched materials and designs to enable the quick adaptation of deep learning techniques to a variety of biomaterial applications, such as drug delivery systems and tissue engineering.

3.4 Ensemble learning techniques for robust prediction

Combining predictions from multiple distinct models, ensemble learning. approaches significantly enhance the accuracy and robustness of machine learning systems. This methodology minimizes the effect of individual model errors by leveraging the complementary strengths of various algorithms, ultimately leading to more reliable predictions. In the context of biomaterial property prediction, ensemble methods effectively integrate outputs from diverse models such as decision trees, support vector machines (SVMs), and neural networks. This fusion enables the detection of complex, multidimensional patterns that may not be captured by a single model alone. Figure 5 depicts the ensemble learning framework, highlighting its ability to synthesize predictions from multiple models to improve performance and reliability in biomaterials research

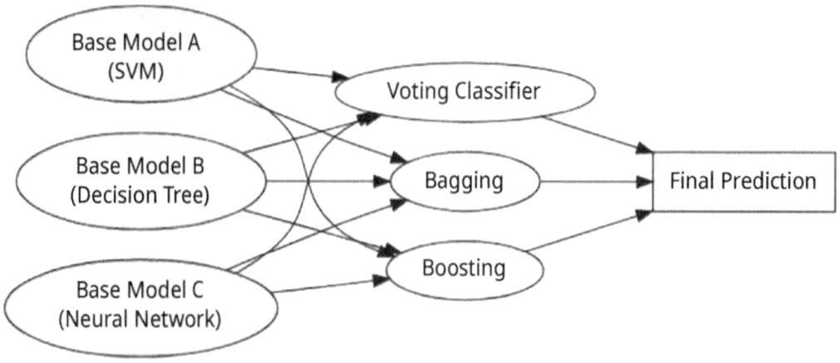

Figure 5: Ensemble learning techniques for robust predictions.

3.4.1 Boosting and bagging in biomaterial property prediction

Two well-known ensemble learning methods that have demonstrated significant effectiveness in biomaterial property prediction are bagging and boosting. Bagging, also known as bootstrap aggregating, is the process of training many models separately on various training data subsets, then combining their predictions, usually by averaging. This method lowers variance and increases the model's stability. On the other hand, boosting concentrates on training models in a sequential manner, with each new model aiming to rectify the mistakes made by the preceding ones. Biomaterial property predictions have been effectively improved with the use of techniques like AdaBoost and gradient boosting, which target hard-to-predict samples and iteratively enhance the model's performance. In the field of biomaterial research, both approaches lead to more consistent and trustworthy prediction results.

3.4.2 Stacking models for enhanced predictive performance

An ensemble learning method called stacking, also known as stacked generalization, integrates multiple models by teaching a meta-model how to optimally aggregate their predictions. Using this technique, a layered architecture is created, with base models, also known as level-0 models, trained on the original data and their predictions serving as inputs for a level-1 model, a higher-level meta-model. The meta-model then determines how best to integrate these forecasts to arrive at a conclusion. Stacking is a useful technique in biomaterial property prediction because it combines various types of models, including neural networks, decision trees, and regression models, to leverage their individual strengths and enhance overall performance. Using stacking to combine the complementary qualities of multiple models can improve prediction accuracy and resilience.

3.4.3 Hybrid approaches combining ML algorithms for biomaterials

Hybrid approaches that combine different ML algorithms can address the limitations of individual models and provide more comprehensive solutions for biomaterial property prediction. For instance, combining regression techniques with clustering methods can enhance the model's ability to capture both global trends and local patterns in biomaterial data. The same can be said about combining linear models with nonlinear ones, such as decision trees, to obtain more detailed insights into the material's behavior. These hybrid models can be tuned to different aspects of biomaterial study, including mechanical properties, by leveraging the complementary natures of the two algorithms. The development of such hybrid approaches can lead to improved predictive accuracy and better insights into biomaterial performance.

Although ensemble learning approaches have many benefits, there are still certain obstacles in their application to the prediction of biomaterial properties. The efficacy of ensemble approaches can be impacted by problems including computational complexity, interpretability of the models, and the requirement for large, high-quality datasets. Additionally, much thought and trial are needed to choose the best base models and determine the optimal combination approach. In order to overcome these obstacles, future ensemble learning research for biomaterials should concentrate on creating more efficient algorithms, enhancing the interpretability of models, and investigating novel approaches for fusing different models together. Large-scale biomaterial datasets and advancements in processing capacity further unlock the potential of ensemble learning approaches, resulting in more reliable and accurate prediction models in the field of biomaterial research.

3.5 Model interpretability and explainability

Especially in the context of biomaterial research, model interpretability is an essential component of ML. It's crucial to comprehend a model's methodology in order to guarantee that the outputs are trustworthy and useful in real-world scenarios. Interpretability offers information on the parameters influencing model predictions in the field of biomaterials, where choices about material composition and design can have a significant impact. Because of this transparency, researchers are better able to verify the model's predictions, identify possible mistakes, and have confidence in the decision-making process. Interpretability demonstrates that the model is based on reliable assumptions and pertinent facts, which promotes confidence among stakeholders and aids in the regulatory approval process.

3.5.1 Techniques for explaining machine learning predictions

There are several methods for improving ML models' interpretability, which make it clearer how predictions are made. Feature significance analysis is a frequently employed technique that pinpoints the most important variables impacting model results. Methods such as LIME and SHAP are types of techniques that inform the user about how various features impact the results in the model. Visualization methods such as individual conditional expectation plots and partial dependency plots can further demonstrate the influence of changes in certain attributes on predictions. When combined, these techniques allow for a more thorough comprehension of model behavior and make it easier to improve biomaterial design strategies based on model findings.

3.5.2 Ensuring transparency in AI-driven biomaterial design

Transparency in AI-driven biomaterial design is essential for building confidence in the use of ML models.

In research and development, transparent practices involve providing clear documentation of the model's architecture, training process, and data sources. Detailed descriptions of the model's assumptions, limitations, and potential biases also contribute to a comprehensive understanding of its functionality. This makes it possible to achieve heightened transparency, which ensures results are reproducible, thereby aiding peer review that is highly beneficial for the progression of the field. Additionally, the disclosure of the model's performance indicators and the results of its validation demonstrates the suitability of the model when applied to real-life contexts.

The second problem is how to achieve a balance between model complexity and interpretability?

Simplicity and interpretability of the models versus model complexity were especially an issue in ML for biomaterials. These can also be very precise, especially deep neural networks, but again, their architectures are very hard to understand. In turn, decision tree models are significantly easier to interpret but do not provide high accuracy levels. Appropriate bias, therefore, means choosing models that achieve reasonable accuracy without compromising the level of interpretability needed by the research. Interpretable complex models can also be adopted, in which both accurate and easily explainable models can be used to obtain high training and performance rates.

The future of model interpretability in biomaterials likely involves advancements in both methodologies and tools. Emerging techniques, such as XAI and integrated model-agnostic approaches, promise to enhance the ability to interpret complex ML models. The development of new frameworks for evaluating interpretability also contributes to more effective and user-friendly explanations of model predictions. As ML models become increasingly integral to biomaterial research, ongoing efforts to improve interpretability will be essential for ensuring that these models are both effective and understandable. Continued research in this area supports the broader adoption of AI-driven techniques and facilitates their application across various biomaterial design challenges.

4 Optimization of biomaterial properties using machine learning

4.1 Multiobjective optimization in biomaterial design

In biomaterial design, multiobjective optimization plays a critical role, as biomaterials often exhibit multiple interdependent properties that can be in conflict with one another. Enhancing one characteristic such as biocompatibility or mechanical strength may come at the cost of compromising another. Techniques like Pareto optimization provide a structured approach to navigating these trade-offs by identifying a set of non-dominated solutions that represent the best possible compromises across all objectives. Through the generation of a Pareto front, researchers can visualize the trade-offs and select optimal solutions that meet diverse design goals. This strategy supports a more holistic assessment of biomaterials, promoting the creation of materials that simultaneously satisfy multiple functional requirements. Figure 6 illustrates the concept of Pareto optimization in biomaterials, showing the balance achieved between conflicting objectives on a Pareto front.

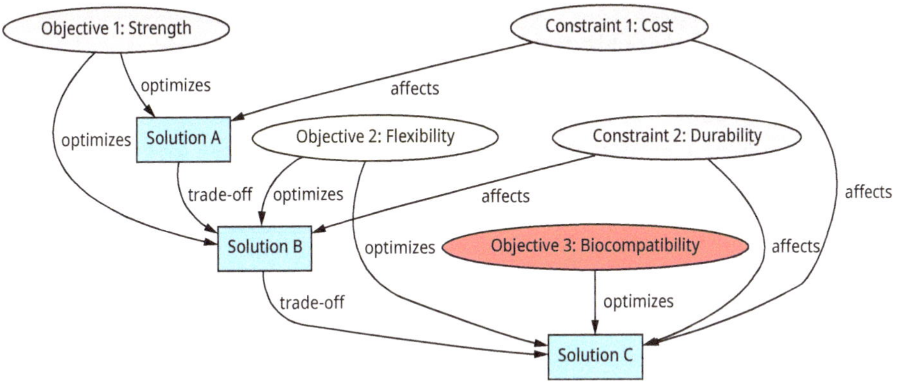

Figure 6: Multiobjective optimization in biomaterials design.

4.1.1 Pareto optimization in biomaterial research

Pareto optimization is a foundational method in multiobjective optimization, offering a way to handle scenarios where trade-offs are necessary. This technique determines the admissible region of solutions, in which no objective function can be optimized without the degradation of another. In the context of biomaterial design, Pareto optimization assists scholars in identifying materials that yield optimal compromises, thus avoiding the sacrifice of one attribute for another, such as strength, durability, and biocompatibility. As an optimization map, the Pareto front helps researchers choose materials depending on an application-associated set of requirements and outcomes. This method is effective in steering the synthesis of next-generation biomaterials that possess desirable characteristics.

4.1.2 A genetic algorithm approach to multiobjective biomaterial design

Hence, the use of genetic algorithms (GAs) for multiobjective optimization in biomaterials is advantageous because of their ability to search complex solution spaces. Like natural biology, GAs use selection, genetic crossover, and mutation in an attempt to evolve toward an optimum solution. For biomaterials design, GAs have the capability to solve multiobjective problems, including adjusting preferences in the search for solutions among a set of objectives. This iterative strategy makes it possible to identify new kinds of biomaterials that possess desired characteristics. Thus, Fischer et al. (2018) highlight the progress in materials' properties and their functionality. The characteristic features of GAs, which make them suitable for solving a wide range of optimization problems, establish them as a powerful tool in the biomaterials field.

4.1.3 Prospects for multiobjective optimization for biomaterials

Multiobjective optimization presents several challenges in biomaterials research, including computational complexity and the difficulty of accurately defining objective functions. Balancing multiple conflicting objectives often requires sophisticated algorithms and significant computational resources. Additionally, the accuracy of optimization results depends on the quality of the input data and the appropriateness of the objective functions. Addressing these challenges involves employing advanced optimization techniques, refining data quality, and ensuring that the objectives reflect realistic material performance criteria. Overcoming these hurdles is essential for achieving effective and reliable multiobjective optimization in biomaterial design.

4.1.4 Case studies of optimized biomaterial properties

Case studies of biomaterial qualities that have been optimized offer important insights into the use of multiobjective optimization approaches. Research has indicated that the mechanical and biological qualities of scaffolds can be enhanced for tissue engineering purposes through the application of GAs. In a similar vein, polymer-based biomaterials utilized in medical implants have had strength and flexibility balanced through the application of Pareto optimization. These case studies highlight how cutting-edge methods result in the creation of biomaterials with improved performance and compatibility for certain applications, demonstrating the useful advantages of multiobjective optimization. Researchers can better grasp how to use these techniques in their own work by examining successful instances.

4.2 Reinforcement learning for adaptive biomaterial design

RL offers a powerful approach for optimizing biomaterial design in dynamic systems. Unlike traditional supervised learning methods, RL algorithms learn optimal strategies through interactions with the environment and feedback from their actions. In the context of biomaterials, RL can be employed to adaptively tune material properties based on real-time performance data and evolving requirements. For instance, RL agents can be used to adjust parameters in the fabrication process of biomaterials to achieve desired properties such as enhanced mechanical strength. By continuously learning from the outcomes of their actions, RL models can identify optimal design solutions that may not be apparent through static approaches.

4.2.1 Training RL agents to optimize biomaterial properties

Training RL agents involves defining a reward function that aligns with the objectives of biomaterial design. This function quantifies the success of various actions taken by the agent, such as changes in material formulation. The training process typically includes interactions with a simulation environment setup, where the agent performs actions and receives feedback in the form of rewards. Through iterative trials and error correction, the RL agent refines its strategy to maximize the cumulative reward, leading to optimized biomaterial properties. Techniques such as Q-learning, policy gradients, and actor-critic methods are commonly used in training RL agents, with the choice of method depending on the complexity of the design problem and the available data.

4.2.2 Case studies: reinforcement learning in biomaterial engineering

Several case studies illustrate the successful application of RL in biomaterial engineering. For example, RL has been used to optimize the synthesis parameters of polymer composites to achieve specific mechanical and thermal properties. Elsewhere, adaptive control of the deposition process through RL was used to optimize the fabrication process in additive manufacturing, improving the quality and homogeneity of printed biomaterials. The RL capabilities revealed in these case studies include handling difficult optimization tasks and the ability to respond to changes in context – key features that suggest the applicability of RL to biomaterial design and manufacturing process improvement. All the case studies demonstrate the potential gains and challenges associated with the RL approach when applying it to actual biomaterial use cases.

4.2.3 Challenges in implementing reinforcement learning (RL) for biomaterials

Implementing RL for biomaterials presents several challenges including the need for accurate simulation models and the high computational cost associated with training agents. Developing reliable simulations that accurately reflect the behavior of biomaterials in various conditions is crucial for effective RL training. The well-known RL predicament of the exploration vs. exploitation dilemma often results in suboptimal solutions if the agent spends too long exploring the space before deciding on the most efficient action. Meeting these needs requires enhancing the simulation realism, making simulations shorter, and ensuring that RL algorithms are ready to adapt between exploitation and exploration if the goal is to reach good design solutions.

4.3 Bayesian optimization for efficient material search

Bayesian optimization is a powerful and efficient strategy for optimizing complex, costly functions, making it especially valuable in materials science for the discovery and refinement of optimal biomaterial properties. This method uses probabilistic surrogate models, often based on Gaussian processes, to estimate how a biomaterial will perform given specific input features. The central concept involves constructing a surrogate model of the objective function and using it to strategically explore the input space. With each iteration, Bayesian optimization updates the model using new data and concentrates the search on the most promising regions. This approach significantly reduces the number of required experiments while still identifying high-performing materials, thereby accelerating innovation and conserving resources. Figure 7 illustrates the Bayesian optimization workflow in biomaterial discovery, showcasing how it guides efficient exploration through iterative learning.

4.3.1 Application of Bayesian methods in biomaterial discovery

In biomaterial discovery, Bayesian optimization was utilized to systematically explore and identify materials with desirable properties. This approach was particularly useful when the objective function was expensive to evaluate, such as in the case of experimental material testing. The objective functions are applied to enhance more than one material property, such as mechanical strength, biocompatibility, and durability through Bayesian optimization. Using Bayesian optimization, an inherently stochastic model was formulated, which accounts for the randomness and variability of material characteristics to select the most suitable candidates for further experimentation in order to reduce the time it takes to arrive at the most effective biomaterials. The method was particularly suitable for the biomaterial design problem because it can moderate between exploitation and exploration within the high-dimensional space.

4.3.2 Case studies: Bayesian optimization in biomaterial property tuning

Several case studies demonstrate the successful application of Bayesian optimization in tuning biomaterial properties. For example, Bayesian optimization has been used to optimize the composition and processing conditions of polymer blends to achieve specific mechanical properties and biocompatibility. In another study, the technique was employed to enhance the performance of metal alloys by identifying optimal alloying elements and concentrations. These case studies illustrate the versatility and effectiveness of Bayesian optimization in addressing various challenges in biomaterial property tuning, highlighting its potential to significantly advance material design and development.

Some of the difficulties experienced in Bayesian optimization for biomaterials.

Nevertheless, there are a number of difficulties associated with using Bayesian optimization in the context of biomaterials. This is mostly due to the computational cost required in the probabilistic modeling and optimization steps as well as the potential difficulty and dimensionality of the material design space. In the same regard, a poor choice of the surrogate model means that the function evaluations used to train the Bayesian optimization may also be inadequate. Imagine having limited data; this reduces the accuracy of the predictions we make and results in a less comprehensive search of the design space. Meeting these challenges thus demands caution in model choice, the sources of data, and computational resources to make Bayesian optimization provide reliable and practicable solutions.

4.3.3 Applications of Bayesian optimization with other methods

To improve the efficiency of biomaterial design by using Bayesian optimization, compared with other optimization approaches, it usually needs to be used in conjunction with other similar methods. For instance, combining Bayesian optimization with GAs can be meaningful because each of the methods used is suitable for overcoming various aspects of the optimization problem. Domain knowledge and constraints considered in the framework can also be integrated with other approaches to improve the optimization results. When incorporating Bayesian optimization with other methods,

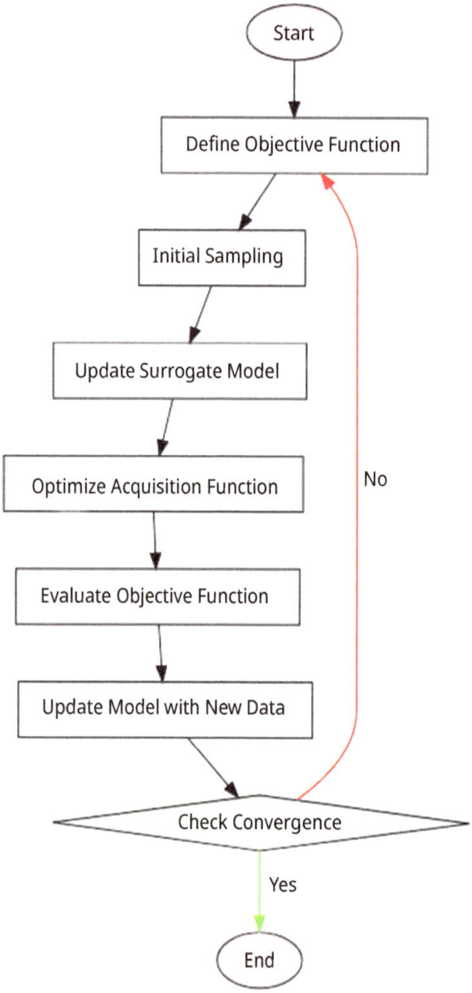

Figure 7: Bayesian optimization for efficient material search.

scientists and engineers can generate more exhaustive solutions for biomaterials and enhance biomaterial performance by almost 50%.

It was found that the next generation of Bayesian optimization in biomaterial research was on the brink of seeing significant progress in computational as well as in modeling strategies. Another emerging trend is the creation of more elaborate probabilistic computational models that simulate the properties and/or behavior of materials with a higher level of accuracy. Enhanced algorithms that reduce computational overhead and improve convergence rates are also anticipated. The integration of Bayesian optimization with high-throughput experimental platforms and advanced data analytics is likely to lead to more efficient and scalable material discovery processes. As these advancements continue, Bayesian optimization plays an increasingly pivotal role in the rapid and efficient development of novel biomaterials with optimized properties.

4.4 Evolutionary algorithms for biomaterial design

Evolutionary algorithms (EAs) are a type of optimization method based on the workings of various natural evolution processes. Some of the well-known metaheuristic algorithms include GAs, evolutionary strategies, and genetic programming (GP), all of which use techniques such as selection, mutation, and crossover to refine solutions. In the context of biomaterial design, EAs are employed to explore and optimize complex design spaces where traditional methods may be insufficient. By simulating evolutionary processes, these algorithms can effectively navigate the vast search space of material properties and fabrication parameters, leading to the discovery of novel biomaterials with desired characteristics. The adaptability and robustness of EAs make them particularly well-suited for solving multiobjective optimization problems in biomaterial science.

4.4.1 Application of genetic programming for biomaterial property optimization

A kind of EA called GP develops computer programs to address optimization issues. GP is used to create prediction models and optimization techniques for biomaterial property optimization that change and adapt in response to performance criteria. In GP, a starting population of feasible solutions is generated and subsequently modified repeatedly by genetic operations, including crossover and mutation. This methodology enables the identification of novel biomaterial compositions and processing procedures that may not be discernible using traditional design methodologies. GP is an effective method for customizing biomaterials to meet certain functional needs because of its adaptability in developing intricate models.

4.4.2 Integrating evolutionary algorithms into machine learning

This work is valuable for researchers in the biomaterials optimization field because the integration of EAs with ML makes the optimization process more effective. The utilization of evolutionary algorithms makes it possible to tune the hyperparameters of machine learning models, perform feature selection, and improve the accuracy of predictions. On the other hand, the use of ML models helps establish the direction of the EA and determine the fitness of prospective solutions much faster. This kind of integrated approach combines the exploratory analysis capabilities of evolutionary algorithms with the analytical power of ML. Examples of such integration include real-time control of biomaterial properties, design adaptation, and enhancing biomaterial discovery.

4.4.3 Case studies: evolutionary approaches in biomaterial engineering

Several case studies illustrate the successful application of EAs in biomaterial engineering. For instance, EAs have been employed to optimize the composition and structure of biodegradable polymers for medical implants, leading to improvements in their mechanical properties and degradation rates. Other studies have utilized EAs to design advanced composites with tailored thermal and electrical properties for electronic and structural applications. Additionally, they demonstrate how EAs can be used to solve intricate design problems and meet target material characteristics, as well as how EAs can likely spur creative biomaterial design by searching for nonconventional solutions.

Future enhancements in the use of EAs for biomaterial design will likely focus on improving the mechanism's efficacy and diversifying its potential uses. It is expected that future developments in parallel processing and related computational mechanisms will enhance the capability and scalability of evolutionary algorithms in computing, enabling them to address even more complex optimization problems. It is noteworthy that EAs are enriched through synergy with other modern technologies, such as quantum computing and advanced analytics. EAs will be essential in driving innovation and identifying novel materials with ideal characteristics as the field of biomaterials continues to develop. Sustaining research and development in this area is imperative to meet the increasing demands of contemporary biomaterial applications.

4.5 Hyperparameter tuning for ML models in biomaterials

Another critical stage in the model development pipeline for biomaterials is the selection of hyperparameters, which must be defined prior to the training process. These hyperparameters including learning rate, number of layers, regularization terms, and more have a profound impact on model performance, generalization, and predictive accuracy. In the context of biomaterials, where precision in property prediction and optimization is essential, the proper tuning of hyperparameters ensures that machine learning models are well-calibrated to deliver optimal results. By optimizing these parameters, researchers can significantly boost model accuracy and extract reliable, actionable insights for material design and application. Figure 8 illustrates the role of hyperparameter optimization, emphasizing its impact on model outcomes in the biomaterial domain.

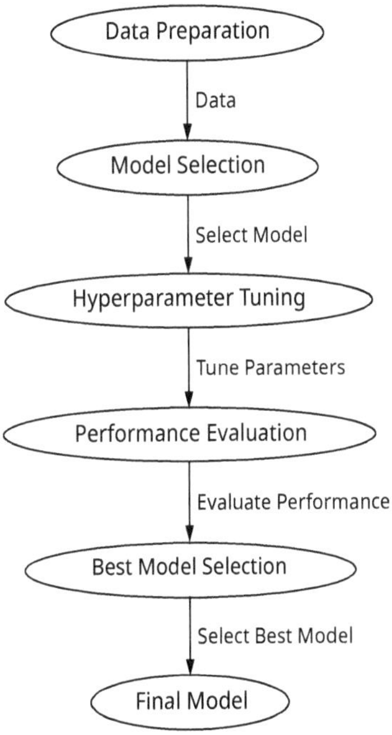

Figure 8: Hyperparameter tuning for machine learning models in biomaterials.

4.5.1 Techniques for hyperparameter tuning

There are several methods for fine-tuning ML models' hyperparameters that are applied to biomaterials. Grid search is a methodical process that tests every conceivable combination of hyperparameter values to determine the ideal configuration by evaluating a predetermined set of values. Contrarily, random search chooses hyperparameter combinations at random, which is more effective than grid search, particularly in high-dimensional environments. In previous evaluations, Bayesian optimization uses a statistical model of the objective function that guides the exploration toward promising areas of the hyperparameter space. In previous evaluations, Bayesian optimization brings a statistical model of the objective function that guides the exploration toward promising areas of the hyperparameters space. In the same regard, there are sophisticated hyperparameter tuning auto-optimization frameworks like Optuna and Hyperopt, among others, that allow for more advanced optimization algorithms because they possess inherent intelligence to optimize based on certain metrics. Altogether, these techniques help enhance the hyperparameter determination process and, therefore, improve the application of ML in biomaterials research. Altogether, these techniques help enhance the hyperparameters determination process and, therefore, improve the application of ML in biomaterials research.

4.5.2 Challenges in hyperparameter optimization for biomaterials

Hyperparameter optimization for biomaterials presents several challenges. The high dimensionality of the hyperparameter space, especially in complex models, can lead to increased computational demands and extended search times. The choice of hyperparameter values can interact in nonlinear ways, complicating the optimization process. Additionally, the need for domain-specific knowledge in biomaterial applications requires careful consideration of hyperparameters that influence model performance and interpretability. Balancing exploration and exploitation during the search process is crucial to avoid overfitting to specific datasets and ensure that the model generalizes well to new biomaterial samples. Addressing these challenges involves employing efficient optimization techniques and leveraging computational resources effectively.

4.5.3 Evaluation metrics for hyperparameter tuning

Finding the ideal configuration requires assessing ML models' performance during hyperparameter adjustment. Recall, accuracy, precision, and *F*1 score are examples of common assessment measures that shed light on the predictive power of the model. Regression models are evaluated for biomaterial applications using additional metrics like MSE to assess how well they predict material qualities. Cross-validation methods, such as *k*-fold cross-validation, are used to guard against overfitting and guarantee that the model performs consistently across various data subsets. Through a methodical assessment of these indicators, scientists pinpoint the hyperparameter configurations that result in optimal performance, guaranteeing the ML models' resilience and dependability for biomaterials research.

Subsequent developments in hyperparameter tuning of biomaterials have therefore aimed at optimizing the performance of optimization algorithms. The incorporation of AutoML frameworks often holds the potential to eliminate the occurrence of the hyperparameter tuning procedure by facilitating the automated identification of these parameters. In the same way, more extensive and faster hyperparameter searches have been made possible by recent improvements in computational power, including parallel processing and educational computers. Research into novel optimization algorithms, such as metaheuristic approaches and reinforcement learning-based methods, further improves the tuning process. As ML models become increasingly sophisticated and integral to biomaterial design, ongoing innovations in hyperparameter tuning will be essential for achieving optimal performance and advancing the field.

5 Machine learning in the fabrication and characterization of biomaterials

5.1 Predictive modeling for additive manufacturing of biomaterials

The use of predictive modeling was critical in the additive manufacturing (AM) of biomaterials because it predicts the effects of different fabrication processes. Conventional manufacturing processes are replaced with different methods of layered manufacturing, of which the most recognized is 3D printing. In predictive models, the details of biomaterials involved in the AM process are incorporated, as well as their properties, structure, and performance potential, broadly speaking, based on ML. These models play an important role in enhancing the process parameters, limiting the time spent experimenting with various possibilities, and aiding in the search for new types of advanced biomaterials for the medical and industrial sectors. Another use of the predictive modeling technique is to understand the impact of various factors on the final product, which enables researchers to achieve improved and optimized methods of fabrication.

5.1.1 Methods and strategies of predictive analytics

Finding the ideal configuration requires assessing ML models' performance during hyperparameter adjustment. Recall, accuracy, precision, and *F*1 score are examples of common assessment measures that shed light on the predictive power of the model. Regression models are evaluated for biomaterial applications using additional metrics like MSE to assess how well they predict material qualities. Cross-validation methods, such as *k*-fold cross-validation, are used to guard against overfitting and guarantee that the model performs consistently across various data subsets. Through a methodical assessment of these indicators, scientists pinpoint the hyperparameter configurations that result in optimal performance, guaranteeing the ML models' resilience and dependability for biomaterials research.

5.1.2 Accommodation of experimental data into machine learning currents

The combination of experimental information with data-scheduled AI methods allows for improving the accuracy and scope of prediction models in additive manufacturing. Process data, such as material aspects, working parameters, and the results of operations, are used to train and calibrate ML algorithms and models as well. This involves collecting experimental data, primarily for integration tasks, preparing this data in the right format, and ensuring its completeness to feed it into the training phase and build predictive models. Validation of these models with other sets of experimentation ensures the reliability of their distribution. In this way, the models can be created and refined until consensus is reached, and changes to the materials and fabrication techniques are incorporated. This work demonstrates how using empirical data and ML tools to generate additional forecasts could help improve additive manufacturing processes for biomaterials.

5.1.3 The issue in applying the various concepts of new structural dynamics for predictive modeling of additive manufacturing

There are challenges associated with using predictive modeling in additively manufactured biomaterials. One of the significant difficulties is the description of factors influencing the final properties of the material of interest, as well as the dependencies between process characteristics and final results. The scattered nature of experimental data, due to variations in materials and conditions, also affects the precision of predictive models. Due to the presence of multiple process parameters and numerous material characteristics as input variables, the input space is quite high-dimensional and thus leads to issues such as overfitting of the model. While attempting to solve these problems, it is necessary to address tasks such as the application of high-level ML methodologies, the selection of sufficient approaches for data collection, and the constant verification of the model to determine whether the created predictive models can be useful in real life.

It was probable that future research about predictive modeling in AM of biomaterials would depend on improvements in specific ML algorithms, as well as computational methods. New approaches that are currently at the prestudy level and which include deep learning and hybrid models hold the key to enhancing both the accuracy and robustness of the predictions, since the models are able to capture additional data complexity beyond the linear model. The combination of data acquisition and the usage of prognosis models allows real-time alterations in the fabrication procedures, thereby improving the flexibility and productivity of AM. Fourth and finally, the richer sources of data that are available, utilizing a broader spectrum of materials as well as the conditions under which manufacturing processes are carried out contribute to building less-specialized, all-encompassing models. In the process of further development of additive manufacturing as a tool for creating high-performance structures, further research into predictive modeling of processes is essential for enhancing the potential of biomaterials in various sectors.

5.2 Machine learning in biomaterial surface characterization

A key component of biomaterial research is surface characterization, which provides essential insights into the chemical and physical properties of material surfaces factors that critically influence biological interactions and functional performance. Traditional analytical methods such as Scanning Electron Microscopy (SEM), Atomic Force Microscopy (AFM), and X-ray Photoelectron Spectroscopy (XPS) are widely used to examine surface topography, roughness, and chemical composition. These techniques generate detailed information about surface morphology and elemental distribution, which is crucial for understanding how biomaterials interact with biological tissues. The high volume and complexity of the data produced by these methods present challenges for data interpretation and analysis. Figure 9 illustrates the common surface characterization techniques and their role in mapping the structural and chemical features of biomaterial surfaces.

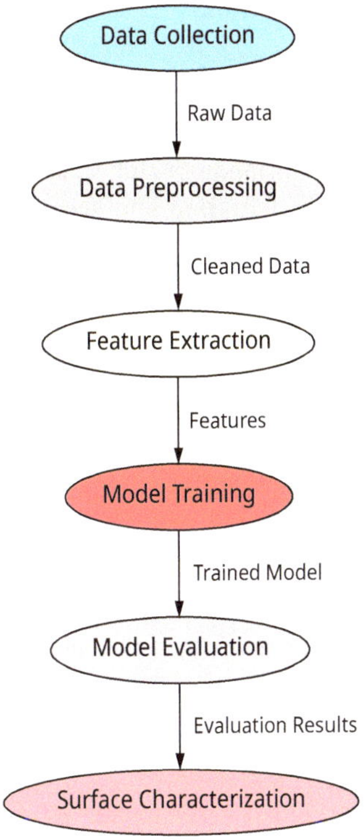

Figure 9: Machine learning in biomaterials surface characterization.

5.2.1 A coupling approach of machine learning with surface characterization methods

ML, when combined with surface characterization techniques, provides a better understanding of the data being obtained. Semantic convergence with other resolutions, like SEM, can be conducted using a CNN model. These algorithms are well capable of scanning data and attribute maps in large dimensions, allowing for higher definition and precision in surface characterization. For example, CNNs can automatically detect and quantify surface roughness, pore size, and texture, which are critical for assessing biomaterial suitability for specific applications. Similarly, SVMs can classify different surface chemical states based on XPS data, facilitating the identification of surface modifications and functionalization.

5.2.1 Enhancing resolution and accuracy through machine learning

Both the resolution and accuracy of surface characterization were also improved through ML, where sophisticated algorithms are applied to isolate and magnify features that are difficult to identify with the naked eye. Certain methods, for instance, image super-resolution and anomaly detection, enhance the definition and accuracy of surface images, thus improving the measurement of features on the surface. Other parameters of the surfaces that are hardly considered in the classificatory schemes used in the conventional approach can be artificially incorporated into the ML models. This capability is particularly valuable for studying nanoscale features and subtle surface modifications, which are crucial for applications in drug delivery, tissue engineering, and biomaterial development.

5.2.2 Applications in biomaterials surface modification and optimization

ML was instrumental in optimizing biomaterial surface properties by analyzing the effects of various surface modifications and treatments. For instance, ML algorithms, as statistical and structural models, can be applied for modeling/estimating the effect of such coatings, roughness, and texture on the biomaterials. These patterns fall into three categories: topography, chemistry, and mechanics; and by associating them with functional consequences such as cell adhesion and protein adsorption, the ML models can determine which characteristics are necessary for an ideal biomaterial for a particular task. It has applications in enhancing the performance and biocompatibility of a broad range of biomaterials for medical devices, implants, and tissue engineering.

5.2.3 Current issues and trend analysis of machine learning in surface characterization

There are still some difficulties with the use of ML in the characterization of biomaterial surfaces. Implementation challenges include data quality, variability in measurements on the epithelial surface, and the requirement for large datasets accompanied by annotations. Meeting these demands entails continuous research into improving the best algorithms to implement, optimizing techniques for data acquisition, and harmonizing protocols for characterizing surface features. Future work involves the use of AI in connection with new characterization techniques that are continuing to develop, such as high-throughput imaging and real-time analysis for the advancement of surface characterization. I conjecture that further developments in the fields of ML as well as surface sciences are likely to enable biomaterial scientists to discover new materials and create the next-generation biomaterials with superior surface characteristics.

5.3 ML-driven techniques for mechanical property prediction

Modern biomaterials science has made significant advances in various areas of material application, including the prediction of mechanical properties through the use of ML technologies. These techniques utilize feed-forward neural networks to estimate a range of mechanical characteristics, such as tensile strength, hardness, and elasticity, among others, based on input features derived from material composition, processing parameters, and structure. ML, as a subdiscipline of AI, is capable of analyzing large datasets of experimentally measured properties along with the features used for their determination – patterns that conventional analytical methods often fail to reveal. This approach enhances the suitability of biomaterial creation and synthesis to meet specific characteristic demands.

5.3.1 Regression models in mechanical properties prediction

In most ML applications, regression models are used to estimate the mechanical properties of biomaterials. Simple interpretability is provided by linear regression models, which create a direct link between input characteristics and mechanical parameters. On the other hand, nonlinear interactions are captured, and prediction accuracy is increased by using more sophisticated regression techniques like support vector regression and polynomial regression. Multiple regression models are used in ensemble approaches, such as random forests and gradient boosting machines, to improve prediction performance and resilience. To provide precise forecasts of mechanical characteristics, these techniques take into account a wide range of input features, such as the composition of the material, the processing conditions, and the structural factors.

5.3.2 Neural networks toward enhanced mechanical property prediction

ML methods, especially deep learning architectures, have become promising techniques for predicting the mechanical properties of biomaterials. CNNs excel at processing data with spatial relationships, such as images of material microstructures, to predict mechanical properties. RNNs and LSTMs can model sequential data, such as processing histories, to forecast how material properties evolve over time. The ability of neural networks to learn complex, hierarchical features from large datasets enables more accurate and nuanced predictions of mechanical properties. Transfer learning, which involves fine-tuning pretrained models on specific biomaterial datasets, can further enhance the performance of neural networks in this domain.

5.3.3 Integrating multimodal data for enhanced prediction accuracy

ML algorithms for mechanical property prediction are more accurate and resilient when multimodal data sources are included. Integrating information from other modalities, such as simulation output, experimental measurements, and material characterization methods, provides a more thorough knowledge of the variables affecting mechanical properties. Heterogeneous data types are integrated using strategies like data fusion and multiview learning, which improve the model's prediction performance and capacity to capture intricate connections. ML algorithms can provide more precise forecasts and better guidance for the design and optimization of biomaterials by utilizing a variety of data sources.

Even with the progress made in ML to forecast mechanical qualities, there are still a number of obstacles to overcome. The amount and quality of data used to train models can affect how accurate the predictions are; noisy data produces incorrect findings. Deep neural networks and other complicated models are difficult to interpret, making it challenging to comprehend the fundamental principles that underlie predictions. The creation of increasingly complex models with enhanced interpretability and the ability to handle a variety of data types is one of the future research objectives. Technological developments in domain-specific ML methods, such as physics-informed neural networks, can provide fresh possibilities for fusing data-driven methods with mechanistic insights. As ML continues to evolve, addressing these challenges will be crucial for advancing the prediction of mechanical properties in biomaterials and enabling their effective design and application.

5.4 Real-time monitoring and control in biomaterial fabrication

Real-time observation during the manufacturing of biomaterials is crucial for ensuring the precision, consistency, and quality of the final products. By actively monitoring fabrication processes such as 3D printing, electrospinning, and additive manufacturing, researchers and engineers can detect anomalies or defects early in the process that might otherwise compromise material properties and performance. This real-time feedback enables immediate adjustments, reducing waste and enhancing reproducibility. As a result, biomaterials produced under monitored conditions exhibit greater reliability, uniformity, and application effectiveness. Figure 10 illustrates a real-time monitoring system in biomaterial fabrication, highlighting the integration of sensors and feedback loops for defect detection and quality control.

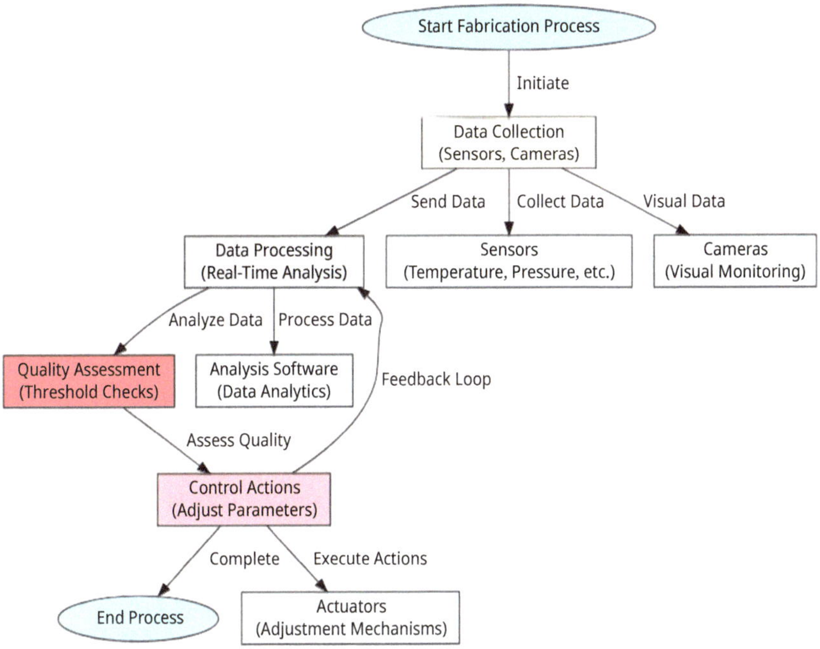

Figure 10: Real-time monitoring and control in biomaterial fabrication.

5.4.1 Methods of real-time data gathering

There are several methods that are used to collect data in real time during biomaterial fabrication. Optical sensors, temperature sensors, and pressure sensors provide information on various aspects throughout the fabrication process. Using optical microscopes and high-speed cameras, the structural evolution and overall morphology of the materials are recorded. Furthermore, data acquisition systems and control algorithms facilitate the integration of sensor data with processing devices used in fabrication to control processes as they occur. The application of these techniques ensures a thorough check on the fabrication process, thereby paving the way for high-quality biomaterials.

5.4.2 Uncertainty in combining machine learning into manufacturing process enhancements

Real-time monitoring and control of biomaterial fabrication significantly depend on the application of ML algorithms. Depending on the data gathered from the sensors and imaging systems, experts apply ML algorithms to establish an understandable relationship between the different parameters and the processes taking place in the plant. Models can also predict problems that may occur in the future, allowing changes to be made in advance to maintain optimal fabrication. Weighted based on the feedback of information gathered from real-time data, reinforcement learning approaches are capable of constantly adjusting the process parameters to enhance the quality of the material. When used together with monitoring systems, ML provides much better control of fabrication procedures, thereby producing superior biomaterials.

5.4.3 Issues related to the deployment of RTMS

There are observed challenges in incorporating real-time monitoring systems in the fabrication of biomaterials. The complexity of fabrication processes, based on the nature of targets (structural and other) and the variability in the parameters to be monitored, presents a major challenge to designing an adequate sensor network as well as the architecture and protocols for data integration. Real-time data analysis becomes very challenging in situations where a huge amount of data is collected concurrently. Additionally, improving the accuracy of sensors and handling data is a technical and computational problem since more data generated from multiple sensors requires more significant processing power. Remediation of these challenges requires the employment of sound data acquisition techniques, sophisticated signal processing, and competent techniques of data control in order to facilitate adequate and effective real-time data control.

Enhancing the accuracy and capacity of the monitoring models is most likely to be the main focus of further development in real-time monitoring and control for biomaterial synthesis. A better and higher level of examination of manufacturing processes should be achievable with such enhancements in probe devices and data compilation methods. Here, it is impossible not to underline the profound positive impact of ML and AI on the improvement of process optimization and, in particular, on forecasting capacities. In addition, smoother and more scalable monitoring solutions can be generated by expanding real-time monitoring with Industry 4.0 concepts like cloud computing and the internet of things. Further advancements in the monitoring and control of biomaterial fabrication should be possible as fabrication technology continues to evolve to enhance the real-time control of material synthesis.

5.5 Integration of ML with characterization techniques

Characterization techniques are fundamental in the investigation of biomaterials' properties and performance. These techniques include methods for evaluating many properties discussed in this chapter such as mechanical, thermal, and chemical properties as well as microstructures. Typical characterization techniques include SEM for observing the morphology of the formed surface features, XRD for identifying the crystalline structure of the material, and AFM for I/O characterization at the nanometer scale. Additionally, methods like Fourier-transform infrared spectroscopy and thermal characterization (e.g., differential scanning calorimetry (DSC)) are used to explain chemical content and thermal characteristics. The combination of these characterization techniques with ML can provide deeper insights into biomaterials by obtaining and utilizing high-powered data sets.

5.5.1 Methods for improving data analysis based on machine learning

Using biomaterial characterization techniques, various methods can be applied from machine learning (ML) to analyze and track the obtained data. Conventional techniques provide inadequate solutions when it comes to the analysis of large and complicated data sets, while ML approaches are well-suited to the analysis of large-scale data for pattern and relationship detection. For instance, with supervised learning, it is possible to train learning algorithms to predict the properties of materials from characterization data, and by applying an unsupervised learning approach, it is possible to discover the structures and clusters in the data. Analyses like PCA and t-SNE are often employed to reduce the dimensionality of a complex dataset; insights derived can be easily obtained from these techniques. By integrating ML with characterization techniques, researchers can expand the capability of handling large datasets, increasing the probability of accurately predicting the behavior of biomaterials and exploring more about them.

5.5.2 Forecast analysis and property enhancement

Predictive models that can anticipate material qualities and performance based on characterization data are created by integrating ML with characterization techniques. These models forecast new biomaterial samples based on sophisticated algorithms and historical data. Regression models, for example, use microstructural data to predict mechanical characteristics, while classification algorithms can distinguish various types of materials based on their spectral fingerprints. Through the identification of optimal property combinations and design guidance, the predictive capabilities of ML models aid in the optimization of biomaterials. This method not only expedites the creation of novel materials but also enhances the effectiveness of methods for material optimization and selection in scientific and industrial settings.

5.5.3 Real-time data integration and analysis

The integration of ML with characterization techniques also supports real-time data analysis and decision-making. Modern characterization methods often produce data at high frequencies, and ML algorithms can be employed to process and analyze this data in real time. For example, real-time monitoring of material properties during fabrication processes can be achieved using ML models that continuously analyze sensor data and adjust processing parameters as needed. This capability enhances control over material properties and enables adaptive optimization during the fabrication process. Real-time data integration and analysis are crucial for applications requiring precise material specifications and quality control, significantly improving the efficiency and effectiveness of biomaterial development.

6 Ethical considerations and future directions

6.1 Ethical implications of AI in biomaterial design

The integration of Artificial Intelligence (AI) into biomaterial design introduces significant ethical challenges, particularly regarding the transparency and accountability of the design and decision-making processes. In addressing the second research question, the study highlights the difficulty in understanding how AI models arrive at decisions, especially when using complex, opaque algorithms. Therefore, it becomes essential to establish transparent documentation of AI processes, including the data sources, model architectures, and decision pathways used during biomaterial development. Researchers are urged to adopt frameworks for evaluating the implications of AI-generated decisions on the accessibility, effectiveness, and safety of biomaterials. This reinforces the importance of synthesizing and applying ethical standards and principles to guide the responsible and trustworthy deployment of AI in this domain. Figure 11 illustrates the ethical framework for AI in biomaterial design, emphasizing transparency, model traceability, and accountability in the decision-making pipeline.

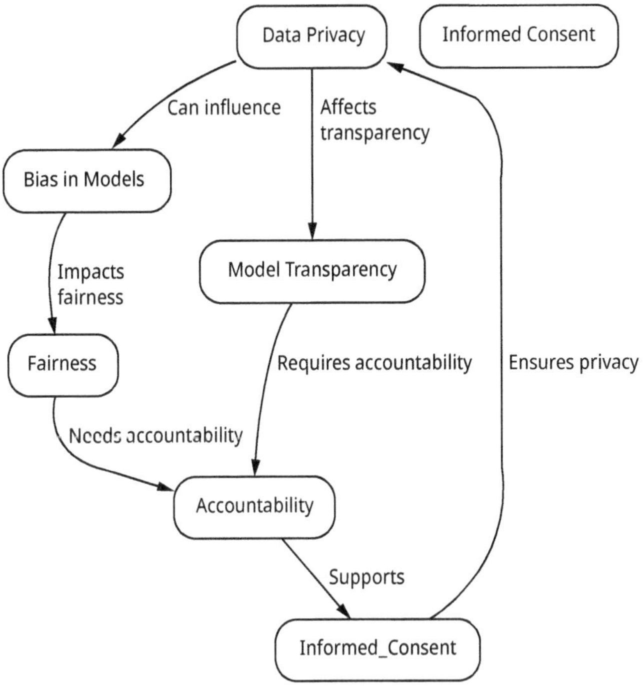

Figure 11: Ethical Implications of AI in biomaterials design.

6.1.1 Addressing bias and fairness

Bias in AI models poses a critical ethical concern, especially in the context of biomaterial design. Bias can arise from various sources, including imbalanced datasets, historical inequalities, and algorithmic limitations. In biomaterial applications, biased AI models lead to unequal performance across different demographic groups. To address these issues, it is necessary to use techniques to collect diverse and inclusive data samples, as well as to employ bias detection and reduction methods. Optimization of biomaterials demands constant monitoring of AI models and efforts to eliminate discrimination risks, ensuring equal opportunities to benefit from the development of new technologies.

6.1.2 Privacy and data security protection

Private health information and material data are key elements that are often collected and processed in connection with the application of AI in biomaterial design. The final ethical concern is the confidentiality and privacy of information. With ever-increasing technological advancements, the privacy and security of information call for maximum precaution to avoid manipulation. Privacy minimization measures include data encryption, limiting physical and digital access to the data, and employing anonymization techniques to minimize the chances of privacy breaches and data extrusion. Additionally, legal requirements that must be followed in the ethical management of data in AI-driven biomaterial research and development include HIPAA and GDPR.

Innovations are not intrinsically good in and of themselves; the objective that underlies this policy is to promote responsible innovation.

In keeping with the RI principles, proper assessment of the social and environmental impact potentials of new technologies was taken into account. Some of the ethical questions are: What are the biomaterial's implications for human beings' overall health? What are the biomaterials' consequences for the environment? And are there any adverse ramifications associated with the universality of biomaterials? The above themes define RI, and these concerns are best handled by ethicists, policymakers, and stakeholders to shape the future of beneficial technology. This is why it is necessary to conduct risk analysis and encourage constructive discussion of the consequences of AI usage to guarantee the optimal correspondence of biomaterial design advancements to ethical standards and, therefore, to the general good.

The future ethical considerations in biomaterial design must be addressed as AI technology advances. New ethical issues emerge alongside the new technologies that serve as constant tools in the modern world, including synthetic biology and complex ML algorithms. Two issues cited are the continual research on potential problems, ethical review processes, and the formation of adaptive guidelines that can accommodate future challenges posed by technological advancements. AI technology can only be utilized responsibly and for the good of society if the biomaterial research community stays ahead of potential problems and cultivates an ethically conscious culture.

6.2 Challenges in data privacy and security

Issues pertaining to data security and privacy are significant when using ML in biomaterials research. Dealing with huge amounts of personal and confidential information, including research outcomes, inventions, and personal data, is a routine process when incorporating complex analytics and ML solutions. To ensure that such data is not leaked, it becomes essential to ensure the privacy, availability, and integrity of such a data set. Since a large amount of data is collected and analyzed, and systems are increasingly interconnected, data privacy and security become more challenging. Solving such problems is highly important to restore people's trust and ensure compliance with legal requirements.

6.2.1 Information security needs to be safeguarded in biomaterials research

In biomaterials research, patient information, the details of the experiment, and the composition of the materials require protection. Since the management of PII and PHI is governed by strict data protection laws and regulations, which offer a standard that is higher than that of ITIL, researchers must adhere to regulations such as HIPAA and GDPR. Reduction of privacy threats is possible if data is kept de-identified before analysis is carried out on it. In addition, there is always a need to limit access opportunities, employ effective types of encryption, and ensure the availability of secure methods for storing information in order to avoid leakage, breaches, and illicit appropriation.

These set measures guarantee that the data is always kept safe, with integrity and accuracy at all times.

Data management was identified as a significant challenge when employing ML for biomaterials because data must be both accurate and complete. Inaccurate or incomplete data can lead to mistakes in model generation and result in incorrect decisions due to the use of compromised data. A considerable improvement in preparing and consolidating data correctness involves implementing validation and verification methods. Measures such as data checksums, version control, and audit trails can detect and mitigate unauthorized alterations. Effective data management methods must also be established to ensure data validity, and essential security measures, such as security audits, should be applied.

6.2.2 Data break-ins and cybersecurity risks

The threats to the security and privacy of any data collected in biomaterials research are significantly compounded by cybersecurity dangers and data leaks. Schemes that threaten identity and disrupt study processes include ransomware and phishing. To combat these attacks, preventive measures must be employed, ranging from applying advanced technologies, such as fire walls, IDS, and MFA. In addition, the development of robust incident response frameworks and ensuring their implementation serve as evidence to businesses that they can effectively tackle and mitigate the impacts of such breaches. Such changes require security processes to be updated and checked periodically, as new threats continue to challenge strong data protection.

6.2.3 Preserving the equity between the possibility of accessing data and the need to protect it

One of the main difficulties when working on biomaterials was how to make the data both more available and more protected at the same time. One couldn't help but agree with the statement that particular sources are needed to provide researchers and potential stakeholders with access while avoiding security threats. Technological measures in the form of role-based access controls, data segmentation, and secure sharing platforms help achieve this balance, ensuring that data is shared appropriately. Additionally, data accessibility, which has been a root cause of most data risks, can be addressed by implementing data governance best practices and entering into secure data-sharing agreements to manage the risks associated with data accessibility.

As for the prospects, it is conjectured that both technological and legal initiatives will develop data privacy and security for biomaterials research. The increase in the use of new and sophisticated technologies, such as blockchain to support data validity and tougher encryption algorithms to enhance data security, are part of the solutions being exhibited globally.

Specifically, the following cannot be overlooked: the development of extensive frameworks that provide solutions for privacy issues and the application of regulations and standards as key factors that contribute to the management of data security concerns. Therefore, it is crucial to establish that the new ongoing process of improving data protection and adapting to new threats is essential for maintaining the privacy of data and the ethical use of ML in biomaterials research.

6.3 Addressing the challenges of model generalization

The integration of Artificial Intelligence (AI) into biomaterial design introduces significant ethical concerns, particularly regarding the transparent documentation of the design process. In line with the second research question, a key challenge lies in understanding how AI models make decisions, as many operate through complex and non-intuitive algorithms. To address this, a sufficient level of transparency is essential detailing not only the data sources used to train the models but also the algorithmic structures and reasoning mechanisms behind their outputs. Researchers are encouraged to implement systematic methods to evaluate and respond to the consequences of AI-driven decisions on the availability, effectiveness, and outcomes of biomaterials. The study underscores the importance of developing and adhering to robust ethical standards and principles that promote the reliable, explainable, and trustworthy application of AI in biomaterial development. Figure 12 presents an ethical oversight framework that emphasizes transparency, accountability, and responsible AI practices in the context of biomaterial design.

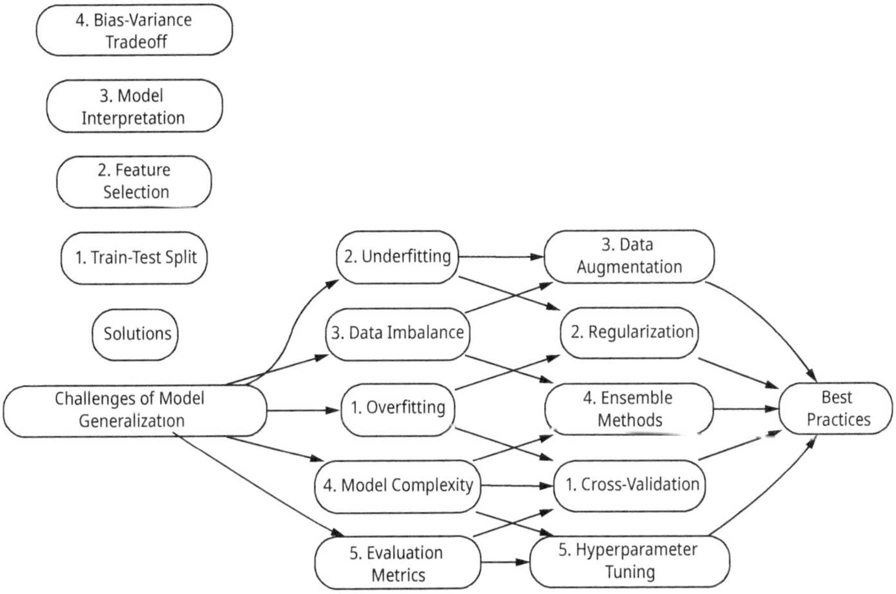

Figure 12: Addressing the challenges of model generalization.

6.3.1 Challenges in achieving effective generalization

Several challenges arise when striving for effective model generalization in biomaterials. One critical difficulty is the presence of fluctuations in the values, which are due to material properties, characterization techniques, and test parameters. This causes overfitting, a situation that makes a model fit well with the training data but perform poorly with unseen data. Another problem is the rising dimensionality, which causes issues with the ability to find important dependencies and possibly generalize from one dataset to another. The presence of noise in the data set can make generalization an even more challenging process. Addressing these challenges requires employing strategies that improve model robustness and ensure reliable performance across various scenarios.

6.3.2 Strategies for improving model generalization

Several tactics are used to enhance model generalization. By imposing penalties on excessively complicated models, regularization approaches like L1 and L2 regularization can aid in the prevention of overfitting. The capacity of the model to generalize is improved by cross-validation techniques, such as *k*-fold cross-validation, which offer a stable framework for assessing model performance on various data subsets. By adding variations, data augmentation approaches can broaden the training dataset's diversity and aid in the model's ability to adapt to various settings. To increase overall performance and stability, ensemble approaches like bagging and boosting integrate predictions from several models. By putting these tactics into practice, one can guarantee a model's efficacy in real-world biomaterial applications and improve its generalization capabilities.

6.3.3 Addressing data quality and representativeness

Achieving successful model generalization requires ensuring representative and high-quality data. Problems with data quality, such as errors and discrepancies, negatively impact the model's capacity to generalize. It is vital to guarantee that the training data is precise, all-inclusive, and representative of the intended purpose. To ensure that the model is exposed to a variety of situations and settings, it is also necessary to take the representativeness of the data into account. This entails populating the dataset with a variety of material types, processing settings, and experimental configurations. Thorough dataset design, consistent source validation, and stringent data cleaning techniques are ways to improve data quality and representativeness.

6.4 Bridging the gap between theory and practice

This raises some challenges in the transformation from theoretical models to practical implementations in biomaterials. Theoretical models, which are used to design practical systems, are in many cases derived by making certain assumptions and considering ideal conditions; therefore, they do not always incorporate all the difficulties encountered when the system is implemented. Some of the problems influencing its practical use are fluctuations in the nature of the material, the conditions under which a part is used, and the limitations of the production process. For example, theoretical models propose certain desired characteristics for a material, but real behavior is affected by spatial imperfections and interactions with living organisms. Overcoming these differences calls for empirical verification and, to a certain extent, the translation of the theoretical models into practice, taking into account the real demands and circumstances.

There was a strong call to incorporate the findings and practices of research with industry practices.

In order to more effectively establish the link between theory and best practice, it was therefore necessary to strengthen the cooperation between theorists in the academic environment and applied practitioners. It was deemed appropriate to state that theoretical knowledge has to be turned into practice-based knowledge by creating mechanisms for industrial research at research institutions and other similar stakeholders. Possibilities for cooperation can include, for example, common projects and studies, as well as initial investigations and mutual trials in which the models and hypotheses are tested and adjusted for applied practice. Another advantage of industry partnerships is that researchers are able to gain a deeper understanding of practical issues and what is needed to solve them, helping them to build better models. Promoting ways of building frameworks that allow for the transfer of information from theory to practice can enhance the process of applying theories to solve practical problems.

6.4.1 Ensuring reproducibility and scalability

The theoretical models had to be reproducible and scalable in order to be used in practical applications. Models used have to be verified under different circumstances and at varying capacities for efficiency and dependability. Reproducibility means performing operations and demonstrations under somewhat different circumstances in an endeavor to ensure that the theoretical behaviors remain valid in other circumstances as well. Scalability reflects the capacity of models to be applied on a larger scale with some loss of their efficiency. In order to ensure that such models can be implemented in the real world, it becomes important to establish standard checklists and specifications for such practices.

6.4.2 Ethical issues: their application in the resolution of real-life problems

The ethical considerations play a central role in bridging the gap between theory and practice. This means that all applications of theoretical models should be undertaken with an awareness of ethical concerns, such as health and safety. It was recommended that strong measures for ethical clearances be implemented to determine the ramifications of applying theoretical models in practical environments. Furthermore, the practice of stakeholder engagement and public consultation helps reveal ethical difficulties, ensuring that solutions are compatible with public ideals. It is critical to strike a positive balance between innovation and applied ethics as the foundation for more reliable biomaterials development.

As for the future of the work on narrowing the gap between theoretical and practical applications, more attention should be paid to the combined work of different disciplines, the improvement of the ways in which the created models are validated, and the creation of more adaptive models. A blurring of lines between traditional disciplines like material science, engineering, and computer simulations enhances solutions beyond pure practice-theory integration paradigms. Better methods for validation, high-throughput experimentation, and real-time monitoring enhance the practicality of theoretical formulations. Furthermore, the creation of more complex models with the ability to adapt to different conditions and applications improves their usability in practice. Their continuous study and development make it easier to implement theory that has been derived from real-life solutions.

6.5 Future trends in machine learning for biomaterials

The future of machine learning (ML) in biomaterials research stands to gain substantially from advances in algorithmic methodologies. Emerging techniques such as deep reinforcement learning (DRL) and neural architecture search (NAS) are set to revolutionize the modeling of complex biomaterial behaviors and the optimization of material properties. DRL introduces a feedback-driven, adaptive design paradigm, where models learn and evolve in response to experimental outcomes, enabling more intelligent and responsive material development. Meanwhile, NAS facilitates the automatic discovery of optimal neural network architectures, leading to enhanced model accuracy and computational efficiency in predicting biomaterial characteristics. Together, these innovations are expected to dramatically improve the precision, scalability, and impact of ML applications in the field. Figure 13 illustrates the integration of advanced ML algorithms like DRL and NAS in the biomaterial development pipeline, highlighting their role in accelerating discovery and enhancing performance.

6.5.1 Integration of multimodal data

Integration of multimodal data sources was anticipated to be a future advance in ML for biomaterials. A more thorough knowledge of biomaterials will be possible by combining information from many characterization methods, such as spectroscopy, microscopy, and mechanical testing, with computational models and experimental findings. The capacity to develop more reliable and generalized ML models, which can predict material qualities under a variety of circumstances and applications, will be improved by multimodal data integration. The successful integration and analysis of many data types will be made possible by methods like data fusion and multiview learning, which produce more precise and useful insights in the field of biomaterials research.

6.5.2 Enhanced interpretability and explainability

In the realm of biomaterials, it was anticipated that the trend toward improved interpretability and explainability of ML models would continue. Understanding ML models' decision-making processes was crucial as they grew in complexity. Improvements in XAI approaches, such as tools for model visualization and feature attribution, made it easier to understand how models make predictions. The validation and adoption of ML models in crucial applications, like clinical usage and regulatory clearances, would be aided by this greater openness. Researchers would be able to learn more about the underlying connections between biomaterial characteristics and design parameters thanks to enhanced interpretability, which helps them make better decisions and create better materials.

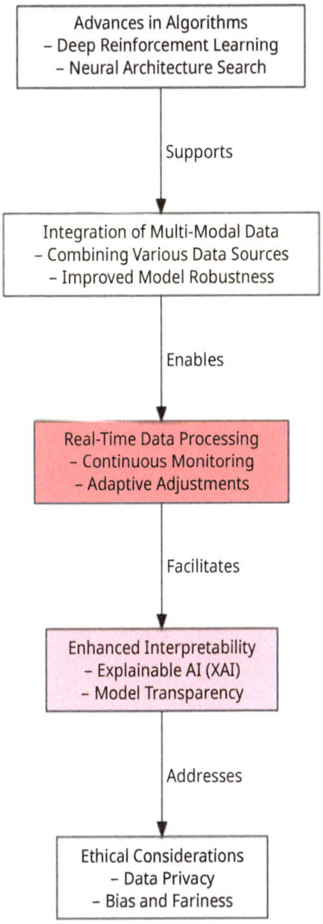

Figure 13: Future trends in machine learning for biomaterials.

6.5.3 Real-time data processing and feedback

Real-time data processing and feedback mechanisms are increasingly being used in biomaterials applications of the future. Continuous monitoring of material characteristics throughout manufacture and testing is made possible by developments in sensor technology and data collection systems. Real-time analysis-capable ML models enable instantaneous modifications to design parameters and fabrication procedures, hence enhancing material performance. This feature is especially helpful in high-precision applications where it's crucial to preserve certain material properties. The manufacturing and research of biomaterials become more productive and efficient when adaptive ML models are combined with real-time data processing.

The ethical and regulatory issues for this project include the following:

Ethical and legal aspects play an even greater part as machine learning (ML) develops in the biomaterials industry. This means that the promotion of responsible use of ML calls for approaches regarding data privacy, model interpretability, and bias. It can be argued that the formation of durable ethical rules and legal regulations is critical for biomaterials research with the help of ML and for defining what is acceptable from a technological point of view while corresponding to the culture of society. This makes it necessary to have constant dialog among researchers, regulators, and other stakeholders to address these challenges and enhance the ethical use of ML technologies in the creation and deployment of biomaterials.

Some of the subthemes included collaborative and interdisciplinary research.

A promising future direction in the development of biomaterials using ML involves the integration of ML into larger collaborative and interprofessional frameworks. This integration of systems requires facets of materials science, computer science, and engineering to solve sophisticated issues and create advancements. It is the cooperation of researchers, industries, and universities that helps in the advancement of new ML methods and their application to biomaterials. Multidisciplinary research improves the capacity to solve complex problems and enables the implementation of developed ML solutions in biomaterials engineering.

7 Conclusion

The combination of biomaterials science with the automotive industry was a major innovation with applications to material selection, design, and property predictions as well as other relevant fields such as medicine and tissue engineering. Additionally, this chapter has outlined how machine learning is evolving the biomaterials research and application field and its implications for the material science domain.

Out of all the different applications in the field of ML, one of the key benefits has been observed in property prediction, where algorithms have helped to increase precision and the rate of success. Some of the highly developed approaches include neural networks and regression, where the behavior of materials under certain conditions is well predicted. These predictive capabilities are invaluable in the process of designing and creating novel biomaterials to be used in new applications for various purposes, which means that the biomaterial innovation cycle can be accelerated and much of the guesswork eliminated. As a result of ML, researchers are now confident in designing and producing materials that conform to multiple performance standards.

Also, the coupling of ML with characterization techniques has transformed the means of analyzing large and complicated datasets. Traditional methods of analysis can be restrictive in terms of their compatibility with computer analysis and their reliance on linear models; ML techniques enhance them. Set measures that include clustering, data dimensionality reduction, and real-time data analysis improve the ability to recognize patterns and associations in large numbers of records. This advancement not only enhances the understanding of how material properties behave but also improves overall decision-making throughout the research and development phase.

In the future prospects of ML in biomaterials, some important trends were anticipated to play significant roles. The integration of multimodal data sources for understanding the properties and behavior of materials became clearer and more comprehensive. Operations involving data integration and analysis in real time promote reactive approaches to research while augmenting the quality of materials produced. Developments in explainable artificial intelligence (XAI) improve the ability to explain ML models, thereby better addressing issues related to model reliability and functionality. The trends presented in this context underline the fact that the application of ML in biomaterials science is still in development and not set in stone.

Ethical aspects play a key role in the further promotion of ML technologies. Using ML in biomaterials requires certain principles regarding how data is to be protected, how samples are to be selected, and how results are to be reported. To establish ethical principles as the core of technological construction, ethical guidelines and appropriate regulatory frameworks must be central to this process.

References

[1] Gokcekuyu, Y., Ekinci, F., Guzel, M. S., Acici, K., Aydin, S., & Asuroglu, T. (2024). Artificial intelligence in biomaterials: A comprehensive review. Applied Sciences, 14(15), 6590.

[2] Soheilmoghaddam, F., Rumble, M., & Cooper-White, J. (2021). High-throughput routes to biomaterials discovery. Chemical Reviews, 121(18), 10792–10864.

[3] Cao, Y., Nakhjiri, A. T., & Ghadiri, M. (2024). Different applications of machine learning approaches in materials science and engineering: Comprehensive review. Engineering Applications of Artificial Intelligence, 135, 108783.

[4] Badini, S., Regondi, S., & Pugliese, R. (2023). Unleashing the power of artificial intelligence in materials design. Materials, 16(17), 5927.

[5] Tahir, F., & Khan, M. (2023). Big data: The fuel for machine learning and AI advancement (No. 10951). EasyChair.

[6] Suwardi, A., Wang, F., Xue, K., Han, M. Y., Teo, P., Wang, P., . . . & Loh, X. J. (2022). Machine learning-driven biomaterials evolution. Advanced Materials, 34(1), 2102703.

[7] McDowell, D. L., Panchal, J., Choi, H. J., Seepersad, C., Allen, J., & Mistree, F. (2009). Integrated design of multiscale, multifunctional materials and products. Butterworth-Heinemann.

[8] Kerner, J., Dogan, A., & von Recum, H. (2021). Machine learning and big data provide crucial insight for future biomaterials discovery and research. Acta Biomaterialia, 130, 54–65.

[9] Suwardi, A., Wang, F., Xue, K., Han, M. Y., Teo, P., Wang, P., . . . & Loh, X. J. (2022). Machine learning-driven biomaterials evolution. Advanced Materials, 34(1), 2102703.

[10] Kerner, J., Dogan, A., & von Recum, H. (2021). Machine learning and big data provide crucial insight for future biomaterials discovery and research. Acta Biomaterialia, 130, 54–65.

[11] Ahmed, E., Mulay, P., Ramirez, C., Tirado-Mansilla, G., Cheong, E., & Gormley, A. (2024). Mapping biomaterial complexity by machine learning. Tissue Engineering, (ja).

[12] Oberkampf, W. L., Trucano, T. G., & Hirsch, C. (2004). Verification, validation, and predictive capability in computational engineering and physics. Applied Mechanics Reviews, 57(5), 345–384.
[13] Di Franco, G., & Santurro, M. (2021). Machine learning, artificial neural networks and social research. Quality & Quantity, 55(3), 1007–1025.
[14] Wang, M., & Pan, N. (2008). Predictions of effective physical properties of complex multiphase materials. Materials Science and Engineering: R: Reports, 63(1), 1–30.
[15] Kerner, J., Dogan, A., & Von Recum, H. (2021). Machine learning and big data provide crucial insight for future biomaterials discovery and research. Acta Biomaterialia, 130, 54–65.
[16] Xue, D., Balachandran, P. V., Hogden, J., Theiler, J., Xue, D., & Lookman, T. (2016). Accelerated search for materials with targeted properties by adaptive design. Nature Communications, 7(1), 1–9.
[17] Basu, B., Gowtham, N. H., Xiao, Y., Kalidindi, S. R., & Leong, K. W. (2022). Biomaterialomics: Data science-driven pathways to develop fourth-generation biomaterials. Acta Biomaterialia, 143, 1–25.
[18] Ahmed, E., Mulay, P., Ramirez, C., Tirado-Mansilla, G., Cheong, E., & Gormley, A. (2024). Mapping biomaterial complexity by machine learning. Tissue Engineering.
[19] Reddy, G. T., Reddy, M. P. K., Lakshmanna, K., Kaluri, R., Rajput, D. S., Srivastava, G., & Baker, T. (2020). Analysis of dimensionality reduction techniques on big data. Ieee Access, 8, 54776–54788.
[20] Ragone, M., Shahabazian-Yassar, R., Mashayek, F., & Yurkiv, V. (2023). Deep learning modeling in microscopy imaging: A review of materials science applications. Progress in Materials Science, 101165.
[21] Darnell, M., & Mooney, D. J. (2017). Leveraging advances in biology to design biomaterials. Nature Materials, 16(12), 1178–1185.
[22] Wang, L., Wang, C., Wu, S., Fan, Y., & Li, X. (2020). Influence of the mechanical properties of biomaterials on degradability, cell behaviors and signaling pathways: Current progress and challenges. Biomaterials Science, 8(10), 2714–2733.
[23] Suwardi, A., Wang, F., Xue, K., Han, M. Y., Teo, P., Wang, P., . . . & Loh, X. J. (2022). Machine learning-driven biomaterials evolution. Advanced Materials, 34(1), 2102703.
[24] Chen-Wiegart, Y. C. K., Campbell, S. I., Yager, K. G., Liu, Y., Frenkel, A. I., Yang, L., . . . & Basham, M. (2020). Multi-modal synchrotron characterization: Modern techniques and data analysis. In Handbook on big data and machine learning in the physical sciences: Volume 2. Advanced analysis solutions for leading experimental techniques, (pp. 39–64).
[25] Hassija, V., Chamola, V., Mahapatra, A., Singal, A., Goel, D., Huang, K., . . . & Hussain, A. (2024). Interpreting black-box models: A review on explainable artificial intelligence. Cognitive Computation, 16(1), 45–74.

Index

https://doi.org/10.1515/9783111503202-015

De Gruyter Series in Advanced Mechanical Engineering

Already published in the series

Volume 13
Raj Kumar Arya, George Verros, J. Paulo Davim (Eds.)
Compressors and Blowers: Maintenance, Practical Guidance, Energy-Efficient Technologies, 2025
ISBN 978-3-11-136335-6, e-ISBN (PDF) 978-3-11-136395-0,
e-ISBN (EPUB) 978-3-11-136417-9

Volume 12
Nishant Kumar Singh, Rajesh Kumar Verma, Virendra Kumar,
J. Paulo Davim (Eds.)
Machining and Tribology of Advanced Materials: From Coatings, Lubrications,
Surface Treatments to Modeling and Simulation, 2025
ISBN 978-3-11-137642-4, e-ISBN (PDF) 978-3-11-137729-2,
e-ISBN (EPUB) 978-3-11-137772-8

Volume 11
Kanak Kalita, Ranjan Kumar Ghadai, J. Paulo Davim (Eds.)
Microfabrication and Nanofabrication: Precision Manufacturing, 2024
ISBN 978-3-11-133543-8, e-ISBN (PDF) 978-3-11-133553-7,
e-ISBN (EPUB) 978-3-11-133559-9

Volume 10
Kanak Singh, Ranvijay Kumar, Vinay Kumar and J. Paulo Davim (Eds.)
3D Printing: Polymer, Metal and Gel Based Additive Manufacturing, 2024
ISBN 978-3-11-108061-1, e-ISBN (PDF) 978-3-11-108163-2
e-ISBN (EPUB) 978-3-11-108206-6

Volume 9
Raj Kumar Arya, Anurag Kumar Tiwari, George Verros, Prashant Malik and J. Paulo Davim (Eds.)
Sustainable Hydrogen Energy: Production, Storage and Transportation, 2024
ISBN 978-3-11-124641-3, e-ISBN (PDF) 978-3-11-124647-5
e-ISBN (EPUB) 978-3-11-124728-1

Volume 8
Tatjana Sibalija and J. Paulo Davim (Eds.)
Soft Computing in Smart Manufacturing: Solutions toward Industry 5.0, 2022
ISBN 978-3-11-069317-1, e-ISBN (PDF) 978-3-11-069322-5
e-ISBN (EPUB) 978-3-11-069325-6

www.degruyter.com

Volume 7
J. Paulo Davim (Ed.)
Nonconventional Machining, 2023
ISBN 978-3-11-058410-3, e-ISBN (PDF) 978-3-11-058447-9
e-ISBN (EPUB) 978-3-11-058415-8

Volume 6
J. Paulo Davim (Ed.)
Manufacturing in Digital Industries, 2020
ISBN 978-3-11-057507-1, e-ISBN (PDF) 978-3-11-057542-2,
e-ISBN (EPUB) 978-3-11-057526-2

Volume 5
Kaushik Kumar, Divya Zindani, J. Paulo Davim
Rapid Prototyping, Rapid Tooling and Reverse Engineering, 2020
ISBN 978-3-11-066324-2, e-ISBN (PDF) 978-3-11-066490-4,
e-ISBN (EPUB) 978-3-11-066345-7

Volume 4
J. Paulo Davim (Ed.)
Additive and Subtractive Manufacturing, 2019
ISBN 978-3-11-054816-7, e-ISBN (PDF) 978-3-11-054977-5,
e-ISBN (EPUB) 978-3-11-054833-4

Volume 3
J. Paulo Davim (Ed.)
Drilling Technology, 2017
ISBN 978-3-11-047863-1, e-ISBN 978-3-11-048120-4,
e-ISBN (EPUB) 978-3-11-047871-6

Volume 2
J. Paulo Davim (Ed.)
Progress in Green Tribology, 2017
ISBN 978-3-11-037272-4, e-ISBN 978-3-11-036705-8,
e-ISBN (EPUB) 978-3-11-039252-4

Volume 1
J. Paulo Davim (Ed.)
Metal Cutting Technology, 2016
ISBN 978-3-11-044942-6, e-ISBN 978-3-11-045174-0,
e-ISBN (EPUB) 978-3-11-044947-1, Set-ISBN 978-3-11-045175-7